Applied Electrostatic Precipitation

Applied Electrostatic Precipitation

Edited by

K. R. PARKER
Consultant in Air Pollution Control
Sutton Coldfield, West Midlands, UK

BLACKIE ACADEMIC & PROFESSIONAL
An Imprint of Chapman & Hall
London · Weinheim · New York · Tokyo · Melbourne · Madras

Published by Blackie Academic & Professional, an imprint of Chapman & Hall, 2–6 Boundary Row, London SE1 8HN, UK

Chapman & Hall, 2–6 Boundary Row, London SE1 8HN, UK

Chapman & Hall GmbH, Pappelallee 3, 69469 Weinheim, Germany

Chapman & Hall USA, 115 Fifth Avenue, New York NY 10003, USA

Chapman & Hall Japan, ITP-Japan, Kyowa Building, 3F, 2-2-1 Hirakawacho, Chiyoda-ku, Tokyo 102, Japan

DA Book (Aust.) Pty Ltd, 648 Whitehorse Road, Mitcham 3132, Victoria, Australia

Chapman & Hall India, R. Seshadri, 32 Second Main Road, CIT East, Madras 600 035, India

First edition 1997

Typeset in 10/12pt Times by Doyle Graphics, Tullamore, Ireland
Printed in Great Britain by St Edmunsbury Press, Suffolk

ISBN 0 7514 0266 4

A catalogue record for this book is available from the British Library
Library of Congress Catalog Card Number: 96–83827

∞ Printed on acid-free text paper, manufactured in accordance with ANSI/NISO Z39.48-1992 (Permanence of Paper).

Contents

List of contributors xiii
Preface xv

1 Why an electrostatic precipitator? **1**
K.R. PARKER

1.1 Introduction 1
1.2 Control system characteristics 2
1.3 Control operating principles 4
1.3.1 Inertial separation 4
1.3.2 Wet scrubbers 4
1.3.3 Fabric filter 5
1.3.4 Electrostatic precipitation 6
1.4 Summary of control system properties 9

2 Milestones in the history of precipitation **11**
K.R. PARKER

2.1 Precipitator installations 11
2.1.1 Early investigations and developments 11
2.1.2 Full-scale precipitator developments 13
2.2 Development of electrical supplies 20
2.2.1 Rectifier types 20
2.2.2 Primary control systems 22
2.2.3 Automatic control systems 22
References 23

3 Basic and theoretical operation of ESPs **25**
C. RIEHLE

3.1 General remarks 25
3.2 Ion production 29
3.2.1 Principles 29
3.2.2 Corona initiation field strength 31
3.2.3 Corona onset voltage 33
3.2.4 Current–voltage relationship 36
3.2.5 Electrical field distribution 44
3.3 Particle charging 52
3.3.1 Charging process 52
3.3.2 Cochet's charging model 52
3.3.3 Time dependence and saturation charge 54
3.4 Particle migration 55
3.4.1 Equation of motion 55
3.4.2 Theoretical migration velocity 57

3.5 Measuring and modelling particle separation 59
3.5.1 Grade efficiency and total efficiency 59
3.5.2 Laminar model 62
3.5.3 Deutsch model 62
3.5.4 Flow field and particle trajectories 76
3.5.5 Diffusivity models 82
3.6 Deposition 85
3.7 Removal 86
References 87

4 Mechanical design considerations for dry precipitators 89
F. KNUTTSEN and K.R. PARKER

4.1 Introduction 89
4.2 Discharge electrodes 89
4.3 Discharge electrode mounting 92
4.4 Collectors 94
4.5 Casings 97
4.6 HT insulators 100
4.7 Rapping 102
4.8 Hoppers 108
4.9 Electrical clearances 110
References 111

5 Aerodynamic factors affecting performance 113
L. LIND

5.1 Introduction 113
5.2 Turbulence and secondary flow 113
5.2.1 Historical resumé 113
5.2.2 Turbulence 118
5.2.3 Secondary flow 122
5.2.4 Numerical flow model 126
5.3 Gas velocity 127
5.4 Gas distribution 129
5.4.1 Standards 130
5.4.2 Residence time 131
5.4.3 Space charge 134
5.4.4 Re-entrainment 136
5.4.5 Erosion 137
5.4.6 Sneakage and sweepage 137
5.4.7 Optimal distribution 138
5.5 Model testing 139
5.6 Computational fluid dynamics 142
5.7 Field testing 148
5.8 Dust build-up and wear 149
References 150

6 The physical and chemical properties of particles and their effect on performance 153
K. PORLE and K.R. PARKER

6.1 Particle size and shape 153
6.1.1 Particle sizing 154
6.1.2 Particle shape and structure 160

6.2 Optical properties 161
6.3 Agglomeration 162
6.4 Cohesivity 163
6.5 Particle electrical resistivity 166
6.6 Chemical compositon and reactivity 172
References 178

7 Performance design considerations 180

C. COTTINGHAM

7.1 Introduction 180
7.2 What are we trying to achieve? 180
7.3 Assessment of the process 181
7.3.1 Typical assessment 182
7.4 Plate spacing 185
7.5 Configuring the ESP 186
7.6 Conclusions 190
References 191

8 Electrical operation of precipitators 192

V. REYES

8.1 Introduction 192
8.2 Precipitator performance and electrical energization 192
8.2.1 Examples 194
8.3 Corona suppression and space charge effects 195
8.3.1 Electrical characteristics with air load 196
8.3.2 Characteristics with dust load 196
8.4 High tension sectionalization 199
8.5 Traditional DC energization 201
8.5.1 Basic principles 202
8.5.2 High voltage power supply ratings 206
8.5.3 Influence of the linear inductor 208
8.6 Intermittent energization 210
8.6.1 Basic principles 211
8.6.2 Comparison with traditional DC energization 212
8.6.3 Collection efficiency 214
8.7 Automatic voltage control and instrumentation 217
8.7.1 Introduction 217
8.7.2 Instrumentation 217
8.7.3 Basic control principles 220
8.7.4 Spark detection and voltage recovery 223
8.7.5 Back-corona detection and corona power control 226
8.8 Pulse energization 230
8.8.1 Introduction 230
8.8.2 Electrical configuration 231
8.8.3 Main features of pulse energization 235
8.8.4 Power consumption 238
8.8.5 Collection efficiency 239
8.8.6 Applications 240
8.8.7 Summary 241
8.9 Supervisory computer control 241
8.9.1 Stand-alone computer 243
8.9.2 Supervisory computer control *via* a gateway unit 244
8.9.3 Advanced control functions 245
Appendix 8A 246
Appendix 8B 247
References 248

9 Precipitator sizing methods and models of electrostatic precipitators **250**
C. PAULSON and M. REA

Editor's note 250

9A Precipitator sizing methods **252**
C. PAULSON

9A.1 Theoretical considerations 252
9A.1.1 Basic dust-collection equation for gas in a duct 252
9A.1.2 Electrostatic precipitation 254
9A.1.3 Improvement of the Deutsch equation 255
9A.1.4 Factors affecting electrostatic precipitation 258
9A.2 Practical considerations 265
9A.2.1 Interpretation of test results 265
9A.3 Precipitator modelling 274
9A.3.1 Mathematical modelling 274
9A.3.2 Practical testing 275
References to 9A 278

9B Models of electrostatic precipitators **280**
M. REA

9B.1 Basic concept 280
9B.1.1 The Deutsch equation 281
9B.1.2 Charging of particles and the modified Deutsch equation 284
9B.2 The modern approach to computer modelling 285
9B.2.1 Early models 285
9B.2.2 Model by Cañadas *et al.* [5] 286
9B.2.3 Modelling at Padova university [6] 288
References to 9B 291

10 Sampling and analysis for particles and heavy metals in gas streams **292**
G.B. NICHOLS and E.B. DISMUKES

10.1 Sampling and analysis 292
10.2 Heavy metals 298
10.2.1 General considerations 298
10.2.2 Sampling methods for multiple types of heavy metals 300
10.2.3 Sampling methods for mercury alone 302
10.2.4 Metal analysis in the laboratory 302
10.2.5 Prospects for real-time monitoring 303
References 303

11 The commissioning of electrostatic precipitators **305**
D.A. STYLER and J.C. WESTBURY

11.1 Introduction 305
11.2 Mechanical commissioning 306

11.2.1 Construction stage 306
11.2.2 Post construction stage 319
11.2.3 Cold commissioning 323
11.2.4 Hot commissioning 326
11.3 Electrical commissioning 326
11.3.1 An overview 326
11.3.2 Managers and commissioning 326
11.3.3 Familiarity revisited 327
11.3.4 The programme – who writes the programme and when? 327
11.3.5 Monitoring 329
11.4 Process commissioning 339
11.4.1 Hot commissioning 340
11.4.2 Back to the real world! 348

12 Dry type precipitator applications 349
K. PORLE and K.R. PARKER

12.1 Introduction 349
12.2 Power generation industry 349
12.2.1 Bituminous coals 350
12.2.2 Anthracite coals 353
12.2.3 Subbituminous coals 354
12.2.4 Brown coals 355
12.2.5 Lignites 356
12.2.6 Oil-based fuels 356
12.3 The cement industry 359
12.3.1 Wet process manufacture 359
12.3.2 Semi-wet processing 360
12.3.3 Dry process production plant 360
12.3.4 Alkali bypass plant 363
12.3.5 Clinker cooler precipitators 363
12.3.6 Cement mill precipitators 364
12.4 General steam-raising plant 365
12.4.1 Moving grate combustors (chain grate, reciprocating and cyclic beds) 366
12.4.2 Fluidised bed units 366
12.5 Biomass-fired steam-raising plants 367
12.5.1 Wood chip combustion 368
12.5.2 Chicken litter, etc. 368
12.5.3 Municipal wastes 368
12.6 Iron and steel works 369
12.6.1 Sinter plants 369
12.6.2 Pelletising plants 372
12.6.3 Steel making 372
12.6.4 Operations involving the casting of hot metal 375
12.7 Non-ferrous industries 375
12.7.1 Copper and nickel recovery 376
12.8 Aluminium smelting 378
12.9 Paper and pulp industry 379
12.9.1 Bark firing 379
12.9.2 Cellulose pulp production 379
12.9.3 Lime sludge burning 380
12.9.4 Magnesium sulphate burning 380
12.10 Conclusions 381
References 381

13 The wet electrostatic precipitator: design and applications 382

K.R. PARKER

13.1 Introduction 382
13.2 Design considerations 383
13.2.1 Dust deposition and removal 383
13.3 Discharge electrodes 386
13.4 HT insulators 387
13.5 Casing/hopper design 387
13.6 Water treatment 390
13.7 Materials of construction 391
13.8 Electrical energisation 393
13.9 Typical applications of wet precipitators 394
13.9.1 Applications in the iron and steel making fields 394
13.9.2 Applications in the chemical industries 397
13.9.3 Applications following acid gas scrubbers 398
13.9.4 Incineration-type processes 399
13.9.5 HAC discharges 400
13.9.6 Glass manufacturing 400
13.9.7 Other applications 400

14 The mist precipitator: design and applications 402

K.R. PARKER

14.1 Introduction 402
14.2 Applications of mist precipitators 407
14.2.1 Collection of sulphuric acid mist 407
14.2.2 Gas detarring 411
14.2.3 Collection of radioactive particles 415
14.2.4 Other mist precipitator applications 416
14.3 Conclusions 417

15 Upgrading of existing precipitator efficiencies 418

K.R. PARKER and H. KRIGMONT

15A Modifications/changes to existing plant 418

15A.1 Assessment of required performance improvement 419
15A.2 Performance improvement modification options 421
15A.3 Alternative solutions 423
15A.3.1 Electrical 423
15A.3.2 Mechanical changes 423

15B Precipitator improvements achieved by changing the electrical resistivity of the particulates 425

15B.1 Change of temperature/relative humidity of the gases 425
15B.2 Flue gas additives to improve performance 426
References to 15A and 15B 428

15C Theory, principles of operation, equipment and applications of flue gas conditioning 429

15C.1 Introduction 429
15C.2 Electrical resistivity 430
15C.2.1 Prediction of fly ash resistivity 432
15C.2.2 Resistivity effects in ESPs 433
15C.3 Flue gas conditioning 436
15.C.3.1 Conditioning by sulfur trioxide 437
15C.3.2 Ammonia conditioning 445
15C.3.3 Dual flue gas conditioning 452
15C.3.4 Balance of plant impact 453
15C.4 Flue gas conditioning equipment 454
15C.4.1 SO_3 FGC systems 454
15C.4.2 Ammonia FGC systems – design features 461
15C.4.3 ESP power consumption 462
15C.4.4 Mixing requirements 463
15C.5 Application of flue gas conditioning in converting hot-side fly ash precipitators to cold-side operation 463
15C.5.1 Introduction 463
15C.5.2 Flue gas conditioning 465
15C.5.3 Installation design 466
15C.5.4 FGC operation 468
15C.5.5 Economics 469
15C.5.6 Conclusions 471
15C.6 FGC systems – optimization 471
15C.6.1 General 472
15C.6.2 Theoretical approach 472
15C.6.3 Procedural approach 474
15C.6.4 Statistical approach 475
15C.6.5 Rapper adjustments 480
15C.6.6 Data collection 481
15C.7 Conclusions 481
References to 15C 481

16 Possible future developments in the field of electrostatic precipitation 483
K.R. PARKER, C. RIEHLE and H. KRIGMONT

16A Electrical developments 487

16A.1 High frequency power conversion or switched mode power supplies 487
16A.2 Nanosecond pulse operation and acid gas control 490
References to 16A 491

16B Use of natural sulphur dioxide as a feed stock for flue gas conditioning systems: flue gas conditioning today and tomorrow 492

16B.1 Background 492
16B.2 'Native' or 'internal' feed stock FGC technologies 494

16B.2.1 'Slip-stream' FGC systems 494
16B.2.2 In-duct FGC systems 495
16B.3 *In-situ* gas conditioning (IGC) approach 496
16B.3.1 Variable exposed area IGC system 496
16B.3.2 Variable catalyst temperature IGC system 497
16B.4 Variable flow IGC system 498
16B.5 Catalyst selection 499
References to 16B 500

16C High temperature/high pressure precipitators for advanced power generation systems 501

16C.1 Fundamentals 503
16C.2 Voltage and current 505
16C.3 Particle charging 509
16C.4 Particle migration 510
16C.5 Grade efficiency 513
16C.6 Open questions 515
16C.6.1 Electrical resistivity 515
16C.6.2 Mechanical stability of material 515
16C.6.3 Rapping 515
16C.6.4 Electrical insulation 516
16C.6.5 Emptying of hoppers 516
16C.6.6 Electrical power consumption 516
16C.7 Symbols 516
References to 16C 517

16D Computer sizing of precipitators 518

Index 519

Contributors

C. Cottingham	Lodge Sturtevant Ltd, George Street Parade, Birmingham, B3 1QQ, UK
E.B. Dismukes	Grady Nichols Enterprises Inc., 400 Kiowa Street, Montevallo, AL 35115, USA
F. Knuttsen	ABB Fläkt Industri AB, S-351 87, Vaxjo, Sweden
H. Krigmont	Allied Environmental Technology, One Pacific Plaza, 7755 Center Avenue, Suite 1100, Huntingdon Beach, CA 92647, USA
L. Lind	FLS Miljo a/s, Ramsingsveg 30, DK2500, Valby, Denmark
G. Nichols	Grady Nichols Enterprises Inc., 400 Kiowa Street, Montevallo, AL 35115, USA
K.R. Parker	17 Somerville Road, Sutton Coldfield, West Midlands, B73 6JD, UK
C. Paulson	CISRO, Division of Coal & Energy Technology, PO Box 136, North Ryde, New South Wales 2113, Australia
K. Porle	ABB Fläkt Industri AB, S-351 87, Vaxjo, Sweden
M. Rea	Dipartimento Di Ingegneria Elettrica, Universita Degli Studi Di Pavoda, Via Gradenigo 6/A, 35131 Padova, Italy
V. Reyes	FLS Miljo a/s, Ramsingsveg 30, DK2500, Valby, Denmark
C. Riehle	Bayer AG, R & D Department, Particle Technology & Fluid Dynamic Group, D51368 Leverkusen, Germany
D.A. Styler	Lodge Sturtevant Ltd, George Street Parade, Birmingham, B3 1QQ, UK
J.C. Westbury	Lodge Sturtevant Ltd, George Street Parade, Birmingham, B3 1QQ, UK

Preface

Increased awareness of the effects of atmospheric pollution and ever tightening legislation have meant that electrostatic precipitators, which have been widely used to separate particulate matter from process gas streams, are now required to achieve collection efficiencies in excess of 99.9% for a number of applications. These changes have challenged the precipitation industry to consider how the equipment can be improved to meet the latest legislation, where control is now focusing on heavy metal and respirable size particulate discharges. In addition to achieving the increased performance at an economic cost, the emissions have to be maintained on a continuous basis, as failure to do so could have serious economic consequences to the plant operators.

These demands have meant that all aspects of technology, engineering and operational concerns have had to be examined, reviewed and in some instances completely modified to meet the present criteria. In spite of commercial precipitators being used for almost 100 years, it has only been in the past two to three decades that the system has changed from essentially a 'black box' art to a scientifically-based technology. Fluid-dynamics, electro-dynamics, solid state electronics and microprocessor disciplines are now the basis for current precipitation theory, design, etc., i.e. the currently accepted 'State of the Art'.

Although computational fluid dynamics and finite element analysis and other computer programs are now widely used by the industry, the sizing of a precipitator to satisfy a certain performance for a specific application is still very dependent on the supplier's knowhow and experience. The original 'black book' approach, although now taking the form of an extensive computerized data bank, is still the basis of sizing. In spite of a great deal of work being carried out using high speed computers and complex programming, there is a reluctance among the suppliers to use this approach, although it is very useful for analysing the field data. This situation is likely to change over the next few years and undoubtedly future precipitator sizing will be computer generated.

Following a review of the early development of electrostatic precipitation, this volume, containing contributions by many of the world's leading experts in the field of electrostatic precipitation, covers the theory of precipitation from both fluid and electrodynamic standpoints, plus the basic practical designs and the gaseous and particulate features which impact on the precipitators' performance.

To assist the non-specialist, Chapters 12, 13 and 14 cover the applications of dry, wet and mist type precipitators, including how the designs are modified to meet a specific duty and an examination of the major process factors which can affect performance. Chapter 11 gives a full description of plant commissioning and Chapter 10 plant testing, for both mass concentration measurements and elemental analysis in terms of particle sizing and chemical make–up. Finally, Chapter 15 shows how it may be possible to improve the performance of an under performing precipitator while Chapter 16 indicates where future developments in precipitation theory, design or application may lead.

As Editor, I am indebted to the publishers for their backing and to the following, without whose experience and expertise, the book could not have been complied – Clive Cottingham, Ed Dismukes. Filip Knuttsen, Henry Krigmont, Leif Lind, Grady Nichols, Colin Paulson, Kjell Porle, Massimo Rea, Victor Reyes, Claus Riehle, David Styler and John Westbury, together with their respective companies for allowing permission to use the material. I would also like to thank Sheila Shepherd for correcting the final manuscript, Ken Darby, who was my mentor and a source of inspiration for some 35 years, and my wife, Maureen, for her encouragement and support in the project.

K.R. Parker
November 1996

1 Why an electrostatic precipitator?

K.R. PARKER

1.1 Introduction

Stricter environmental legislation in many countries is producing ever-tightening regulations and standards governing the emission of fine particles to the atmosphere from all sources. With the ease and rapid means of international communication, many believe that the control of pollution is a modern concept; history, however, indicates that the first recorded measure was in the UK, when, in the 1600s, Parliament prohibited the burning of bituminous sea coal in London, to avoid what the Clean Air Act of 1956 referred to as 'smog'.

It was not until the nineteenth-century UK Industrial Revolution, when water power gave way to steam produced by the burning of carbonaceous fuels to evaporate water, that pollution became a serious threat to those living and working in the vicinity of the new works. The problem was exacerbated by the rapid development of the blast furnace for making iron for the ever expanding needs of industry, and also by the large-scale production of chemicals.

The first attempt to control emissions in the world, certainly in the UK, was the 1863 Alkali and Works Act, aimed specifically at controlling hydrogen chloride emissions released during the manufacture of sodium carbonate using the Leblanc Process. The Alkali Act was revised and extended in 1906 to cover most chemical works discharges and probably formed the basis of most current environmental legislation.

In recent years there has been worldwide recognition of the problems of environmental pollution and most industrialised countries have enacted legislation covering all uncontrolled emissions. The most stringent measures are associated with wealthy countries having high population density and heavy industrialisation, e.g. Japan, North America and Western Europe. For the developing countries, since much of the heavy equipment tends to be imported from the industrialised Nations, a great deal of the plant is supplied fitted with some form of pollution-control device, dependent on the country of supply, proposed or existing regulations and the finance package.

There are many forms of emission which are of worldwide concern, e.g. noise, water, gas and particulate discharges, which are the subject of control and whose discharge or emission levels are steadily reducing with each phase of legislation. In the UK, the allowable particulate emission rate from power-station chimneys has seen, over the past 30 years, a 10-fold decrease and the need for minimising emissions is steadily being recognised by the developing countries in their own right. These changes in legislation do not however signify that the problems of pollution have been overcome and there is still a need for the major technological countries to further improve the 'State of the Art' for equipment designed to reduce pollution. Any developments, in addition to improving performance efficiency, should also increase plant availability and make any plant more cost effective, either by changing the design of existing equipment or the application of new concepts.

There are many aspects of pollution control; this publication, however, will restrict itself to the control of entrained particulates, both solid and liquid phase, in mainly gas-borne streams using electrostatic precipitation, the principle of which is readily applicable to the collection of particles in liquid-phase streams, provided the carrier medium has electrically insulating properties.

1.2 Control system characteristics

There are numerous methods of separating particles from process streams using different principles, e.g. gravity or inertial separation as used in cyclones, impaction and diffusion as applied to fabric filtration, electrical means as applied to electrostatic precipitation and contacting, impingement and impaction in the case of wet scrubbers. Because of the different basic principles used by each form of separator, they each have different properties regarding collection efficiency, process and application suitability.

The effectiveness of the above forms of device is indicated in Figure 1.1, a dust spectrum, which shows the particle size range over which reasonable collection efficiencies can be achieved. While all the various approaches can be very efficient in collecting the large particles, i.e. greater than 10 μm ($1\,\mu m = 1 \times 10^{-6}$ m), the current legislative emission levels mean that effective separation of particles of 1 μm or less is now essential for a large number of processes if they are to comply with legislation.

In theory, while the effective range of inertial separators can be extended towards the submicron range, the required centrifugal force can only be obtained by reducing the diameter of the device, which restricts gas throughput, unless a large number of individual separators are operated in parallel. This becomes impractical for many industrial processes.

Although the high energy wet scrubber, e.g. venturi type, can effectively separate submicron particles, its power consumption, typically 1000 mm w.g., to capture 1 μm particles at 99% efficiency not only means high operating costs but, as the particles will be contained in a liquid

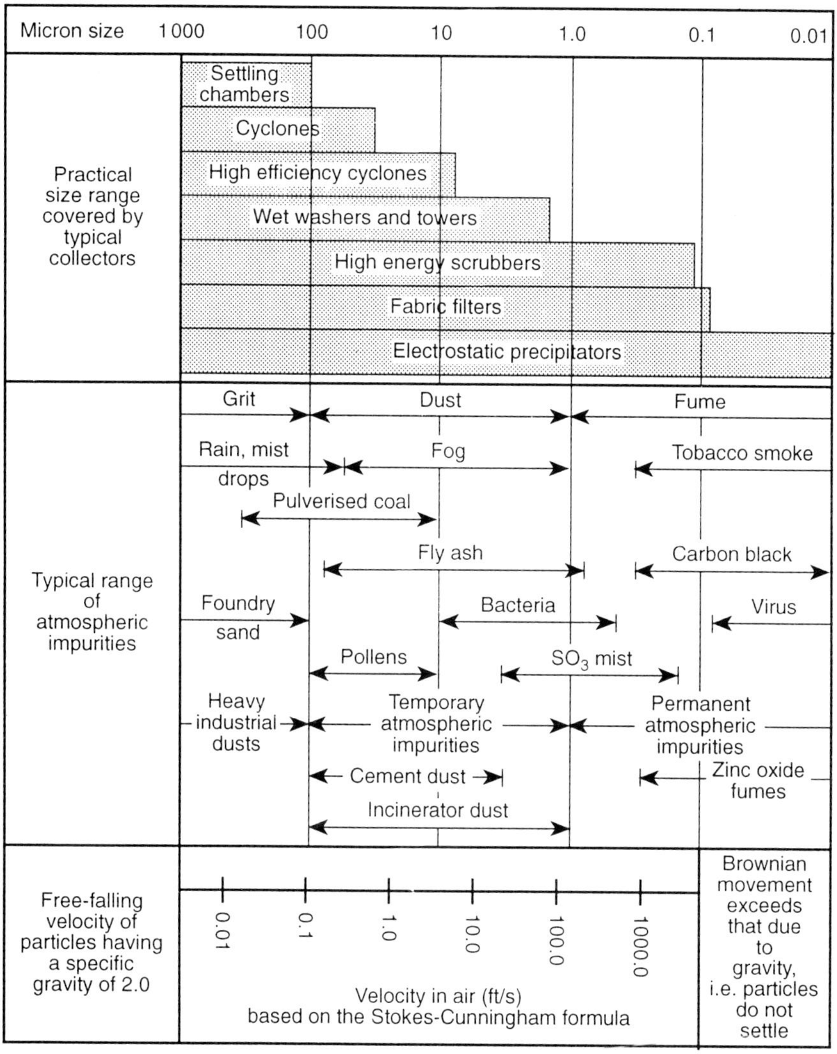

Figure 1.1 Dust spectrum. The size of any particle is defined as the diameter of a sphere having the same free-falling velocity as the particle, usually termed the Stokes diameter. $1\,\mu m = 1 \times 10^{-6}\,m$.

effluent, this needs to be treated to avoid changing an air pollution problem into a more difficult water pollution problem.

The choice of equipment having the capability to effectively collect submicron particles at high efficiency from large process gas streams ($50\,m^3/s$ upwards) tends to be limited to the fabric filter and electrostatic precipitator as cost-effective approaches.

1.3 Control operating principles

For completeness, the operating principles of the various major types of equipment, outlined above, can be summarised as follows.

1.3.1 Inertial separation

There are many different forms and arrangements of collectors using this principle, usually termed cyclones or centrifugal collectors, where the entraining stream is caused to spin or rotate rapidly within a cylindrical vessel. Because of the much higher mass of the particulates compared with the gas molecules, the resultant centrifugal force causes the particles to migrate across the flow to the wall of the containing vessel, where they become disentrained in the low flow region of the device.

The cyclone is a very simple device having a wide range of operating temperatures, but requires a driving force, e.g. a pressure drop of some 100 mm w.g., for effective separation. For most practical applications its use for high efficiency collection is limited to particulates having a particle diameter greater than 10 μm.

1.3.2 Wet scrubbers

All wet scrubbers operate by contacting the particles with large quantities of liquid in a fully turbulent area of the device. The particles are allowed to impact or impinge on the liquid droplets, normally water, such that their size and mass increase. As larger water-wetted particles they can be removed from the entraining gas stream in a simple impingement or cyclonic form of separator.

As a contact/impaction device, the effectiveness of particle droplet collision is dependent on the total energy expended in the device. This can be either as pumping power, to provide small liquid target particles, or as pressure drop, to develop the high relative particle/droplet velocities necessary to ensure that impingement occurs. For submicron particles, venturi scrubbers, having efficiencies in the 99% region, have total operating power requirements equating to some 1500 mm w.g.

The scrubber design is relatively simple; having a small footprint, it can handle a full range of gas temperatures and is insensitive to sticky dusts, but can suffer from erosion and corrosion. Its usage is normally limited, for high efficiency collection duties, to processes having small gas flows, because of the high power demand, large water usage, probable effluent problems and poor plume buoyancy unless gas reheat is practised. The wet scrubber, because of its excellent mass transfer characteristics, however, is widely used to control gaseous emissions, i.e. acid gases from many processes.

1.3.3 Fabric filter

The principle is that the particulate-laden gas stream is passed through a porous membrane which filters off the particulates and allows the clean gas to pass through. As the retained particles deposit and build up to form a layer, the efficiency of separation, which is basically, in its simplest form, an interception and diffusion mechanism, increases, such that particles having a much smaller diameter than the membrane pore size are collected. In the original design of plant, the media comprised a plain woven cloth formed into a bag or sock; the filter material is now usually felted and surface treated to reduce particle penetration while still maintaining high gas porosity.

In theory, such a device should be close to 100% in capturing particles as the layer on the media builds up. In order, however, to maintain operating pressure drop, e.g. 100–150 mm w.g., it is necessary to periodically remove the layer from the media. Assuming that the media can be satisfactorily freed of the dust layer, by means of vibration or air reverse purging to regain a low pressure drop, on returning the filter to service there is always a degree of slippage of the finer particles, which automatically limits the ultimate efficiency or emission level.

From the early days, when the media were manufactured from natural fibres in the form of a bag or sock, the operating temperature range has been considerably extended by using synthetic fibres, woven and sintered metals, or ceramics in various forms. In the latter, the operating temperatures can, dependent on the media selected, approach 1000 °C. Whilst the actual form of the filter can be different for the extreme temperatures, the principle of separation remains unaltered.

Simultaneous development of special media coatings and finishes, to simplify cleaning and general operation, has extended the expected life of the media. This has enabled suppliers to consider firm commitments/guarantees on the life of the media, in addition to performance. This is particularly important for large applications where, in the case of a major power plant, some 35 000 conventional bags would be used to treat the output from each 500 MW boiler installation and, to avoid major emission excursions and legislation violations, a strict maintenance/bag replacement programme needs to be implemented.

Compared with an electrostatic precipitator installation handling a 'difficult fly ash', the cost of the equivalent bag filter would be somewhat cheaper, but its operating power, the maintenance of compressors and cleaning mechanisms, plus the cost of a complete media (bag filter) replacement, probably every 3 years or so, to ensure emission compliance, must be borne in mind and added to the overall cost comparison.

Figure 1.2 compares the installed capital and 15 year total cost of various arrestment systems for a 250 MW power plant designed to meet an emission

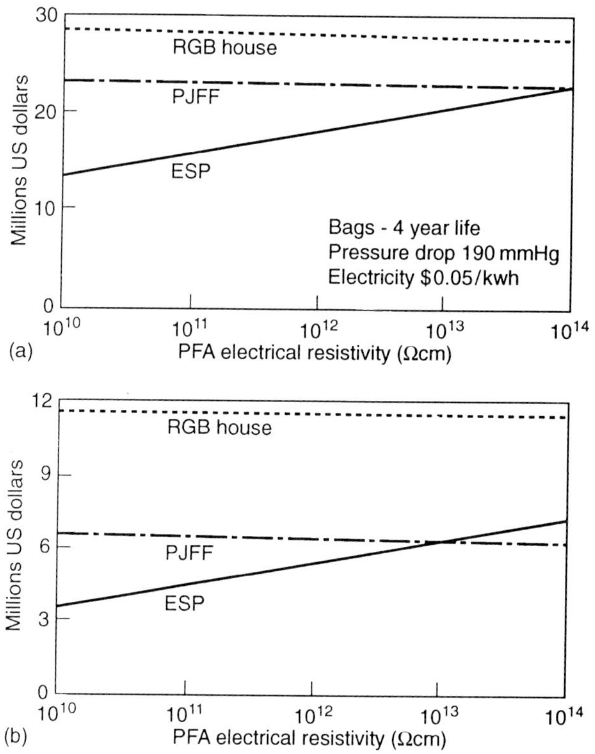

Figure 1.2 Cost comparisons for electrostatic precipitator (ESP), pulse jet fabric filter (PJFF) and reverse gas bag (RGB) house. 1994 costs based on a 250 MW power plant designed to meet an emission of 50 mg/Nm^3 at 6% O_2. (a) Total capital installed cost. (b) 15 year total cost.

of 50 mg/Nm^3 at 6% O_2 dry. The costings indicate that for either a reverse gas bag house or pulse jet fabric filter the cost remains constant, irrespective of the pulverised fuel ash (PFA) resistivity, whereas the electrostatic precipitator (ESP) increases proportionately to the ash resistivity. Overall the ESP has the economic advantage taking all costs into consideration.

Another point, which needs to be carefully considered for some applications, is that if the ash becomes difficult to remove from the bag filter media and the pressure drop continues to increase, the plant process gas throughput and hence output will drop as the induced draught (ID) fan reaches its maximum pressure capability or limiting motor power.

1.3.4 Electrostatic precipitation

The basic theory of operation is that the gas-borne particles are passed through a corona or charging field where they receive an electric charge, usually negative in the case of industrial precipitators, and then as charged particles are deflected by the electric field producing the charging regime. The charged particles then move across the gas stream from the negative electrode to be deposited on the positive electrode, which for convenience is normally earthed or grounded. From the collectors, the particles are removed into receiving hoppers or troughs, either by mechanical shock impulse rapping in a dry application or by water washing in the case of wet precipitators, i.e. those plants operating close to or at water dew-point temperature conditions.

Commercially, electrostatic precipitators have been used for almost a century for the collection of dust, fume and mist particles from all types of processes. The initial applications from mainly chemical/metallurgical based applications were both for reducing air pollution and for the recovery of valuable byproducts, for example the cleaning of combustible gases, or material losses from metal smelting processes. The ability to size precipitators for a specific efficiency was ideal in this respect and economics dictated mid-90% efficiencies for this recovery type of application, since the collected material normally had a positive value.

More recently, the major application has been to control air pollution to meet specific legislative emissions and, consequently, design efficiencies can now approach, and sometimes exceed, 99.9%.

Figure 1.3 illustrates some recently measured emissions for various process applications, together with their relevant legislative emission levels. This indicates that a well-designed and operated electrostatic precipitator can produce emissions well below 10 mg/Nm^3 for a wide range of duties and, as such, can comply with the most stringent regulations.

With the most recent legislation demanding more stringent controls with regard to toxics and heavy metals, many existing in the submicron fume range, then, provided these are in a solid or liquid phase at the normal plant operating temperature, the electrostatic precipitator is an ideal capture vehicle.

For compounds which are in a gaseous phase at the normal process discharge temperature, then, provided the gases can be further cooled to condense thc material, a wet precipitator can be used for very efficient secondary cleaning, e.g. following a sulphur dioxide scrubbing system on a power station. An alternative approach, provided the toxics or gaseous phase material can be adsorbed by an injected target material, e.g. activated carbon, means that the precipitator can operate above any dew-point temperature and hence can be constructed from normal carbon steels at much reduced cost.

The latter method is a preferred option, particularly in the UK, since under Integrated Pollution Control (IPC) legislation, changing air pollution to water pollution is not a viable option since the cleaning of dirty process water can prove to be very expensive. In some instances, such as sticky dusts

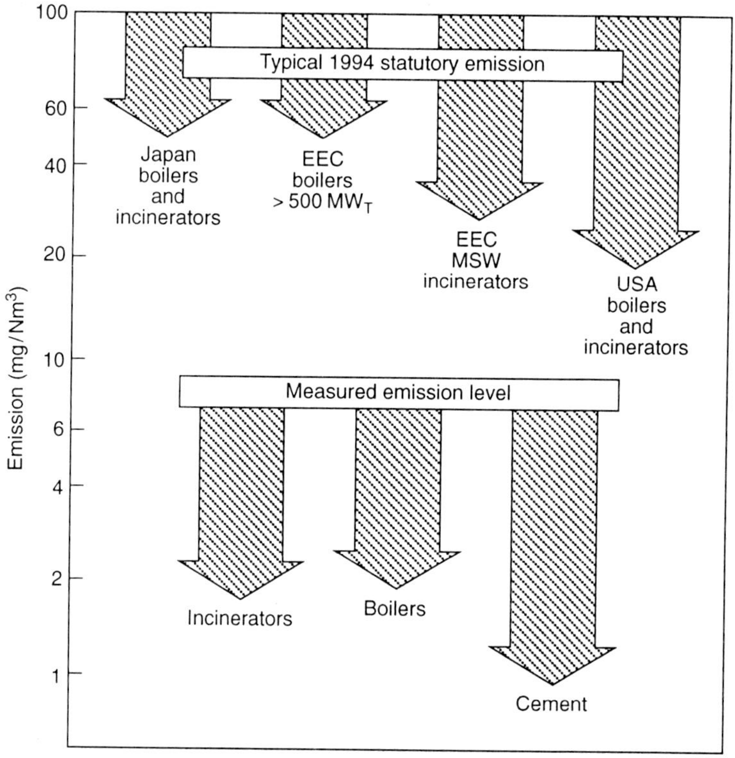

Figure 1.3 Comparison of some measured emissions with statutory regulations. Lodge Cottrell brochure 'An Introduction to Electrostatic Precipitation'.

or dew-point conditions, there is no alternative to using a wet precipitator, but often as this is associated with some form of scrubber or quench system, the water clean-up is an economic part of the total process.

Although it is possible to collect these submicron toxics and heavy metals with either a high energy venturi type scrubber or bag filter, these suffer from very high energy costs in the case of the scrubber or possible 'bleed-through' in the case of the fabric filter, unless expensive precoating systems are adopted; even then there is still a risk. Consequently, there are commercial and performance advantages in using electrostatic precipitators in order to comply with legislation.

The electrostatic precipitator can be summarised as having the following advantages compared with other forms of device for particulate collection or control:

1. can be sized for any efficiency;
2. can operate at temperatures up to 850 °C;

3. can operate at any positive pressure condition but limited to suctions that will maintain corona without the discharge developing into a plasma;
4. can operate with fully saturated gas;
5. has a low pressure loss;
6. acceptable electrical operating costs;
7. not particularly particle-size sensitive;
8. proven long life;
9. excellent reliability;
10. low maintenance requirements.

1.4 Summary of control system properties

The purpose of this initial chapter is to list the major types of device that are available for the removal of particulates, fumes and mists met in industrial air pollution control applications and to attempt to identify the advantages and disadvantages of each. While one can argue that all have

Device	Advantage	Disadvantage
Inertial separator	Simple construction, low cost, low maintenance, no temperature limitation, both wet and dry applications	Limited effective particle size range, high pressure loss, typically 100 mm w.g., possible erosion
Wet collector	Simple construction, low capital cost, small footprint, sticky particulates, effective on wet gas and dusts	High pressure loss up to 1500 mm w.g., large water usage, wet effluent, low plume buoyancy, possible plume odour
Fabric filtration	Intrinsic high efficiency 99.5% plus, reasonable footprint	Pressure loss, average 150 mm w.g., media temperature limitation, fire/melting risk, media of limited life-span, unsuitable for sticky/adhesive dust
Electrostatic precipitator	Sized for any efficiency, full particle size range, low pressure loss, low maintenance, long life 20+ years, wet and dry collection	High capital cost, efficiency sensitive to dust resistivity, large footprint

their own specific applications, for all-round efficient and effective collection of all particulate forms, the electrostatic precipitator ranks probably the highest, if one accepts the cost and ground space implications.

The properties of the devices, outlined in section 1.4, as effective high efficiency particle collectors, are summarised in the table on the previous page.

Please note that the foregoing table should not be used for deciding which of the various devices should be used, but is useful in that it identifies the advantages of each and can be used as a first cut.

As each application is specific, it will be necessary to complete a full costing exercise, considering all the relevant factors, to determine which collection system is most cost effective.

The subsequent chapters will cover the history of the development of the electrostatic precipitator, from its early beginnings through to the latest and possible future applications. In addition, the theory of precipitation and the factors which affect the efficiency and sizing will be covered in detail, together with practical hints and assistance to engineers faced with operating and optimising their electrostatic precipitation plant.

2 Milestones in the history of precipitation

K.R. PARKER

2.1 Precipitator installations

2.1.1 Early investigations and developments

While the first commercially successful precipitators were installed in the early part of this century, historically, the first report of the electrical attraction of smoke appears in 1600, when William Gilbert [1] describes the phenomenon: 'Everything rushes towards electricks excepting flame and flaming bodies and the thinnest air... yet they entice smoke sent out by an extinguished light.' The next report was in 1824, when Dr M. Hohlfeld at Leipzig [2], performed an experiment of clearing fog in a jar containing an electrified point. This work was repeated by C. F. Guitard in 1850 [3].

The first significant investigations using electrostatic precipitation were by Professor O. J. Lodge and Mr J. W. Clark, working at Liverpool University in the early 1880s. Following publication of this work [4,5], Lodge was approached and collaborated with Mr Alfred Walker of Parker, Walker & Co. of Chester, in attempting to remove fumes from gases arising during the smelting of lead at their Works in Deeside.

This was the first attempt to use electrostatic precipitation industrially for the collection of particulates. In 1884 a UK Patent, No. 11 120 [6], was filed by Alfred Walker describing how the apparatus was arranged (Figure 2.1). At this stage it should be appreciated that the only means of developing high voltages was by Wimshurst or Voss electrostatic friction generators, which stored their charge in Leyden Jar capacitors. It should be noted that unlike later plants, Walker's drawing indicates positive energisation of the electrode system. The adoption of negative energisation of the discharge electrodes follows Cottrell's investigations and his patent of 1908.

The plant at Bagillt in North Wales was energised from two 1524 mm (5 ft) diameter Wimshurst machines driven from a 1 hp steam engine. This method of energisation proved ineffective; while an electrostatic machine can provide the necessary voltage, it cannot simultaneously deliver the required particle charging current and, consequently, the plant failed to live up to expectation.

It should be remembered that in the early 1880s electric motors and generators were still being developed and were not readily available outside the laboratory. Even had high voltage equipment been available, electrical insulating materials, such as glass, mica, ebonite and similar materials would not have been entirely satisfactory for precipitation duties.

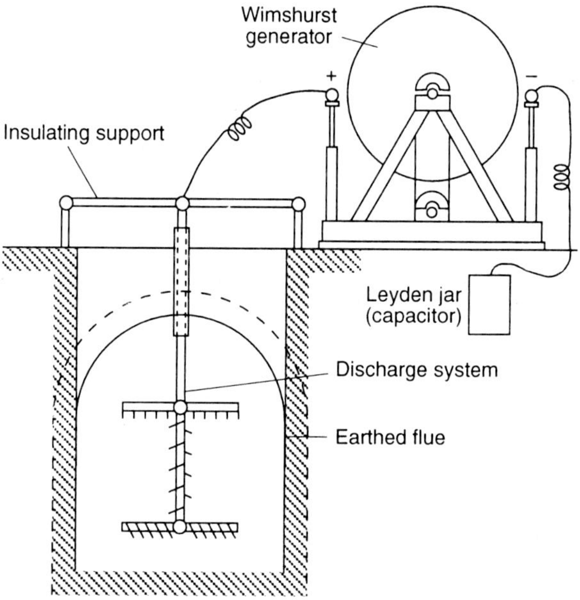

Figure 2.1 Illustration from first US patent on lectrostatic precipitation. A.O. Walker, No. 342 548 (1886).

This early failure did not prevent Walker from realising the potential of electrostatic precipitation and he took out additional Patents in Europe and the USA for the collection of fumes and dust from all types of chemical plant.

Almost in parallel with Lodge and Walker's work in the UK, Dr Karl Moeller [7] in Germany rediscovered the art of precipitation and took out a German Patent, No. 31 911, 'Rohrenformiges Gas and Dampfilter', describing his findings, in 1884.

Following the installation and failure of the plant in North Wales, Lodge appears to have temporarily lost interest in precipitation and tended to concentrate on other experimental work, such as meteorological and atmospheric electrification for the removal of fog and creating rain artificially, electro-magnetic radiation as a precursor to wireless telegraphy and also X-rays for medical use. In 1903, however, he obtained a UK Patent, No. 24 305 [8], covering a high voltage rectifier bridge arrangement, using Cowper–Hewitt mercury arc vapour lamps for the deposition of smoke, dust, mist and the like from gases. The patent describes a recognisable precipitator using barbed wire electrodes and plate collectors. The high tension (HT) equipment employed a gas engined driven dynamo feeding a Ruhmkorff induction coil, then the mercury arc rectifiers in full wave configuration.

In 1905, another UK Patent, No. 25 047 [9], was issued under the names of Lodge, Muirhead and Robinson, covering an improved vacuum type rectifier. These were manufactured by A. C. Cossor who, in their sales literature, describe the device as 'Sir Oliver Lodge's High Tension Valve' for obtaining unidirectional current for use with X-ray tubes, etc. These valves were used extensively by Lodge in the following years for precipitation duties.

Work in the USA, at this time, was being carried out by Dr F. G. Cottrell, a physical chemist, who was studying various ways of controlling air pollution from the Californian Smelters; he made the technological breakthrough for providing high voltage and charging currents simultaneously by using a high voltage AC transformer coupled to a synchronous mechanical switch rectifier. This device proved vastly superior to the earlier forms of equipment and led to the successful development of an electrostatic precipitator for the collection of sulphuric acid mist.

Cottrell also recognised that negative corona had advantages over positive and, in 1908, took out a US Patent No. 895 729 [10], which describes the early apparatus in some detail. Figure 2.2 illustrates how the equipment was arranged. It should be noted that the discharge electrode is described as pubescent and was made of semi-conducting fibrous material. This proved to be important in the success of obtaining a uniform corona over long lengths of wire at relatively low voltage. Because of insulation limitations, the high voltage transformers available were only capable of supplying 10–15 kV.

2.1.2 *Full-scale precipitator developments*

With Cottrell's success in collecting sulphuric acid fumes in the Laboratories of the University of California, the first commercial precipitator was applied in 1907/8 to collect fumes and dust from the Powder Works of the DuPont de Nemours Plant at Pinhole, California [11]. This plant was quickly followed by another installation on the Lead Smelter Facility at Selby, which was embroiled in acute air pollution difficulties over their gaseous discharges. This precipitator was successful in significantly reducing particulate emissions and collected some 2 gal/min (0.15 l/s), of sulphuric acid mist, at a gas flow rate of some 5000 cubic feet per minute (cfm) (2.4 m^3/s).

The next precipitator to be installed was in 1910, at the lead smelter at Balaklala in California [12]. This plant was to remove lead fume, etc., at a gas flow rate of 250 000 cfm (118 m^3/s). Although the plant design had improved with the experience gained from the earlier installations, the precipitator did not live up to expectations, and in spite of extensive investigation, only achieved some 90% collection efficiency. At this stage, it had not been established that to successfully handle lead dust and fumes, it is important to optimise temperature and flue gas moisture to control

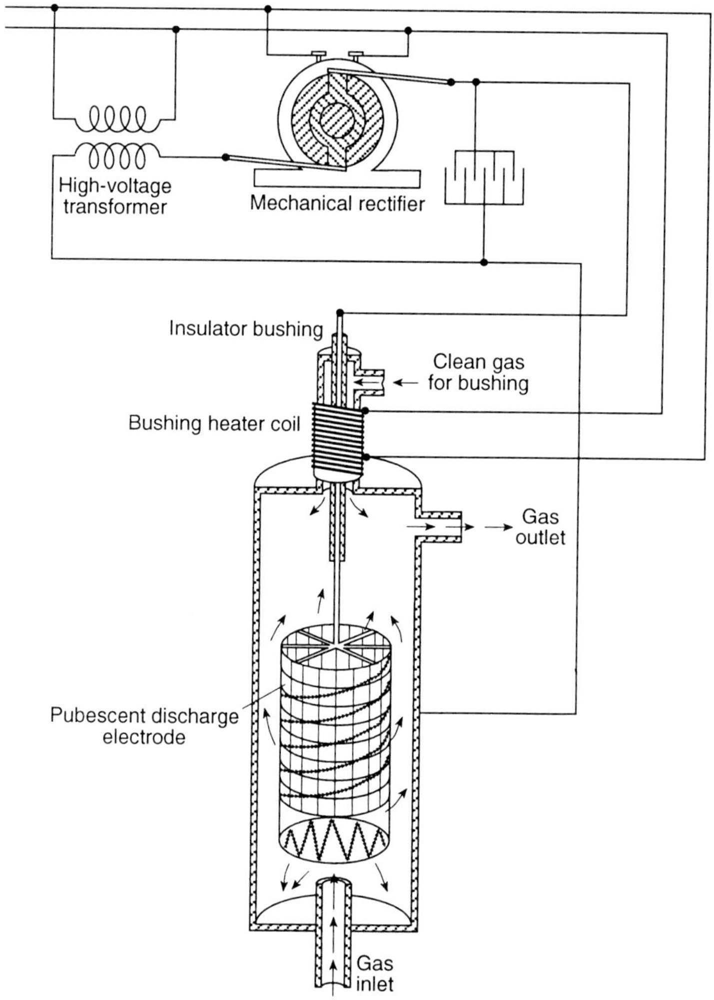

Figure 2.2 Illustration from Cottrell's first electrostatic precipitation patent, No. 895 729 (1908).

particulate electrical resistivity. This later finding, together with the problems of inadequate power supplies, probably halted the 1880s developments by Lodge *et al.*

In spite of the only partial success of the Balaklala plant, further investigations and development work was carried out in the US, since it was recognised that electrostatic precipitation was a most effective means of controlling and removing dust and fumes from many industrial processes.

Walter Schmidt, a student of Cottrell's, developed and patented the fine wire discharge electrode. This breakthrough, enabled larger plants to be constructed, as illustrated by the precipitator installation at the Cement Works of the Riversdale Portland Cement Works in 1912 [13], where the precipitator was designed to handle 1 000 000 cfm (472 m^3/s) of process gas to remove cement and other dusts from their kilns. The plant was highly successful and remained in operation for some 50 years.

Dr Cottrell had, with magnificent generosity, in 1912 decided to relinquish all rights and claims to his US patents, east of the Rockies, with the exception of the cement industry, and gave them in trust. An endowment fund was set up under the auspices of the Research Corporation, so that the income could be used for the further advancement of scientific knowledge. His remaining patents and interests covering the Western US and foreign rights were assigned to Walter Schmidt.

The application of precipitators to the cement industry to collect kiln dust gave rise to the American Potash Industry. This allowed the US to be quickly self-sufficient, not only for its use as a fertiliser but also in pig iron and ferromanganese production, where large quantities of potash are used for fluxes. Although the quantity of potassium in the basic raw cement feed material is small, the processing drives off the low temperature volatile compounds, which after condensing, can be readily captured at a much higher concentration [14]. On some installations, Feldspar and/or sodium salts were added to the raw material to preferentially distil off the potassium salts as halogens, thus enhancing the value of the collected material.

The first precipitator applications worldwide were mainly to combat the worst effect of particulate air pollution. At the time there were no specific emission levels to be met, so the size and installation cost of the electrostatic precipitator was based on the recovery value of the collected material and hence the collection efficiency was generally in the mid-nineties except for easy to collect particulates such as sulphuric acid mist and the like.

Much of the development work in the smelter industry was carried out in California, where severe air pollution, as a result of acid gas and particulate emissions from the treatment of ores, slimes and concentrates, was being experienced by all operating companies who were in litigation with the fruit farmers. To counter this, precipitators were built by the operators under licence from the Research Corporation, so different designs were to be found, although most employed the 5 inch diameter tube, as used by Cottrell *et al.* during their early work at Berkley, California.

A good example of the use of 5 inch tubes, is the precipitator installation on the Garfield Smelter, Salt Lake City, Utah, where the builder, Mr W. H. Howard [15], describes the precipitator, designed to handle 200 000 cfm (94.4 m^3/s) of converter flue gas, as comprising 2520, vertical 5 inch diameter by 10 ft long steel pipes, having No. 10 gauge iron wire as the discharge electrodes, operating from a 25 kV rectifier arrangement. This plant operated satisfactorily and collected several tonnes of fume per day containing some 50% lead, plus various precious metals, which had, up to then, been lost to the atmosphere.

In 1914, because of pollution problems at the giant Anaconda Smelter, investigations to optimise space and cost were carried out using tubes between 12 and 36 inch diameter, which were powered by rectifiers of up to 150 kV. Reports of tubes up to 48 inch diameter, using 240 kV power supplies, are also mentioned [11]. These large diameter tubes, however, had their own operational problems and the final installation at Anaconda, in 1916, was a vertical flow arrangement having corrugated plate collectors. The unit handled a gas flow rate of 700 000 cfm (330 m^3/s), and even by today's standards was large.

Another smelter type installation was at Trail, British Columbia, where one precipitator treated the gases from a lead blast furnace, at a design gas flow rate of 100 000 cfm (47.3 m^3/s), using 13-inch diameter tubes and another similar unit on a lead roasting furnace.

These installations are of historic interest in that people tend to consider wide spacing between the electrode systems as a fairly new concept, whereas, in reality, the early investigators recognised the cost advantages of larger spacings and were not averse to trying them, in spite of the higher voltage requirements.

As indicated earlier, not only were the problems of high resistivity dust, and the use of temperature and moisture conditioning to overcome these effects, recognised by operators in the smelting industry, but the investigators also found that by operating one precipitator at high temperature, around 350 °C (650 °F), to remove the metallic elements, then after cooling and operating a second unit at around 105 °C (220 °F), it was possible to collect high purity arsenic trioxide commercially. This investigation gave rise to the principle of selective material collection using precipitators.

Another area of special mention in the smelter industry is that of the recovery of gold, silver and other bullion metals. In refining these are distilled off as vapour, at the processing temperature, to consequently condense on cooling as fume-size particles. These fumes were virtually impossible to collect by conventional means, such as cyclones and scrubbers, and were lost to the atmosphere. With the high commercial value attached to these materials, a large number of precipitators were installed to collect these fumes, their value being so high as to make the precipitator installation costs insignificant.

A surprising early application of the basic precipitation process can be found in the Oil Field Industry, where Hydrofiners, or Dehydrators, for the removal of water and the breaking of crude oil/water emulsions can be traced back to 1915. Trials using both DC and AC energisation were found to reduce 65% water/crude oil emulsions down to 0.5% water [16]. This application has now been extended to cover other similar type applications where there is a significant difference in the insulation properties of the various phases being processed.

The earliest application in the iron and steel industry was the cleaning of blast furnace gases for reuse. The gases from the blast furnace have a reasonable calorific value/heat content, but contain a fairly high dust loading. This dust, which is rich in terms of potassium salts and hence a valuable source of potash, produces difficulties with duct blockages, rapid fouling of the burners, and unstable combustion. To overcome these problems, the gases were originally cleaned at high temperature to retain their sensible heat.

Possibly the earliest recorded successful hot dry precipitator installation is that at the UK Skinningrove Ironworks, commissioned in 1917. This unit is of the vertical flow, tube type and was energised by the Lodge valve form of rectifier. Although many dry-type hot gas precipitators were installed for cleaning blast furnace gases in the US, UK and Europe, problems with high electrical resistivity of the particulates meant that performance was variable and, subsequently, the wet electrostatic precipitator was found to be more effective overall, in spite of losing the sensible heat content of the gases by operating the plant at a much lower temperature.

Another early application was that of detarring illuminating (Town's Gas) retort or iron works coke oven gases. During the production of town's gas, or the manufacture of coke for use in the blast furnace, reasonably high volatile coal is heated under a reducing atmosphere to distil off the hydrocarbons and leave the coke behind.

The gases given off during this processing have a higher calorific value than blast furnace gas and, in addition to a small quantity of dust, contain a high level of tars, which condense on cooling the gases. To eliminate downstream problems, these gases can be effectively detarred using precipitators [17]. The earliest forms of precipitator were of the tubular vertical flow type, but other designs, such as horizontal and vertical flow plate types, can now be found.

An interesting development occurred in 1915, when the Hooker Electrochemical Company at Niagara Falls was under pressure from the local authority to reduce the emissions from their bleach manufacturing plant. The main problem arose over the discharge of chlorine gas. As the electrostatic precipitator cannot collect gas molecules, it was decided to treat the precipitator inlet gases with finely divided slaked lime; this reacted with the chlorine to produce calcium chloride, which was satisfactorily collected. This could be the first application of acid gas emission control using a dry scrubbing technique; similar approaches are now widely used to abate many acid gas discharges.

Although the sectors first tackled were the metallurgical smelters, followed by cement and iron and steel plants, all producing material having a significant commercial value, it was not until the early 1920s that precipitators were applied to the power industries.

These followed the need for electric power and higher steam rate production, which led to the development of the pulverised coal combustor. These combusters gave serious air pollution, since up to 90% of the ash can be carried over with the furnace gases, as compared with some 10% from hand fired, chain grate and similar type stokers previously employed. With pulverised coal firing, as the coal fly ash has only a limited commercial value, the precipitator installation was primarily for air pollution control. This is still true today, where precipitators are designed, dependent on the fuel ash content and local legislation to collect in excess of 99.9% of the fly ash. The collection of power station fly ash forms the largest single application, in terms of both cost and number, of electrostatic precipitators.

While the foregoing includes some of the original installations, it is not intended to be a complete history of events. With the formation of Research Corporation as a worldwide licensing organisation by 1917, Linn Bradley [18], reports: 'there are many plants in commercial operation on the continent of North America, several on the continent of Europe, in England, some in Japan, Africa and South America, and the work is growing very rapidly'.

The following table lists some of the pioneer installations in, as far as possible, chronological date order:

1884	Lead fumes – roaster	Bagillt, Wales, UK
1907	Sulphuric acid mist – contact process gases	Pinhole, CA, USA
1910	Lead zinc – smelter	Balaklala, CA, USA
1912	Cement kiln dust	Riverside, CA, USA
1912	Lead fume – converters	Garfield, UT, USA
1913	Bullion recovery-slimes	Niagara Falls, NJ, USA
1914	Lead – Dwight Lloyd sinter	Tooele, UT, USA
1914	Lead – blast furnace	Trail, B.C., Canada
1914	Cleaning roaster off gases – sulphuric acid production	Germany
1915	Lime inj. chlorine collection	Niagara Falls, USA
1915	Tar removal – towns gas	Portland, OR, USA
1915	Carbon from calcium carbide furnaces	Germany
1915	Ventilation air cleaning	New Haven, CT, USA
1915	Dehydration crude oils	CA, USA
1916	Paper pulp alkali salt	Canada
1917	Hot BF gas cleaning	Skinningrove, UK
1918	Copper smelter plants [19]	Japan
1919	Copper smelter – central gas cleaning. 2 000 000 cfm	Anaconda, MT, USA

By 1920, the art of electrostatic precipitation had made great strides, from the initial laboratory investigations of a scientific phenomena to becoming an established industrial approach to collecting all forms of particulates and fumes on many diverse applications, as indicated in the foregoing table.

Although many of the initial plants adopted the vertical tube and axial wire configuration, H. D. Braley, in his '*Notes on Electrostatic Precipitation*' 1919 [16], indicates the following:

1. Plant designs
   Gas-type treaters
   (a) Tube
   (b) Plate type
   (c) Water film plate type
   Liquid treaters
   (a) Stationary electrodes
   (b) Rotating electrodes.

The water film plate arrangement, where water is allowed to continually flow over the collectors, was developed for the collection of 'difficult dusts', which had been recognised by the early investigators, or alternatively, for use as a secondary series low temperature unit for selective dust precipitation, i.e. following an upstream dry unit or scrubber for removing the heavy coarser materials.

In the treatment of liquids, the idea of rotating electrodes was to break up coalesced water chains, which on some designs shorted out the electrode system when handling water/oil emulsions.

2. Operational temperatures
Investigations of operating practices have indicated that the following represent a fair average of operating conditions.

300–500 °C	Cement plant and pyrites roasters
180–200 °C	Copper convertors and metallic dusts
Below 125 °C	Condensible arsenic trioxide
80–120 °C	Lead smelters and similar
Below 100 °C	Water or wet film treaters

3. Gas velocity

Majority of plants operating	5–6 fps (1.5–1.8 mm/s)
Lowest velocity found	2.5 fps (0.76 m/s)
Highest velocities used	8–10 fps (2.4–3.0 m/s)

4. Direction of flow
For tube type, both up and down flow can be found, dependent on the application and site conditions. With plate types, horizontal as well as vertical approaches have been used. Both pressure and suction designs have been installed to suit a particular application.

5. Capacity of rectifier equipment
Most precipitators in service are energised from rotary rectifiers and high voltage transformers providing a maximum operating voltage of 60 kV. Figures quoted for specific powers are interesting, in that, for cement plants the specific power is given as 175 W per 1000 cfm (269 W per m^3/s) or 1.25 W per ft^2 (13.5 W per m^2) of collector, whilst for copper and lead smelters, the specific power is 600 W per 1000 cfm (922 W per m^3/s) and 3.0 W per ft^2 (32 W per m^2) of collector. This would explain one of the dilemmas facing some of the early investigators and

designers, in terms of correctly sizing the rectifier equipment, particularly in association with the very primitive control methods of handling power arcs within the precipitator.

Most industrial precipitators at this stage were sized and designed by 'rule of thumb' approaches, based on the broad foundations established by Dr Cottrell *et al.* Although the basic physics of particle charging and migration were known, it was not until 1919/1920 that Evald Anderson, a co-worker of Walter Schmidt, experimentally established the exponential relationship between efficiency and gas flow, later translated to plant size [20]. Walter Deutsch, a mathematical physicist, working for Metallgesselschaft in Germany, theoretically proved the logarithmic relationship between efficiency gas flow and collector area in 1922 [21].

The equation developed by Deutsch in 1922, later revised in 1926, was used for almost 50 years by precipitation engineers for sizing plants having similar dusts, process applications and design efficiencies, but different flow rates. In the 1960s, when new and more stringent legislation called for emissions to be reduced by an order of magnitude, the Deutsch Equation was modified by Matts and Ohnfeldt [22] amongst others, into a more operational form. All present day precipitator equipment suppliers have adopted some version of the modified Deutsch Equation for plant sizing duties and the success of these modern plants can be traced back to the findings of the early workers in the precipitation field.

2.2. Development of electrical supplies

2.2.1 Rectifier types

The ultimate success of electrostatic precipitators for industrial applications can be attributed and paralleled to the development of suitable electrical supplies. Mention has already been made of the need to simultaneously provide high voltage and corona current for particle charging and the reason for the failure of the initial plants using electrostatic-type generators.

The contribution of Dr Cottrell, in developing the synchronous switch rectifier, was possibly the greatest technological breakthrough and allowed the precipitator to be taken from the laboratory into the field.

Although Lodge's high tension valve rectifier was widely used in the UK, by the then recently formed Lodge Fume Deposit Co., the system, however, was not very efficient as a rectifier, had power limitations and, being fragile, was not readily acceptable by most users.

Several DC generators have been investigated and developed over the early part of the century, the earliest probably being that from the Girvin Electric Co. of Philadelphia, where the machine was a vertical belt driven device, having a rotating field and provided with intermediate commutating poles. The armature coils were oil immersed and each connected to the commutator and wired in such a manner as to develop the required HT voltage. The machines were some 33 inch in diameter and 62 inch in height, and could deliver voltages of up to 75 kV at 10 kW.

In the late 1950s, Prof. Fellici proposed a DC generator working on the Van der Graff principle, and although successful in small field trials, the rapid development of silicon technology supplanted its use industrially.

In order to progress the art of precipitation worldwide, a strong friendly relationship developed between the main investigators, Sir Oliver Lodge in the UK, Dr Cottrell in the US and Dr Moeller in Germany. Archive evidence shows that there were biannual meetings between the parties to review their latest developments, patents and findings. This close academic and free interchange of ideas between what was to become Lodge Cottrell, Research Corporation and Metallbank und Metallurgische Ges. (Lurgi), lasted until the US Sherman Anti-Trust Laws made such Associations illegal.

One result of these discussions was that Dr Cottrell allowed Sir Oliver Lodge to use the rotary switch rectifier for precipitator duties. In recognition Lodge, in 1921/1922, renamed his company Lodge Cottrell Ltd. In Germany, Lurgi also used the rotary switch rectifier during this period of major and very varied plant installation.

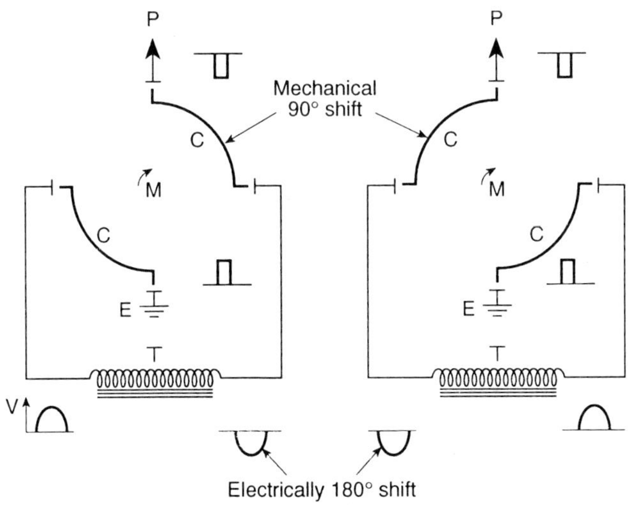

Figure 2.3 Basic principle of mechanical switch rectifier. C, edge connections on insulating disc; E, earth or ground connection (positive); M, synchronous motor (half mains frequency); P, precipitator HT connection (negative); T, transformer secondary winding (AC).

While the rotary rectifier was mechanically robust and reasonably efficient, it was noisy and needed a large room because of insulation requirements and ventilation since the arcing produced both NOx and Ozone. The operation of the rectifier is shown schematically in Figure 2.3 and, although dating from 1908, this form of rectifier was still being supplied in an updated configuration by Lodge Cottrell Ltd in the 1960s for Utility Plant applications.

To reduce some of these difficulties there was a trend, particularly in the US, to use thermionic high voltage diodes or Kenetrons as the valve was called. In Europe development in the selenium rectifier field in the 1940s and early 1950s, tended to replace the rotary rectifier for new installations.

The advent of silicon technology in the late 1950s and early 1960s changed the whole of the precipitator electrics scenario. Virtually all new installations from the mid-1960s onwards were fitted with silicon rectifiers because of their advantages over other forms of rectifier. Not only are they vastly superior in terms of efficiency and are much smaller in size, but the control of the primary input and automatic optimisation systems now incorporate the latest silicon technology.

2.2.2 *Primary control systems*

For many years the only method of controlling the voltage, and hence corona current, was by means of tap changing on the transformer input. This simple tap changing was later modified to an autotransformer or moving coil regulator, where the incoming voltage was modulated, initially by hand wheel, then by a motor drive, as voltage optimisation systems were introduced.

The next type of control was by the use of magnetic saturable reactors or transductors located in the primary circuit, where the output from the saturable reactor is varied by altering the impedance of the device. This is achieved by a DC current passing through a separate winding on the core of the reactor.

Although reasonably successful, these devices were rapidly supplanted by the silicon controlled rectifier or thyristor in the early 1960s. These devices control the power into the transformer by modulating the firing angle of the incoming supply voltage and are superior to previous methods of control.

2.2.3 *Automatic control systems*

With the need to optimise precipitator performance at all times, the development of the automatic voltage control system dates from the 1950s. It was recognised, by this stage, that the optimum precipitator collection efficiency was obtained at the maximum operating voltage as predicted from theory. Once the maximum operating voltage had been achieved, any further increase in primary supply caused the secondary voltage to fall and the current to rise as a result of increased arcing within the precipitator.

The early automatic voltage controllers were electro-mechanical systems, where either the kV or current was monitored and action taken if a certain value was reached or exceeded. These systems were fairly basic in concept and led to discussions as to whether voltage or current control was the best approach. Modern systems monitor both current and voltage to fully optimise performance.

Several forms of wholly electrical device were developed using thermionic valve or magnetic amplifier approaches, but with the rapid development of silicon technology in the 1960s, these were quickly superseded. Initially simple analogue designs were used, then digital and finally microprocessor based, silicon high speed switching systems, which operate in conjunction with the silicon controlled rectifiers, or thyristors, used for primary power modulation.

The latest controller systems are of the stand-alone type, using sophisticated microprocessor units, having programmes which can make the precipitator completely automatic in terms of operation. Facilities can be included to provide complete start-up and shut-down of the precipitator, rapping control, data and fault logging and some provide even a fault finding menu aid, as well as back ionisation detection, pulse modulation and an interruptable power supply to suit a particular set of site conditions. Generally these controllers interface with the plant DCS system for overall simplicity of control.

References

It is not intended for this chapter to be a complete review of the development of electrostatic precipitation, but only to indicate the major milestones in the development of the present high performance plant widely used in industry for the collection of all types of particulate. For those interested in obtaining further insight into the early development of electrostatic precipitators, thc following reference works should be examined.

1. Gilbert, W. (1900) *De Magnete.* (English Edit.) Thompson S.P., pp. 24–5, London, UK.
2. Hohlfeld, M. (1824) Das Niederschlagen des Rauches durch Elektricitat. *Kastner Arch. Gesammte Naturl.*, **2**, 205–6.
3. Guitard, C.F. (1850) Condensation by electicity. *Mech. Mag.* (*London*), **53**, 346.
4. Lodge, O.J. (1884) Dust Free Spaces. Lecture to the Royal Dublin Society, April 2nd (see *Transactions of Society*).
5. Lodge, O.J. (1886) The electrical deposition of dust and smoke with special reference to the collection of metallic fume and to a possible purification of the atmosphere. *J. Soc. Chem. Ind.*, **5**, 572–6.
6. Walker, A.O. (1884) British Patent No. 11 120.
7. Moeller, K. (1884) German Patent No. 31 911.
8. Lodge, O.J. (1884) British Patent No. 24 305.

9. Lodge, O.J., Muirhead, A. and Robinson, E.E. (1905) British Patent No. 25 047.
10. Cottrell, F.G. (1908) US Patent No. 895 729.
11. Cottrell, F.G. (1916) Recent progress in electrical smoke precipitation. Paper Presented at the Second Pan-American Scientific Congress, Washington, Dec. 1915–Jan. 1916. *Eng. Min. J.*, **101**, 385–92.
12. Cottrell, F.G. (1911) The electrical precipitation of suspended particles. *J. Ind. Eng. Chem.*, **3**, 542–50.
13. Schmidt, W.A. (1912) The Control of Dust in Portland Cement Manufacture by the Cottrell Precipitation Processes. Proc. Eighth International Congress. *Appl. Chem.*, **5**, 117–24.
14. The Times Engineering Supplement (1917) October 27th.
15. Howard, W.H. (1914) Electrical precipitation at Garfield. *Trans. Am. Inst. Min. Eng.*, **49**, 540–60.
16. Braley, H.D. (1919) Notes on electrostatic precipitation. 25th General Meeting American Electro-Chemical Society New York. April 3–5 1919, pp. 13–43.
17. White, A.H. *et al.* (1914) The electrical separation of tar from coal gas. *Am. Gas Light J.*, **101**, 177–80.
18. Bradley, L. (1917) The Cottrell process in practice. Abstract of Paper Presented at American Institute of Mining Engineers, and American Electro-Chemical Society, January 26 1917. *Electr. Rev.*, **80** (2066).
19. Hirota, R. and Shiga, K. (1920) Electrical precipitation in Japanese smelters. *Chem. Metall. Eng.*, **22** (6), 276–7.
20. Anderson, E. (1925) Some factors and principles involved in the separation and collection of dust, mist and fume from bases. *Trans. Am. Inst. Chem. Eng.*, **16**, 69.
21. Deutsch, W. (1922) Bewegung und Ledung der Elektricitatstrager in Zylinder Kondensator. *Ann. Physik*, **68**, 335.
22. Matts, S. and Ohnfeldt, O.P. (1963–1964) Efficient gas cleaning with the SF electrostatic precipitator. *Flakt Rev.*, **6, 7**, 105–22.

3 Basic and theoretical operation of ESPs

C. RIEHLE

3.1 General remarks

In electrostatic precipitators (ESPs) the separation of particles from flue gases is based on electrical means. The separation process can be divided into five essential steps which are schematically illustrated in Figure 3.1:

1. generation of charge carriers (section 3.2);
2. charging of the particles (section 3.3);
3. deflection and separation of the particles (sections 3.4, 3.5);
4. dust deposition (section 3.6);
5. dust removal (section 3.7).

In most applications the steps are not as isolated as represented here, but the steps can happen all in one operational section. These are the so-called single-stage precipitators, which are the most important configurations for industrial processes (Figure 3.2a). In so-called two-stage precipitators steps 1/2 are executed separately from steps 3/4/5, i.e. in the first stage the ions are produced and the particles are charged while in the second stage the particles are collected and removed (Figure 3.2b). The two-stage configurations are only used in small-scale applications as, for example, clean room technology. Since single-stage precipitators have a far greater industrial significance than two-stage ESPs, the discussions will be confined to the first. However, most of the presented subjects and equations can be equally applied to two-stage configurations.

In single-stage electrostatic precipitation, two principal different designs exist: tube-type (Figure 3.3a) and plate-type precipitators (Figure 3.3b). In plate-type precipitators a row of discharge wires – orientated along the gravitational axis – are positioned between parallel collecting plates forming a duct. Dust is deposited on the collecting plates due to electrical and Van der Vaals forces. The plates are cleaned by mechanical impact and the dust layer or large agglomerates fall down into the hoppers. An alternative to removing the dust layer by mechanical impact occurs in wet precipitators, where the dust is removed by a flushing liquid flowing over the plates.

In industrial precipitators the total length of the collecting plate is divided into series and parallel fields. A 'field' is characterized by an independent power supply unit. Therefore, the fields of a precipitator can be

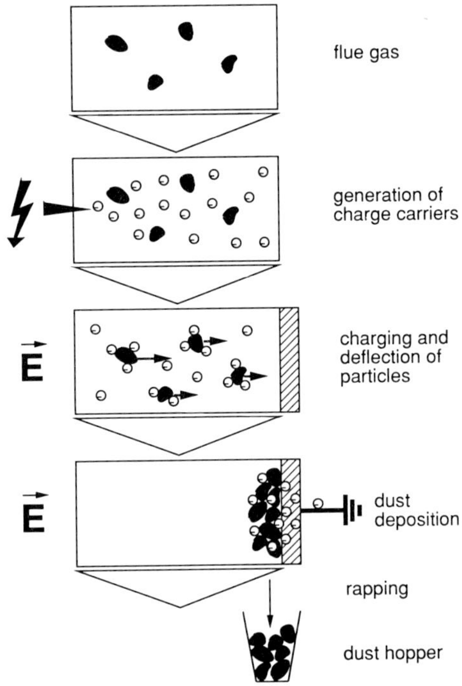

Figure 3.1 Schematic illustration of the separation process in ESPs.

operated with different electrical conditions leading to different transport situations for the particles. Thus the operating conditions can be optimized to different dust loadings and different particle sizes as the gas passes through the precipitator.

In tube-type precipitators the discharge wire and the tube are orientated along the gravitational axis and, in most current applications, the collecting plate is cleaned by liquid films running down into a sump at the precipitator base. For this reason tube-type ESPs are often applied to processes with self-draining liquid particles, e.g. acid mists or tar. In the past, tube-type precipitators were used for dry applications, where the dust was removed by mechanical rapping as in the plate-type approach previously mentioned.

In contrast to the plate-type, in the tube-type, the gas flow and the discharge wires are aligned. Apart from the precipitator arrangement in question, the gas flow conditions should be carefully designed, since re-entrainment of particles from high gas velocities and sneakage are often severe problems preventing high efficiencies being obtained. Therefore, special care should be taken in obtaining a homogeneous gas flow distribution at the precipitator inlet.

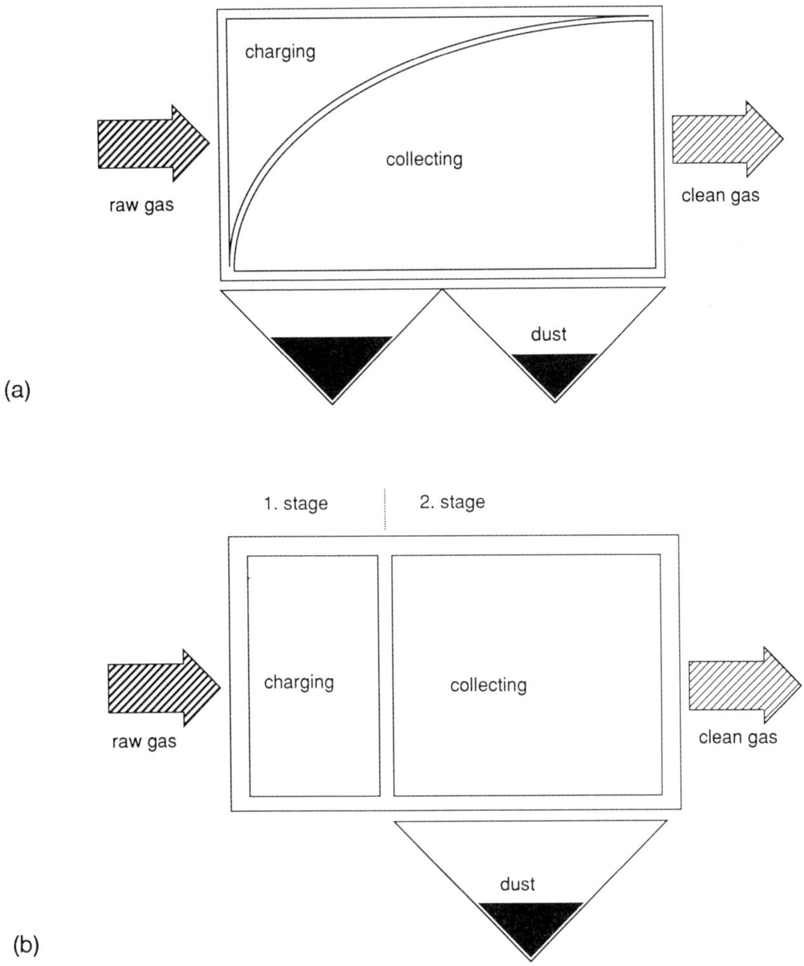

Figure 3.2 (a) Single-stage precipitator; (b) two-stage precipitator.

Besides the flow field, the precipitator's state of operation is determined by the electrical conditions and the dust which is to be collected. Figure 3.4 overviews the influencing parameters and their interactions with respect to particle transport. The interaction between the electrical state and the flow field is called electrohydrodynamic. For example the turbulence intensities of the flow field change when switching on the high voltage [1–3]. The interaction between the dust particles and the electrical state governs charging and discharging mechanisms and space charge effects. The interaction between the dust particles and the flow field governs the particle

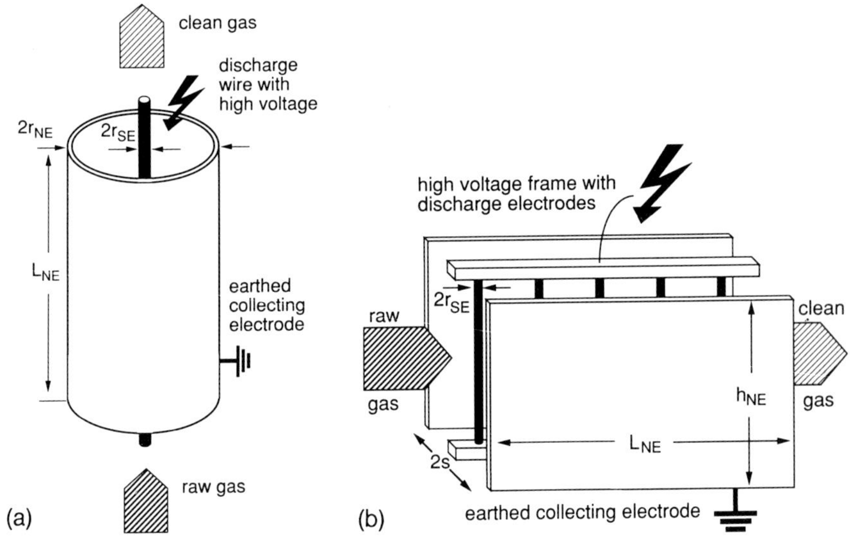

Figure 3.3 (a) Tube-type electrostatic precipitator; (b) plate-type electrostatic precipitator.

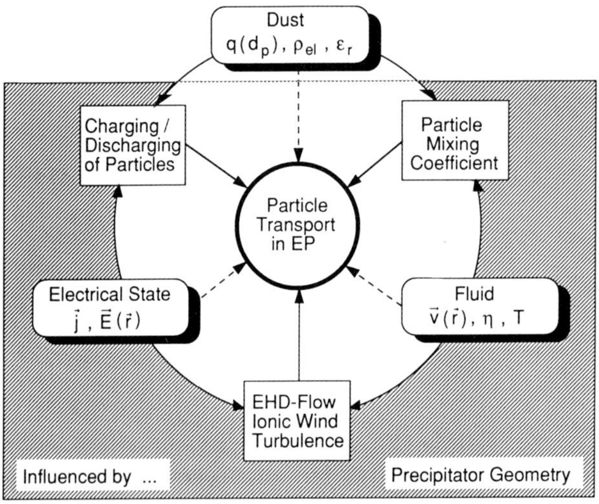

Figure 3.4 Overview of influencing parameters and their interactions with respect to particle transport in a precipitator.

dispersion and the existence of particle concentration profiles. Figure 3.4 illustrates that particle transport is governed more by the interactions than by the influencing parameters themselves. Furthermore, apart from the dust, all influencing parameters and interactions are a function of the precipitator's geometry. Therefore, similar transport conditions for particles on different scales of precipitator size provide at least geometrical similarity. The problem of electrical similarity in ESP operation was investigated recently [4, 5].

3.2 Ion production

3.2.1 Principles

To produce large numbers of charge carriers in a gas phase, a critical electrical field strength has to be overcome. To create this high electrical field strength, high voltage is applied to a special electrode design. When the applied voltage exceeds a distinct value, an electrical current between the two electrodes can be measured indicating a corona discharge. This is the corona onset voltage as illustrated in Figure 3.5. A further increase in voltage will lead to a progressively increasing current until spark-over occurs. The spark-over marks the electrical breakdown of the gas. Figure 3.6a shows a photograph of corona discharges along five parallel smooth wires, while Figure 3.6b shows a successful photograph of a spark-over in a laboratory ESP taken by J. Miller [6]. This shot also visualizes the corona discharges at the ends of the emitters, in this case, spiked-type electrodes.

What is the characteristic of a corona discharge compared with a spark-over and how does it start? Usually a gas contains in the order of 10^{19} neutral molecules per cm^3. Because of natural radioactivity and cosmic radiation some molecules become ionized and recombine immediately afterwards. If an electrical field is present during ionization, the electron will be accelerated and separated quickly from the remaining positive ion (Figure 3.7). After a short distance the electron will hit another neutral gas

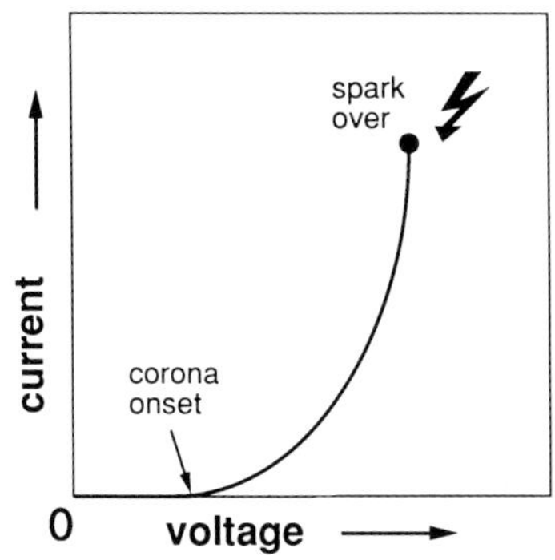

Figure 3.5 Typical current–voltage relationship.

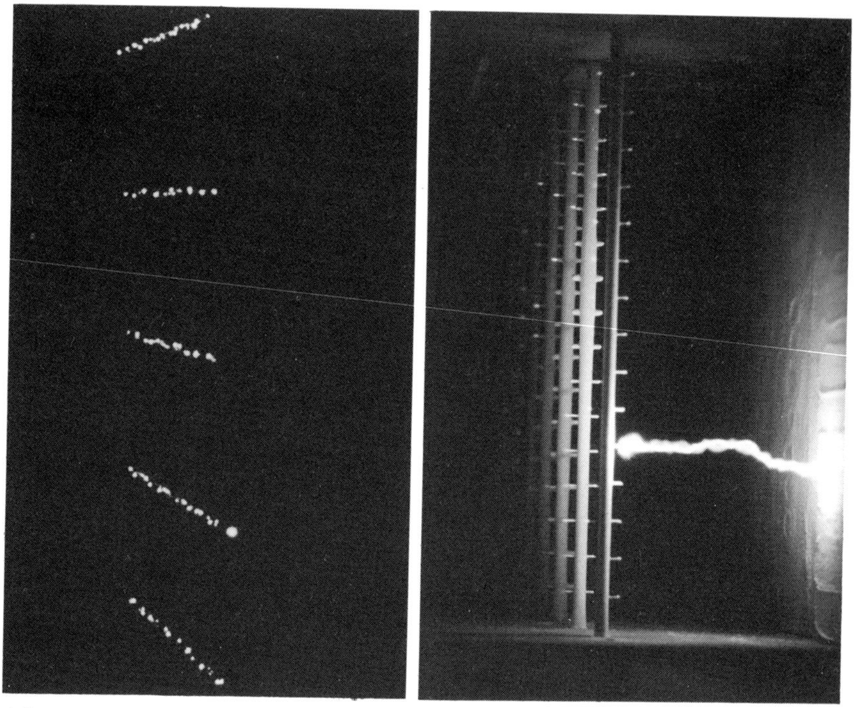

(a) (b)

Figure 3.6 (a) Corona discharges along five parallel smooth wires [24]; (b) spark-over in a laboratory ESP (photograph taken by J. Miller [6]).

molecule and an additional, second electron will be produced, provided the kinetic energy has been high enough for ionization. Thus, an avalanche effect is able to start in regions where the electrical field strength is high enough to bring the electrons to ionization energy levels. Recombining ions and highly activated molecules emit photons and the weak blue glow is therefore characteristic for a corona discharge in addition to a crackling noise. An inhomogeneous electrical field occurs at curved surfaces, e.g. wires, which means that the high field values are restricted to regions close to the wire. This region, where ionization processes occur, is called the active zone. When the moving electrons come into the region of lower electrical field strength, the passive zone, they are not able to ionize any further molecules, but they will attach themselves to an electronegative gas molecule, such as O_2, SO_2, Cl_2, thereby forming negative gas ions. The spatial extension of the active zone is only a few percent of the passive zone, i.e. particle separation occurs almost completely in the passive zone.

The drifting charge carriers, which are continously produced by the so-called 'trichel pulses' of the corona discharge, represent the electrical

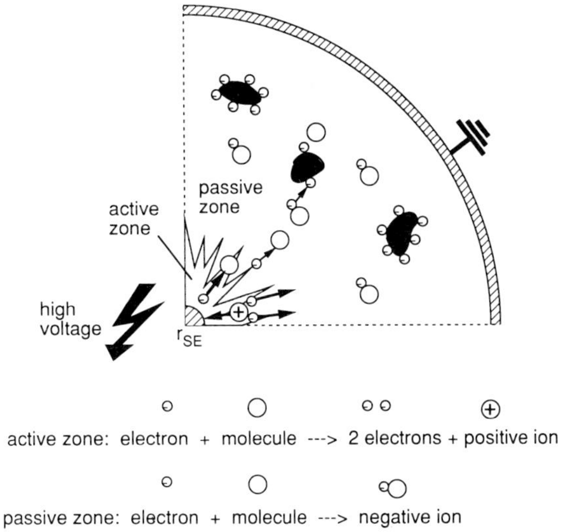

Figure 3.7 Principle of a corona discharge.

discharge current. If the applied voltage, however, exceeds a certain level, a discharge begins to pass straight through the gas as a 'streamer'. Suddenly electrical breakdown of the gas sets in and spark-over occurs.

In general, the discharge electrode can be operated on negative or positive polarity. For a given geometry, however, the corona initiation voltage and the electrical breakdown of the gas occur at higher voltages for negative energization than for positive. Because of these higher electrical field strengths, most industrial applications prefer negative corona, i.e. discharge electrodes are energized with negative high voltage and the collecting plates (positive) are grounded. Readers who are interested in more details of corona discharge with respect to the ESP application may refer to [7, 8].

3.2.2 Corona initiation field strength

From section 3.2.1, it should be clear that the precipitator configuration, the gas state and especially the gas composition, determine an essential part of the ESP's electrical properties. The electrical field required to start a continous ionization process will obviously depend on the ionization energies of the gas species present and the mean free path between collisions. Since the mean free path is related to the gas state, the corona initiation field strength will be a function of gas density. Furthermore, the curvature of the wire determines the inhomogeneity of the electrical field. A pure theoretical description, however, is still missing because the analysis of a complex mixture of gases is difficult. Peek, however, has proposed a semi-empirical relationship (equation 3.1) where the initiation field strength E_0 is expressed as a function of the relative gas density δ (equation 3.2), the discharge wire radius r_{SE} and two empirical constants, A and B, characterizing the gas and corona polarity [9].

$$E_0 = A\delta + B\sqrt{\delta / r_{SE}} \tag{3.1}$$

$$\delta = \frac{\rho_2}{\rho_1} = \frac{p_2}{p_1} \cdot \frac{T_1}{T_2} \tag{3.2}$$

Usually temperature and pressure are related to normal conditions, i.e. $T_1 = T_0 = 273\,\text{K}$ and $p_1 = p_0 = 1$ bar. For typical ESP conditions (negative corona), Robinson [10] recommends as empirical constants:

$$A = 3,2 \cdot 10^6\ \text{V/m}, \qquad B = 9 \cdot 10^4\ \text{V/m}^{1/2}$$

Figure 3.8 shows the corona initiation field strength E_0 as a function of wire radius. The thick line represents the relationship for the recommended values; the thin lines were calculated with the constants A, B for pure SO_2 and CO_2, respectively, in order to show the sensitivity to gas composition. In general an increase of wire radius leads to a decrease in electrical field strength, this trend reducing for radii > 1 mm.

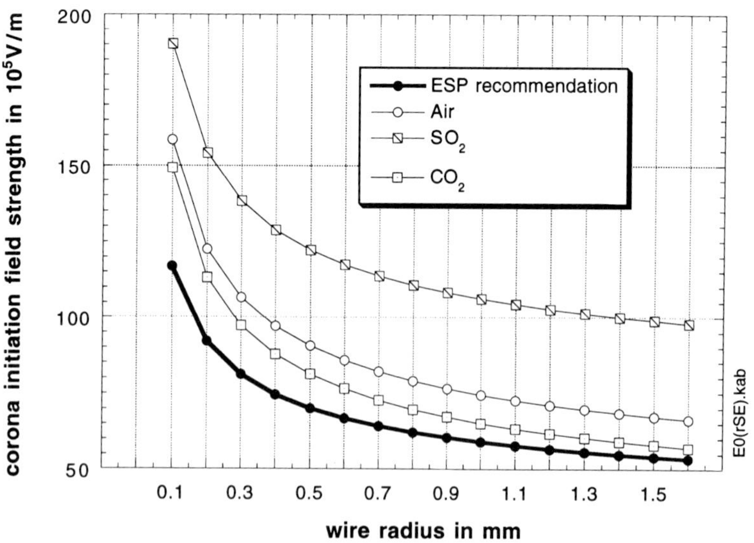

Figure 3.8 Corona initiation field strength E_0 as a function of wire radius.

3.2.3 *Corona onset voltage*

The conditions for the corona initiation field strength are essentially set by the design of the discharge electrode. The voltage necessary to overcome this critical field strength, however, is set by the complete configuration of discharge and collecting electrodes. For this reason the corona onset voltage has to be additionally a function of the collecting electrode's design. Therefore, a distinction between the two main ESP designs must be considered (see Table 3.1 for typical characteristic values of ESP operation).

3.2.3.1 Tube-type. The collecting electrode of a tube-type ESP is simply described by the tube's radius r_{NE}. From electrostatic field theory the following relationship can be deduced for coaxial electrode configurations:

$$V_0 = E_0(\delta, r_{SE}) \cdot r_{SE} \cdot \ln \frac{r_{NE}}{r_{SE}} \tag{3.3}$$

In Figure 3.9 the corona onset voltage is plotted as a function of the collecting tube radius for three different wire radii. The curves hold for a relative gas density of 1.0 (this dependence will be discussed in chap. 16). Generally, increasing tube size needs higher voltages for corona onset. Alternatively, the onset voltage can be lowered by reducing the radius of the discharge wire.

Table 3.1 Typical characteristic values of ESP operation

	Tube	Plate
v_0 (m/s)	1.5–2.5	1.0–2.0
$2s$ (mm)	150–250	250–400
E (kV/cm)	5–5.5	3.5–4
j (mA/m^2)	~1.0	~0.5
Particle concentration in raw gas (g/Nm^3)	Up to 10	Up to 1000
Efficiencies (%)	99–99.5	99.5–99.9+
Applications	Mist collection, e.g. SO_3, hydrocarbons, tars	Dry dusts, e.g. P.F., cement, iron ore; wet dusts e.g. BF gas, BOS gas, DESOX gas
Flow rates (m^2/s)	Up to 20	Up to 1000
Δp	20	10–15
Energy consumption		
Efficiency (%)	**DC power (W/m^3/s)**	**DC power[a] (W/m^3/s)**
99.0	500	170
99.5	1000	300
99.75		450
99.90		1000

[a]Depends on type of discharge electrode.

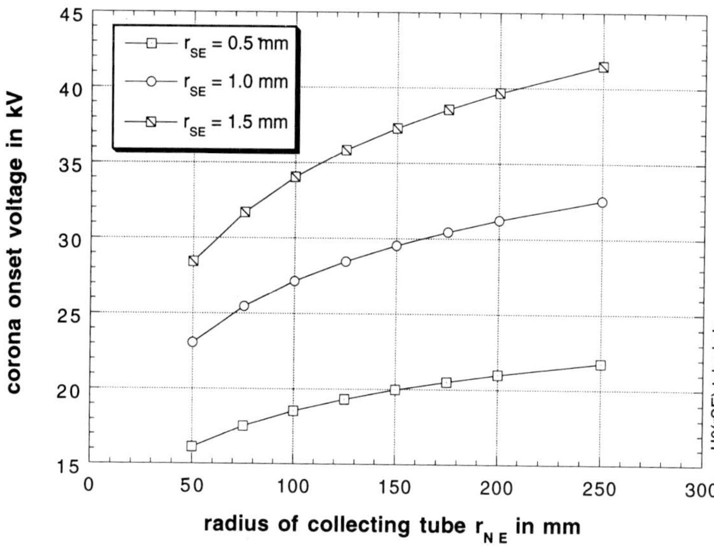

Figure 3.9 Plot of corona onset voltage as a function of the collecting tube radius for three different wire radii.

3.2.3.2 Plate-type. Beside the wire radius r_{SE} and the wire-to-plate distance s for the determination of plate-type configurations another parameter is necessary, i.e. the distance between the discharge wires, $2c$ (Figure 3.10). When describing plate-type configurations, a characteristic length d is introduced, which is a function of the wire–plate distance s and the wire-to-wire distance $2c$. The corona onset voltage for plate-type ESPs is given by equation (3.4) while the characteristic length ratio is determined by equations (3.5a,b,c).

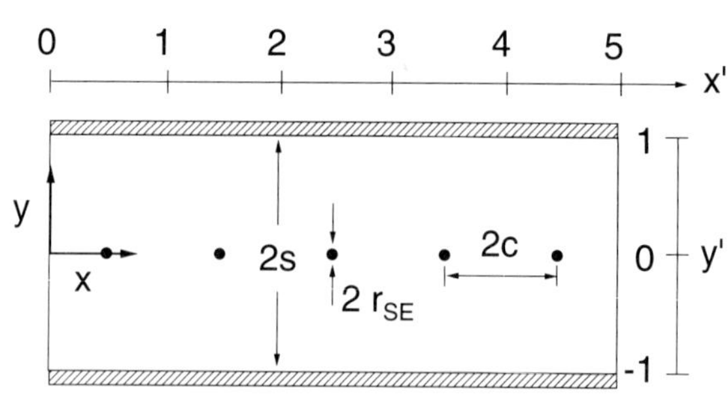

Figure 3.10 Nomenclature for plate to plate configuration.

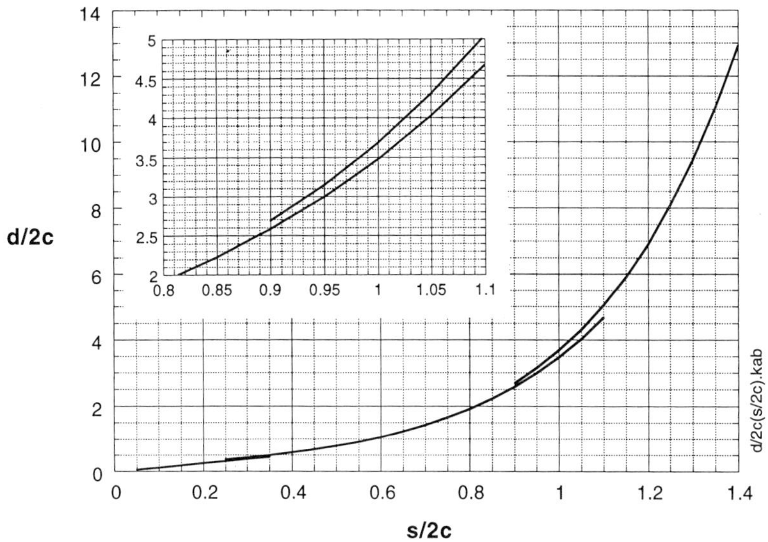

Figure 3.11 Plot of values for $d/2c$, dependent on the $s/2c$ ratio.

$$V_0 = E_0(\delta, r_{SE}) \cdot r_{SE} \cdot \ln \frac{d(s, 2c)}{r_{SE}} \tag{3.4}$$

$$\frac{d}{2c} = \frac{4}{\pi} \cdot \frac{s}{2c}; \qquad \text{for} \qquad s/2c < 0{,}3 \tag{3.5a}$$

$$\frac{d}{2c} = 0{,}18 \cdot \exp\left(2{,}96 \cdot \frac{s}{2c}\right); \quad \text{for} \quad 0{,}3 < s/2c < 1{,}0 \tag{3.5b}$$

$$\frac{d}{2c} = \frac{1}{2\pi} \cdot \exp\left(\pi \cdot \frac{s}{2c}\right); \qquad \text{for} \quad 1{,}0 < s/2c \tag{3.5c}$$

Dependent on the $s/2c$ ratio, the values for $d/2c$ can be calculated according to the different relationships (3.5a,b,c) and are plotted in Figure 3.11, from which it will be seen that wire–plate spacings larger than the wire-to-wire distance progressively dominate the characteristic length. Higher characteristic lengths mean that higher onset voltages are required. In other words, a very short wire-to-wire distance interferes with the corona discharge, while wide wire-to-wire distances need lower corona onset voltages (the electrical field in the space between them and the collecting plate, however, will decrease, resulting in deteriorating particle collection).

Table 3.2 gives examples for the value range of the characteristic length d for typical wire–plate distances and some variations in wire-to-wire distance. Figure 3.12 illustrates the corona onset voltage for plate-type

Table 3.2 Examples of typical size values in plate-type design with characteristic length d

s/mm		150		200		250	
$s/2c$	$d/2c$	$2c$/mm	d/mm	$2c$/mm	d/mm	$2c$/mm	d/mm
0.5	0.79	300	237	400	316	500	395
1.0	3.57	150	535.5	200	714	250	892.5
1.5	17.72	100	1772	133	2357	166	2941

configurations as a function of the characteristic length d for different electrode radii. The behaviour is analogous for tube-type ESPs, where an increasing d means higher corona onset voltages. Again the onset voltage can be lowered by reducing the radius of the discharge wire.

3.2.4 *Current–voltage relationship*

The electrical current is formed by gas ions passing through the passive zone as mentioned above (section 3.2.1). The drift velocity of these gas ions is proportional to the electrical field applied and the constant of propor-

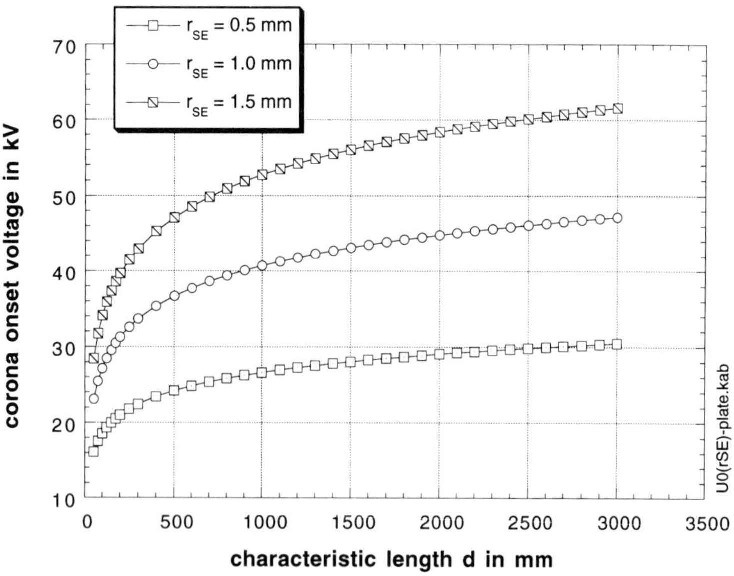

Figure 3.12 Corona onset voltage for plate-type configurations as a function of the characteristic length d for different electrode radii.

Table 3.3 Mobilities of single-charged gas ions at 0 °C and 1.0 bar [8]

Gas	Mobility in $10^{-4}\,m^2/Vs$	
	Negative ions	Positive ions
Air (dry)	2.1	1.36
Air (very dry)	2.5	1.8
CO_2 (dry)	0.98	0.84
H_2O (100 °C)	0.95	1.1
SO_2	0.41	0.41

tionality is called the electrical mobility, b_i, of gas ions. Since theoretical expressions for mobilities, derived by kinetic theory of gases, are still questionable, experimentally determined values are recommended whenever these are available. Table 3.3 gives a few mobilities for positive and negative gas ions [8]. It has been found experimentally that ion mobility is almost inversely proportional to the relative gas density δ (equation 3.2) over a wide temperature–pressure range, thus equation (3.6) holds. Since most ESPs operate with negative corona and a gas composition similar to air, the following mobility value for normal conditions is assumed (equation 3.7).

$$b(p, T) = \frac{b(p_0, T_0)}{\delta} \tag{3.6}$$

$$b(T_0, p_0) = 2 \cdot 10^{-4}\,\frac{m^2}{Vs} \tag{3.7}$$

For typical electrical field strengths e.g. $1 - 5 \times 10^5$ V/m, the gas ion's velocity ranges between 20 and 100 m/s.

The operating condition of an ESP is mainly governed by its electrical state. The relationship between the applied voltage and the resulting current is called the current–voltage characteristic. This curve describes all possible electrical operating states of a given ESP configuration. Usually the total electrical current i_{tot} is related to the total collecting area $A_{NE,tot}$. Thus the electrical current density j_{NE} results (equation 3.8), having the unit of current per unit area. Sometimes the total electrical current is related to the total length of discharge wires $L_{SE,tot}$. Thus an electrical current per unit length results (equation 3.9).

$$j_{NE} = \frac{i_{tot}}{A_{NE,tot}} \tag{3.8}$$

$$j_{SE} = \frac{i_{tot}}{L_{SE,tot}} \tag{3.9}$$

In practice, however, the electrode designs are not always as simple as illustrated in Figures 3.3a and 3.3b. Especially the discharge electrodes,

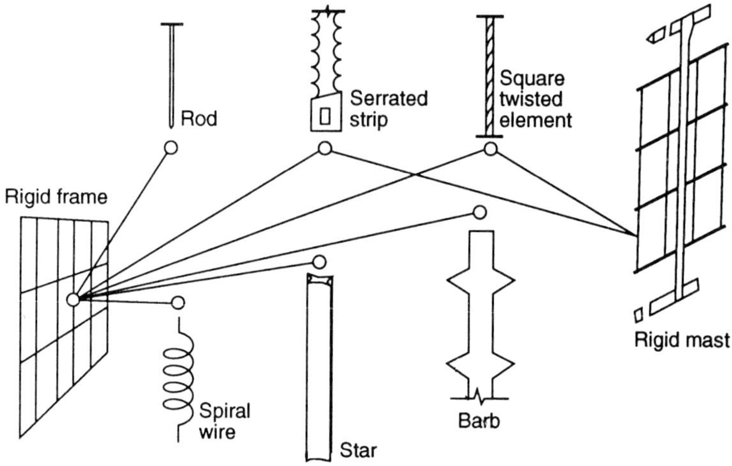

Figure 3.13 Discharge electrode elements.

which are rarely round wires. All sorts of sharp edges, points, barbs, etc., are common and Figure 3.13 gives only a few examples. For most of these discharge electrodes, a theoretical description for the current–voltage relationship is impossible. Thus the characteristic of a certain electrode has to be measured in the laboratory. A great variety of collecting plate designs also exist and some examples are shown in Figure 3.14. The background of collecting plate profile design is directed more towards the improvement of particle collection and retention and less by electrical properties. Most collector profiles can be considered flat in relation to the electrode/collector spacing.

When designing industrial precipitators an estimation of the electrical power is needed and the dependence on geometry might be helpful. Therefore, the next sections introduce the theoretical approaches to current–voltage relationships for idealized geometries.

3.2.4.1 Tube-type. In tube-type ESPs, the number of tubes N_t equals the number of discharge elements N_{SE} and the tube length L_{NE} is approximately equal to the wire length L_{SE} (see Figure 3.3a). Therefore, equation 3.10 holds for the current per discharge wire j_{SE}. When the total current is related to the collecting area j_{NE}, equation (3.11) results, wherein, i_t represents the current per tube and i_{tot} the total current.

$$j_{SE} = \frac{i_{tot}}{L_{SE}^{tot}} = \frac{N_t i_t}{N_t L_{SE}} \approx \frac{i_t}{L_{NE}} \tag{3.10}$$

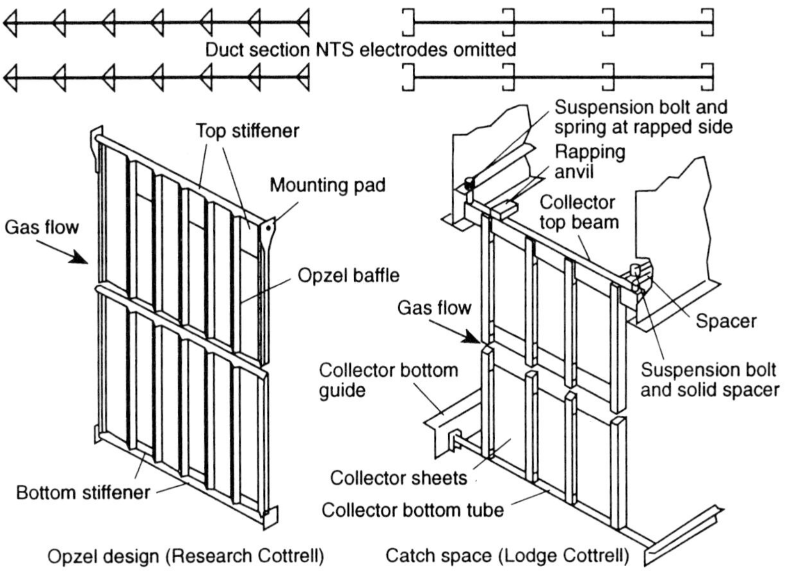

Figure 3.14 Collecting plate designs: Opzel Design (Research Cottrell); Catch Space (Lodge Cottrell).

$$j_{NE} = \frac{i_{tot}}{A_{NE}^{tot}} = \frac{N_t i_t}{N_t 2\pi r_{NE} L_{NE}} = \frac{i_t}{2\pi r_{NE} L_{NE}} \tag{3.11}$$

Considering the last two equations a general transformation rule for both kinds of current density in tube-type ESPs can be given:

$$j_{NE} = \frac{j_{SE}}{{}^{2}\pi r_{NE}} \tag{3.12}$$

With the assumption of low current flow, a quite simple relationship can be deduced from the theory of electrostatic fields (equation 3.13). This approximation was originally published by Townsend in 1915 [11]. It is applicable as long as equation 3.14 holds. The permittivity of vaccuum ε_0 is 8.86e12 As/Vm.

$$j_{NE} = \frac{j_{SE}}{2\pi \cdot r_{NE}} = \frac{4\varepsilon_0 \cdot b(\delta) \cdot U \cdot (U - U_0(r_{NE}, r_{SE}, \delta))}{(r_{NE})^3 \cdot \ln \dfrac{r_{NE}}{r_{SE}}} \tag{3.13}$$

$$\left(\frac{r_{NE}}{E_0 \cdot r_{SE}}\right)^2 \cdot \frac{r_{NE} \cdot j_{NE}}{\varepsilon_0 \cdot b(\delta)} \ll 1 \tag{3.14}$$

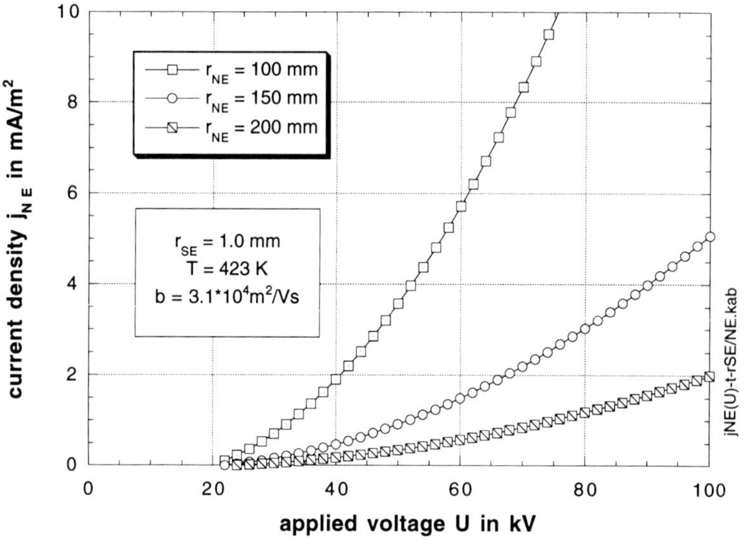

Figure 3.15 Dependency of current density, j_{NE}, on the collecting tube radius.

Figure 3.15 illustrates the dependence of current density j_{NE}, according to equation (3.13), on the collecting tube radius. The wider the duct, the progressively lower the current density becomes for a given voltage. Compared with the collecting tube radius, the influence of the wire radius is negligibly small.

In the case of higher currents, a simple calculation becomes impossible. The corresponding expression is given in equation (3.15), according to Townsend [11], where Ψ_t and Φ_t are defined by equations (3.16) and (3.17), respectively. Equation (3.15) is plotted in Figure 3.16, therefore for practical purposes only equations (3.16) and 3.17) have to be used.

$$\Psi_t = \sqrt{1+\Phi_t} - 1 - \ln\frac{1+\sqrt{1+\Phi_t}}{2} \tag{3.15}$$

$$\Psi_t = \frac{U-U_0}{U_0} \cdot \ln\frac{r_{NE}}{r_{SE}} \tag{3.16}$$

$$\Phi_t = \left(\frac{r_{NE}}{E_0 \cdot r_{SE}}\right)^2 \cdot \frac{r_{NE} \cdot j_{NE}}{\varepsilon_0 \cdot b(\delta)} \tag{3.17}$$

3.2.4.2 Plate-type. Plate-type ESPs consist of a number of parallel ducts N_d (see Figure 3.3b). The number of discharge electrodes per duct is N_{SE}. The current per unit wire length j_{SE} is given by equations (3.18), where i_{tot}, i_d and i_{SE} represent the total current, the current per duct and the current per

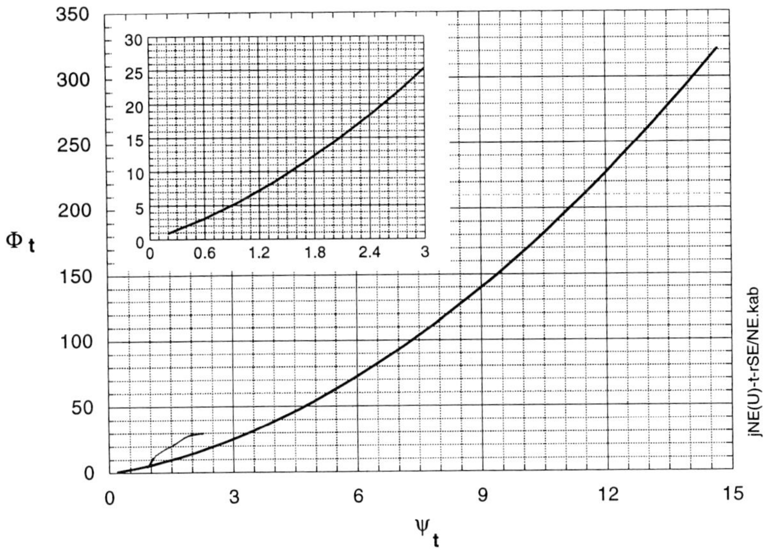

Figure 3.16 Plot of equation (3.15).

discharge wire, respectively. For the current density per unit collecting area j_{NE}, equation (3.19) holds.

$$j_{SE} = \frac{i_{tot}}{L_{SE}^{tot}} = \frac{N_d i_d}{N_d N_{SE} h_{SE}} = \frac{N_{SE} i_{SE}}{N_{SE} h_{SE}} = \frac{i_{SE}}{h_{SE}} \tag{3.18}$$

$$j_{NE} = \frac{i_{tot}}{A_{NE}^{tot}} = \frac{N_d i_d}{N_d A_{NE}} = \frac{i_d}{2L_{NE} h_{NE}} \tag{3.19}$$

Combining the last two equations, a general transformation for both types of current density can be deduced (equation 3.20). This can be approximated to equation (3.21) (where $2c$ = discharge element spacing, Figure 3.10) because the relationships (3.22) and (3.23) hold for most plate-type designs.

$$j_{NE} = j_{SE} \frac{N_{SE} h_{SE}}{2L_{NE} h_{NE}} \tag{3.20}$$

$$j_{NE} \approx \frac{j_{SE}}{4c} \tag{3.21}$$

$$h_{NE} \approx h_{SE} \tag{3.22}$$

$$\frac{L_{NE}}{N_{SE}} \approx 2c \tag{3.23}$$

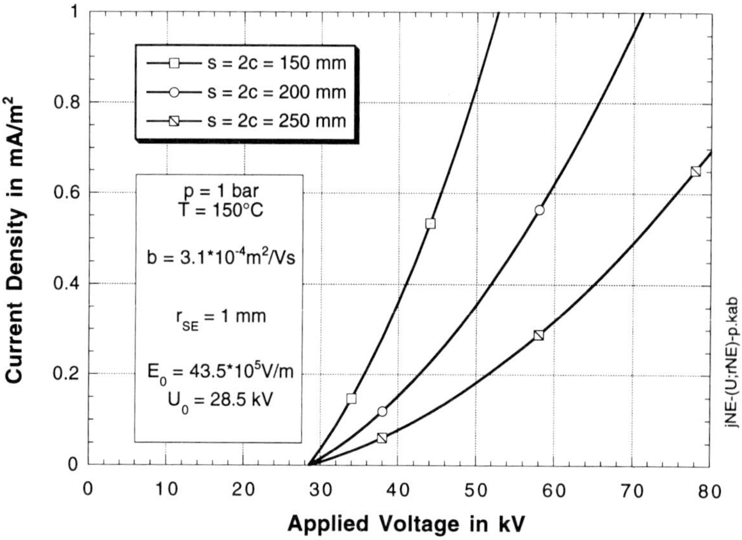

Figure 3.17 Plot of current–voltage relationship for a range of parameters according to equation (3.24).

The wire–plate arrangement is much more difficult to handle theoretically than the wire–tube configuration. However, P. Cooperman derived an analogous solution for the current density (equation 3.24) by assuming low currents, which is fulfilled as long as equation (3.25) holds [12].

$$j_{\mathrm{NE}} = \frac{j_{\mathrm{SE}}}{4c} = \frac{\pi\varepsilon_0 \cdot b(\delta) \cdot U \cdot (U - U_0(d, r_{\mathrm{SE}}, \delta))}{c \cdot s^2 \cdot \ln \dfrac{d}{r_{\mathrm{SE}}}} \tag{3.24}$$

$$\left(\frac{s}{E_0 \cdot r_{\mathrm{SE}}}\right)^2 \cdot \frac{1}{\ln \dfrac{d}{r_{\mathrm{SE}}}} \cdot \frac{c \cdot j_{\mathrm{NE}}}{\pi \cdot \varepsilon_0 \cdot b(\delta)} \ll 1 \tag{3.25}$$

The current–voltage relationship according to equation (3.24) is plotted in Figure 3.17, for a range of parameters.

For high currents, G. Cooperman proposed a different approach [13]. He applied an approximative solution for high currents given by White [7] to certain parts of the precipitator duct. The result and definitions are presented in equations (3.26) to (3.28). Herein $E_{\mathrm{NE},j=0}$ characterizes the highest electric field strength at the collecting plate achievable in the electrostatic case.

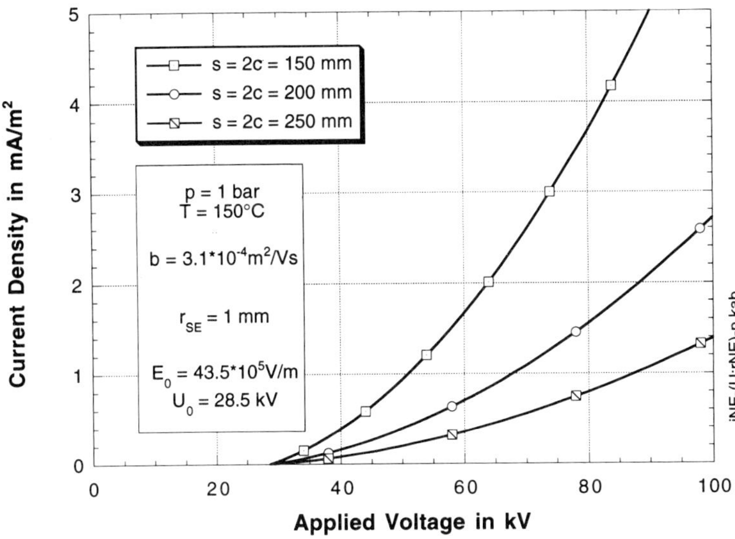

Figure 3.18 Plot of current–voltage relationship for the same parameter settings as in Figure 3.17, according to equation (3.26).

$$j_{\mathrm{NE}} = \frac{\varepsilon_0 b}{16 s^3} \left[\alpha + \sqrt{\alpha^2 + 192(sE_{\mathrm{NE},\,j=0})^3 (U - U_0)}\right] \tag{3.26}$$

$$\alpha = 9(U - U_0 + sE_{\mathrm{NE},\,j=0})^2 - 12(sE_{\mathrm{NE},\,j=0})^2 \tag{3.27}$$

$$E_{\mathrm{NE},\,j=0} = \frac{\pi U_0}{2c \ln \dfrac{d}{r_{\mathrm{SE}}}} \tag{3.28}$$

The current–voltage relationship according to equation (3.26) is plotted in Figure 3.18 for the same parameter settings as Figure 3.17, allowing comparison of both.

Since, in the case of higher currents, a simple calculation becomes impossible for plate-type designs, an analogous graphical procedure can be carried out, as for the tube configuration. The corresponding expression is given in equation (3.29) according to Robinson [10] while Ψ_{p} and Φ_{p} are defined by equations (3.30) and (3.31), respectively. Equation (3.29) is plotted in Figure 3.19; therefore, for practical purposes only equations (3.30) and (3.31) have to be used.

$$\Psi_{\mathrm{p}} = \frac{1}{2}\left\{a \cdot \ln\left[\left(\frac{a-1}{a+1}\right) \cdot \left(\frac{a+b}{a-b}\right)\right] - \ln\left[\frac{\Phi_{\mathrm{p}}}{2} \cdot \left(\frac{b+1}{b-1}\right)\right]\right\} \tag{3.29a}$$

$$a = \sqrt{2\Phi_{\mathrm{p}} + 1} \tag{3.29b}$$

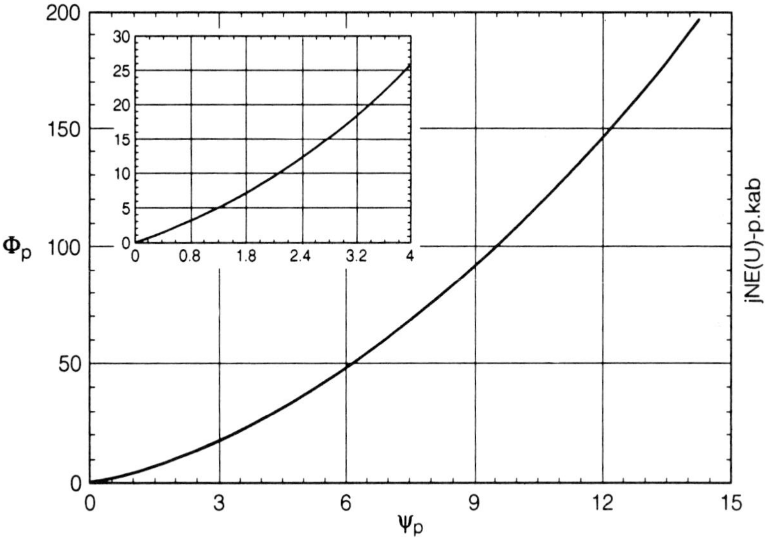

Figure 3.19 Plot of equations (3.29a), (3.29b) and (3.29c).

$$b = \sqrt{\Phi_p + 1} \tag{3.29c}$$

$$\Psi_p = \frac{U - U_0}{U_0} \cdot \ln \frac{4s}{\pi r_{SE}} \tag{3.30}$$

$$\Phi_p = \left(\frac{s}{E_0 \cdot r_{SE}}\right)^2 \cdot \frac{16s \cdot j_{NE}}{\pi^3 \cdot \varepsilon_0 \cdot b(\delta)} \tag{3.31}$$

When looking at common plate-type designs the ratio $s/2c$, i.e. the half duct width to the wire-to-wire spacing, is normally around 1. This does not happen accidently, but the current density reaches a maximum value for that configuration. This can be demonstrated by taking equation (3.24) and calculating current densities as a function of wire-to-wire spacing for given duct widths and given wire radius. This relationship is shown in Figure 3.20, for $s = 200$ mm and $r_{SE} = 1.0$ mm, an applied voltage of 40 kV and an ion mobility of 3.1×10^{-4} m^2/Vs. If the wire-to-wire distance is very small, the corona will be completely suppressed.

3.2.5 *Electrical field distribution*

Ions are generated as soon as the electrical field strength overcomes the critical value E_0. However, the distribution of the electrical field in the tube,

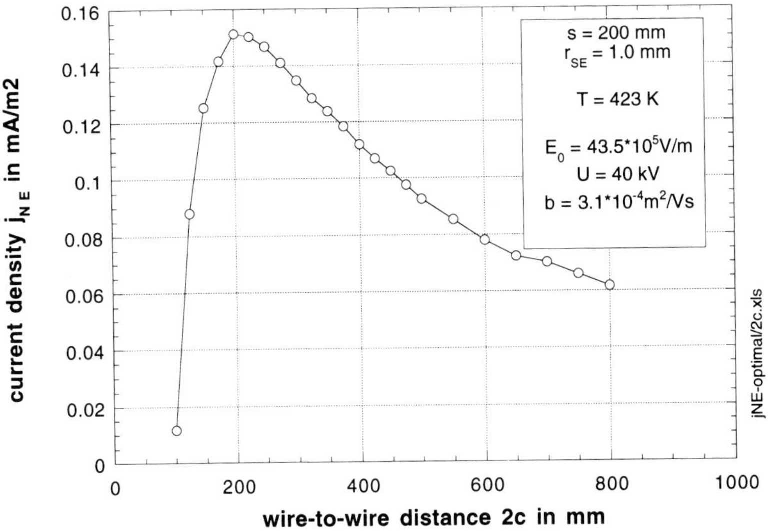

Figure 3.20 Current density as a function of wire-to-wire spacing for given duct widths and given wire radius.

or in the duct, governs the distribution of the current on the collecting plates, the charging of the particles and their migration. Electrical field strength is therefore a local function and the field distribution depends on the geometry and the operating state of a precipitator. Quite often the electrical state of a precipitator is simply described by an electrical field strength according to equation (3.32). However, the value E_{ps}, which is U/s for plate-type and U/r_{NE} for the tube-type designs, represents only a rough approximation of the real state, because it characterizes the electrical field only for parallel plate electrodes. For that reason E_{ps} is called the pseudo-homogeneous electrical field strength.

$$E_{ps} = \frac{\text{applied voltage}}{\text{distance wire-collecting plate}} \tag{3.32}$$

The field distribution is obviously affected by the distribution of electrical charge. Therefore, in order to get the electrical field distribution, the distribution of the gas ions has to be known and vice versa. This is why a correct solution of the problem needs, for example, an iterative procedure to solve Poisson's equation and the equation of continuity of electrical charge as outlined in Figure 3.21 [5]. However, such detailed computations are restricted to simple electrode geometries. For practical applications,

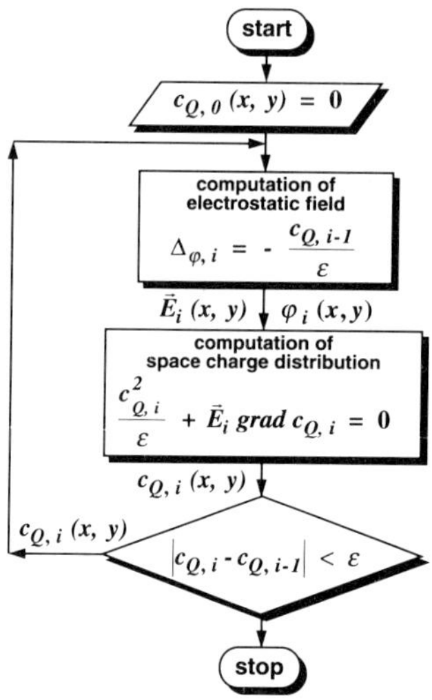

Figure 3.21 Outline of procedure to solve Poisson's equation and the equation of continuity of electrical charge.

simpler approaches, such as neglecting space charge or assuming a homogenously distributed charge, are often preferred.

3.2.5.1 Tube-type. In a coaxial geometry the radial position is described by r. Taking the electrical field equation, the distribution for the electrostatic case can easily be calculated (equation 3.33a). Considering a homogenously distributed ionic space charge, Robinson derived the distribution as given by equation (3.34a) [10]. In Figure 3.22 both relationships are plotted in dimensionless form. As can be seen, the space charge, resulting from a voltage higher than corona onset, leads to an increasing electrical field strength near the collecting plate.

$$E(r) = \frac{U}{r \cdot \ln \frac{r_{NE}}{r_{SE}}} \tag{3.33a}$$

$$E(r) = \sqrt{\frac{r_{NE} \cdot j_{NE}}{\varepsilon_0 \cdot b(\delta)} + \frac{r_{SE}}{r}\left[E_0^2 - \frac{r_{NE} \cdot j_{NE}}{\varepsilon_0 \cdot b(\delta)}\right]} \tag{3.34a}$$

The dimensionless forms, equations (3.33b), (3.34b), result from introducing dimensionless quantities according to equations (3.35) to (3.39). The dimensionless current density (equation 3.36) and the dimensionless voltage (equation 3.37) have proved to be helpful when comparing current–voltage

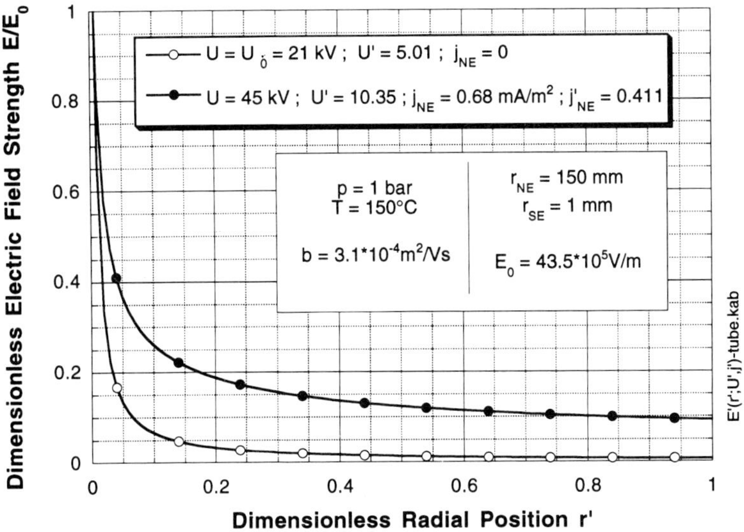

Figure 3.22 Plot of radial position and electric field strength in dimensionless form.

relationships for precipitators with different absolute size [4].

$$E'(r') = \frac{1}{r' \cdot \ln \dfrac{1}{r'_{SE}}} \tag{3.33b}$$

$$E'(r') = \sqrt{j'_{NE} + \frac{1}{r'}\left[\frac{1}{r'_{SE} \cdot U'^2} - r'_{SE} \cdot j'_{NE}\right]} \tag{3.34b}$$

$$E' = \frac{E(r)}{E_{ps}} = \frac{E(r)}{U/r_{NE}} \tag{3.35}$$

$$j'_{NE} = \frac{j_{NE}}{\dfrac{\varepsilon_0 \cdot b(\delta)}{r_{NE}^3} \cdot U^2} \tag{3.36}$$

$$U' = \frac{U}{E_0 \cdot r_{SE}} \tag{3.37}$$

$$r' = \frac{r}{r_{NE}} \tag{3.38}$$

$$r'_{SE} = \frac{r_{SE}}{r_{NE}} \tag{3.39}$$

3.2.5.2 Plate-type. For wire-plate geometries an analytical solution can be derived for the electrostatic case [14, 15]. The vector components $\boldsymbol{E}_{x,j=0}$, $\boldsymbol{E}_{y,j=0}$ are given in equations (3.40) and (3.41), respectively. In order to allow simple programming on a computer, some abbreviations have been used according to the equations (3.42) to (3.47). The electrostatic solution is a superposition of the electrical field of m wires between two parallel plates. The coordinates are used as introduced in Figure 3.10. The coordinates and all length scales occurring are related to the half duct width s (equations 3.48 to 3.51) in order to derive relative quantities.

$$\boldsymbol{E}_{x,j=0}(x', y') = \frac{U}{s} \cdot \pi \cdot \cos\alpha \cdot \frac{\sum_m \dfrac{\sinh\beta}{N_1}}{N_2} \tag{3.40}$$

$$\boldsymbol{E}_{y,j=0}(x', y') = \frac{U}{s} \cdot \pi \cdot \sin\alpha \cdot \frac{\sum_m \dfrac{\cosh\beta}{N_1}}{N_2} \tag{3.41}$$

$$\alpha = \pi y'/2 \tag{3.42}$$

$$\beta = \pi(x' - 2mc')/2 \tag{3.43}$$

$$N_1 = \cosh^2\beta - \cos^2\alpha \tag{3.44}$$

$$N_2 = \sum_m \ln\left\{\frac{\cosh\gamma + \cos\delta}{\cosh\gamma - \cos\delta}\right\} \tag{3.45}$$

$$\gamma = \pi mc' \tag{3.46}$$

$$\delta = \pi r'_{SE}/2 \tag{3.47}$$

$$x' = x/s \tag{3.48}$$

$$y' = y/s \tag{3.49}$$

$$2c' = 2c/s \tag{3.50}$$

$$r'_{SE} = r_{SE}/s \tag{3.51}$$

In Figure 3.23a, a three-dimensional plot of the spatial distribution of the $\boldsymbol{E}_y$-component in the electrostatic case is plotted in dimensionless form (equation 3.52) as a function of the relative coordinates for a precipitator duct with five discharge wires (the $\boldsymbol{E}_y$-component is responsible for the transport towards the collecting plates). As can be seen, $\boldsymbol{E}_y$ is very high only

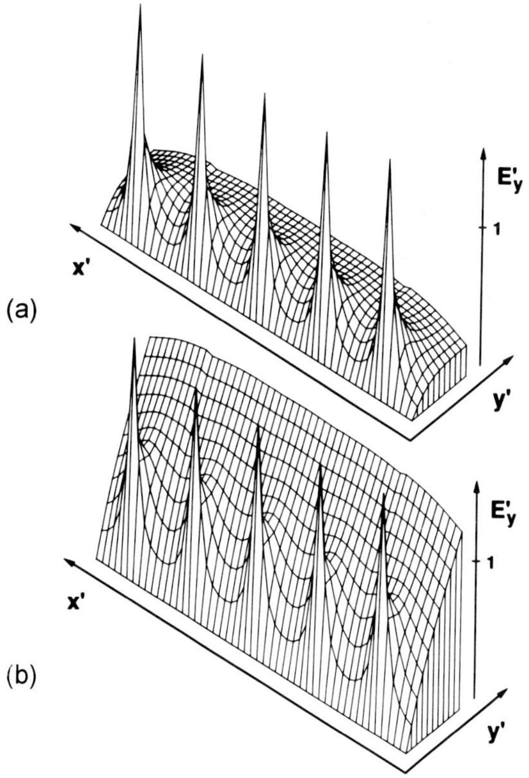

Figure 3.23 Three-dimensional plot of the spatial distribution of the E_y-component (dimensionless form) in the electrostatic case (a), in the case of a homogeneously distributed ionic space charge (b), as a function of the relative coordinates for a precipitator duct with five discharge wires.

close to the wires and reaches a constant level for $y' > 0.5$. Between the wires it decreases down to zero.

$$E_i'(x', y') = \frac{E_i(x', y')}{E_{ps}} = \frac{E_i(x', y')}{U/s} \tag{3.52}$$

In considering the ionic space charge, Oglesby and Nichols [8] proposed to take the constant space charge deduced by P. Cooperman [12]. Thereby, the current density is assumed to be only a function of the y-coordinate. The resulting field is a superposition of the electrostatic field plus the field of a homogenously distributed ionic space charge. The dimensionless electrical field strength is given in equations (3.53) and (3.54), while Figure 3.23b demonstrates the increase of the electrical field near the collecting plate

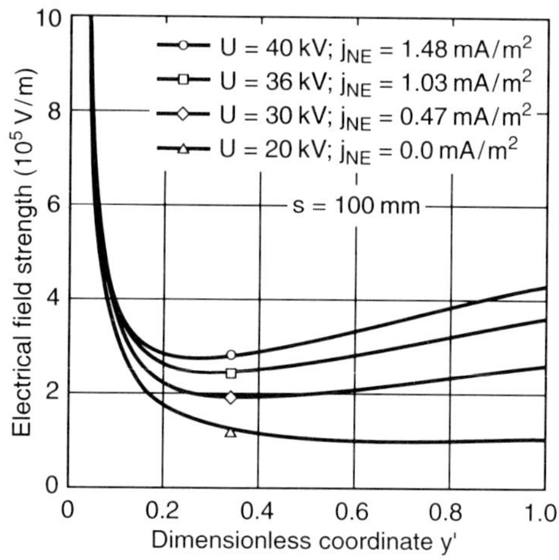

Figure 3.24 Numerical results for a plate-type ESP with a duct spacing of $2s = 200$. The increase in the electrical field strength with increasing y' depends on the current density.

caused by considering the space charge for the same conditions as in Figure 3.23a, but a voltage higher than corona onset.

$$\boldsymbol{E}'_x(x', y') = \boldsymbol{E}'_{x,j=0}(x', y') \tag{3.53}$$

$$\boldsymbol{E}'_x(x', y') = \boldsymbol{E}'_{x,j=0}(x', y') + j'_{\mathrm{NE}} \cdot 2c' \cdot \ln\left(\frac{d}{r_{\mathrm{SE}}}\right) \cdot \frac{y'}{\pi} \tag{3.54}$$

$$j'_{\mathrm{NE}} = \frac{j_{\mathrm{NE}}}{\dfrac{\varepsilon_0 \cdot b(\delta)}{s^3} \cdot U^2} \tag{3.55}$$

As already mentioned, a correct solution to the problem needs an immense computational effort. An iterative solution for electrical field and charge distributions has been calculated by various investigators, for example by [5, 8, 16]. Figures 3.24 and 3.25 show such numerical results for a plate-type ESP with a duct spacing of $2s = 200$ mm according to [5]. Obviously the increase of the electrical field strength with increasing y' depends on the current density (Figure 3.24). At the collector surface, this increase in field strength can be 200 or 300% compared with the electrostatic case. Of further concern is the local concentration of ionic space charge c_{iQ}, which is continuously decreasing with increasing y' (Figure 3.25).

In an electrical field, the space charge concentration is distributed in the interelectrode space. An example of a calculated distribution of ionic space charge according to Miller [17] is depicted in Figure 3.26. This shows that

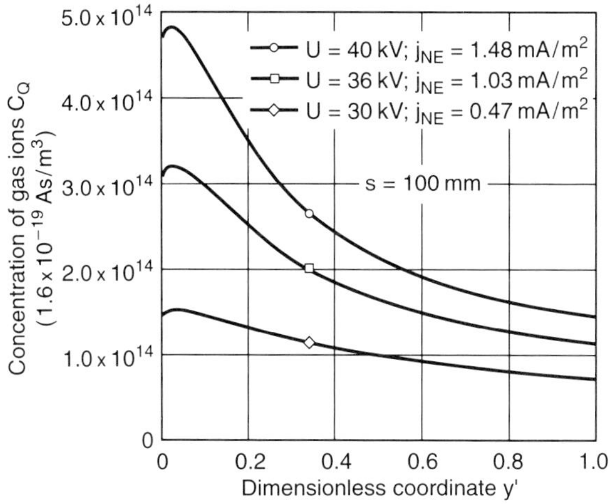

Figure 3.25 Numerical results for a plate-type ESP with a duct spacing of $2s = 200$. The local concentration of ionic space charge c_{iQ} is continuously decreasing with increasing y'.

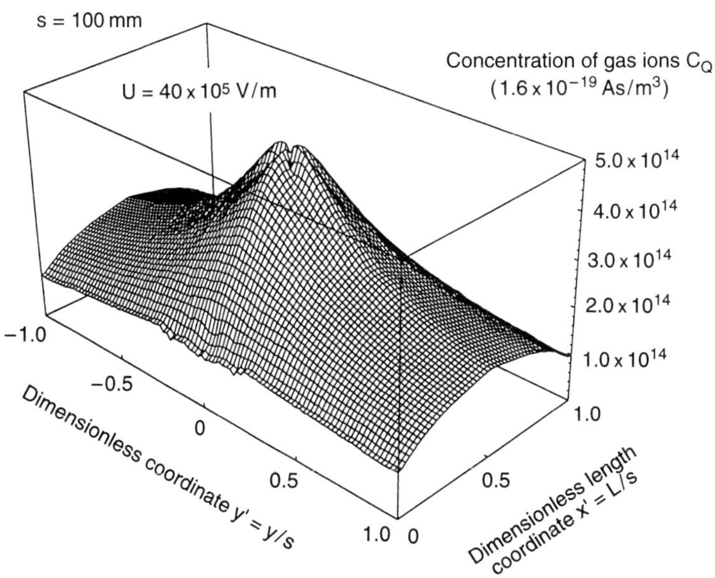

Figure 3.26 An example of a calculated distribution of ionic space charge according to Miller [17].

the highest ion concentrations occur close to the discharge wire and decrease towards the collecting plates. The highest value occurs at the collecting plate just opposite the wire.

Both distributions of electrical field and space charge will influence the particle charging process and the particle transport towards the collecting plates.

3.3 Particle charging

3.3.1 Charging process

Usually the charging process is divided into a field charging region, for particles $>1\ \mu m$, and a diffusion charging region for particles $<0.1\ \mu m$ [7, 8, 10]. While field charging requires the presence of an electrical field, which drives the free movable charge carriers, the diffusion process is based on randomly moving gas ions caused by temperature and described by the kinetic gas theory, i.e. Brownian motion.

Obviously, in an ESP, particles of all sizes experience both situations simultanously, as particles $<0.1\ \mu m$ are also driven by the electrical field. A simple addition of charges resulting from both mechanisms is sometimes made [18]; however, Oglesby and Nichols argue that both electric currents towards the particles have to be superimposed [8]. Initial approaches considering this were made by Murphy *et al.* [19] but they didn't succeed with a complete solution. Based on their work, Liu and Yeh [20] published a simplified theory which led to reasonable agreement with measured results. Smith and McDonald [21] extended this model to all particle sizes, thus including field and diffusion charging as limiting cases. The model of Liu and Kapadia [22], which was published later, gives better approximations in some cases. All these models however, cannot be solved analytically, i.e. they need numerical effort.

For practical work it seems to be reasonable, therefore, to look for charging theories describing charging processes continuously from small to larger particle sizes. A reasonable alternative to models based on numerical solution is Cochet's analytic equation [23]. This allows an easy calculation and the correlation to actual ESP conditions is quite reasonable in the critical size range, as Figure 3.27 demonstrates.

3.3.2 Cochet's charging model

The particle saturation charge (i.e. after infinite time) according to Cochet is given by equation (3.56). For calculations the electrical field strength E

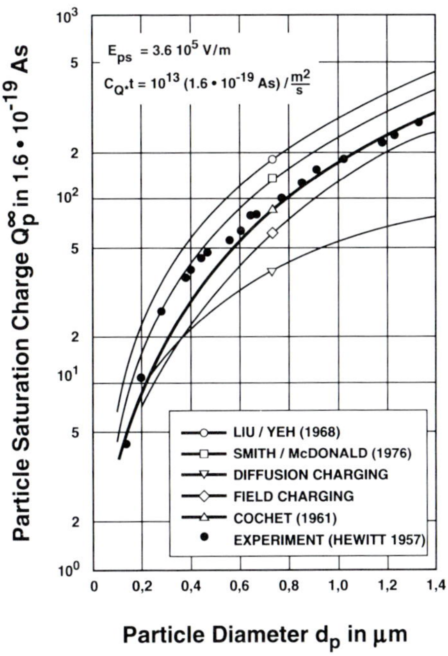

Figure 3.27 Plot of particle diameter *versus* particle saturation charge based on Cochet's analytic equation [23].

and the electrical permittivity of the particle material ε_r have to be specified.

$$Q_p^\infty = \left\{(1 + 2\lambda/d_p)^2 + \left(\frac{2}{1 + 2\lambda/d_p}\right) \cdot \left(\frac{\varepsilon_r - 1}{\varepsilon_r + 2}\right)\right\} \cdot \pi\varepsilon_0 d_p^2 E \qquad (3.56)$$

In Figure 3.28 calculations were made for typical electrical field strengths and a temperature of 150 °C, which changes the mean free path of the molecules from 0.065 μm to 0.1 μm. The electrical permittivity of the particles was assumed to be $\varepsilon_r = 10$. As can be seen from the figure, a 0.1 μm particle carries 5 elementary charge units in an electrical field of 3.0×10^5 V/m, while a 9.0 μm particle reaches 10 000 electrons under the same conditions. For larger particle sizes the saturation charge is proportional to particle surface and increases linearly with the electrical field strength.

Generally, the electrical field strength is a local function in a precipitator (see Figures 3.22/3.24) and therefore this influences the particle saturation charge. However, these local functions have not, to date, been considered in common ESP models. Such a solution only becomes possible when particle trajectories are calculated (as demonstrated in 3.5.5).

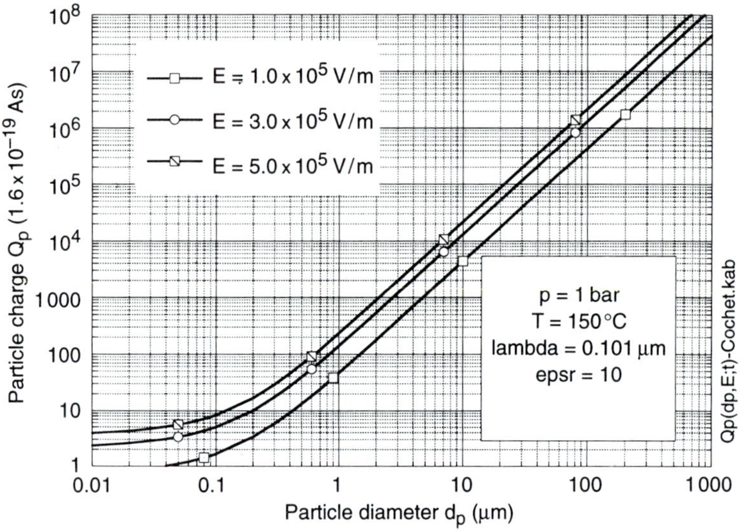

Figure 3.28 Plot of particle diameter *versus* particle charge: calculations for typical electrical field strengths and a temperature of 150 °C, which changes the mean free path of the molecules from 0.065 μm to 0.1 μm.

3.3.3 *Time dependence and saturation charge*

The time dependence of the charging process is described by equation (3.57), using the time constant τ_Q for the charging process (equation 3.58) [10]. The dynamic charging behaviour is independent of particle size if a homogenous electrical field is considered.

$$Q_p(t) = Q_p^{\infty} \cdot \frac{t}{t + \tau_Q} \tag{3.57}$$

$$\tau_Q = \frac{4\varepsilon_0}{c_Q b} \approx \frac{4\varepsilon_0}{j_{NE}/E} \tag{3.58}$$

Before the time dependence can be studied, the time constant τ_Q has to be specified from the electrical conditions. Referring to Ohm's law, the concentration of gas ions c_Q can easily be expressed in terms of current density, mobility and electrical field (approximation in equation 3.58). For typical electrical states, τ_Q has values $\leqslant 10$ ms. The dynamic behaviour of the charging process was calculated and is presented in Figure 3.29 for time constants 1, 10 and 100 ms. As can be seen, the particles will reach about

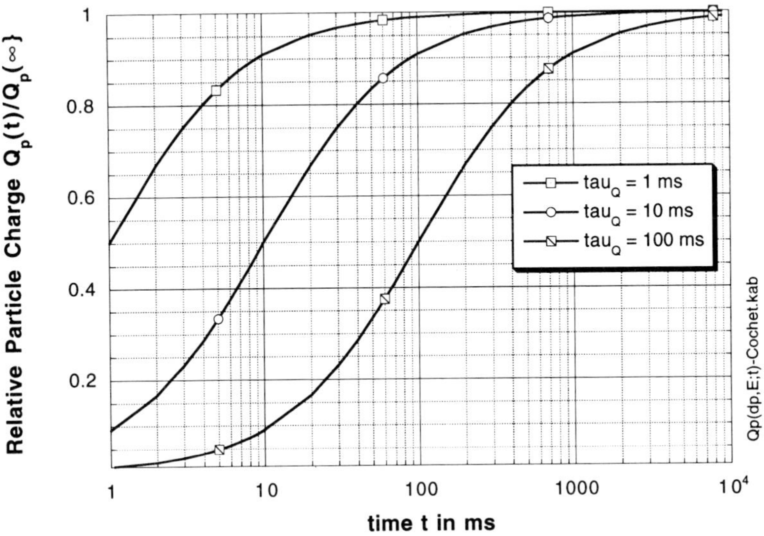

Figure 3.29 Plot of relative particle charge versus time under typical electrical conditions.

90% of their saturation charge within some 10 ms under typical electrical conditions (i.e. $\tau_Q \leqslant 10\,\text{ms}$).

3.4 Particle migration

3.4.1 Equation of motion

In order to characterize the state of particle motion the balance over all forces acting upon the particle is needed. For a particle in an ESP, these forces are the momentum force $\vec{F}_T$ (equation 3.59), the electrical force $\vec{F}_{el}$ (equation 3.60) and the drag force $\vec{F}_w$ (equation 3.61).

$$\vec{F}_T = -m\vec{a} \tag{3.59}$$

$$\vec{F}_{el} = Q_p\vec{E} \tag{3.60}$$

$$\vec{F}_w = c_w(\text{Re}) \cdot A_{pr} \cdot \frac{\rho_F}{2} \cdot v_{rel} \cdot \vec{v}_{rel} \tag{3.61}$$

The sum over all acting forces has to be zero (equation 3.62).

$$\vec{F}_T + \vec{F}_{el} + \vec{F}_w = \vec{0} \tag{3.62}$$

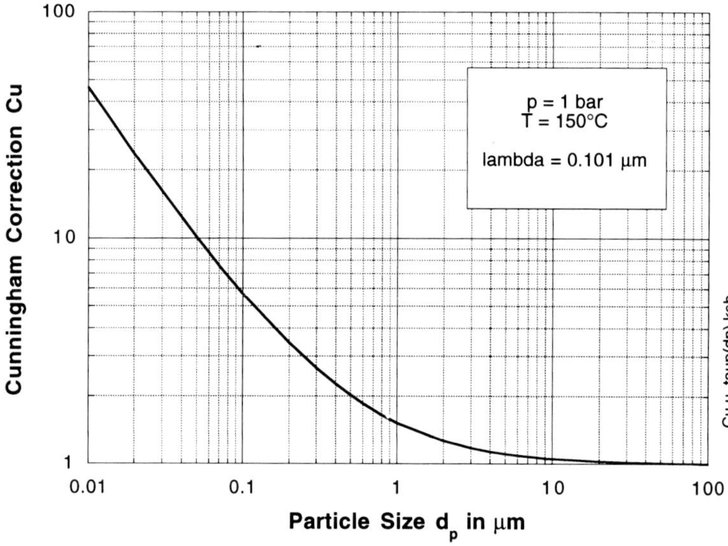

Figure 3.30 Cunningham correction factor applied to Stokes' law when particle size reaches the region where the fluid loses its continuum characteristic.

Before solving this differential equation, however, the drag force must be specified. In the case of low Reynolds numbers i.e. Re ≪ 1 (equation 3.63), the drag coefficient c_w is given by equation (3.64), which is fulfilled in typical ESP conditions for particle sizes less than 20 μm. The drag relationship in this regime for spheres is given by Stokes' law (equation 3.65).

$$\mathrm{Re}_\mathrm{p} = \frac{d_\mathrm{p} \cdot |\vec{v}_\mathrm{rel}|}{\nu} = \frac{d_\mathrm{p} \cdot |\vec{v} - \vec{w}|}{\nu} \tag{3.63}$$

$$c_\mathrm{w} = \frac{24}{\mathrm{Re}_\mathrm{p}} \tag{3.64}$$

$$\vec{F}_\mathrm{w} = 3\pi\eta d_\mathrm{p} |\vec{v} - \vec{w}| \left(\cdot \frac{1}{\mathrm{Cu}} \right) \tag{3.65}$$

If the particle size reaches the region where the fluid loses its continuum characteristic (mean free path of the molecules λ), then Stokes' law needs correction by the Cunningham factor Cu, given by equation (3.66) and plotted in Figure 3.30 for 150 °C.

$$\mathrm{Cu} = 1 + 1.246 \cdot \frac{2\lambda}{d_\mathrm{p}} + 0.42 \cdot \frac{2\lambda}{d_\mathrm{p}} \cdot \exp\left(-0.87 \cdot \frac{d_\mathrm{p}}{2\lambda} \right) \tag{3.66}$$

Assuming the fluid to have no component towards the collecting plate and that the particles reach their saturation charge, equation (3.67) results as a differential equation, characterizing the motion of a charged sphere in an electrical field E.

$$\frac{\mathrm{d}w}{\mathrm{d}t} + \frac{3\pi\eta d_{\mathrm{p}}}{m(\cdot\,\mathrm{Cu})} w = \frac{Q_{\mathrm{p}}^{\infty}}{m} E \tag{3.67}$$

3.4.2 *Theoretical migration velocity*

Taking $w(t = 0) = 0$ as the initial condition, the solution of equation (3.67) can easily be found. This describes the time dependence of the particle migration velocity (equation 3.68). Herein, the relaxation time τ_{p} characterizes the dynamic behaviour of the particle (equation 3.69) and w_{th} is the theoretical migration velocity, which is the steady state velocity of the particle (equation 3.70) [24].

$$w(t) = w_{\mathrm{th}}\left\{1 - \exp\left(-\frac{t}{\tau_{\mathrm{p}}}\right)\right\} \tag{3.68}$$

$$\tau_{\mathrm{p}} = \frac{m_{\mathrm{p}}(\cdot\,\mathrm{Cu})}{3\pi\eta d_{\mathrm{p}}} = \frac{\rho_{\mathrm{p}}\cdot d_{\mathrm{p}}^2}{18\eta}(\cdot\,\mathrm{Cu}) \tag{3.69}$$

$$w_{\mathrm{th}} = \frac{Q_{\mathrm{p}}^{\infty} E}{3\pi\eta d_{\mathrm{p}}}(\cdot\,\mathrm{Cu}) \tag{3.70}$$

The theoretical migration velocity is plotted in Figure 3.31 as a function of particle size for three different electric field strengths, a temperature of 150 °C and an electrical permittivity of $\varepsilon_{\mathrm{r}} = 10$. The theoretical migration velocity shows a minimum located at 0.35 μm. For larger particles the increase of w_{th} proceeds linearly with particle size (equation 3.70 and equation 3.56). For smaller particles the increase in w_{th} is more pronounced. From Figure 3.31 the strong influence of the electrical field becomes obvious (w_{th} increasing with E^2).

The transient behaviour of particle motion is characterized by the relaxation time τ_{p}, which is plotted in Figure 3.32 as a function of particle size. It is worth noting that the relaxation time does not depend on the electrical conditions. The particle velocity at different time intervals, in relation to the steady state value, can easily be calculated with equation (3.68). In order to illustrate the acceleration phase, this velocity ratio is plotted in Figure 3.33 for different time steps. Thus, after 1 ms, particles smaller than about 6 μm have already reached steady state velocity, while a 10 μm particle has reached only 80% and a 20 μm particle only about 30%.

When the migration velocities of all particles in an ESP are known, then the separation process can be calculated.

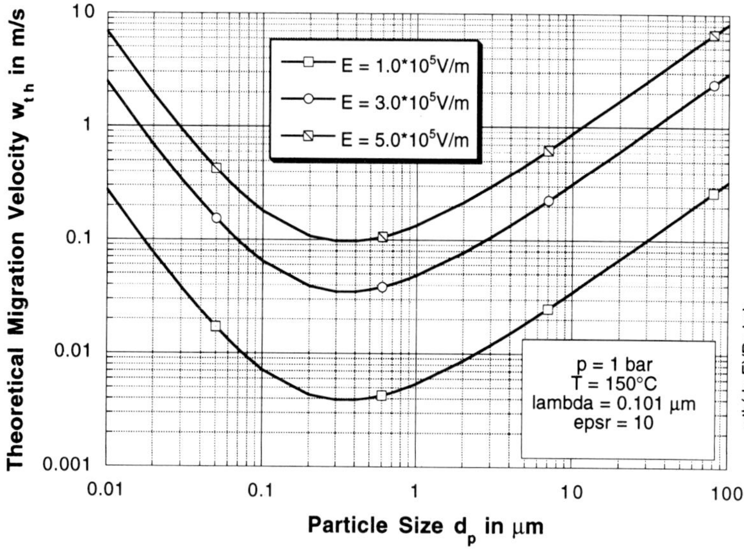

Figure 3.31 Plot of theoretical migration velocity as a function of particle size for three different electric field strengths, a temperature of 150 °C and an electrical permittivity $\varepsilon_r = 10$.

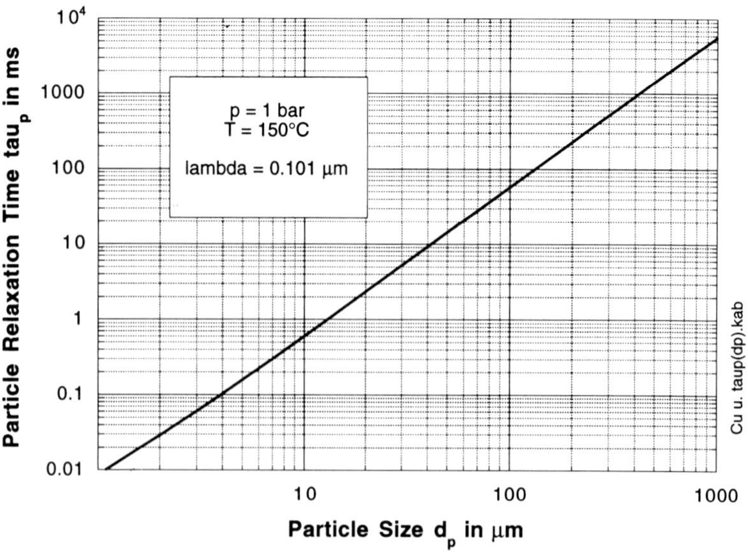

Figure 3.32 Plot of particle relaxation time as a function of particle size.

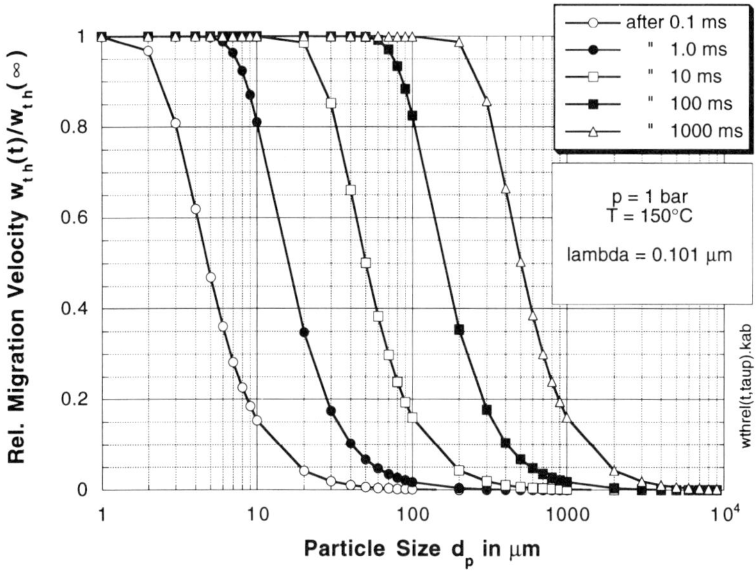

Figure 3.33 Plot of particle velocity, at different time steps, in relation to the steady state value as an illustration of the acceleration phase.

3.5 Measuring and modelling particle separation

3.5.1 Grade efficiency and total efficiency

If a particle collection process is to be characterized, a method for measuring particle 'size' is necessary. Although particles of all sizes might occur in a disperse system, a measuring method always produces 'size classes'. The borders of the size classes (let's say $i - 1$) are termed by x_i, the width of the size class (interval) by Δx_i, while the median of a size class is termed by $d_{p,i}$. An illustration of the nomenclature is given in Figure 3.34.

Let's assume we analyse the particles collected in a hopper with a sieving tower consisting of six sieves plus the bottom. Thus, we have six given borders x_2 to x_7 and an estimate of the smallest particle size x_1. After sieving and weighing the sieves and the bottom, the relative mass parts are determined according to equation (3.71); these, of course, refer to the median of the size interval. In particle size analysis, a cumulative mass distribution is often used, where x_i is related to the upper end of a size class (see definitions 3.72, 3.73 and Figure 3.34).

$$\Delta Q(d_{p,i}) = \frac{M(d_{p,i})}{M_{tot}} \tag{3.71}$$

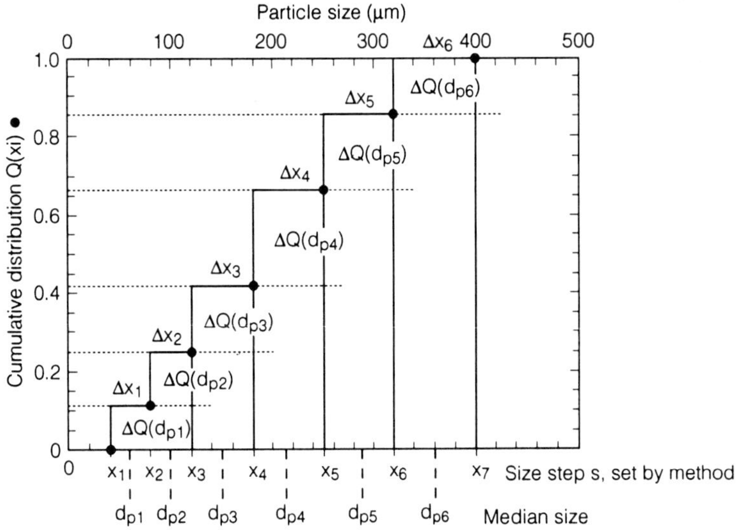

Figure 3.34 Nomenclature of particle 'size' measurement.

$$\Delta Q(d_{p,i}) = Q(x_{i+1}) - Q(x_i) \tag{3.72}$$

$$Q(x_j) = \sum_{i=1}^{j-1} \Delta Q(d_{p,i}); \quad \text{for } 2 \leqslant j \leqslant N \tag{3.73}$$

An example with numbers is given in Table 3.4, together with the corresponding density distribution as given according to equation (3.74).

$$q(d_{p,i}) = \frac{\Delta Q(d_{p,i})}{\Delta x_i} \tag{3.74}$$

We now have a formal method to characterize the particle sizing of dust, at the precipitator's inlet, the fine fraction at the outlet or the coarse fraction in the hopper.

Table 3.4 Example of particle size distribution and derived quantities

i	$x_i/\mu m$	$d_{pi}/\mu m$	$M(d_{pi})/g$	$\Delta Q(d_{pi})$	$Q(x_i)$	$\Delta(x_i)/\mu m$	$q(d_{pi})/1/\mu m$
1	40	60	25.927	0.11	0	40	0.00275
2	80	100	32.998	0.14	0.11	40	0.00350
3	120	150	40.069	0.17	0.25	60	0.00283
4	180	215	58.925	0.25	0.42	70	0.00357
5	250	285	44.783	0.19	0.67	70	0.00271
6	320	360	32.998	0.14	0.86	80	0.00175
7	400	—	—	—	1.00		
			235.7				

Grade efficiency or fractional efficiency means we evaluate the efficiency for each particle size class. For size class $d_{p,i}$ for example, when the collected coarse amount is related to the inlet amount, the result is a fractional efficiency (equation 3.75); when the emitted, fine fraction is related to the inlet amount the result is called fractional penetration (equation 3.76). Obviously equation (3.77) has to hold.

$$T(d_{p,i}) = \frac{M_c(d_{p,i})}{M_e(d_{p,i})} = \frac{c_c(d_{p,i})}{c_e(d_{p,i})} \tag{3.75}$$

$$P(d_{p,i}) = \frac{M_f(d_{p,i})}{M_e(d_{p,i})} = \frac{c_f(d_{p,i})}{c_e(d_{p,i})} \tag{3.76}$$

$$T(d_{p,i}) + P(d_{p,i}) = 1 \tag{3.77}$$

If an integral balance is made for total particle collection, the result is a total efficiency, or a total penetration P, holding for the quantity balanced, e.g. mass generally (number or surface can also be balanced). The total mass efficiency can also be expressed in mass concentrations as demonstrated in equation (3.78). Again equation (3.79) has to hold.

$$E = \frac{M_{c,\,tot}}{M_{e,\,tot}} = \frac{M_{e,\,tot} - M_{f,\,tot}}{M_{e,\,tot}} = 1 - \frac{c_{f,\,tot}}{c_{e,\,tot}} \tag{3.78}$$

$$E + P = 1 \tag{3.79}$$

Generally, a fractional efficiency curve gives much more information about a separator than the overall efficiency E. In order to derive a fractional efficiency curve, particle size distribution measurements are needed. Combining equations (3.75) and (3.71) leads to a simple relationship for calculating fractional efficiency (equation 3.80) from measured collected particles and inlet distribution.

$$T(d_{p,i}) = \frac{M_{c,tot} \cdot \Delta Q_c(d_{p,i})}{M_{e,tot} \cdot \Delta Q_e(d_{p,i})} \tag{3.80}$$

From equation (3.80) the calculation procedure for the total (mass) efficiency can be deduced (equation 3.81).

$$E = \sum_i T(d_{p,i}) \cdot \Delta Q_e(d_{p,i}) \tag{3.81}$$

When total mass efficiency is the only interest, then, by using an isokinetic sampling device before and after the ESP, an alternative calculation can be done. If the mean velocity in the ESP and the cross-sections at both measuring locations is identical equation (3.82) holds, which considers the different fluxes (measuring time and cross-section) for sampling.

$$E = 1 - \frac{\Delta M_f}{\Delta t_f \cdot A_f} \cdot \frac{\Delta t_e \cdot A_e}{\Delta M_e} \tag{3.82}$$

Measuring total mass efficiencies may often be the simplest way to characterize a separation process. However, when considering modelling the separation processes, fractional efficiency is certainly the base and this is why selective particle size measurements are essential.

Needless to say, all standard models refer to the ideal precipitation process, i.e. where all negative influences are known and ignored, e.g. back-corona of the dust layer, re-entrainment of precipitated dust and field bypassing, called sneakage. Correctly speaking, this means that these idealized models need correspondingly idealized experiments for confirmation or failure – a problem when taking plant results, an advantage with lab-scale experiments.

3.5.2 Laminar model

The word 'laminar' in this case means that there is no driving force for the particles in the opposite direction to the electrical field. In other words, there is no remixing force arising, e.g. from turbulence, so that each particle follows a trajectory only determined by gas flow and electrical conditions. It can be shown by different methods [24, 25] that, in the case of a fluid velocity profile, fully charged particles and a homogeneous electrical field, grade efficiency is determined by straight particle trajectories. This situation is illustrated in Figure 3.35, which makes clear that grade efficiency is given by equation (3.83). This is the ratio of the mean gas residence time in the precipitation zone to the time a particle of size d_p needs to travel the distance s.

$$T(d_p) = \frac{w_{th}(d_p, E)L_{NE}}{v_0 s} = \frac{t_{res}}{t_s(d_p, E)} \tag{3.83}$$

3.5.3 Deutsch model

In contrast to the laminar model, the Deutsch model assumes an infinite remixing force, e.g. caused by 'turbulence', which redistributes the remaining particles at each downstream location homogeneously over the precipitator duct. Apart from the boundary layer, the particles in the downstream direction have the mean velocity of the fluid, are fully charged and migrate in a homogeneous electrical field. The situation is illustrated in Figure 3.36 and was firstly published in 1922 by W. Deutsch [26].

3.5.3.1 Grade efficiency. Taking a balance over the control volume sketched in Figure 3.36, the differential equation (3.84) results (written for

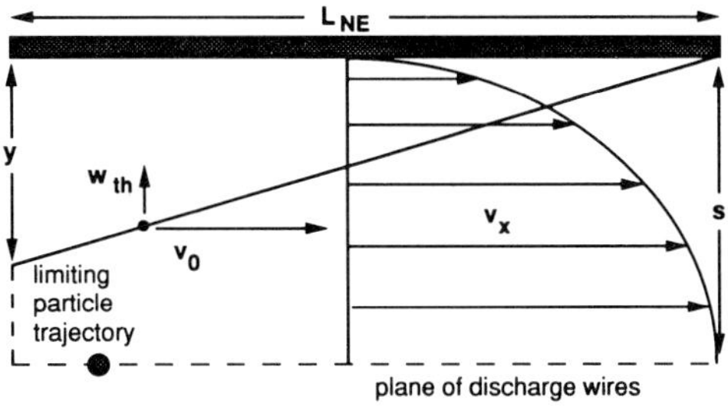

Figure 3.35 Illustration of the laminar model. Grade efficiency determined by straight particle trajectories in the case of a fluid velocity profile, fully charged particles and a homogeneous electrical field.

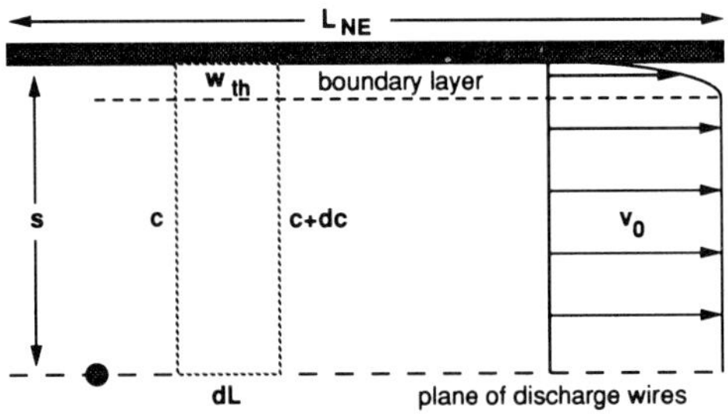

Figure 3.36 Illustration of the Deutsch model.

plate-type), which can be integrated quite easily in order to get the concentration of particles with size d_p, as a function of downstream location L (equation 3.85).

$$\frac{dc(d_p)}{c(L, d_p)} = -\frac{w_{th}(d_p)}{v_0 s} \cdot dL \tag{3.84}$$

$$c(L, d_p) = c(0, d_p) \cdot \exp\left\{-\frac{w_{th}(d_p, E)L}{v_0 s}\right\} \tag{3.85}$$

Remembering equation 3.75 for the definition of grade efficiency, the final formula is equation (3.86) for plate-type ESP:

$$T(d_p) = 1 - \exp\left\{-\frac{w_{th}(d_p, E) \cdot L_{NE}}{v_0 \cdot s}\right\} \tag{3.86}$$

and equation (3.87) for tube-type ESP:

$$T(d_p) = 1 - \exp\left\{-\frac{2 \cdot w_{th}(d_p, E) \cdot L_{NE}}{v_0 \cdot r_{NE}}\right\} \tag{3.87}$$

When the total collecting area and the total flow rate is used, a general

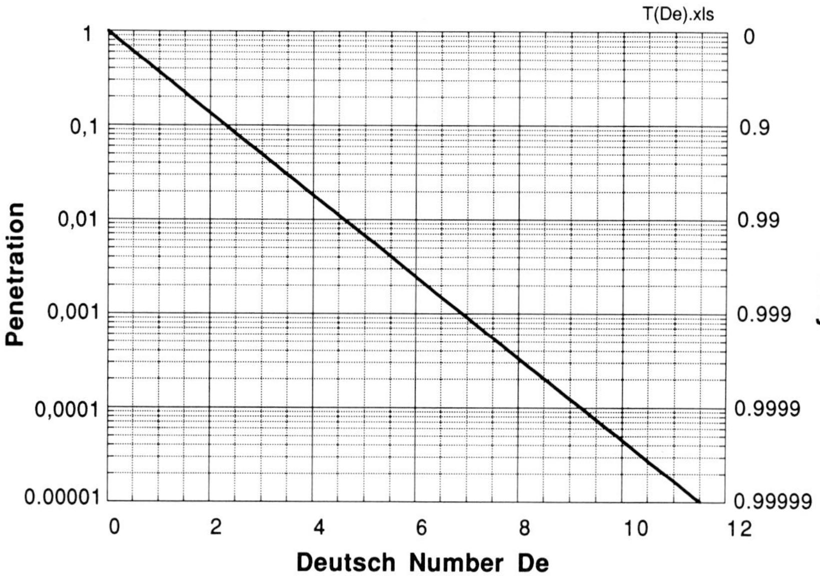

Figure 3.37 Grade efficiency plotted as a function of the Deutsch number.

formulation is possible (equation 3.88).

$$T(d_p) = 1 - \exp\left\{- w_{th}(d_p, E) \cdot \frac{A_{NE}}{\dot{V}}\right\} \tag{3.88}$$

Some years ago, Self defined the exponent in equations (3.86) to (3.88) as the 'Deutsch number' [25]. In Figure 3.37, the grade efficiency is plotted as a function of the Deutsch number. Obviously the exponent can also be interpreted as dimensionless residence time (equation 3.90) or as the product of a particle drift parameter Ω, which includes operational conditions (v_0, E), and a dimensionless length describing geometry (L_{NE}, s) [4, 24].

$$T(d_p) = 1 - \exp\{- \mathrm{De}(d_p; E, v_0)\} \tag{3.89}$$

$$\mathrm{De} = \frac{w_{th}(d_p; E)}{s} \cdot \frac{L_{NE}}{v_0} = \frac{t_{res}(v_0; L_{NE})}{t_{drift}(d_p; E; s)}$$

$$= \frac{w_{th}(d_p; E)}{v_0} \cdot \frac{L_{NE}}{s} = \Omega(d_p; E, v_0) \cdot L'_{NE}(s, L_{NE}) \tag{3.90}$$

In Figure 3.38, showing the grade efficiency for a plate-type precipitator zone of 5 m length, the strong influence of electric field strength on performance is demonstrated. The typical minimum efficiency at around 0.3–0.4 μm is shifted from 5% up to 70%, by increasing the field from 1.0 to 5.0 $\times$ 10^5 V/m.

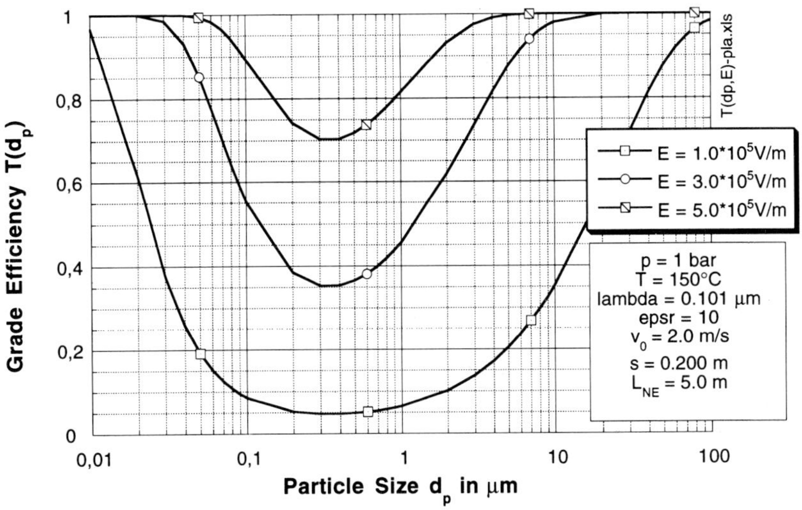

Figure 3.38 The grade efficiency for a plate-type precipitator zone of 5 m length demonstrates the strong influence of electric field strength on performance.

Figure 3.39 compares measured and calculated grade efficiencies and illustrates the influence of mean gas velocity. The measurements were performed in a lab-scale ESP with ambient air under ideal precipitation conditions (no sneakage, no rapping, no back-corona) by using an optical particle counter, which is able to analyse the particle size distributions '*in situ*' very quickly, thus guaranteeing constant conditions [24]. From Figure 3.39 it can be seen that around 10 μm the compliance between measurements and calculations is quite good. For smaller particle sizes ($<2\,\mu$m), however, all measured efficiency values are far higher than the calculated ones. A quantitative explanation for this behaviour is still missing. Maybe agglomeration happens or the fine particles can be charged more efficiently than assumed.

Furthermore, Figure 3.39 demonstrates that grade efficiency increases in wider ducts as long as they are operated with the same pseudo-homogeneous electrical field strength. This increase can be explained with an increasing space-averaged electrical field strength (see Table 3.5). A detailed

Table 3.5 Space-averaged electrical field strength $\langle\langle E\rangle\rangle$ (kV/cm)

s (mm)	100	71	50
E_{ps} kV/cm	5.49	5.14	4.83
j_{NE} mA/m^2	2.9	3.4	4.5
j_{NE} mA/m^2	5.49	4.68	3.86
E_{ps} kV/cm	5.0	4.8	4.4

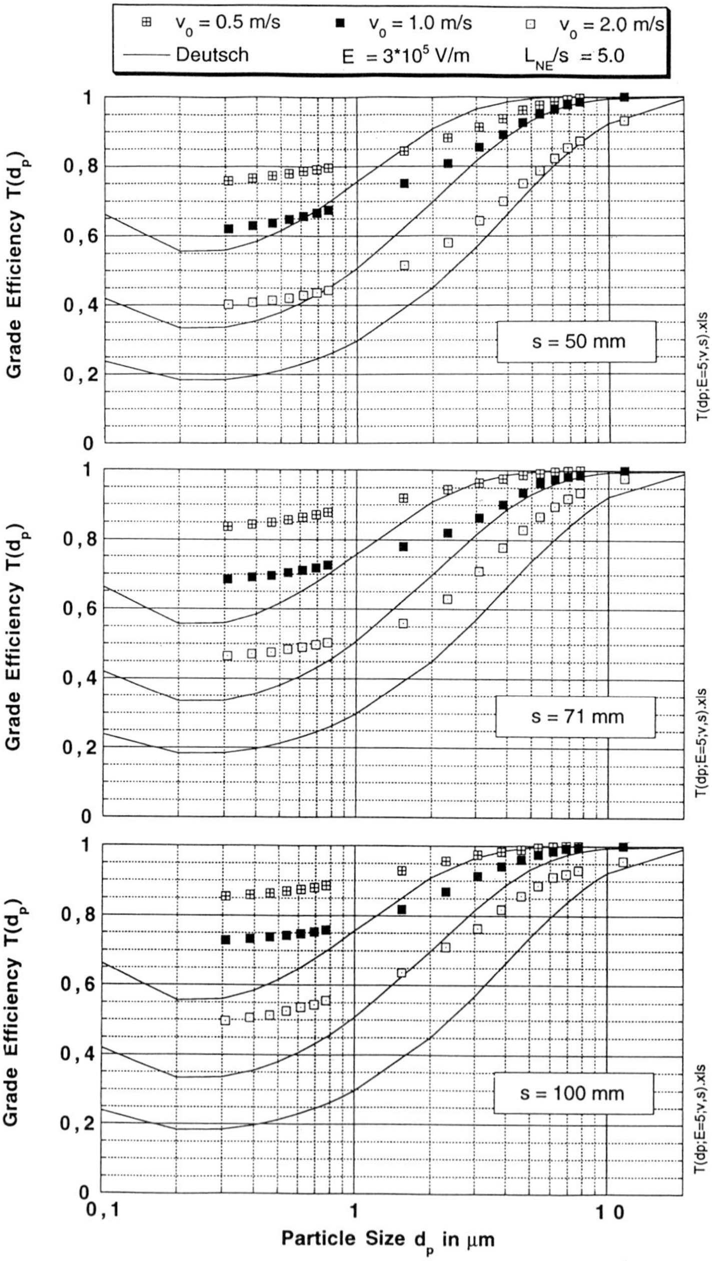

Figure 3.39 Comparison of measured and calculated grade efficiencies and illustration of influence of mean gas velocity.

discussion concerning the comparison of electrical operating conditions in precipitators of different size is given in [4, 24].

3.5.3.2 Effective migration velocity. The ESP performance is often characterized by the 'effective migration velocities' w_{eff} or w_k, which are defined by equations of the Deutschian type (equations 3.91a/b and 3.92a/b). While the Deutsch model predicts the grade efficiency $T(d_p)$ by means of an exponential function (see equation 3.88), wherein the drift velocity $w_{th}(d_p)$ of particles in the electrical field occur, equations (3.91) and 3.92) define a velocity value by taking the exponential function to describe the total mass efficiency (E_3). Therefore, effective migration velocities represent overall averages, characterizing performance as total efficiencies do.

$$w_{eff} \equiv -\frac{\dot{V}}{A_{NE}} \cdot \ln(1 - E_3) \tag{3.91a}$$

$$E_3 = 1 - \exp\left(-w_{eff} \cdot \frac{\dot{V}}{A_{NE}}\right) \tag{3.91b}$$

$$w_k = \frac{\dot{V}}{A_{NE}} \cdot \{-\ln(1 - E_3)\}^{1/k} \tag{3.92a}$$

$$E_3 = 1 - \exp\left[-\left(w_k \cdot \frac{\dot{V}}{A_{NE}}\right)^k\right] \tag{3.92b}$$

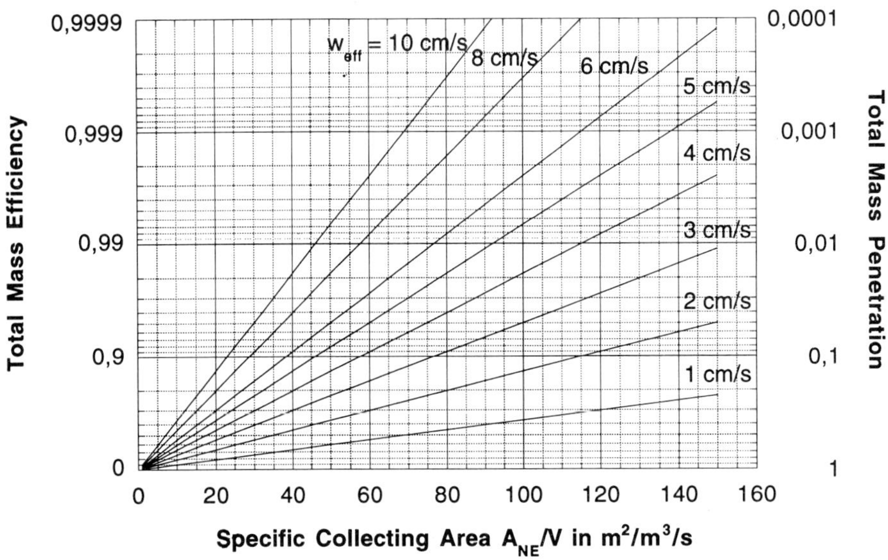

Figure 3.40 Dependence of total mass efficiency on specific collecting area for different w_{eff} values.

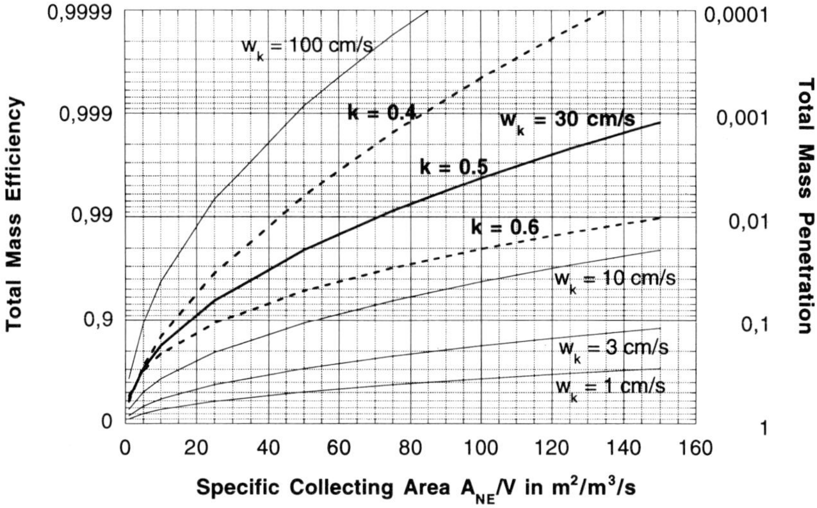

Figure 3.41 Dependence of total mass efficiency on specific collecting area for different modified effective migration velocity (w_k) values.

Figure 3.40 shows how total mass efficiency depends on specific collecting area for different w_{eff} values according to equation (3.91). The plotted straight lines overpredict performance, when for an existing ESP the emission needs to be further improved. The reason for this overprediction is that mass transfer along the collecting plates is not constant. It decreases with locations further downstream, because smaller, and therefore lighter particles, have lower transport velocities, i.e. they need more time to be collected.

In order to overcome this overprediction caused by neglecting particle size distribution effects, Matts and Öhnfeldt [27] introduced a modified effective migration velocity w_k. Their equation (3.93) is plotted in Figure 3.41 for different w_k values. Obviously, for the same specific collecting area and the same total efficiency, the w_k value is much higher than the corresponding w_{eff} value. Furthermore it is demonstrated for $w_k = 30\,\text{cm/s}$ that the empirical parameter k has a strong influence on the slope of the curve. A typical value for k lies within 0.4–0.6 depending on the actual problem.

These w_{eff} and w_k values do not represent particle velocities (but they characterize mass transfer; see next section). This is why it is not surprising that effective migration velocities (typically a few cm/s) are often one order of magnitude smaller than theoretical migration velocities, i.e. different characteristics should not be compared. There is, of course, only one unrealistic case, where w_{eff} is identical with a particle velocity, and that is in the case of a monodispersion, i.e. all particles have exactly the same size.

Finally, Figure 3.42 illustrates the relationship between the total mass efficiency and the effective migration velocity, and in particular, their

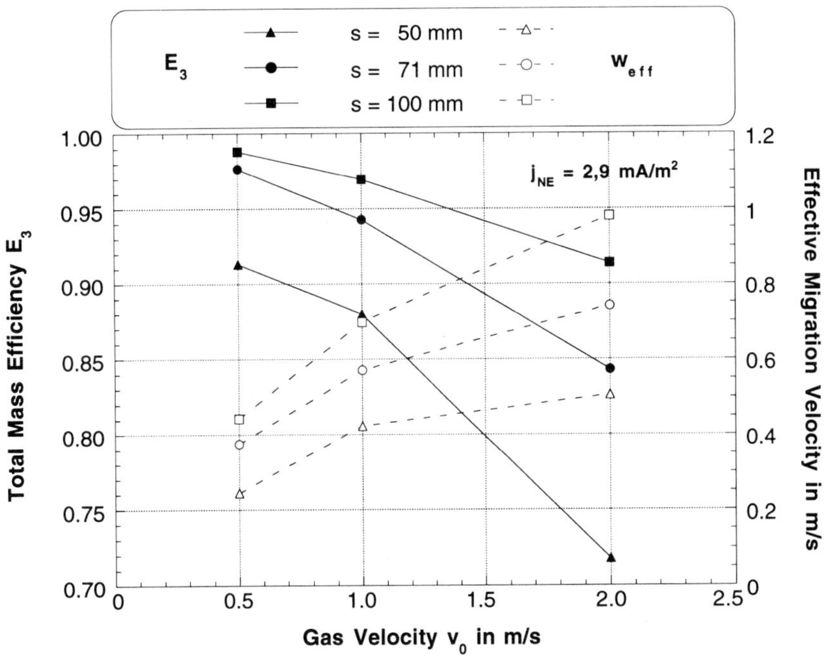

Figure 3.42 Relationship between the total mass efficiency and the effective migration velocity with particular reference to gas velocity and duct spacing.

dependence on gas velocity and duct spacing. The total mass efficiency (left y-axis) stems from gravitional efficiency measurements in three geometric similar lab-scale precipitators of different absolute size ($s = 50/71/100$ mm and $L_{NE}/s = 5.0$). Figure 3.42 holds for a constant current density $j_{NE} = 2.9\,\text{mA/m}^2$ at the collecting plates, i.e. the DC voltages have been adjusted according to the different duct widths.

It is worth noting that the effective migration velocities (calculated according equation 3.91a) in Figure 3.42 are much larger (0.25 m/s to 1.0 m/s) than those usually observed in large-scale precipitators (0.04 m/s to 0.20 m/s). There are two main reasons for this behaviour: firstly, the longer a precipitator becomes, the lower the effective migration will be in total, because the local mass flux continously decreases with the length coordinate (see next section); secondly, all secondary influences, usually occurring in practice, e.g. re-entrainment, sneakage and back-corona, all causing deteriorating efficiency and thereby reducing the effective migration velocity, were excluded from the investigations by the design and the performance of the experiments.

What are the trends of the measured results plotted in Figure 3.42 where an increase in the mean gas velocity leads to a progressive decrease in the total efficiency? From equation (3.91a) it follows that the observed decrease in total efficiency correlates with an increase in the effective migration velocity. This contradicts the interpretation of the effective migration velocity as a quantitative measure for the particle transport velocity, since the transport conditions perpendicular to the main flow direction, determined by the electrical state, have not changed. To be more precise, the local transport velocities of individual particles have not changed, but the mass flux towards the plates will increase because of a higher volume flow. The increase of the effective migration velocity with increasing gas velocity points to a correlation of w_{eff} and mass flux.

When comparing the precipitators with respect to constant specific collecting areas (where $v =$ const. since $L_{\mathrm{NE}}/s = 5.0$), the wider duct shows higher total efficiencies. Here, a decrease in total efficiency correlates with a decrease in the effective migration velocity. This means that an increase in the effective migration velocity corresponds to both a decrease and an increase in the total efficiency. This observation expresses the fact that the mass flux towards the collecting plates can be increased in two ways: first, by increasing the individual drift velocity of the particles in the electrical field where the number of individual particles transported per unit time remains constant. However, because of an increased drift velocity the particles are deposited earlier on the plates or the same number of particles are collected in a shorter precipitator. On the other hand, an increase in the volume flow leads to an increase in the number of particles transported per unit time towards the plates, but without a change in the individual drift velocity. Obviously only the first case can represent a real precipitator improvement.

3.5.3.3 Mass flux. The total collected mass flow (particles per unit time) or the total collected mass flux (particles per time and area unit) can simply be derived from fundamental considerations enabling a clearer representation of particle transport in precipitators [28]. If $i_{\mathrm{m,e}}$ represents the total particle mass flow at the precipitator entrance and $i_{\mathrm{m,f}}$ the fine particle mass flow at the outlet, then the total mass flow $i_{\mathrm{m,c}}$ of collected particles is described by equation (3.93). Additionally, the definition of the total mass efficiency was considered (equation 3.78).

$$i_{\mathrm{m,c}} = i_{\mathrm{m,e}} - i_{\mathrm{m,f}} = i_{\mathrm{m,e}} \cdot E_3 \tag{3.93}$$

The incoming particle mass flow $i_{\mathrm{m,e}}$ can be calculated by using the particle mass concentration at the inlet $c_{\mathrm{m,e}}$ and the volume flow $\dot{V}$ according to equation (3.94). When relating the total mass flow of collected particles $i_{\mathrm{m,c}}$, to the collecting area A_{NE}, a particle mass flux results, which is called 'total mass flux of collected particles', characterizing the total precipitation capability of the apparatus (equation 3.95). 'Total' means here a two-fold integration indicated by the double bracket: the first bracket indicates the integration over particle size distribution, the second bracket indicates the integration over the precipitator length.

$$i_{m,e} = c_{m,e} \cdot \dot{V} \tag{3.94}$$

$$\langle\langle j_{m,c} \rangle\rangle = c_{m,e} \cdot E_3 \cdot \frac{\dot{V}}{A_{NE}} \tag{3.95}$$

$$= c_{m,e} \cdot E_3 \cdot \frac{v \cdot s}{L_{NE}}$$

With the normal measurements of total efficiencies, this total flux of collected mass $\langle\langle j_{m,c} \rangle\rangle$ can be calculated as simply as the widely used effective migration velocity w_{eff}. However, the total flux of collected mass has two advantages. On one hand this quantity is based on a proper physical background. Further, it is linguistically clearly distinguished from the steady state theoretical drift velocity of the particles w_{th} in the electrical field, thus avoiding misleading interpretations of w_{eff} as a particle velocity. On the other hand $\langle\langle j_{m,c} \rangle\rangle$ is an explicit function of particle mass concentration in the raw gas: an influence which is well known in practice but not explicitly considered in w_{eff}.

When evaluating total efficiencies, the mass concentrations in the raw and clean gas have to be calculated; therefore, the total flux of collected mass can readily be calculated with known quantities according to equation (3.95). This procedure was carried out with the efficiencies in Figure 3.43 where the resulting total flux of collected mass are plotted as open dots (y-axis on the right-hand side). The mass flux values range from 10 to 40 mg/m^2 s and, as is to be expected, they increase with increasing gas velocity.

A limiting straight line exists for the total flux of collected mass, since $\langle\langle j_{m,G} \rangle\rangle$ has to be zero at gas velocity zero. If a precipitator is operating with total efficiencies close to 1.0, the slope of the limiting straight line is obviously solely a function of the ratio, precipitator length to half duct width L_{NE}/s (see equation 3.95). Because of geometrical similarity, the precipitators investigated belong to a configuration characterized by $L_{NE}/s = 5.0$, leading to the following slope of the limiting straight line (see Figure 3.45):

$$\frac{50\,\dfrac{\text{mg}}{\text{m}^2\text{s}}}{2.5\,\dfrac{\text{m}}{\text{s}}} = 20\,\frac{\text{mg}}{\text{m}^3}$$

Obviously, precipitators with different L_{NE}/s-values have different limiting straight lines.

Furthermore, Figures 3.42 and 3.43 show differences with respect to precipitator size. These wide-spacing effects can be explained by higher

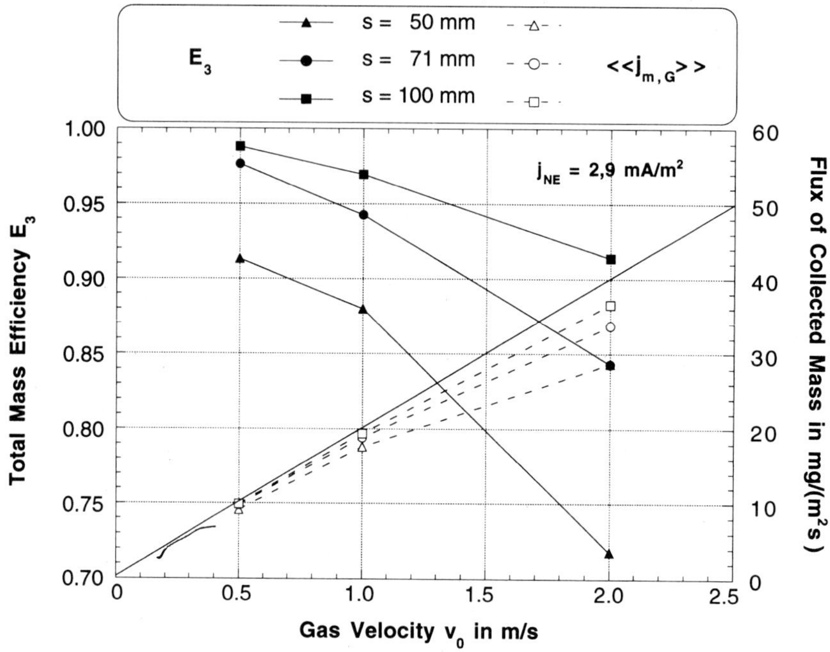

Figure 3.43 Plot of total flux of collected mass with corresponding total mass efficiencies as a function of increasing gas velocity.

space-averaged electrical field values as before (see Table 3.5, $j_{\mathrm{NE}} = 2.9\ \mathrm{mA/m^2}$).

It should be pointed out that in the deduction of equation (3.95) no model for particle transport was required, only a calculation. A model becomes necessary when the total mass efficiency E_3 or the grade efficiency $T(d_{\mathrm{p}})$ is to be predicted.

The relationship between effective migration velocity and the total flux of collected mass results when combining equation (3.95) with equation (3.91). Thus, equation (3.96) allows an easy transformation from effective migration velocities into total fluxes of collected mass and vice versa. The only things required to be known are the total efficiency and the total particle mass concentration in the raw gas, both quantities required to be measured when w_{eff} is to be found.

$$\langle\langle j_{\mathrm{m,c}} \rangle\rangle = -\frac{E_3 \cdot c_{\mathrm{m,e}}}{\ln(1 - E_3)} \cdot w_{\mathrm{eff}} \tag{3.96}$$

Generally, a flux can be represented as a product of a transport velocity and a concentration. Similarly, the total flux of collected mass can be split up into two factors, a 'velocity' w_{eff} and a ratio representing a 'concentration'. Obviously, the ratio represents a fictitious concentration value, since the mass concentration in the raw gas is connected in a physically uninterpretable way with the total efficiency – therefore w_{eff} is a fictitious velocity.

However, there might be a practical argument for a representation with effective migration velocities: when comparing Figures 3.42 and 3.43, a different sensitivity of w_{eff} and $\langle\langle j_{m,c}\rangle\rangle$ to changes in duct width and changes in gas velocity becomes obvious. For the w_{eff}-representation changes of gas velocity and changes of duct width are of the same order, but in a $\langle\langle j_{m,c}\rangle\rangle$-representation, changes in duct width are far smaller compared with changes in gas velocity.

Besides the total collected mass flux, the mass flux concept can be applied locally, i.e. locally collected mass flow of particles of size d_p, towards the collecting plate at location L, is the product of the mass concentration of these particles $c_m(L, d_p)$ at location L and the theoretical migration velocity $w_{th}(d_p)$ (steady state transport velocity) of particles of size d_p (equation 3.97). The mass concentration of particles of a given size $c_m(d_p)$ in relation to the total mass concentration is equivalent to $\Delta Q_3(d_p)$ and can therefore be represented by means of the corresponding particle size distribution function (equation 3.98).

$$j_{m,c}(L, d_p) = c_m(L, d_p) \cdot w_{th}(d_p) \tag{3.97}$$

$$c_m(d_p) = c_m \cdot \Delta Q_3(d_p) = c_m \cdot q_3(d_p) \cdot \mathrm{d}\,d_p \tag{3.98}$$

$$T(d_p) \equiv 1 - \frac{c_m(L, d_p)}{c_{m,e}(d_p)} \tag{3.99}$$

If the validity of a certain particle transport model is assumed, e.g. that of Deutsch, then both the particle size dependent mass concentration and the change of particle size distribution (which is caused by the classifying effect when passing through the electrical field) can be calculated. Figure 3.44 shows, for example, how the average theoretical migration velocity changes when passing through a precipitation zone assuming a log-normal number distribution for the dust (for details refer to [29]). Analogously, a local mass flux of collected particles $\langle j_{m,c}(L)\rangle$ can be calculated by using the local number concentration $c_N(L)$ and the number density distribution $q_0(d_p, L)$ at that location (equation 3.100).

$$\begin{aligned}\langle j_{m,c}(L)\rangle &= \int_{d_{p,min}}^{d_{p,max}} \mathrm{d}\,j_{m,c}(d_p, L)\\ &= \int_{d_{p,min}}^{d_{p,max}} c_N(L) \cdot m_p(d_p) \cdot w_{th}(d_p) \cdot q_0(d_p, L) \cdot \mathrm{d}\,d_p\end{aligned} \tag{3.100}$$

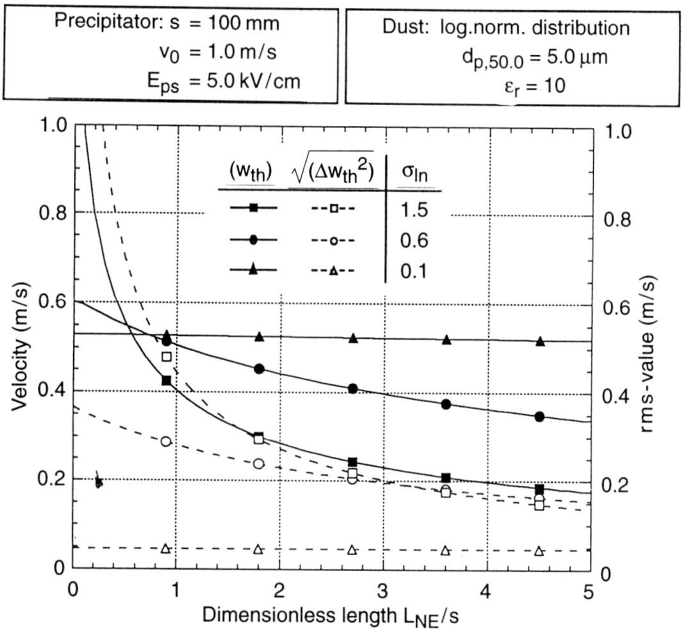

Figure 3.44 Plot to show how average theoretical migration velocity changes when passing through a precipitation zone, assuming a log-normal number distribution for the dust.

Integrating this locally collected flux over the complete precipitator length L_{NE} leads to the total mass flux $\langle\langle j_{\mathrm{m,c}}\rangle\rangle$ already introduced (equation 3.101).

$$\langle\langle j_{\mathrm{m,c}}\rangle\rangle = \frac{1}{L_{\mathrm{NE}}} \cdot \int_0^{L_{\mathrm{NE}}} \langle j_{\mathrm{m,c}}(L)\rangle \cdot \mathrm{d}\,L \tag{3.101}$$

This total mass flux, calculated by means of a transport model, can be compared with the total mass flux derived from measurements according to equation 3.95. Depending on the level of correspondence, the transport model could be validated.

The locally collected mass flux can also be simulated by means of particle tracking in a computed flow field (the method is described in 3.5.4). For a precipitator with half duct width $s = 200\,\mathrm{mm}$, a length of 4 m and a mean gas velocity (air, temperature 150 °C) of 1 m/s the flow field was calculated by means of a commercial CFD (computational fluid dynamic) package [30]. The turbulent flow was modelled by using the $k - \varepsilon$ model with an inlet turbulence intensity of 10%. For the simulation of locally collected mass fluxes, the particle size distribution represented in Figure 3.45 was

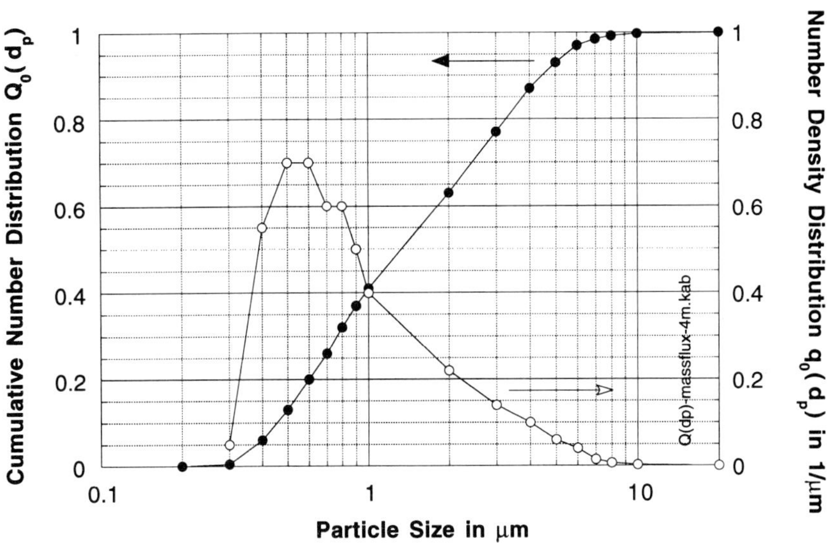

Figure 3.45 The particle size distribution represented here was assumed, as at the precipitator inlet, for the simulation of locally collected mass fluxes.

assumed. In order to consider the stochastic walk of particles in turbulent flows for particle tracking, a random walk model was used [30]. The particles were homogeneously distributed across the inlet and the number of calculated particle tracks corresponds to the relative frequency of size d_p in Figure 3.45. In total, 10^5 particle tracks were calculated and the locations of collection were registered, or alternatively the particles escaped.

Figure 3.46 shows, for each particle size, how the precipitated mass is distributed along the collecting electrode. The locally precipitated mass of a size class is related to the total mass of the size class at the inlet. As expected the large particles with high drift velocities w_{th} are precipitated early in the ESP. Each size class shows a curve with a maximum, which forms in the entrance region ($L' < 5$), then the curve decreases with increasing precipitator length. For particles $<2.5\,\mu$m almost no profile for the collected mass is readily observed (presumably, the little increase at $L' > 15$ is an artefact caused by the flow field conditions and the particle tracking algorithm).

When the results for the size classes are summated at each location, the total, locally collected mass or the collected number of particles can be plotted. In Figure 3.47 these masses or numbers are related to their total at the ESP inlet. This simulation of collected particles shows a strong maximum when represented as mass within $L' < 5$. This situation is analogous to the real behaviour where most of the dust can be found in the first hopper. The sharp maximum is a result of the mass representation compared with the number representation. The number curve of collected

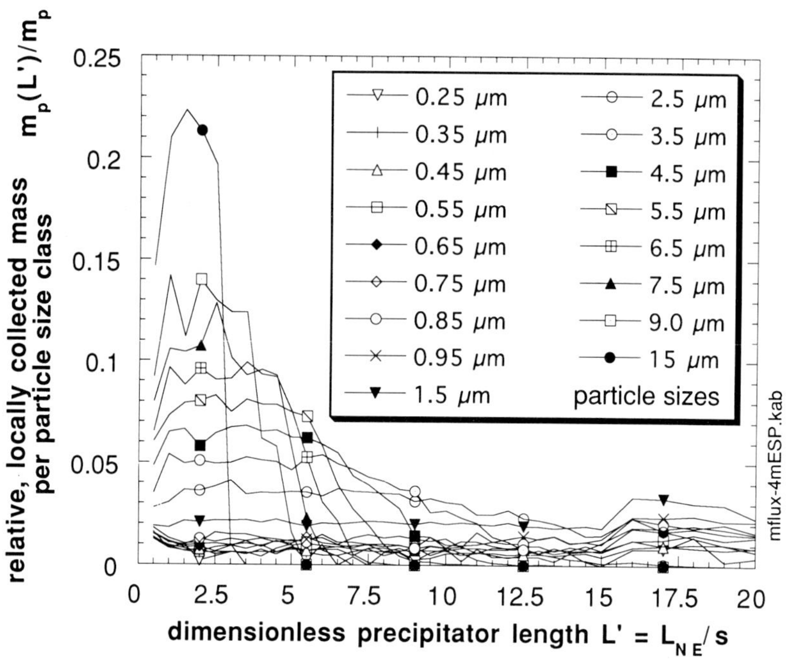

Figure 3.46 Diagram showing how the precipitated mass is distributed along the collecting electrode for each particle size.

particles shows a less pronounced maximum which is to be found slightly further downstream in the ESP.

Although particle tracking models (PTM) are in an early stage of development, they could have a practical application as shown by the results presented in Figures 3.46 and 3.47, which are quite promising for future analytical work. If, for example, for a given ESP configuration and a given dust, the curves in Figure 3.47 can be simulated, then an approximation function can be found, which should allow excellent scale-up in the case of precipitation improvement by adding further collection zones. Variations in operating conditions, such as changing gas velocity or electrical field strength, can also be easily investigated.

3.5.4 Flow field and particle trajectories

Observing particle transport in an ESP reveals (Figure 3.48) that, on the one hand, turbulent flow structures exist, while on the other hand, the particles are not homogeneously distributed over the duct. As Figure 3.48 and video tapes could show [31], dust free zones are clearly observed in the duct under idealized transport conditions. Obviously, this observation is more congru-

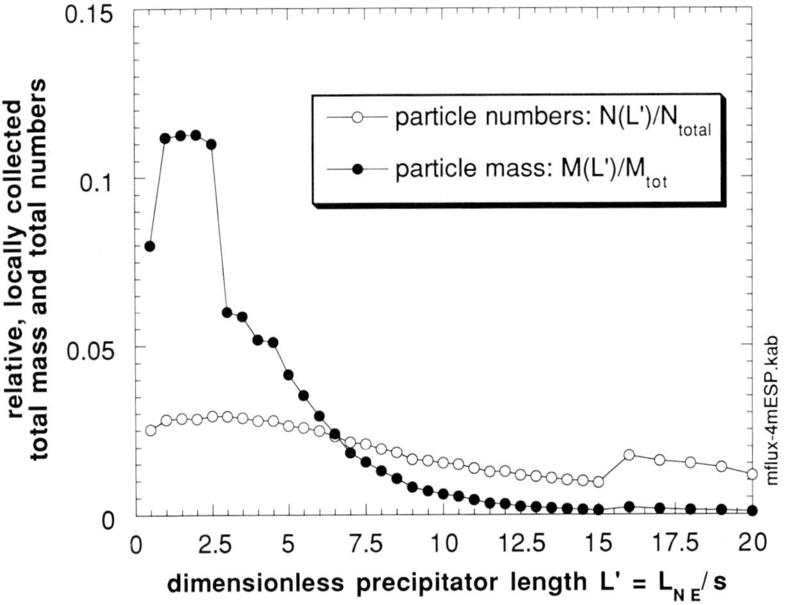

Figure 3.47 The total, locally collected mass or the collected number of particles plotted in relation to their total at the ESP inlet.

ent with the assumptions of the laminar model than with those of the Deutsch model.

For that reason, this section considers an alternative way of modelling particle transport in ESPs. Particle tracks can be calculated by solving the equation of motion provided the flow field and the electrical field are known. A simplified version of the equation of motion is given by equation (3.102), where the drag force is assumed to be Stokesian, which holds for particles $<20\,\mu$m under typical ESP conditions [24].

$$m_p \frac{d\vec{w}}{dt} = 3\pi\eta d_p(\vec{v} - \vec{w})\left(\cdot\frac{1}{Cu}\right) + Q_p^\infty \vec{E} \tag{3.102}$$

In turbulent flows, each local velocity component is described by its time-averaged value and a time-dependent fluctuation according to equation (3.103).

$$\vec{v}(\vec{r};t) = \begin{pmatrix} v_x(\vec{r};t) \\ v_y(\vec{r};t) \end{pmatrix} = \begin{pmatrix} \bar{v}_x(y) + v'_x(y;t) \\ \bar{v}_y(y) + v'_y(y;t) \end{pmatrix} \tag{3.103}$$

The simplest approach for the fluid flow assumes a time constant turbulent velocity profile and a corresponding profile for the turbulence intensity,

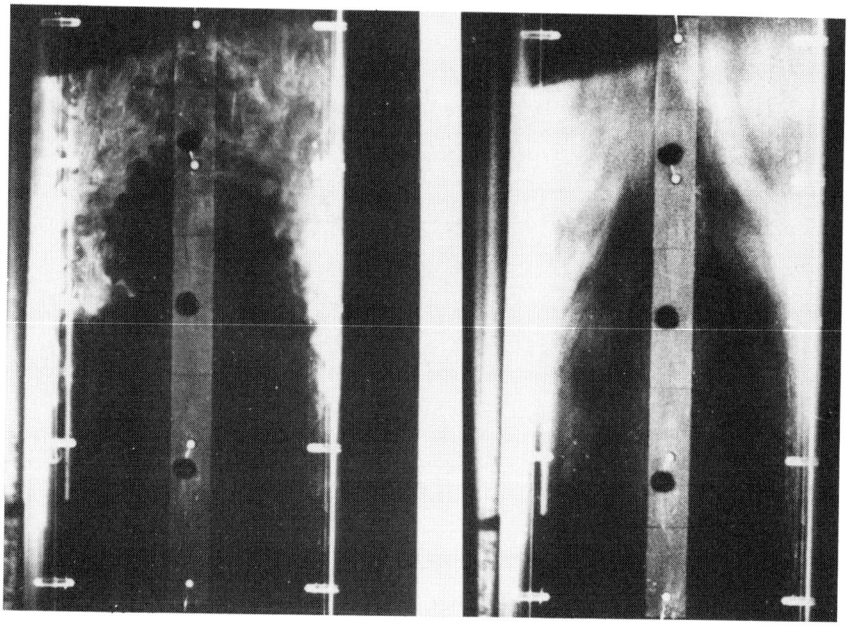

Figure 3.48 Particle transport in an ESP showing that, on the one hand, turbulent flow structures exist, while, on the other hand, the particles are not homogeneously distributed over the duct [4].

both only a function of y' as depicted in Figure 3.49 (in the near wall region a laminar sublayer is assumed). The velocity profile holds for a volume flux of 1.0 m/s and the profile of turbulence intensity is rather arbitrary; however, values of 10% in the middle of the channel, increasing in near wall regions, are typical for turbulent flows (for details see [24]).

When looking at the time averaged process, the mean particle tracks can easily be calculated by common numerical methods (e.g. Runge–Kutta–Fehlberg). Examples of typical particle tracks are given in Figure 3.50 for particles of sizes 1.0 and 10 μm. A lab-scale ESP ($s = 100$ mm, $v = 1.0$ m/s, $U = 30$ kV) has five discharge electrodes and an inhomogeneous distribution of the electrical field strength, together with the space charge according to 3.2.5, has been considered in the simulations if the particles enter the ESP with zero charge.

Figure 3.50 demonstrates how particles of different size are deflected in different ways. Particles entering the ESP on the symmetry line of the discharge electrodes get a kick by the first electrode towards the collecting wall. For electrode numbers >2, however, this kick and the electrodes' influence, because of inhomogeneity, has already disappeared. Furthermore, the particle charging process, in general, cannot be neglected, especially for

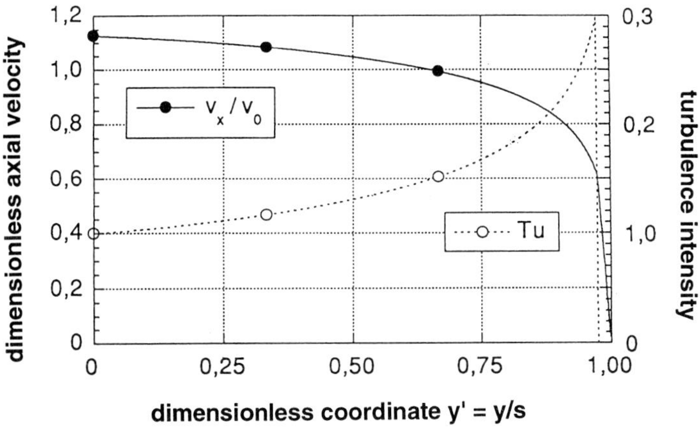

Figure 3.49 A time constant turbulent velocity profile and a corresponding profile for the turbulence intensity, both a function of y', comprise the simplest approach for the fluid flow.

fine particles. The 1.0 μm particles entering the ESP away from the plane of discharge wires will not reach 90% of their saturation charge within the considered precipitation zone.

When the stochastic character of turbulent flow is to be considered, the time-dependent part of the velocity is expressed in terms of an rms value of the x-component (isotropic turbulence) according to equations (3.104) and (3.105). The rms value is expressed by means of the assumed velocity and

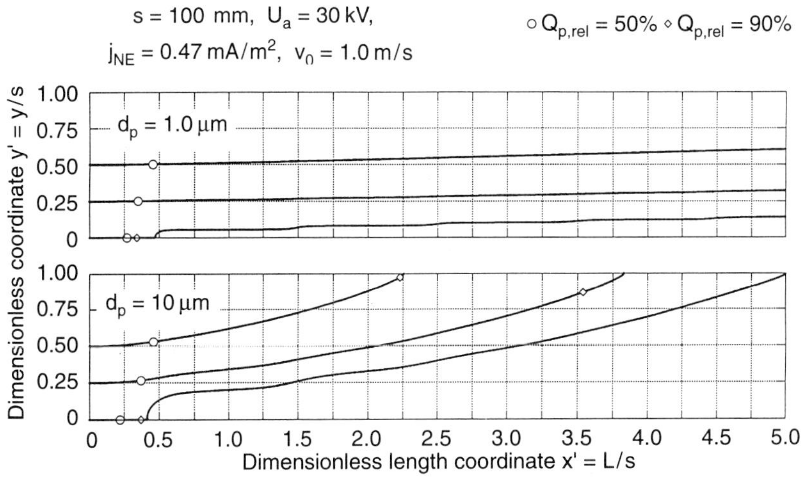

Figure 3.50 Examples of typical particle tracks for particles of sizes 1.0 and 10 μm.

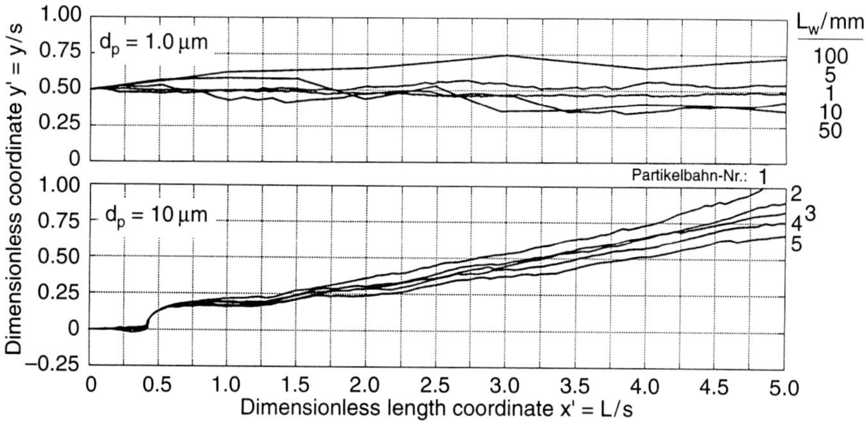

Figure 3.51 Top: plot to show how the structure of calculated particle tracks depend on eddy size (turbulent structure). Bottom: Five stochastic particle tracks with square starting location and an eddy size of 5 mm.

turbulence profiles (Figure 3.49) according to equation (3.106). The average y-component is zero.

$$v'_x(y';t) = \chi v_{x,\mathrm{rms}}(y') \tag{3.104}$$

$$v'_y(y';t) = \chi v_{x,\mathrm{rms}}(y') \tag{3.105}$$

$$v_{x,\mathrm{rms}}(y') = \bar{v}_x(y') \cdot \mathrm{Tu}(y') \tag{3.106}$$

The actual, time-dependent velocity value is simulated by means of a random number χ, which has to be revised from time to time. This time step can be interpreted as the life time of turbulent eddies or, when expressed in a distance, as a scale characteristic for the turbulent structure (eddy size L_w).

Figure 3.51 (top) shows how the structure of calculated particle tracks depends on this eddy size. Compared with the observations depicted in Figure 3.48, an eddy size of 5 mm leads to particle paths structured realistically. For this eddy size, Figure 3.51 (bottom) shows five particle tracks started at the same locations and same conditions. As in real turbulent flows, the simulated randomness leads to a spread in the particle paths.

With modern computational fluid dynamics (CFD) tools, however, the calculation of the flow field and even the tracking of particles have become more comfortable. Although an inhomogeneous electric field cannot be considered in standard versions, an homogeneous field can be taken into account quite easily. An example is given in Figure 3.52.

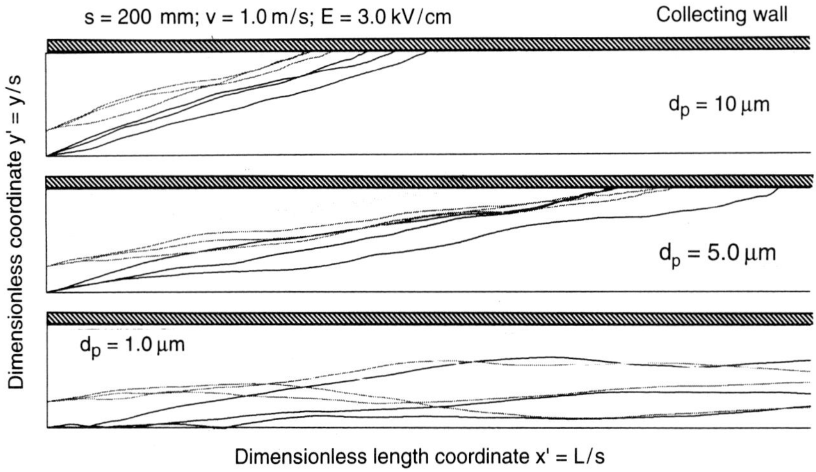

Figure 3.52 Example of how the homogeneous field can be taken into account in the calculation of the flow field with modern CFD.

The method of particle tracking can be used to simulate grade efficiencies of separators. For the lab-scale ESP discussed ($s = 100\,\text{mm}$, $L = 5\,\text{s}$, $v = 1.0\,\text{m/s}$, $U = 50\,\text{kV}$), grade efficiencies of different models together with experimental values are plotted in Figure 3.53.

Obviously, for particle sizes around $10\,\mu\text{m}$ the Deutsch model and the measured grade efficiencies are in good compliance. For decreasing particle size, however, measured particle precipitation is much better than the Deutschian values. While for sizes $< 1\,\mu\text{m}$, the performance is even better than the values predicted by the laminar model. The results of the particle tracking models introduced above are also plotted in Figure 3.53. It is interesting to note that both results are rather different. Thus, the PTM model based on a simple flow profile, but considering the inhomogeneous structure of the electrical field with a homogeneously distributed space charge, leads to an enormous upshift in efficiency for particle sizes $< 1\,\mu\text{m}$ – although the values are less than the measurements. This might indicate that fine particles are much better transported than assumed up to now, originally caused by the commonly neglected inhomogeneous distribution of the electrical field and space charge effects. This explanation is further supported by the results of another PTM model which uses a simulated turbulent flow field and a simple homogeneous electrical field strength. Herein, without electrical inhomogeneities, the particles in the fine particle

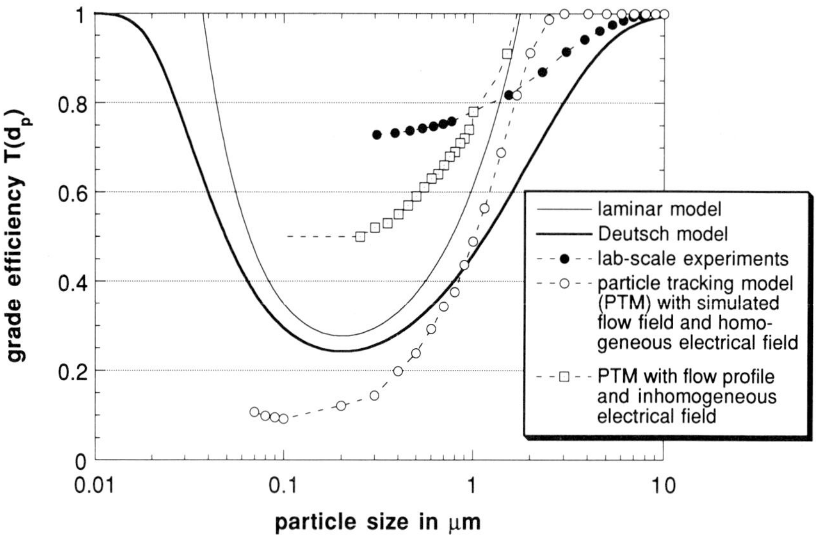

Figure 3.53 Plot of grade efficiencies of different models together with experimental values.

region are collected even lower than the predictions according to Deutsch. For particle sizes $> 1\ \mu m$ the simulated efficiencies lie in between the laminar and the Deutschian curve.

Looking at these results, there seems to be promise in combining the inhomogeneous electrical conditions with the simulated fluid flow; even so, such a simulation technique could still show a deviation from the measurements. It is presumed that the agglomeration effects, which occur under ESP operation [32], will have to be considered further to obtain compliance

3.5.5 Diffusivity models

Sometimes particle transport in ESPs is described as a combination of convective and diffusive processes [15, 25]. Assuming that the average flow field and the electrical field are of vector type as given in equation (3.107), then the so-called convective diffusion equation looks as equation (3.108).

$$\vec{v} = \begin{pmatrix} v_0 \\ 0 \end{pmatrix}; \qquad \vec{E} = \begin{pmatrix} 0 \\ E_{ps} \end{pmatrix} \tag{3.107}$$

$$v_0 \cdot \frac{\partial \bar{c}}{\partial x} + w_{th} \cdot \frac{\partial \bar{c}}{\partial y} - D_p \cdot \left(\frac{\partial^2 \bar{c}}{\partial x^2} + \frac{\partial^2 \bar{c}}{\partial y^2} \right) = 0 \tag{3.108}$$

Dedimensionalizing equation (3.109) with half plate-to-plate distance s, gas velocity v_0 and particle concentration at the entrance $c(x' = 0)$ reveals that the convective diffusion equation (equation 3.109), for clarity the bars have been omitted) is governed by two dimensionless numbers: the ratio of the transverse particle velocities Ω (equation 3.110), which is called the transport parameter (compare [4]), and the ratio of two counter-current transport effects, namely electrical transport against eddy diffusivity, the Peclet number, Pe (equation 3.111).

$$\frac{1}{\Omega} \cdot \frac{\partial c'}{\partial x'} + \frac{\partial c'}{\partial y'} - \frac{1}{\mathrm{Pe}} \cdot \left(\frac{\partial^2 c'}{\partial x'^2} + \frac{\partial^2 c'}{\partial y'^2} \right) = 0 \tag{3.109}$$

$$\Omega = \frac{w_{\mathrm{th}}}{v_0} \tag{3.110}$$

$$\mathrm{Pe} = \frac{w_{\mathrm{th}} \cdot s}{D_{\mathrm{p}}} \tag{3.111}$$

The method of determining the analytical solution was comprehensively described by Leonard *et al.* and a detailed solution is published [33]. The particle concentration is given by a sum of m eigen functions (equation 3.112) and, unfortunately, the complex relations for the eigen values Θ_m and the constant C_m are difficult to handle. When solving equation (3.112) on a computer, the effect of different eddy diffusivity values is easily shown as

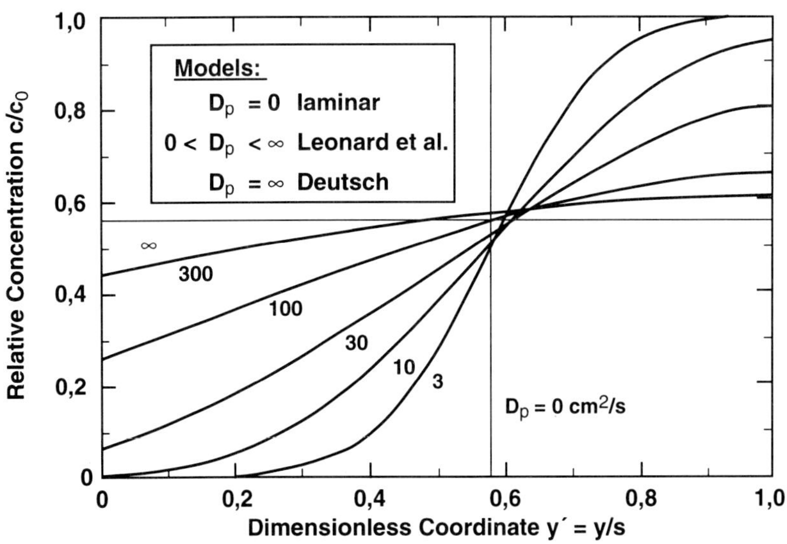

Figure 3.54 When using a computer to solve equation 3.112 the effect of different eddy diffusivity values is easily shown.

Figure 3.54 demonstrates. For low diffusivity values $D_p \to 0$ the model tends to the laminar case, while for high diffusivity values $D_p \to \infty$ the Deutsch model emerges.

$$c(x', y') = c(x' = 0) \cdot \sum_{m=1}^{\infty} C_m \cdot \exp(-\Omega \cdot x' \cdot F_m) \cdot \exp\left(\frac{\mathrm{Pe} \cdot y'}{2}\right)$$

$$\cdot \left\{\cos(\Theta_m y') + \frac{\mathrm{Pe}}{2\Theta_m} \cdot \sin(\Theta_m y')\right\} \tag{3.112}$$

$$C_m = \frac{C_Z}{C_N} \qquad \text{with:} \tag{3.113}$$

$$C_Z = \frac{\exp\left(-\frac{\mathrm{Pe}}{2}\right)}{\Theta_m^2 + \frac{\mathrm{Pe}^2}{4}} \cdot \left\{-\mathrm{Pe} \cdot \cos(\Theta_m) + \left(\Theta_m - \frac{\mathrm{Pe}^2}{4\Theta_m}\right) \cdot \sin(\Theta_m) + \frac{\mathrm{Pe}}{\exp\left(-\frac{\mathrm{Pe}}{2}\right)}\right\}$$

$$C_N = \frac{1}{2} + \frac{\mathrm{Pe}}{4\Theta_m^2} + \frac{\mathrm{Pe}^2}{8\Theta_m^2} + \frac{1}{4\Theta_m} \cdot \left(1 - \frac{\mathrm{Pe}^2}{4\Theta_m^2}\right) \cdot \sin(2\Theta_m) - \frac{\mathrm{Pe}}{4\Theta_m^2} \cdot \cos(2\Theta_m)$$

$$F_m(\Omega, \mathrm{Pe}) = \frac{1}{\Omega^2} \cdot \frac{\mathrm{Pe}}{2} \cdot \left\{\sqrt{1 + \Omega^2 \cdot \left[1 + \left(\frac{2\Theta_m}{\mathrm{Pe}}\right)^2\right]} - 1\right\} \tag{3.114}$$

$$\tan \Theta_m = \frac{-2\left(\frac{2\Theta_m}{\mathrm{Pe}}\right)}{1 - \left(\frac{2\Theta_m}{\mathrm{Pe}}\right)^2} \tag{3.115}$$

An analytical integration of the concentration profile at precipitator's exit will readily lead to the grade efficiency (equation 3.116).

$$T(d_p) = 1 - \sum_{m=1}^{\infty} C_m \cdot \exp\left(-\Omega \cdot \frac{L_{NE}}{s} \cdot F_m\right) \cdot \frac{\sin \Theta_m}{\Theta_m} \cdot \exp\left(\frac{\mathrm{Pe}}{2}\right) \tag{3.116}$$

For practical aspects the grade efficiency curves are important to know, but the particle size is hidden behind the transport parameter, Ω, and the Pe-number. This is why the values for both parameters cannot be varied independently. In Figure 3.55 grade efficiencies for a precipitator of $2s = 200$ mm, $L_{NE} = 500$ mm, $v_0 = 1.0$ m/s and $U = 50$ kV is illustrated, calculated according to equation 3.116.

The efficiency minimum in all curves is caused by the Cochet equation (see section 3.3) considering diffusional effects in the field charging process of particles. It may be surprising that the difference in efficiency for particles below 1 μm is rather small, while the largest difference is to be observed

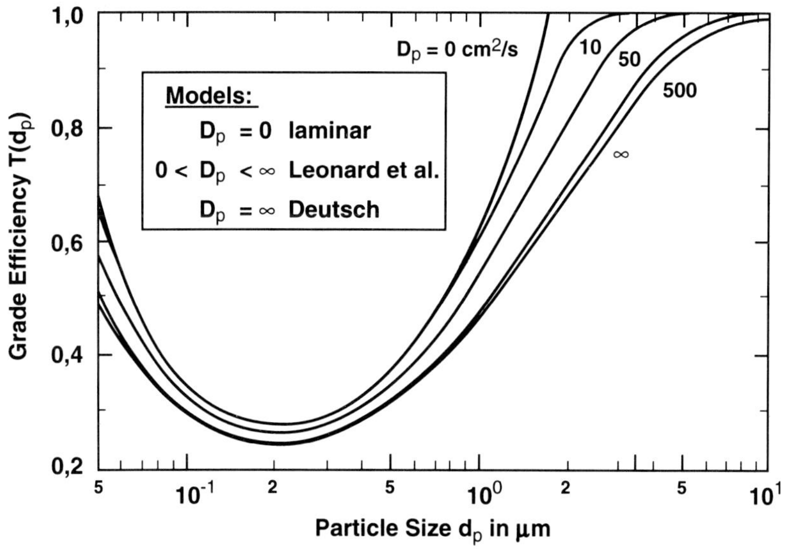

Figure 3.55 Grade efficiencies for a precipitator of $2s = 200\,\text{mm}$, $L_{NE} = 500\,\text{mm}$, $v_0 = 1.0\,\text{m/s}$ and $U = 50\,\text{kV}$, calculated according to equation 3.116.

around $2\,\mu$m. Of course the absolute trends depend on the operating conditions used for calculation.

Since the differences range between the laminar and the Deutschian limit, neither of which match reality (see Figure 3.53), and because of the complexity of the equations, the finite diffusivity models are not recommended for practical work. (When the local character of the electrical field is considered, which requires a numerical iterative solution process, there still remains some contradictions when simulating grade efficiencies [34].)

3.6 Deposition

When the drifting, electrically charged particles touch down on the collecting electrode, the electric charge has to flow from the particles into the ground electrode. Since the electrical charges have to pass through the dust layer, dust resistivity plays an important role in ESP operation (Figure 3.56). The transport of electrical charge occurs by two mechanisms, surface and volume conductivity, which are discussed in detail elsewhere.

Because the numbers of electrical charge carriers created by the corona are usually orders of magnitude more than needed for charging all particles to saturation [24], not only the charges on the particles have to pass through the dust layer but *all* charges created by the corona. Therefore,

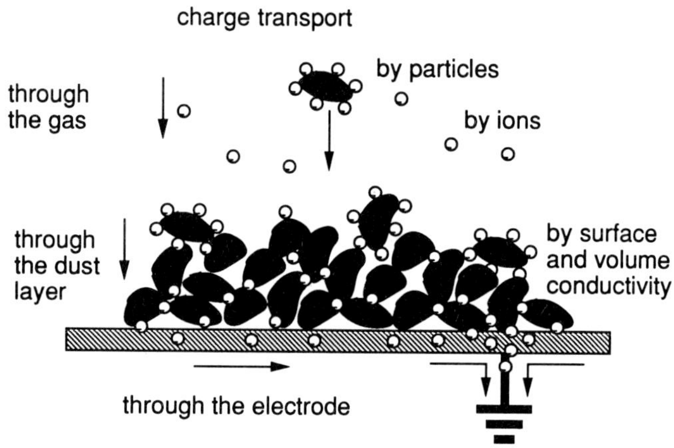

Figure 3.56 Illustration of the important role played by dust resistivity in ESP operation since the electrical charges have to pass through the dust layer.

operating with high currents can have advantages when considering particle charging. On the other hand, it can be critical with respect to the initiation of back-corona in the dust layer, which dramatically deteriorates particle precipitation. For this reason pulsing high voltage systems, resulting in lower total charge carrier flow, can be very effective when applied to high resistivity dusts. These lower currents lead, in addition, to less energy consumption [35–37].

3.7 Removal

The dry removal of the dust layer is made by mechanical impact which is achieved by hammers in certain time steps. In most common applications these cleaning cycles are made on-line, i.e. the gas flow and electrics are still in operation, while the dust layer breaks up into large agglomerates which settle into the hoppers. Because of the ongoing gas flow, finer and already collected particles can be readily re-entrained into the gas stream. This re-entrainment, which depends for example on turbulence levels, can severely reduce the ESP efficiency.

On the other hand, the electrical current, which is still flowing through the dust layer, holds the dust layer like an electrical glue. This electrical force, caused by the charge transfer between touching particles, increases the common adhesive force and is called an electric clamping force. This force plays an important role in the cleaning process of the collecting electrode, but has been poorly investigated to date.

References

1. Riehle, C. and Löffler, F. (1993) Particle dynamics in an electrohydrodynamic flow field investigated with a two-component laser-Doppler velocimeter. *Part Part. Syst. Charact.* **10**, 41–7.
2. Shaugnessy, E.J. and Davidson, J.H. (1986) Turbulence generation by electric body forces. *Exp. Fluids*, **4**, 17–26.
3. Shaugnessy, E.J. and Davidson J.H. (1985) The fluid dynamics of electrostatic precipitators. *Aerosol Sci. Technol.*, **4**, 471–6.
4. Riehle, C. and Löffler, F. (1993) Reflections on similarity laws concerning particle transport in electrostatic precipitators. *Powder Technol.*, **77**, 201–8.
5. Miller, J., Riehle, C., Schwab, A.J. and Löffler F. (1994) Numerische Feldberechnungen in Elektrofiltern im Hinblick auf elektrisch ähnliche Betriebszustände. *J. Electrostatics*, **33**, 213–28.
6. Miller, J. (1994) Private communications.
7. White, H.J. (1963) *Industrial Electrostatic Precipitation*, Addison-Wesley, Pergamon Press.
8. Oglesby, S. and Nichols, G.B. (1978) *Electrostatic Precipitation, Pollution Engineering and Technology*, Vol. 8, Marcel Dekker, New York, Basel.
9. Peek, F.W. (1929) *Dielectric Phenomena in High Voltage Engineering*, 3rd ed., McGraw Hill, New York.
10. Robinson, M. (1971) *Electrostatic Precipitation*, Ed. E. Strauß, *Air Pollution Control*, Part 1, John Wiley, New York.
11. Townsend, J.S. (1915) *Electricity in Gases*. Oxford University Press, pp. 375–6.
12. Cooperman, P. (1960) A theory for space charge limited currents with application to EPs. *Trans. AIEE*, **79**, 47–50.
13. Cooperman, G. (1981) New current-voltage relation for duct precipitators valid for low and high current densities. *IEEE Trans. Ind. Appl.*, **A-17 (2)**, 236–9.
14. Spiegel, M.R. (1974) *Complex Variables* (Schaum Outline Series), McGraw-Hill, New York, pp. 252–3.
15. Self, S.A., Kihm, K.D. and Mitchner, M. (1987) Comparison of wire-plate and plate-plate EPs in turbulent flow. *J. Electrostatics*, **19**, 21–32.
16. Leutert, G. and Böhlen, B. (1972) Der räumliche Verlauf von elektrischer Feldstärke und Raumladungsdichte im Plattenelektrofilter. *Staub-Reinhalt. Luft*, **32**, 297–301.
17. Miller, J. (1995) Private communications.
18. Sato, T. (1987) Charging process of fine particles in unipolar ion flow. *Trans. I.E.E. Jpn.*, **107**, 155–61.
19. Murphy, A.T., Adler, F.T., and Penney, G.W. (1959) A theoretical analysis of the effects of an electrical field on the charging of fine particles. *Trans. AIEE*, **78**, 318–26.
20. Liu, B.Y.H. and Yeh, H.C. (1968) On the theory of charging aerosol particles in an electrical field. *J. Appl. Phys.*, **39**, 1396–402.
21. Smith, W.B. and McDonald, J. R. (1976) Development of theory for the charging of particles by unipolar ions. *J. Aerosol Sci.*, **7**, 151–66.
22. Liu, B.Y.H. and Kapadia, A. (1978) Combined field and diffusion charging of aerosols in the continuum regime. *J. Aerosol Sci.*, **9**, 227–42.
23. Cochet, R. (1961) Lois Charge des Fines Particules (Submicroniques) Etudes Théoriques – Controles Récents Spectre de Particules, *Coll. Int. la Physique des Forces Electrostatiques et Leurs Application*, Centre National de la Recherche Scientifique, Paris, **102**, 331–8.
24. Riehle, C. (1992) Bewegung und Abscheidung von Partikeln im Elektrofilter. Dr.-Ing. Thesis, University of Karlsruhe.
25. Self, S.A., Kihm, K.D. and Mitchner, M. (1987) Comparison of wire-plate and plate-plate EPs in laminar flow. *J. Electrostatics*, **19**, 21–32.
26. Deutsch, W. (1922) Bewegung und Ladung der Elektrizitätsträger im Zylinderkondensator. *Ann. Phys.*, **68**, 335–44.
27. Matts, S. and Öhnfeldt, P.O. (1963/64) Efficient gas cleaning with SF electrostatic precipitators. *Fläkten*, **1-12**, 93–110.
28. Riehle, C. (1995) Mass flux and effective migration velocity in electrostatic precipitators. *Powder Technol.* (December issue).

29. Riehle, C. (1993) Zur Messung und Beschreibung des korngrößenabhängigen Stofftransports im Elektrofilter. *Staub-Reinhalt. Luft*, **53**, 383–8.
30. FLUENT Inc. (1995), User guide.
31. Riehle, C. and Löffler, F. (1991) Revision of the Deutsch model based on video tapes of particle transport in electrical wire-plate precipitators, *EPRI 9th Particulate Control Symp.*, Oct. 15–18, Williamsburg/USA EPRI TR 100471, 2, Palo Alba, CA, USA.
32. Riehle, C. and Wadenpohl, C. (1995) Electrically stimulated agglomeration at an earthed surface, *Powder Technol.* (December issue).
33. Leonard, G., Mitchner, M. and Self, S.A. (1980) Particle transport in electrostatic precipitators. *Atmosph. Environ.*, **14**, 1289–99.
34. Riehle, C. and Löffler, F. (1995) Grade efficiency and eddy diffusivity models. *J. Electrostatics*, **34**, 401–13.
35. Peterson, H.H. (1981) New trends in electrostatic precipitation: wide duct spacing, precharging, pulse energization. *IEEE Trans. Ind. Appl.*, **IA-17**, 496–501.
36. Darby, K. (1984) Pulse energization – an alternative to conditioning for highly resistivity dusts. *Proc. 2nd Int. Conf. ESP*, Kyoto, Japan, pp. 495–501, APCA, Pittsburgh, USA.
37. Mayer-Schwinning, G. and Neulinger, F. (1985) Sind Pulsgeneratoren tatsächlich die Lösung aller Entstaubungsprobleme?, *Zem. Kalk-Gips*, **38**, 405–11.

4 Mechanical design considerations for dry precipitators

F. KNUTTSEN AND K. R. PARKER

4.1 Introduction

The major operational components of any precipitator, as will be appreciated from chapter 3, are the discharge electrodes, which provide the necessary corona current for charging the particles, and the collectors or receiving electrodes, on which the charged particles are deposited. Once the particles have been deposited they will need to be removed, in order that precipitation will continue satisfactorily, either by mechanical rapping in the case of a dry precipitator, or by liquor washing if the application is wet.

These components are housed in a completely stable and gas tight casing, usually thermally insulated, the lower end of which is formed into hoppers or troughs for receiving the particulates removed from the collectors. In order that the discharge electrodes can be electrically energized, to the required voltage to produce the corona, they are isolated from the casing by some form of electrical lead through insulator arrangement, which is coupled to the transformer rectifier equipment providing the precipitator's voltage and current needs.

While the foregoing identifies the necessary components for precipitation, the design of the individual items have special requirements if the precipitator is to not only perform completely satisfactorily, but give long-term consistent performance with the minimum maintenance needs. This chapter will review the mechanical items of equipment in some detail, in order to meet these objectives.

4.2 Discharge electrodes

The basic requirement of any discharge element used in an electrostatic precipitator, is, when electrically energized, the field intensity adjacent to the element surface is such that the molecules in the vicinity have their work function raised to promote ionization of the gas, together with the release of some electrons. The production of these ions and electrons is covered in detail in chapters 3 and 8, so will not be discussed further in this section.

To produce the necessary field intensity at the surface of the electrode, it is essential that the radius of curvature of the element is small compared

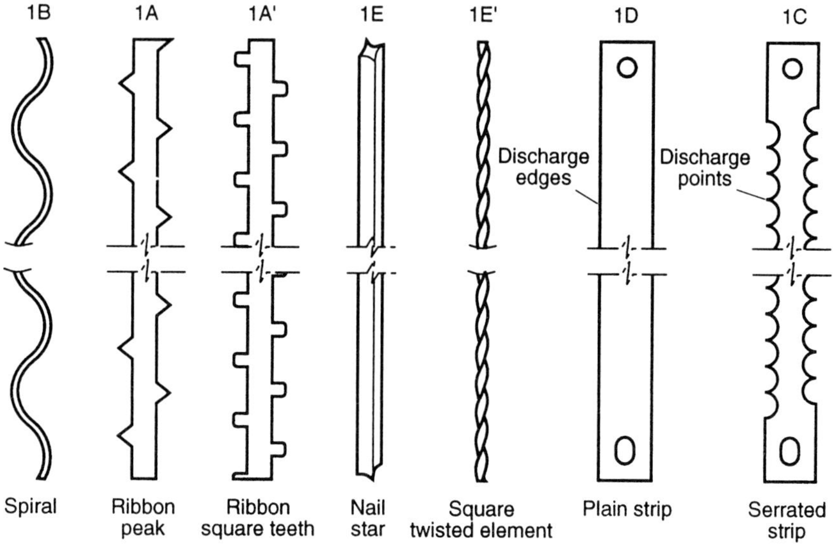

Figure 4.1 Discharge electrode profiles.

with that of the positive collector electrode. In most practical applications this does not present any problem, since the collector, particularly for a plate precipitator, can be considered as having an infinite radius of curvature.

Over the years, the discharge electrode has taken many different forms, from a simple round, square or barbed wire, through to so-called controlled emission electrodes for specific duties, some units being claimed as 'unbreakable'. As a modern power plant precipitator can contain the equivalent of 50 km of wire electrode, the cost, in a competitive bid situation, means that a simple readily available shaped material can offer an economic advantage, so a round wire formed into a spiral or twisted square section wires can often be found in many installations.

Some typical electrode profiles are illustrated in Figure 4.1; these can be split into two different categories, uncontrolled and controlled corona forms. In the controlled corona type, these have specific positions or areas where the emission develops, such as the spikes on electrodes 1A, or in the case of the spiral electrode 1B, the position closest to the collector, i.e. the point of maximum field intensity. For the square or round wire format, the corona develops somewhat randomly and, if observed, the highest or brightest corona points move around the electrode to coincide with the maximum field intensity which changes with dust conditions, either on the electrode system or in the gas phase itself.

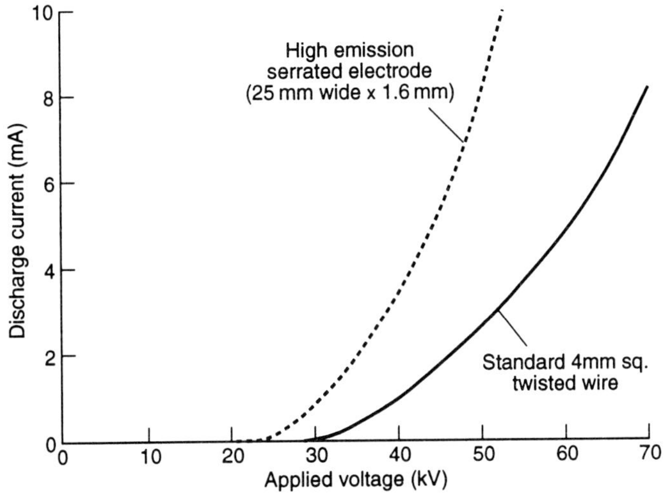

Figure 4.2 Laboratory emission characteristics. Square *vs.* serrated electrodes, collectors at 305 mm spacing.

Each precipitator supplier will emphasize the advantages of their specific electrode form but, in the absence of space charge or corona suppression effects, all electrodes can provide acceptable performance figures. One of the disadvantages of the controlled emission electrode is that without corona suppression or space charge effects, the emission is considerably greater than for a simple wire electrode, as indicated in Figure 4.2 [1]. These data were obtained under laboratory conditions in air, for a square wire electrode 1E and a controlled emission electrode 1C. In actual operation a similar characteristic is obtained, but the shape of the curves is modified to reflect the space charge effect resulting from the dust particles.

This increase in power consumption, under free running operation, can have a considerable effect on the operating costs of a large precipitator installation. Where, however, 'power saving' measures are practised, using modern microprocessor control systems (see chapter 8, section 8.9), this increased power consumption can be minimized without significantly increasing the emission to any great extent.

Where severe space charge or corona suppression is to be expected, then a high emission discharge electrode is essential to allow the particles to be effectively charged. Under these conditions the corona emission characteristic of the high emission electrode reduces to equate roughly with that of a simple wire profile, hence the power consumption is limited automatically [2]. There is, however, a significant increase in precipitator performance, as the space charge or corona suppression effects are negated by the higher corona current flow, resulting in a higher 'drift' velocity.

Theoretically, only the first field needs to have high emission electrodes, since once this field performs efficiently, the particles causing the corona suppression or space charge effects, are largely removed and the downstream fields operate normally. In many installations, it is normal to find the same form of electrode fitted throughout the unit unless the plant is very large and a change in electrode profile can be economically justified.

4.3 Discharge electrode mounting

In general there are two main approaches, dependent on the design philosophy of the supplier, as to how the electrode systems are mounted within the precipitator. The first system can be considered as the outboard of the field proper, i.e. the discharge electrode support system is carried from an insulated frame located in the interfield area, and the second inboard, where the suspension is carried from an insulated frame located above the collectors.

For the outboard suspension, individual electrode elements are normally mounted in a rigid frame, often referred to as a 'bedstead', where the electrodes are fixed between horizontal members forming the frame. The total discharge system itself is typically carried from four support insulators

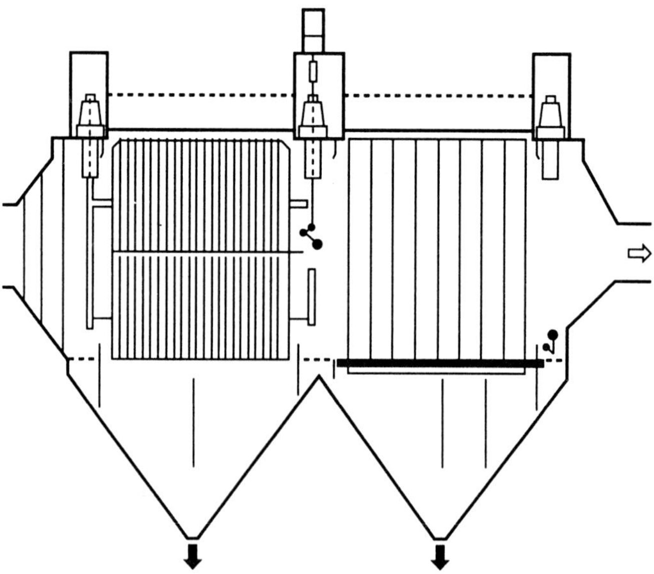

Figure 4.3 Schematic arrangement of complete precipitator. Wide interfield spacing, bedstead frame (spiral electrodes), rolled channel collectors.

located above the interfield area, and the individual frames are normally positioned within the height and length of the collector duct, as illustrated in Figure 4.3.

The inboard type of suspension, being above the field, enables the electrodes to be hung from a top support frame to pass between the collectors. The early installations were fitted with weighted wire electrodes, where the tensioning weight was located beneath the bottom of the collectors. More modern designs using the inboard approach have the electrode elements carried in a mast, or use the 'unbreakable' full-height electrode, both with the extremities being located above and below the collector, as indicated in Figure 4.4.

The outboard approach, where the insulators are positioned over the interbank area, means that the casing is longer than the inboard system, but with this, the casing is taller to cater for the top and bottom electrode support systems.

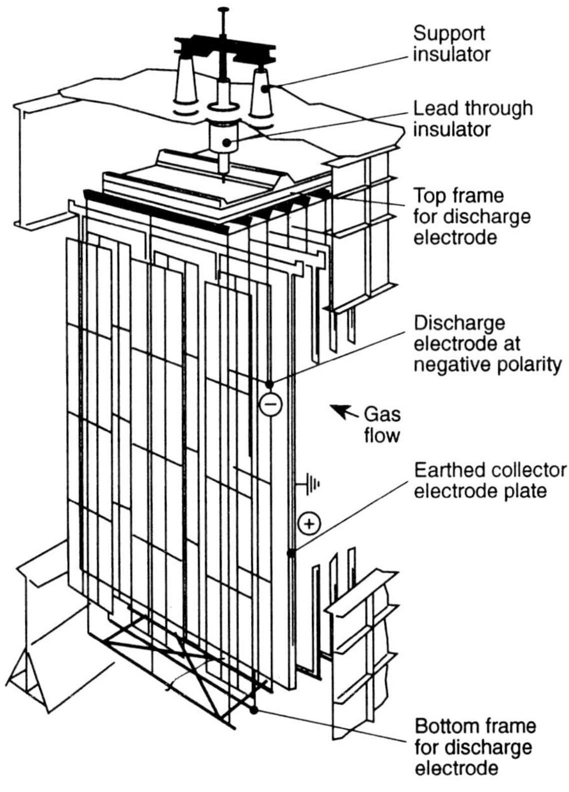

Figure 4.4 Schematic arrangement of complete precipitator field. Narrow interfield spacing, mast-type discharge electrodes (twisted wire), catch space collectors (courtesy Lodge Sturtevant Ltd).

Obviously one of the main criteria for plant availability is the freedom of electrode failure, as a broken or swinging electrode can electrically short out the field and hence impact on performance. To minimize this, it is important that the choice of electrode design must be such that swinging must be prevented. Many designs are arranged to have only short unsupported lengths of electrode elements, so that the amplitude of element vibration, arising either through electrical energization or mechanical rapping, is reduced.

To further increase the potential electrode life, large section elements are used instead of smaller round wires, but as the corona emission is reliant on a small radius of curvature, the electrode edge must be as sharp as possible. For example, under a given situation, a sharp 4 mm square wire has the same emission characteristics as a round wire of 2.5 mm diameter, but has some four times the cross-section and hence is mechanically superior.

4.4 Collectors

The prerequisite for any collector is that it provides a relatively flat profile to the discharge electrode which gives a uniform electric field, and is mechanically stable under all operating conditions of temperature, gas flow and rapping.

Over the years, the collector like the discharge electrode has taken many different forms, from an initial simple flat or corrugated plate to the present large fabricated designs.

During the 1950s and 1960s, there was a vogue for so-called 'catch pocket' profiles where, in theory, the dust, upon being dislodged by the rapping, passed through the pocket into a vertical quiescent zone, such that re-entrainment was minimized [3]. In practice, as the pockets caused local electric field disturbances at the surface, and the maximum field was at the pocket, this resulted in the greatest dust deposition adjacent the pocket, so that when rapped, the dislodged dust tended to fall outside the protected area. Their use has been discontinued; not only was the theory unsound, but mainly because of cost, since to provide the quiescent zone, the collector plate comprised two separate sheets.

Most installations today, which can have collector plates up to 5 m by 15 or 16 m high, employ either full-height cold-rolled strips having inbuilt stiffening ribs, joined together to form the collector, or alternatively, some type of fabricated collector, employing pressed or rolled sheets connected between vertical stiffening members, e.g. channels. These approaches (Figures 4.5 and 4.6) have been found to provide the requisite mechanical strength and stability to maintain the degree of electrode/collector alignment necessary to meet the low emission levels presently demanded. Particularly, so as to minimize costs, collector metal thicknesses are typi-

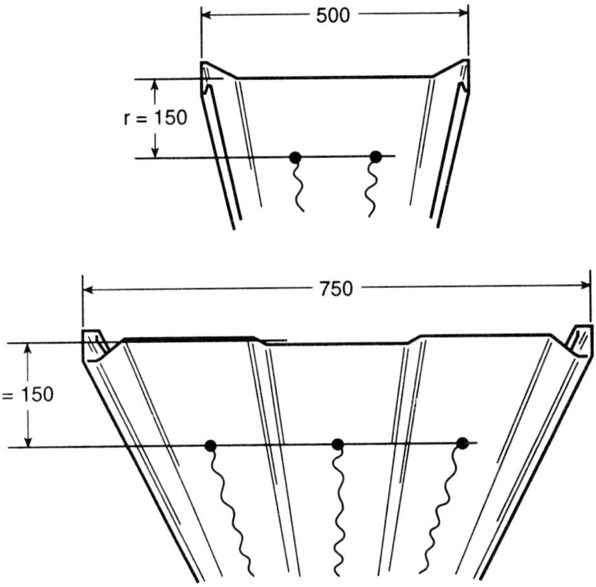

Figure 4.5 Collector design – cold-rolled strip profiles.

cally around 1.5 mm, which without the stiffeners, would be totally unsuitable.

In either approach, the elements are carried from a suitable top support member sometimes rapped, carrying the main load, plus a lighter frame for bottom end location purposes, or a heavier bottom frame, which can be alternatively mechanically rapped. Again, like the discharge electrode system, the format depends on the suppliers' design concepts.

The actual profile of the collector depends on the availability of fabricators to produce a suitable profile in the lengths and tolerances required. Each supplier will have a preferred profile for his particular cold-rolled sheet, but essentially all forms have raised upstands at each edge and various stiffening rib arrangements to balance out rolling stresses, such that twisting or 'banana-ing' is eliminated. Some have slots and fingers built into the edge stiffeners, which interlock, to provide additional stability and alignment. Some rolled channel profiles are illustrated in Figure 4.5.

There are a number of different fabricated forms of collector having flat sheets connected between vertical stiffening members. The so-called 'catch space' collector is illustrated in Figure 4.6; this comprises a hollow top beam carrying full-height rolled channels, 75 mm × 50 mm, having 1.5 mm pressed or rolled sheets connected to them forming the collector proper.

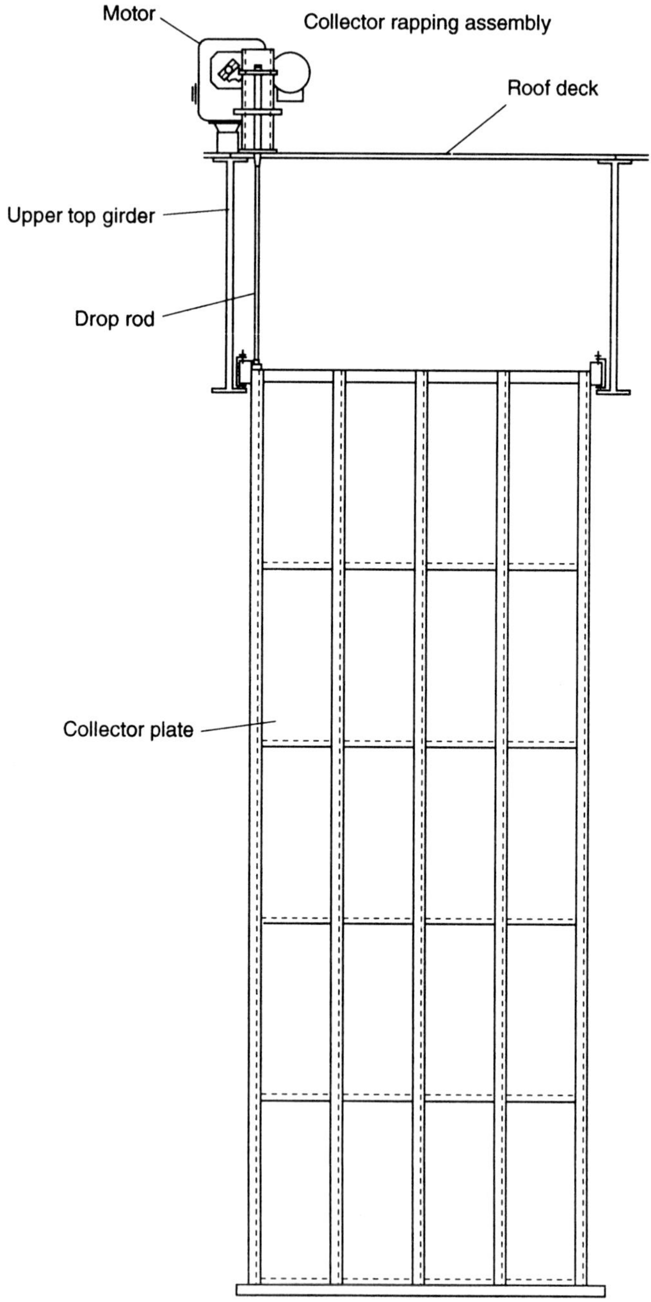

Figure 4.6 Collector assembly – catch space design (Lodge Sturtevant Ltd).

On some smaller wet precipitators, e.g. with a maximum height of 8 m, a simple thick 5 or 6 mm hot-rolled plate collector can be found; again this would be supported from a heavy top member and located at the bottom by a lighter frame. Other forms of collector can be found in practice, particularly on mist precipitators, where tubular metal or fabric reinforced plastic (FRP) thick collectors can be found. These alternative forms of collector are detailed in chapter 14.

4.5 Casings

The casing, containing all the precipitator internals, can weigh several hundred tonnes and is an important part of the overall design. Not only must it be completely gas tight, but must also withstand all the various loads imposed on it without distortion, in order to maintain electrode collector alignment. The loads would include wind, snow and possibly seismic conditions, in addition to the dead load of the internals, plus dust deposits on the internals and hoppers.

Casings are generally designed to international standards utilizing the ‘allowable stress’ in the material, for load, gas pressure and temperature, etc.; alternatively, some casing designs are now based on ‘deflection under load’. As any precipitator structure must be rigid enough to maintain electrode alignment and to contain the gas under all conditions, the design based on ‘deflection’, rather than ‘allowable stress’, can result in an optimum weight/cost arrangement.

Typically each precipitation field would be separated by a vertical column member and portal, which carries the collector and discharge electrode (DE) loads. For the outboard insulator suspension arrangement, this portal comprises a double roof girder forming a hollow beam or tunnel in which the discharge electrode suspension insulators can be located. This hollow beam/tunnel tends to produce wider interbank access areas than the inboard arrangement, but in this case, the discharge electrode load must be carried by structural cross-beams connected back to the collector support bridge girder spanning the unit (see Figures 4.3 and 4.4), showing typical side elevations.

Most large precipitators incorporate a main lower ring beam system upon which the precipitator casing is built; this ring beam is, in turn, supported from a substructure having sliding bearings situated between the substructure and ring beam stubs. This arrangement (Figure 4.7) allows the complete casing to expand with temperature changes, which not only eliminates mechanical stresses, but ensures that alignment is maintained. Figure 4.8 illustrates how, on a large casing, one point on the substructure is fixed and all other locations can move.

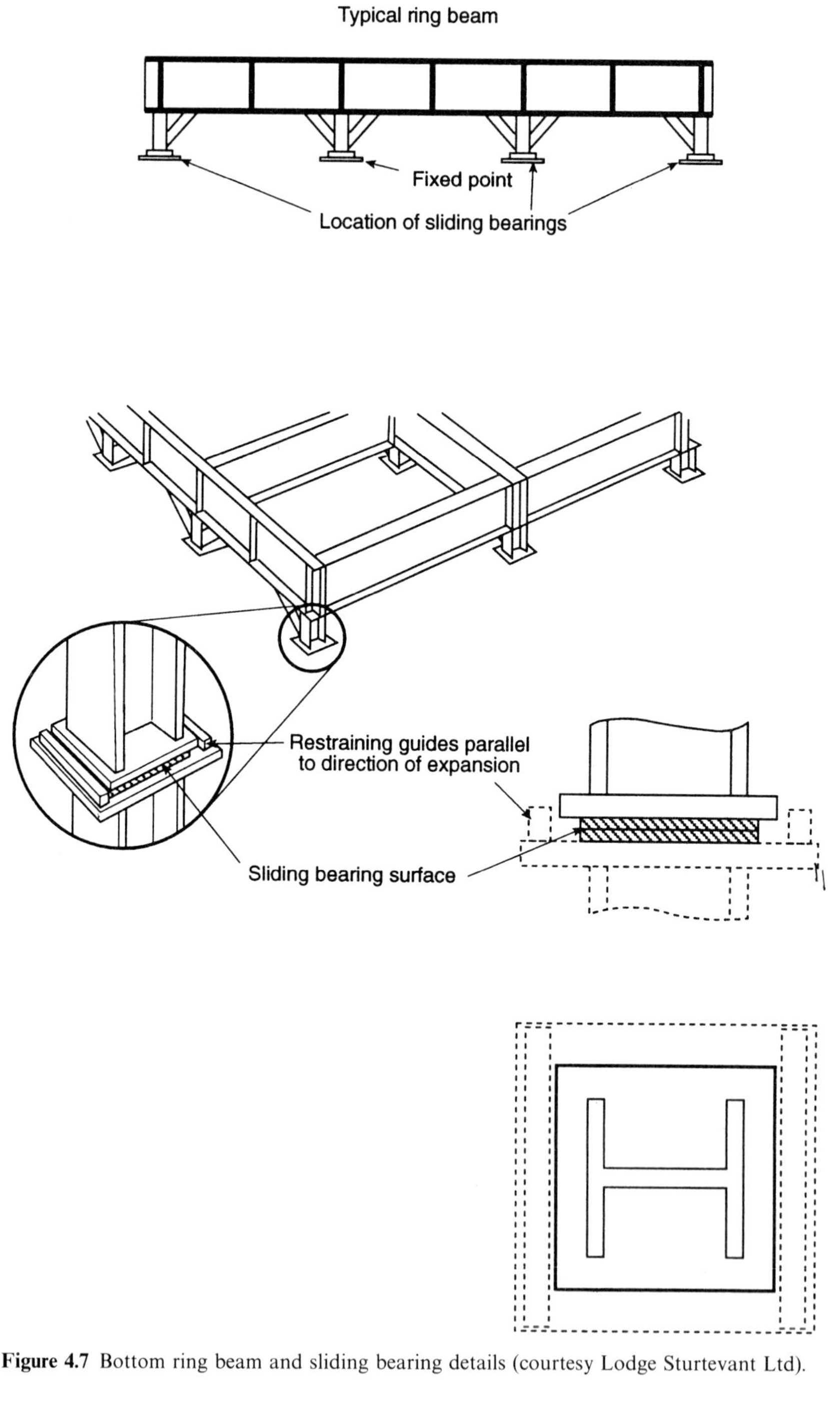

Figure 4.7 Bottom ring beam and sliding bearing details (courtesy Lodge Sturtevant Ltd).

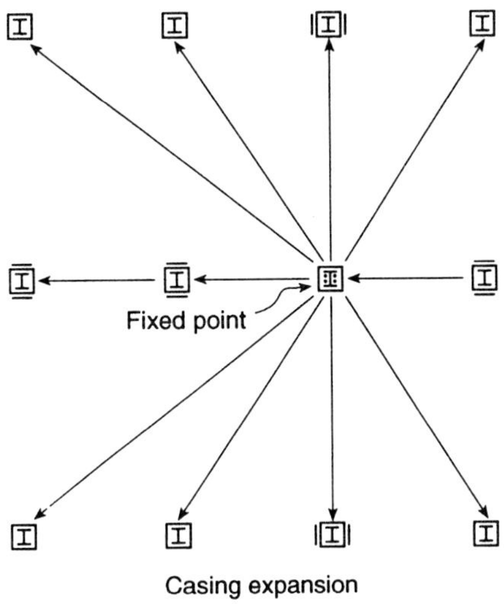

Figure 4.8 Substructure/casing expansion conditions.

The lower end of the casing terminates in a hopper or trough mounted directly onto the ring beam; sometimes the side walls are extended vertically downwards to provide additional storage capacity for the dust. The specified design condition for the hopper connection is usually the hopper dead load plus the hopper completely full of dust, calculated with the bulk density of the appropriate dust in question.

In the past, casings have been constructed from reinforced concrete and other building materials, but the present trend, because of speed of erection and cost, is to employ steel fabrications, usually partly prebuilt off-site, for example field side walls, hoppers, ring beams and possibly mouthpieces, depending on the size of installation. FRP and similar corrosion-resistant materials can also be found on wet or mist precipitators.

For large installations, e.g. power plants, to reduce costs and to help maintain complete rigidity, interfield support props and cross-ties can often be employed in the design of the precipitator casing.

For smaller precipitators, some designs use a monolithic approach where, instead of sliding bearings, the casing expansion is reflected directly into the column legs, which are arranged as an extension to the side wall columns.

Substructures both in steel and reinforced concrete can be found; with the latter, the precipitator design would always incorporate sliding bearings between the ring bcam column stubs and substructure, whereas, with a monolithic approach, the substructure must be of steel and would form part of the casing design. The foundations of the substructure must be carefully designed to carry all the loads reflected back from the precipitator itself.

To ensure that the casing is gas tight, any access doors and cut-outs used for rapping drives/insulators, etc., or instrumentation, must be carefully considered and designed such that they can be sealed effectively. Any air ingress into an operating precipitator will not only result in electrical disturbances but invariably causes local corrosion from the gas being cooled below dew-point. Any door used for man access, in addition to having effective sealing, should also be electrically key interlocked for safety reasons, as the operating voltage on any precipitator is lethal.

On the majority of dry precipitators the casing is thermally insulated to reduce corrosion and give a long life. The thermal insulation, or, at least its fixing to the steel work, needs to avoid all thermal bridges; otherwise corrosion can occur on the inside of the casing, which could ultimately affect its mechanical integrity and possibly, in the short-term, electrode alignment.

4.6 HT insulators

Possibly one of the limiting items affecting plant availability and performance could be the HT insulators used for electrically isolating the discharge electrode system from the casing. Not only must the insulator material be capable of withstanding the maximum voltages imposed upon it during all operating conditions but, on some precipitator designs, the insulator must withstand the full weight of the complete discharge system and possible deflection loads.

To minimize potential problems of electrical tracking, because of deposited materials, e.g. dust, carbon, moisture or sulphuric acid, it is usual to find the insulators heated and/or purged. This is essential to comply with some recent legislative measures demanding that the precipitators are electrically energized immediately the process plant is switched on from cold so as to minimize all possible emissions. Under this condition, the internals, particularly the insulators, which are usually mounted partly outside the gas stream, can, unless preheated, encourage water/acid deposition from the gases with potential resultant tracking. If tracking occurs, not only will the precipitator efficiency be reduced, but the insulator will fail electrically, due to uneven heating or arcing causing thermal stresses. In this situation the insulator would require replacing completely to return the precipitator to optimum performance.

Generally the insulator material best suited for this duty is high alumina porcelain, although special high temperature applications have used 99.9% alumina, which, from Figure 4.9, offers better high temperature insulation/tracking properties [4].

Although many designs mechanically support the discharge electrode system directly on a cone-shaped (flower pot) insulator, others use post insulators outside the gas stream and use the HT insulator as an unloaded

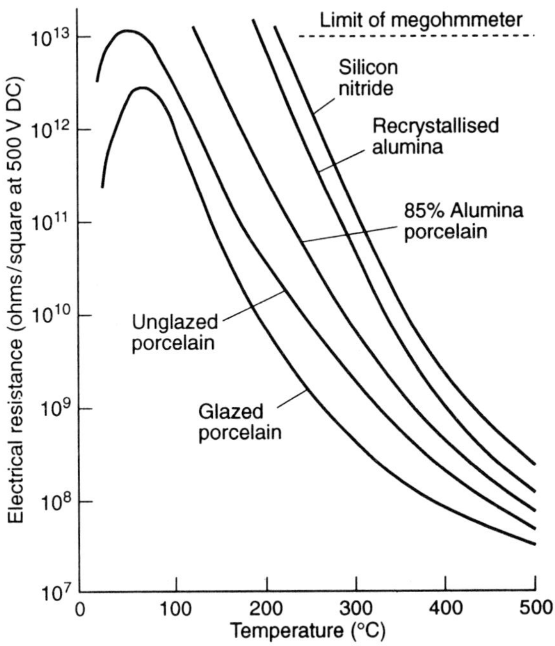

Figure 4.9 Insulator materials, temperature *vs.* resistance curve (courtesy Lodge Sturtevant Ltd).

gas seal. Until the recent legislation measures, the post insulator support design was used with dirty gas insulators having a tracking length of 1 m or so. With this approach it was necessary, before energizing the precipitator electrically, to allow sufficient time for the gas temperature to heat the system up to give dry conditions which, with the long tracking length, has been found sufficient on most installations to overcome dust tracking problems.

In addition to the hollow portal beam spanning the precipitator, in which the insulators can be mounted, two further insulator arrangements can be considered: firstly, the use of thermally insulated and heated boxes mounted on the roof and, secondly, a complete top housing covering the whole roof area. Figure 4.10 shows typical insulator box and top housing arrangements.

The latter approach is normally found where the discharge electrodes are carried from post insulators and the rapping is externally mounted, as will be described later. The advantage of this approach is that the top housing forms a weatherproof self-heated room over the casing, which enables both electrical and mechanical components to be visually inspected during plant operation and any required maintenance can be carried out, after isolating the electrics, in warm dry conditions.

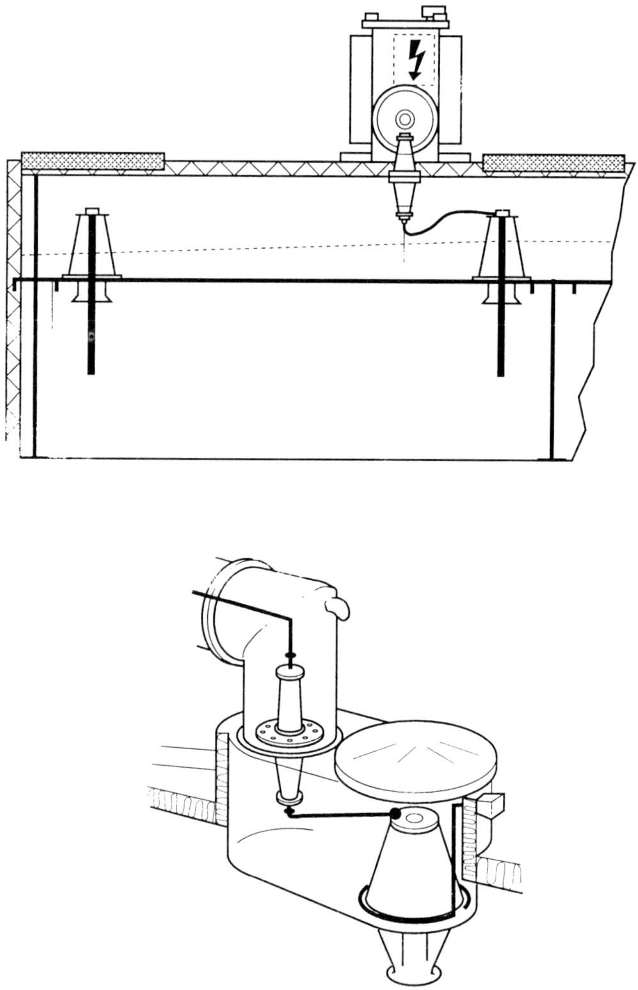

Figure 4.10 Insulator box and top housing designs.

4.7 Rapping

So that a precipitator always functions at optimum efficiency, it is necessary for any deposits on the collector and discharge system to be periodically removed, so that any build-up has minimal impact on the electrical operating conditions. This removal is usually achieved by mechanical impact rapping in the case of dry precipitators and by liquor washing for wet applications. (The different washing systems employed are fully described in chapter 13.)

In modern dry installations, dependent on the supplier, the rapping is achieved either by a rotating tumble hammer or by dropping a heavy rod onto an anvil connected to the component being rapped. The tumble hammer system is normally operated through a motor-driven shaft and the drop rod by either a motor-driven cam shaft or a magnetic lifting mechanism. Older systems, employing vibrators or mechanical lifting and dropping of the collectors, have been superseded by the foregoing, but can sometimes be found.

Although, theoretically, cleaning to the metal surface would be ideal, the energy required to dislodge a 10 μm particle from a metal surface is around 1000 g [5], which, while possible, would result in not only complete disintegration of the deposited material, causing severe re-entrainment and a very much reduced efficiency, but such impact intensity would also invariably result in fatigue and mechanical damage to the components being rapped.

In order that the particles reach the hopper, the size of the agglomerate, when dislodged, must be some 1000 μm in effective diameter to overcome the horizontal component of gas flow [6]. This means that a compromise has to be established on any operating plant, such that the rapping provides sufficient energy to dislodge the deposited layer from the surface, without causing the agglomerated layer to shatter and become re-entrained, and also limits mechanical damage to the components being rapped. Because the rate of deposition varies along the precipitator, the rate or frequency of rapping is reduced from the inlet to the outlet fields; this is necessary to enable the deposited layer to achieve sufficient thickness to produce a 1000 μm agglomerate when sheared from the surface by the rapping.

Although the frequency of collector rapping is relatively slow, it nevertheless is mechanically searching, at a rate of 12 raps per hour; the components undergo 1 million impact cycles in approximately 10 years, and to eliminate premature failure most suppliers carry out full-scale laboratory investigations. These are not only to ensure that the system is capable of long-term reliability, by equivalent 10 year life testing, but also that the energy distribution is sufficient and reasonably uniform across the collector or discharge system. An energy distribution across a collector, rapped at the lower end by a 5 kg tumble hammer, is presented in Figure 4.11; this shows a typical format which has proven satisfactory in the field. The results also indicate the potentially damaging energy levels imparted around the anvil area, which have to be suitably designed to ensure long trouble-free life.

Figures 4.12 and 4.13 indicate typical arrangements for a tumble hammer bottom and collector rapping system and a cam lifted drop rod arrangement at the top of the collector. While the tumble hammer system is invariably mounted within the casing, as shown, the drop rod can be either external, as shown, or internal. An advantage of the external arrangement is that it operates in a dust-free environment, but requires a number of roof penetra-

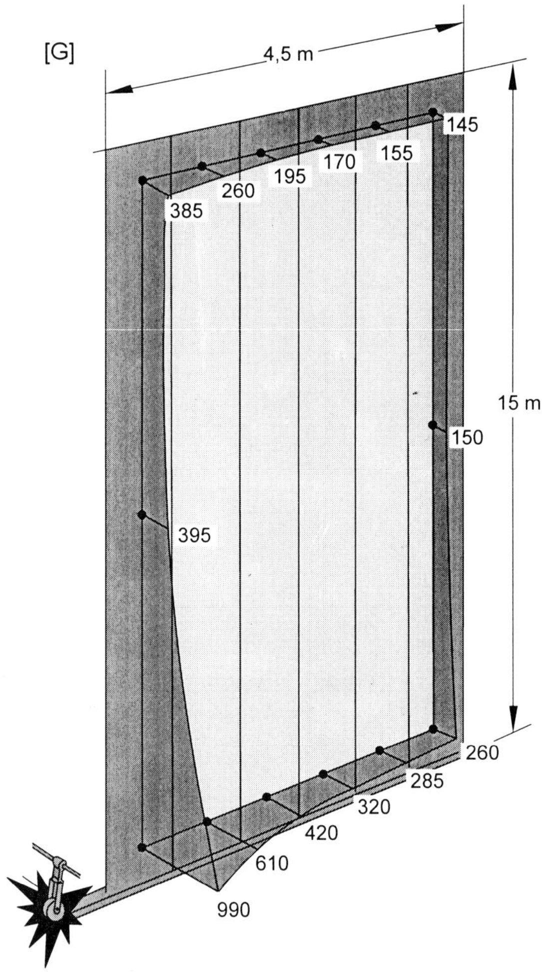

Figure 4.11 '*g*' distribution over a tall collector.

tions. On some installations additional rows of tumble hammer rapping can be found in the collector height, to further improve the distribution profile, when handling sticky or 'difficult' dusts [7].

While the greatest rate of dust deposition occurs on the collectors, even small accumulations on the discharge electrodes can have a deleterious effect on the electrode emission characteristics, since the radius of curvature is increased (see chapter 3), and it is equally, if not more important, to ensure that the discharge electrodes are effectively rapped. As the system contains very much less material than the collectors, transference of energy through

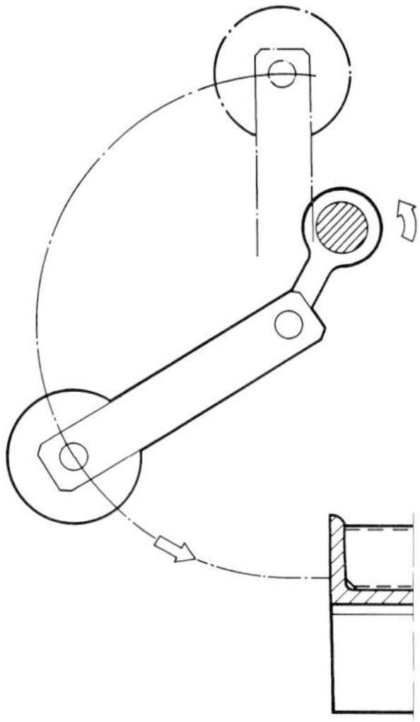

Figure 4.12 Tumble hammer collector arrangement (Flakt).

the discharge system is more difficult. Some 50% of the total applied energy is lost through every 90° change of direction; nevertheless, both internal tumble hammer and external drop rod rapping systems are employed. The internal tumble hammer raps either an individual frame or a pair of frames, whereas the external drop rod form of rapping can only impart energy to a specific area of the support frame. By limiting the size of individual support frames, very effective rapping can be achieved, albeit at the cost of additional insulators and support frames.

It is generally accepted that, if rapping intensities of 100 g (0 to peak) plus can be achieved on the collectors and a similar value on the discharge electrodes, then the resultant cleanliness will be satisfactory in terms of maintaining performance. While 'g', the acceleration due to gravity, is used as a measure of rapping effectiveness, this is very much an oversimplification, since the value determined will be dependent on the size and thickness of the element being rapped as shown by the following formula [6]:

$$\text{acceleration} = 4 \cdot \pi^2 \cdot f^2 \cdot d$$

where f is the fundamental frequency of vibration (Hz) and d is the displacement (m), and

$$f = \frac{K}{2} \cdot \left(\frac{D \cdot g}{wA^4}\right)^{1/2}$$

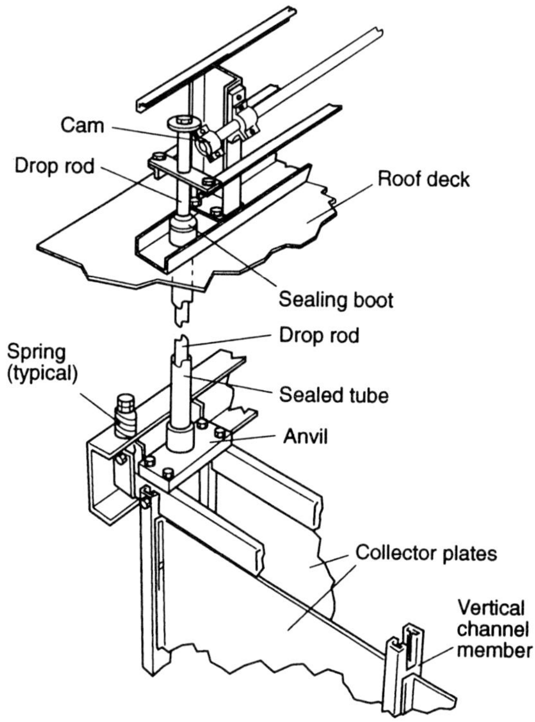

Figure 4.13 Drop rod rapping arrangement (courtesy Lodge Sturtevant Ltd).

where K is a constant, $D = Et^3(1 - V^2)$, t is plate thickness, V is Poisson's ratio, E is modulus of elasticity, w is weight per unit area and A is short edge dimension.

Since the measurement of g to determine the effectiveness of rapping is not the complete story, various manufacturers often quote lower values than the above 100 g, for their particular system as being adequate to enable their plant to perform satisfactorily. In this instance, one must consider the reputation and track record of the supplier, prior to accepting the statement covering lower g values, which might be correct for the vast majority of particulates met in practice.

Arguments have also arisen as to how the g value should be measured on the collector or discharge element, e.g. transverse to or in the plane of the rapping; many specifications overcome this by stating that the acceleration will be normal to the surface, and whether the reported value is to be peak to peak or zero to peak. Another factor to be borne in mind when measuring g values is that the size and weight of the accelerometer and the

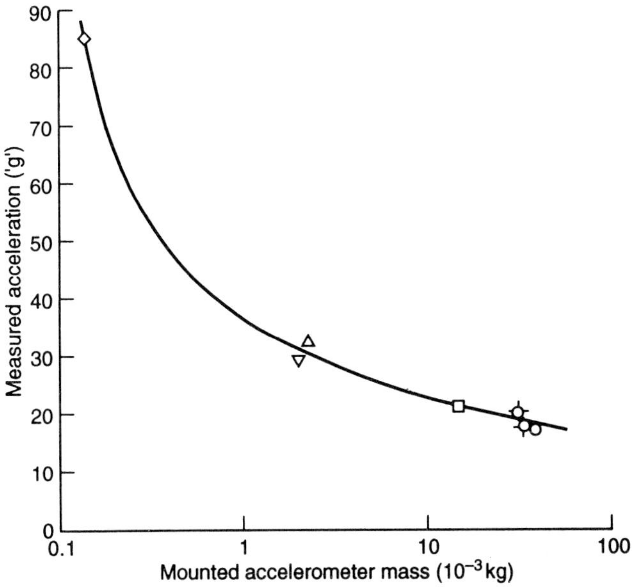

Figure 4.14 Effect of accelerometer mass on measured '*g*' value.

frequency cut-off range can affect the measured value, thus further complicating the situation. This is illustrated in Figure 4.14, which shows how the measured value for a 5 m × 15 m catch space collector top rapped by a 15 kg rod dropped through 100 mm, is influenced by the mass of the accelerometer [7]. Ideally, since the collector is relatively thin, the measuring device should have zero mass to eliminate local disturbances and changes in the mode of plate vibration.

In practice, a lightweight accelerometer of 0.2 *g*, is possibly the nearest approach to the ideal and, to avoid 'ringing', and hence false high readings, some suppliers advocate that a practical approach is to use a measuring cut-off frequency of 10 kHz, which is well below the collector's natural frequency of vibration, but there is no adequate theory that has been adopted worldwide.

Rapping re-entrainment, which is inevitable on mechanically rapped dry units, is minimized on most modern installations by using microprocessor control systems to ensure that collectors in successive fields are not simultaneously rapped. Some installations have the kV and power slightly reduced on the field immediately before the blow is delivered, to eliminate any flashover induced re-entrainment which may arise because of the dust disturbance. In other plants, power-off rapping is employed, i.e. all electrostatic clamping forces are removed to assist in dust dislodgement.

Other installations minimize re-entrainment by completely isolating the flow during rapping by means of inlet and outlet isolating dampers, so-called off-flow rapping; all fields are then rapped simultaneously, usually with the power off. The isolated flow is then left for some period of time to allow the dust to reach the hoppers prior to reconnecting the flow back into service. An alternative to complete flow isolation is to only blank-off the duct being rapped with an air curtain or damper system [8]. The economics of these approaches need careful evaluation, since the cost of the additional dampers and duct seals may outweigh the gain from a potentially reduced plate area precipitator.

4.8 Hoppers

Examination of data log records indicates that at least 70% of the reported dry precipitator problems are associated with hopper dedusting difficulties of some sort, so one cannot overemphasize the importance of satisfactory hopper and dedusting system design in any precipitator installation.

The particulates dislodged from the collector system have to be initially contained in a receptacle in order that they may be removed from the process gas stream. These receptacles usually take the form of pyramidal hoppers or troughs located beneath the fields of the precipitator. To ensure that the particulates can be extracted from these, it is essential that the discharge opening in the base, leading to the final evacuation system, should be no smaller than 300 mm diameter, regardless of the amount of material in the hopper. The corner valley angle depends on the characteristics of the dust but is not normally less than 55° to the vertical.

The bottom third or so of the hopper must be heated to maintain the dust above the gas dew-point temperature; heating can be accomplished by thermostatically controlled electrical tapes or pads, rated at $1.5\,kW/m^2$, or by low pressure steam coils or jackets. Other appendages, which are often incorporated, are some form of manual poke rod and a hopper level indicator.

To assist in the emptying of particularly cohesive materials, some hoppers have vibrators or hammers fitted to the side walls of the hopper. The use of vibrators can, however, sometimes worsen the situation by compacting the dust, so their use must be carefully considered. For potentially adhesive particulates, such as those from paper mills operating the older Kraft process, the base of the precipitator is a continuation of the vertical side walls to form a full-width trough, the particulates being extracted by a full-width scraper chain conveyor feeding into a narrower end conveyer fitted with a sealing device.

To optimize the performance of any precipitator, it is essential to eliminate any gas bypass of the field, as the gas expands over the hopper region. This can be effectively achieved by full-height vertical interfield baffles in the case of trough hoppers, while for pyramid-type hoppers adequate deflection plates located so as to divert the flow upwards will be fitted. With designs using a frame-type discharge electrode carrier, centre hopper baffles extending up to the bottom of the collectors can be found; for plants having the lower end of the discharge electrodes terminating beneath the collectors, the height of any potential centre baffle would not assist in controlling bypass and, in this instance, it is imperative that the small inlet and outlet hopper deflector baffles are fully effective. This is normally checked on large, even full-scale models of the area.

There are numerous devices on the market for hopper level indication/detection, employing one of the following principles: capacitance change, change in a tuning fork frequency, change in pressure/suction, change in rotational speed of a driven disc, ultrasonic and radioactive sensors. As any device will only determine a local dust situation, overfull hoppers can still arise, in spite of an apparent 'empty' hopper indication.

Should the hopper overfill with dust, not only will the dust short-out the electrics, but could also result in severe mechanical damage if the level should reach high enough into the field area. If 'rat holing' occurs during the emptying from a very overfull hopper situation, since the dust is fluid, if only one duct empties, the hydrostatic pressure exerted by the dust in the adjacent filled ducts can result in local distortion of the lower part of the collector and sometimes the discharge system itself, with a deleterious effect on precipitator performance.

To reduce this risk, or at least give further early warning, in addition to the hopper level devices, chains are sometimes hung from the four corners of the discharge frame. The dust on reaching the chains usually causes the electrics to trip on reduced kV output, thereby preventing further dust being precipitated. One should remember, however, that although dust will not be precipitated, it will still tend to be deposited under gravity, particularly at the inlet, so that action to remedy the overfull hopper situation is still a high priority if damage to the internals is to be avoided.

As many precipitators operate under suction, it is imperative that air inleakage into the hopper is prevented to eliminate possible re-entrainment of collected particulates. Even an unmeasurable inleakage flow can re-entrain sufficient dust in passing through the hopper area to significantly affect the overall emission from the plant, particularly the outlet hoppers. For example, a precipitator operating with an emission of $50\,mg/Mm^3$, will, if mixed with an outlet hopper inleakage, equating to only 0.25% of the total flow but carrying $10\,g/Nm^3$ of hopper dust, result in a final emission approaching $75\,mg/Nm^3$, i.e. a 50% increase.

Although hopper evacuation systems fall outside the scope of this publication, the precipitator designer/supplier must be fully conversant with the proposed evacuation system capacity and design, since this will not only impact on the choice of hopper, but also on the method selected to prevent air inleakage during emptying sequences.

For volumetric systems, e.g. conveyors of various types, rotary valves are often used sited below the hopper take-off point to meter the dust flow into the conveyor. As the rotary valve acts as a pump on the return empty half-cycle, its position in the down leg and operation must be such as to provide a positive head of dust to eliminate inleakage. For either dense or lean phase pneumatic conveying systems, both rely on slide-type gate valves for isolation and sealing, so the above 'pumping' problem will not arise; however, it is most important that these gate or slide valves seal, particularly when positive pressure systems are used.

4.9 Electrical clearances

The electrical clearances that must be carefully considered, if the precipitator is to operate efficiently and as designed, fall into two main categories; the first are those associated with the discharge electrodes and collectors and the second, the peripheral clearances outside the field itself. Any electrical clearance which is too small will limit the operating voltage of the unit and so affect the efficiency of precipitation.

In general, the electrodes are suspended so that they hang exactly mid-way between adjacent collectors. In practice, although the alignment is critical, with erection and fabrication tolerances, one must accept a misalignment of up to $\pm$ 10 mm on a field height of 10 m and above, but a tighter tolerance for smaller plants.

The position of the discharge element within the collector and the distance between elements is also important to maximize efficiency; generally, a minimum electrode element–electrode element distance of half the collector/collector spacing is used to give satisfactory corona formation and current distribution over the collector surface (see chapter 3). For collectors, the clearance between the stiffening channel edges and any emitting point should be greater than the collector half-spacing by some 25% to provide satisfactory emission and voltage conditions. This distance applies in both the horizontal and vertical planes, since the electrodes or collectors may not hang exactly plumb.

Another area of concern with the discharge electrodes is the position where any support or carrier members pass through the collectors. In this case, not only should the member be low emitting, e.g. have a large radius of curvature, but must also pass centrally between adjacent collectors.

For clearances outside the actual field area, the positioning of any live discharge electrode support members must be well clear of any earthed structural components. For general safety, the following 'rule of thumb' minimum clearance distances should be adopted for any design. As positively energized components have a lower breakdown strength than negatively energized elements, any edges on the earthed collector, casing or support members can result in premature breakdown. For example, for components having sharp edges opposing one another, a separation of three times the discharge electrode/collector distance should be used, even with reasonably curved surfaces, such as those on rolled angles or joists; the minimum separation should be 1.5 times the discharge electrode/collector distance, with intermediate separations being adopted for differing profiles and edge conditions.

A special area of concern is where the discharge electrode support tubes pass through the casing and insulator. To minimize the quantity of heated purge air used with the cone or flower pot insulator, the open area at the base is significantly reduced by a restrictor, which would take the form of a profiled electrical stress cone or bus ring. This, with a large diameter suspension tube, e.g. 50 mm or so, effectively minimizes the risk of breakdown in this area, in spite of the actual clearance approaching, or being sometimes less than, the discharge electrode/collector separation. The location of the suspension tube must be truly central within the bus ring to be fully effective.

Where bus ducts, rather than HT cable, are used to feed power to the electrodes, their diameter and profile must be such as to eliminate the risk of breakdown within the duct itself. If insulators are employed to centralize, or carry the bus connection, these should have adequate tracking length to prevent breakdown because of atmospheric dust deposition, etc.

On large precipitators having discharge electrode frames spanning two or more hoppers, the actual distance between the frame and any cross-hopper apex member, should take into account any potential dust deposition on the member to minimize the risk of breakdown.

Similar comments would apply to internal tumbling hammer discharge electrode rapping systems, where the 'live' shaft and hammers could give rise to suspect areas, which would be difficult and expensive to overcome.

Other components, which may be built into the system, should adopt the above 'rule of thumb' minimum clearance distances in order that the maximum applied voltage is across the discharge electrode/collector system and not elsewhere in the plant.

References

1. Unpublished Lodge Cottrell Ltd Research Report.
2. Parker, K.R. (1980) The precipitation of difficult dust. *J. Electrostatics*, **8**, 355–67.
3. Forrest, J.S. and Lowe, H.J. (1957) Present performance and scope for improvement in power station electrostatic precipitators. Proc. Conference on the Mechanical Engineers' Contribution to Clean Air, pp. 42–59, Inst. Mech. Engs, London, UK
4. Unpublished Lodge Cottrell Ltd Research Report.
5. Lowe, H.J. and Lucas D.H. (1953) The physics of electrostatic precipitation. *Br. J. Appl. Phys., Suppl.* 2, 840.

6. Baylis A.P. and Russell-Jones A. (1982) Collecting electrode rapping designed for high efficiency electric utility boiler electrostatic precipitators. Proc. 4th EPA/EPRI Symposium on the Transfer and Utilization of Pasticulate Control Technology. Houston, Texas, Oct 11–14th – EPRI, Palo Alto, CA.
7. Darby, K. *et al.* (1990) The rapping systems for cleaning the discharge electrodes and collectors of electrostatic precipitators. Proc. 4th International Conference on Electrostatic Precipitators, Beijing, China, Chap 7, pp. 498–513, International Academic Publishers, Beijing, 1992.
8. Falaki, H.R. (1993) Experimental study of flow diversion during the rapping of collector plates inside an ESP. Proc. 10th Particulate Control Symposium and the 5th International Conference on ESPs, 5–8 April, Washington, USA, pp. 37.1–15, EPRI TR LOS048 V2, Palo Alto, CA, USA.

5 Aerodynamic factors affecting performance

L. LIND

5.1 Introduction

In any precipitator the gas flow carrying the particles which have to be precipitated plays an important role. Gas volume rates between parallel connected chambers should not deviate too much from the average value used for sizing, and the gas flow distribution across the face of each individual chamber should be within certain limits. Furthermore, it is important to include the turbulence and the secondary flow, generated from internal structures, and from the corona wind itself. It is the intention of this chapter to explain these matters in detail.

5.2 Turbulence and secondary flow

5.2.1 Historical resumé

The flow between the electrodes has been known to be turbulent right from the very early days of precipitation but, due to the complexity of the flow, a more quantitative analysis was not performed until the appearance of the computer, half a century later. The flow which is influenced by gravity, electric field and viscosity controls the movement of the electrons, positive and negative ions, neutral gas molecules and particles. This is indeed a very complex flow, difficult to conceive, to measure and very difficult to calculate theoretically.

Going back more than 70 years to 1922, Deutsch put forward his expression for the electric particle migration velocity [1]:

$$\mathbf{v}_E = q\boldsymbol{E}_p/(3\pi d_P \mu) \tag{5.1}$$

the nomenclature being $\mathbf{v}_E$ the electric migration velocity, q the particle electrical charge, $\boldsymbol{E}_p$ the precipitating field strength vector, d_P the particle diameter, μ the dynamic viscosity of the gas.

Deutsch took the forces acting on spherical particles to be the electrical Coulomb and the Stokes' fluid drag forces, presuming a quiescent gas. The determination of the charge was his prime concern, and his charging theory was extended down to particles of less than 0.1 μm diameter.

The diffusion charge and the field charge were later treated by, e.g. Cochet [2] and McDonald *et al.* [3], by introducing the concept of particle mobility

$$m_P = q/(3\pi d_P \mu) \tag{5.2}$$

Substituting in equation (5.1) gives:

$$v_E = m_P E_p \tag{5.3}$$

This is the Coulomb–Stokes value, which Deutsch introduced in his expression for the efficiency

$$\eta = 1 - \exp[(-v_E/v)(L/d)] \tag{5.4}$$

where v is average axial gas velocity, L is precipitator length and d is half the duct width. $(v_E/v)(L/d)$ is the so-called Deutsch number D.

In the derivation of equation (5.4), Deutsch presumed the particle concentration profile to be flat over the electrode space due to 'whirling motion, electric wind etc.' In other words, he assumed full turbulent mixing and thereby the diffusion or degree of turbulence to be infinite.

According to Mayer-Schwinning [4], the influence of the electric wind was discussed as early as 1924, and in 1931, Deutsch stated [5] that the ion wind had a positive as well as a negative influence on precipitation. He meant that 'the gas flow was only responsible for the uniform distribution of the particles within the precipitator' (quote from [4]). The physical mechanism Deutsch looked upon was a core of flow with infinite turbulence intensity, and a boundary layer close to the collecting wall, where the Coulomb force overcomes the Stokes' force.

In his 1963 book, [6] White demonstrates how precipitator efficiency with laminar flow becomes 100% after a length $L_{Lam} = vd/v_E$. The efficiency expression is:

$$\eta_{Lam} = 1 - D \tag{5.6}$$

In the late 1970s Professor White expressed his great interest in achieving realistic values of the turbulence level inside the precipitator, being well aware of the fact that equation (5.4) might be pessimistic.

Since Deutsch, many specialists have tried to 'crack the nut' by including a finite turbulence intensity in an appropriate manner. In 1968, Myron Robinson [7] published a review of articles dealing with the influence of turbulence. His own interest was the 'electric wind' and its contribution to turbulence and the resultant secondary flow and its influence on precipitator performance. From his review the essence from a few contributors follows.

(a) In 1959, Friedlander [8] introduced a particle vorticity diffusion and a migration velocity in the two-dimensional differential equation for the particle concentration in both axial and transverse directions.

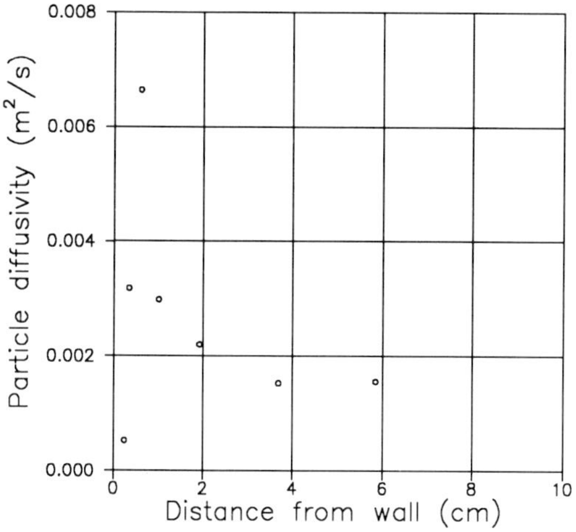

Figure 5.1 Effective particle diffusivity measured with LDA at 2.5 m/s and −38 kV. Duct width is 250 mm. Redrawn from [13].

(b) P. Cooperman used a similar model in the years from 1960 to 1966 [9] adding re-entrainment to advance the problem.

(c) In 1967, Myron Robinson published a paper [10] with a modified Deutsch equation inspired by Inuyshkin and Averbukh and P. Cooperman.

(d) Crowe and Stock published, in 1974, a paper which included a calculation on the electrodynamic flow in a tube precipitator including an effective viscosity for a pipe flow without an electrical field [11].

(e) In 1979, Bernstein and Crowe [12] described observations using a laser light sheet in a wire plate precipitator. The same paper deals with numerical calculations of electric field and gas flow, the latter described by a k-ε turbulence model. They concluded that the interaction between electric field and gas flow increases the migration in certain regions, but decreases migration and augments re-entrainment in other regions.

(f) The same year, Harald Gross of Stuttgart published results from measurements in a laboratory precipitator [13]. He measured the effective particle diffusivity Γ, using a laser doppler anemometer, and found a peak close to the collector wall (Figure 5.1). Furthermore, the particle migration velocity was recorded and he considered that the particles must overcome a diffusion barrier at the wall before precipitation. As this barrier is independent of duct width, this is a fluid mechanical argument for wider ducts having higher migration velocities than corresponding narrower ones.

(g) In the late 1970s a group of scientists at the High Temperature Gas Dynamic Laboratory at Stanford University took up the topic, continuing through the 1980s [14,15]. They contributed powerfully to the understanding and the solution of the influence of turbulence on performance. The

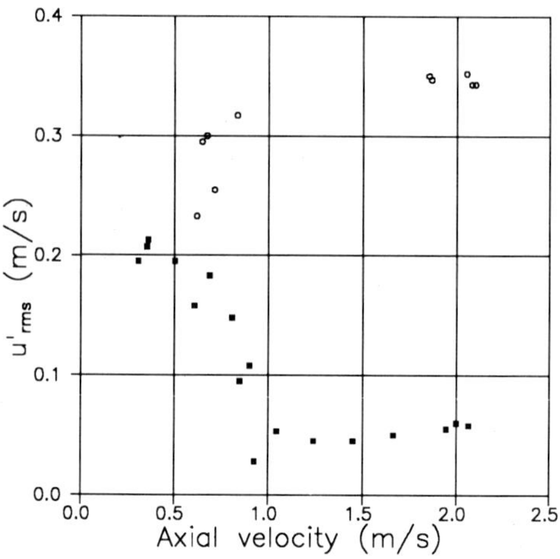

Figure 5.2 u'_{rms} measured in laboratory precipitator using LDA between electrodes. Circles and squares are from two different horizontal levels. Redrawn from [20].

work concluded in a theoretical and practical demonstration of the expectation of a higher efficiency electrofilter with a finite level of turbulence. This they illustrated in a graph showing penetration P in dependence of the Deutsch number, $D = v_{\mathrm{E}}L/(v \cdot d)$ with Péclet number as a parameter, $P_{\mathrm{E}} = v_{\mathrm{E}}d/\Gamma$. Suffix E stands for 'electrical', v_{E} being determined from equation (5.1). P_{E} is ∞ for laminar and 0 for fully turbulent flow. The diffusivity $\Gamma \approx 0.2 \cdot u'_{\mathrm{rms}} \cdot L_{\mathrm{EU}}$, where u'_{rms} is the rms value of the turbulent fluctuating axial velocity and L_{EU} is an Eulerian length scale. Typical turbulent velocity

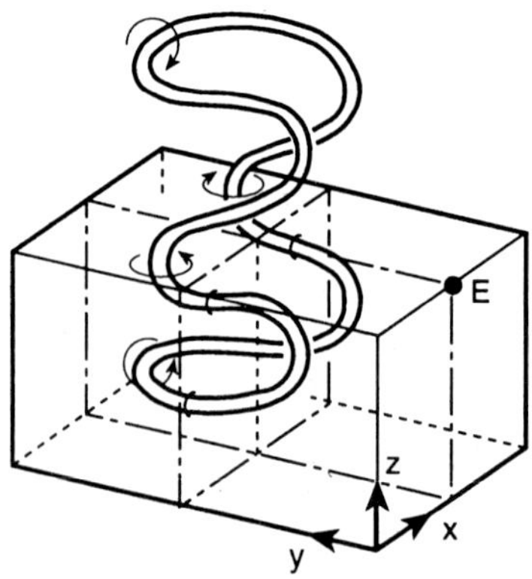

Figure 5.3 Vortex structure topology at zero bulk velocity. Emitter point at E. From [21].

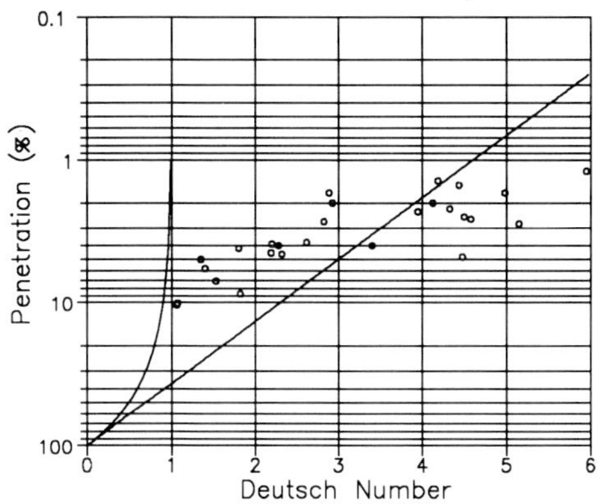

Figure 5.4 Penetration *versus* Deutsch number. Curve is laminar theory, straight line is Deutsch's theory, while circles are from laboratory precipitator measurements. Redrawn from [22].

fluctuations are shown in Figure 5.2 from [20]. Calculated and measured concentration profiles, performed at Stanford, clearly demonstrate the claims by former researchers of increasing concentration towards the collector wall.

Y. Yamamoto *et al.* in 1980 published a numerical study of secondary flow interaction [16], followed in 1981 by an experimental and theoretical study ($2-D$) of the secondary flow in a positive corona single stage precipitator [17]. In 1982 Thomsen *et al.* repeated the experiment but with negative tuft-corona [18].

Later, in 1984, Shaugnessy *et al.* from Duke University, North Carolina, published their version of the influence of the electrical body forces on the flow creating a secondary flow pattern [19].

In 1986. Larsen and Sørensen on the basis of measurements in a laboratory precipitator, described an analytical procedure for simulating the secondary rolls generated by the current [20]. Larsen and his group published papers on secondary flow and turbulence in precipitators through the 1980s. They demonstrated that the vortex picture begins having a complicated course (Figure 5.3) at zero axial velocity, changing to vortices being stretched by the bulk flow at higher axial velocities [21]. One of the findings was that the performance might decrease below the Deutsch limit at higher Deutsch numbers, because particles trapped in the middle of the vortices escape the electrode zone without reaching the collecting walls. This might happen with laminar flow as well. The phenomenon appears to be confirmed from tests on a small laboratory precipitator (see Figure 5.4, from [22]). In the figure the laminar curve and the Deutsch line are seen together with the measuring points shown as circles. Above a Deutsch number of approximately 4, the points drop below the Deutsch line.

5.2.2 Turbulence

A description of the influence of turbulence upon precipitator efficiency, based primarily on theories and measurements from Stanford University, California, and the Technical University of Denmark, is given below.

Turbulence is generated from the inlet screens, the frame members and structural members inside the casing. The discharge electrodes themselves produce eddies or vortices, which degenerate to turbulence and finally into heat. The boundary layer, extending from the leading edge of the first collecting plate down through the duct of the given field, creates turbulence due to wall friction, being enhanced from any protrusions on the collecting plates, the so-called antire-entrainment baffles. Normally the field length is too small to make the boundary layers of opposite walls meet, hence real 'duct flow' with fully developed velocity profiles is unlikely to occur.

The electrons of the corona become attached to gas molecules converting them to gas ions, which move with high velocity due to the electric field. The ions and the neutral gas molecules, thus carried away, form the so-called 'ion wind', which is a major source of turbulence.

There is a clear difference between positive and negative corona for smooth wires; the positive corona is a rather stable glow along the wire, while the negative corona forms corona tufts. The tufts are randomly distributed and move with time, and the number increases with current density. For discharge electrodes with corona spikes (high emitters) the difference in appearance between positive and negative polarity is thought to be smaller. As positive corona is hardly used today in industrial

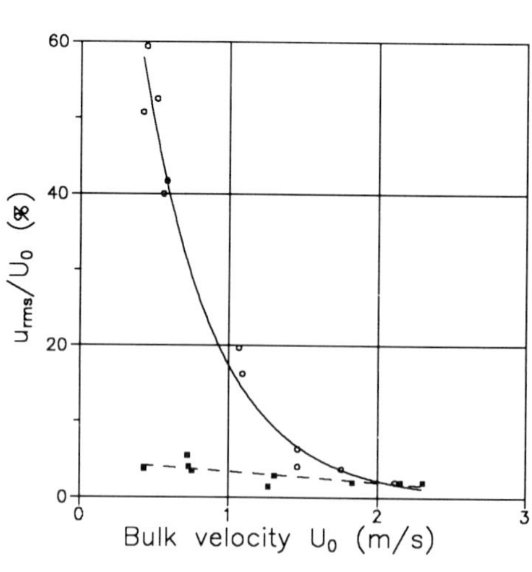

Figure 5.5 Turbulence intensity *versus* bulk velocity. Black squares without tension and circles with -50 kV and 0.65 mA/m^2. Redrawn from [18].

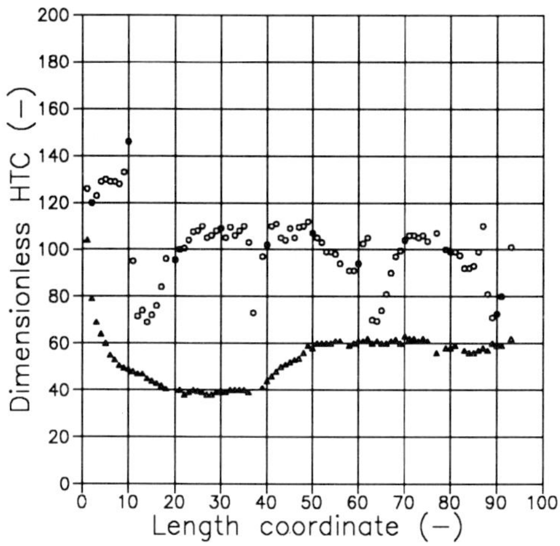

Figure 5.6 Heat transfer coefficients from heated plate to air. Triangles from flat plate, circles from plate with transversal baffles. Redrawn from [24].

precipitators, it will not be treated here, but the potential of positive corona with respect to high efficiency precipitation due to low turbulence might be worth exploiting in the future. Therefore, this chapter deals only with negative corona.

From simple theory, we know that turbulence has a negative influence on precipitator efficiency, as laminar flow should give 100% efficiency for a given precipitator length, while the Deutsch theory, assuming 100% turbulence, should produce an upper limit with respect to precipitator length.

In Figure 5.5 we see the turbulence intensity as a function of axial gas velocity [18]. This reveals the fact that too low an axial velocity would be expected to have an unwanted effect on efficiency, due to turbulence, and it should be avoided.

Often, in the literature, it has been claimed that low turbulence is created by the shielding of collecting plate protrusions or baffles. This is not the case; in [23] it is shown, by laboratory measurements and theory, that baffles have a negative effect on efficiency, and [24] demonstrates the negative effect of baffles using similarity between high heat transfer coefficients downstream of a baffle and high turbulence (Figure 5.6). The superiority of baffle-free plates was demonstrated by field mcasurements on collectors formed from horizontal plate strips without baffles.

The laminar molecular kinematic viscosity of air, at standard conditions, is of the order 15×10^{-6} m^2/s, whereas, the turbulence particle diffusivity Γ, predicted from concentration profile measurements in a parallel plate laboratory precipitator with precharged particles, is of the order 60 to

$120 \times 10^{-6}\,m^2/s$, i.e. four to eight times higher. With baffles an even higher level was found, $30 \times 10^{-4}\,m^2/s$, i.e. about 20 times higher [14].

An equation describing the concentration balance, assuming constant turbulent particle diffusivity, is:

$$(u + u_E)\partial n/\partial x + (v + v_E)\partial n/\partial y + (w + w_E)\partial n/\partial z$$
$$= \Gamma(\partial^2 n/\partial x^2 + \partial^2 n/\partial y^2 + \partial^2 n/\partial z^2) \qquad (5.7)$$

where $u(x, y, z)$ is the time-averaged velocity, $n(x, y, z)$ is the time-averaged particle concentration, u_E, v_E and w_E are the electrical migration velocities, and x and y are coordinates in the axial direction and perpendicular to the collecting wall, respectively; z is the third Cartesian coordinate.

Initial conditions and boundary conditions are:

$$n = n_0 \qquad \text{at } x = 0$$
$$\partial n/\partial y = (v + v_E)n/\Gamma \qquad \text{at } y = 0$$
$$\partial n/\partial y = 0 \qquad \text{at } y = d$$
$$\partial n/\partial z = 0 \qquad \text{at } z = 0 \text{ and } L$$

d is half the duct width, and L the precipitator length.

Equation (5.7) can be solved numerically, as in [25] for two dimensions, x and y. Figure 5.7 shows a typical measured concentration profile at $0.4\,mA/m^2$, redrawn from [25]. This profile differs from the flat 'Deutsch' profile only in the core region, i.e. ± 0.20 times half-duct width.

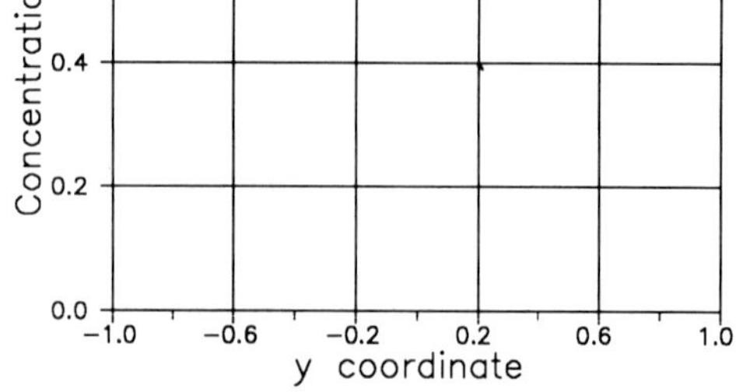

Figure 5.7 Measured concentration profile in a wind tunnel precipitator at approx. $0.4\,mA/m^2$. Redrawn from [25].

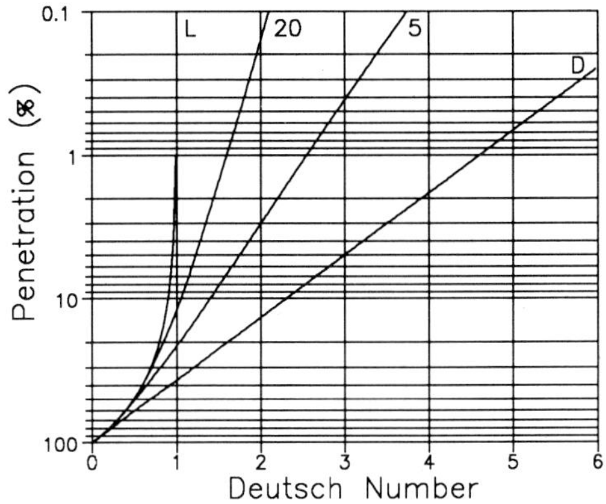

Figure 5.8 Penetration *versus* Deutsch number for various electrical Péclet numbers. $L \sim P_E = \infty$ and $D \sim P_E = 0$, curves are for $P_E = 5$ and 20. Redrawn from [26].

Integrating the concentration profiles across the duct width gives the mean concentration $n_m(x)$ and the efficiency along the precipitator $\eta(x) = 1 - n_m(x)/n_m(0)$. Instead of x, $(v_E/v)(x/d)$, the dimensionless length or the Deutsch number, is used in Figure 5.8 as abscissa with penetration, $P = 1 - \eta$ as ordinate. In solving equation (5.7) P_E, the electrical Péclet number, appears as a parameter: $P_E = v_E d/\Gamma$.

Figure 5.8 shows that a finite turbulence level, i.e. P_E greater than 0, gives less penetration as compared with the penetration found by the Deutsch

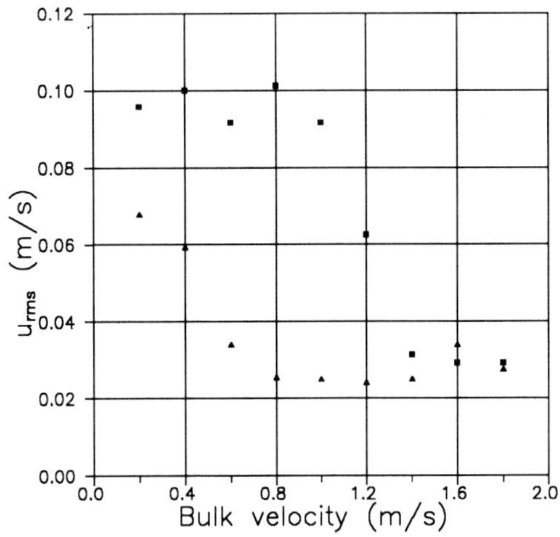

Figure 5.9 u_{rms} *versus* bulk velocity. Black squares with DC and triangles with pulse. Current density 0.05 mA/m^2. Redrawn from [27].

equation. Notice that a disadvantage of the graph is the fact that v_E is part of both the abscissa and the parameter. This has been modified [26] by introducing a dimensionless number equal to the Péclet number divided by the Deutsch number, i.e. $vd^2/(\Gamma L)$.

In [25] the reason for the measured migration velocity enhancement in the entrance region to the precipitator is believed to be caused by increased particle charging due to turbulent motion. This leads the authors to conclude that relatively high turbulence ahead of the first field and moderate and low turbulence further down through the precipitator is desirable.

In the case of pulse operation, with approx. 150 μs pulse width, it has been shown on a wind tunnel laboratory precipitator that the turbulence level is more moderate than for the case of DC operation. This is due to the fact that the impulsive electrical body forces deliver a smaller momentum to the flow than the continuous body forces of direct current (Figure 5.9, from [27]).

5.2.3 *Secondary flow*

Similar to the turbulence phenomena described, the effect of secondary flow on the precipitation rate is limited to a description of the effect of the non-uniform electric field created by negative corona, which is dominant in industrial precipitation.

Many commercial discharge electrodes are fitted with sharp points or pins acting as specific or 'controlled' corona emitting points. The ion current emitted from these entrains surrounding gas, creating an ion wind, blowing towards the plate at the level of the pin and back again between pins in order to maintain continuity. Even for a discharge electrode of the wire type, negative corona tend to be generated at discrete points, where they make tufts which move up and down along the wire. Basically the interaction between electrical field and gas flow creates tufts that are positioned at almost constant pitch.

This non-uniformity in electrical field and current gives rise to body forces acting on the fluid, creating vorticity, which drives the secondary flow, and the stretching of the vortices is a source of turbulence.

In [17], Yamamoto and Velkoff revealed secondary flow with mainly vertical axes using a smoke visualizing technique. In [21], the structure for low bulk velocity is interpreted from calculations of the flow field, suggesting a vortex topology structure as seen from Figure 5.3. For low velocities, 0.1–0.2 m/s, the flow field is reported to fluctuate at a frequency corresponding to the bulk velocity divided by axial electrode pitch. As bulk velocity increases for a given current density, the secondary flow pattern, vortices with almost vertical axes, changes and at realistic velocities their axes tend to turn horizontal (provided the discharge electrodes are vertical). Thus, we end up with secondary vortices like those indicated in Figure 5.10 from [20].

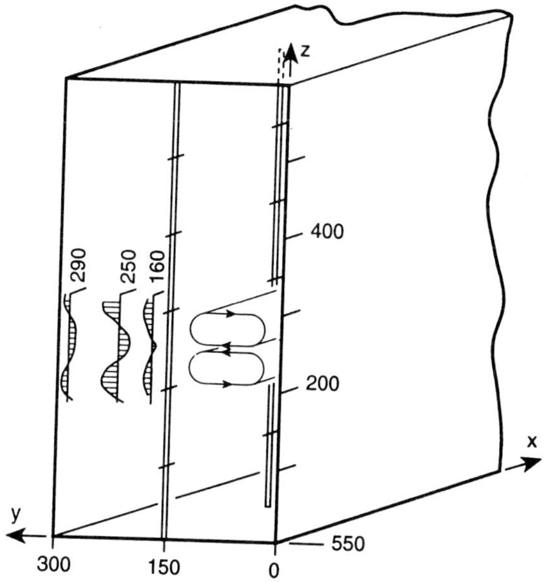

Figure 5.10 Secondary rolls generated by negative tuft corona. Velocity profiles indicated, too. From [20].

See also [28] for other suggested vortex structures. At even higher velocities, the vortices break down to turbulence, which is dissipated fairly quickly into heat.

Larsen [29], introduced analytical expressions for the secondary flow with velocity components satisfying mass conservation:

$$v = v_0 \sin(\pi y/d) \cos(\pi z/d) \tag{5.8}$$

$$w = -(L/d) v_0 \cos(\pi y/d) \sin(\pi z/d) \tag{5.9}$$

y being the coordinate perpendicular to the collecting surface, z the coordinate along the discharge electrode body, d half the duct width, L half the pin pitch, and v_0 the maximal roll velocity or 'roll strength'. The 'roll strength' is defined as the maximum transversal mean velocity at the level of pins, normally found half-way between the emission electrode and the collector wall.

If the v and w expressions are inserted in the two-dimensional particle concentration balance equation (5.7) we get:

$$U_0 \partial n/\partial x + (v + v_E)\partial n/\partial y + w \partial n/\partial z = \Gamma(\partial^2 n/\partial y^2 + \partial^2 n/\partial z^2) \tag{5.10}$$

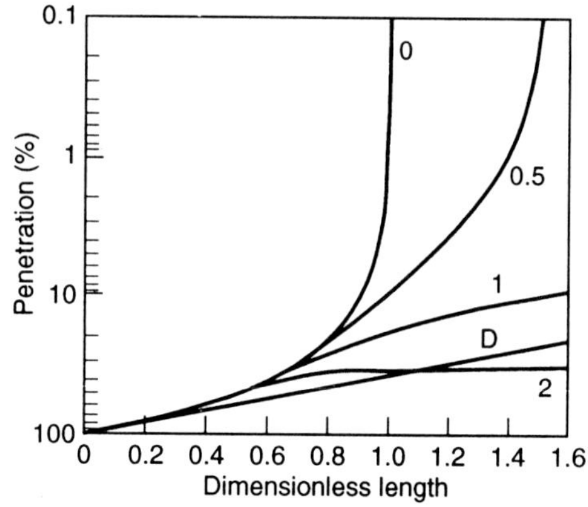

Figure 5.11 Penetration *versus* dimensionless length (= Deutsch number) for roll strengths $v_0/v_E = 0$, 0.5, 1 and 2 for infinite Péclet number (laminar). Deutsch's theory gives the straight line marked D. Redrawn from [30].

By assuming a constant axial velocity U_0, constant drift velocities v_E and u_E, putting w_E equal to zero, and ignoring axial diffusion, then using the proper boundary conditions from equation (5.7), the efficiency can be found by integration:

$$\eta(X, P_E, v_0/v_E) = 1 - 1/n_0 \int_0^d \int_0^L n(x, y, z) \cdot \mathrm{d}y \cdot \mathrm{d}z \qquad (5.11)$$

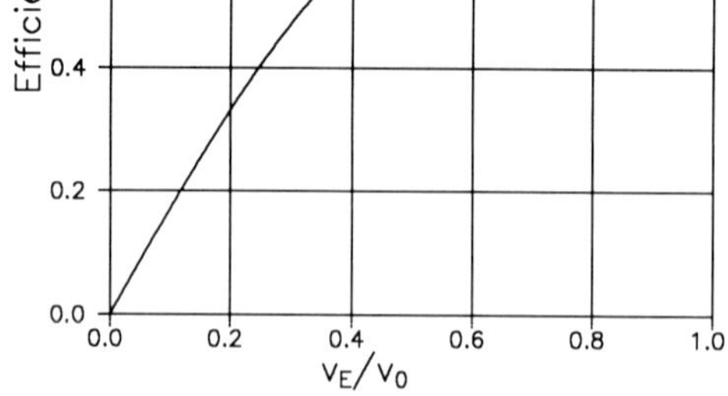

Figure 5.12 Maximum efficiency *versus* inverse roll strength for infinite Péclet number (laminar). Redrawn from [30].

X being a dimensionless axial position, $X = (v_E/U_0)(x/d)$, P_E the electrical Péclet number, $P_E = v_E d/\Gamma$ and v_0/v_E the dimensionless 'roll strength'.

Figure 5.11 shows penetration $(1 - \eta)$ *versus* X for $P_E \to \infty$, corresponding to zero diffusion. It is obvious that the expected laminar efficiency is markedly reduced with $v_0/v_E > 0$ and for a dimensionless roll strength of 2, the Deutsch limit is intersected at $X \approx 1.1$. The achievable maximum efficiency for laminar flow *versus* inverse roll strength is shown in Figure 5.12.

Roll strength has been found from measurements in a wind tunnel precipitator to increase with increasing velocity up to a maximum, dependent on current density, and to decrease again at higher bulk velocities, possibly due to the fact that the rolls are incompletely developed at any given axial position.

For a given bulk velocity, the roll strength increases with DC current density almost linearly as seen from Figure 5.13. Experiments also show that the roll strength increases as the distance between corona points over half-duct width approaches unity.

In reference [29], a thorough physical and mathematical treatment of the secondary flow is given. The non-uniformity in electrical field and current gives rise to body forces f_e acting on the fluid. The vorticity drives the secondary flow and the stretching of the vortices causes turbulence. Mathematically the time-averaged Navier–Stokes equations including f_e can be expressed as:

$$V \cdot \nabla\omega = \omega \cdot \nabla V + \nu\nabla^2\omega + \nabla \times f_e/\rho \tag{5.12}$$

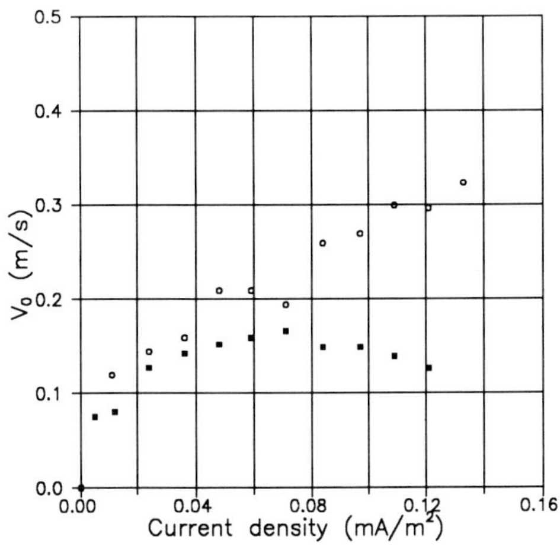

Figure 5.13 Roll strength *versus* current density. Circles are with DC and squares with pulse operation. Redrawn from [27].

where V is the velocity vector $u(x)$, $v(y)$, $w(z)$, ρ is gas density and ν is gas kinematic viscosity. ω is the vorticity vector $\omega_x, \omega_y, \omega_z$. The body force $f_e = \rho_i E$ with ion density ρ_i and electrical field E. Assuming the ion space charge $\rho_i \gg$ the dust space charge ρ_d, which is normally justified, the electrodynamic equations and the flow equations can be solved separately. A unidirectional and periodic field $f_e = (0, f_y, 0)$ is introduced in equation (5.12), which is solved numerically. The author presents a perturbation analysis using a periodic, unidirectional field to establish the effect of current density and electrode geometry on the fully developed roll strength, including an estimate of the roll development length.

The roll strength is found as:

$$v_0(x) = (\rho_i E)d^2/(\nu 2\pi(\alpha + \alpha^{-1})^2) \cdot \{1 - \exp[-(\pi^2(1 + \alpha^{-2}))x/(\mathrm{Re} \cdot d)]\} \quad (5.13)$$

where $\mathrm{Re} = U_0 d/\nu$ and $\rho_i E = i_m/m$, the rotational part of the electrical force field driving the ionized gas motion, m is ion mobility, $\alpha = L_z/(2d)$ the roll aspect ratio and L_z is the vertical distance between corona peaks.

The question, how pulse operation influences the secondary flow, in cases where the pulse frequency exceeds the characteristic frequency of rolls, has to some extent been answered by investigations performed by the same group of researchers. Experiments referenced in [27], using approx. 150-μs pulses, show rolls of almost the same strength as for DC up to about 0.08 mA/m^2, at bulk flow velocities of 1 m/s (Figure 5.13). At higher current densities, pulse roll strength decreases while DC roll strength increases. Though the exact current limit is specific to a given arrangement, one might conclude from Figure 5.13 that there is no real difference between pulse and DC energization with respect to roll intensity, particularly as pulse operation is normally used with high resistivity dust, demanding low current densities.

5.2.4 *Numerical flow model*

A full model consists initially, in drawing up the Maxwell equations for electric field and space charge in three dimensions and implementing a proper particle charging theory; secondly, the full Navier–Stokes equations in three dimensions for the flow field including the electric body forces and the necessary turbulence model with the corresponding equations; and finally the three-dimensional particle transport equations. The boundary conditions, which form the solution of the partial differential equations; are generally straightforward; only the corona region still needs some empiricism. It is a demanding task working out the proper codes, finding an effective and fast solution procedure and having access to sufficient computer power. The difference equations are coupled and need to be solved iteratively demanding significant computer space and time.

Jamshid Zamany [30] tried, but had to limit the problem. He worked with the electrical equations in three dimensions combined with a two-dimensional turbulent flow model. He used a constant diffusivity in the calculation domain proportional to the square of axial velocity times duct width, $\Gamma \sim U^2d$. In order to determine the electrical operation point with dusty gas, only the corona starting voltage is needed for a given clean gas state. The discharge electrodes can have any form, including fully three-dimensional spike types, and the collecting plate surface can be corrugated or provided with baffles. The space charge distribution is three-dimensional and the current density pattern on the collector surface is predicted. Corona quenching is calculated, together with axial current density averaged in the vertical coordinate direction. Efficiency as a function of axial length can be predicted for each class of particle size.

Further improvement of the model includes three-dimensional flow field, local particle diffusivity, a turbulence model reflecting the physics inside the precipitator and re-entrainment at the near wall region. In a later improved version the local field strength and the local current density on the dust surface and across the dust layer are calculated and used for prediction of back corona.

5.3 Gas velocity

The Deutsch migration velocity, presuming a quiescent gas, only depends on the molecular gas viscosity. Through the years a large number of efficiency measurements has been performed showing that there is a variation in the migration velocity parameter with bulk velocity.

Among the papers dealing with this subject, the contribution from Dalmon and Lowe from the 1950s has been chosen [31]. They used a pilot precipitator where dust burden, temperature and velocity could be controlled. Figure 5.14 shows the calculated migration velocities for 3.5, 11 and 51 μm particles, increasing with increasing velocity up to about 2 m/s and a decrease at higher velocities, presumably due to re-entrainment from scouring.

By means of the theoretical code from [30], the efficiency has been determined and the migration velocity parameter calculated using the Deutsch expression. The result is seen in Figure 5.15 revealing an increase with velocity, as measured by Dalmon and Lowe. The code does not take re-entrainment into consideration, so the constant level above approximately 2 m/s does not reflect reality. The data scatter at high velocities is due to numerical instability, not to physics.

The conclusion is that an optimally designed precipitator should operate in the range where the migration velocity is almost independent of axial velocity. If a precipitator is to operate at a reduced velocity, the variation should be included in the sizing. It should be remembered that the reduced

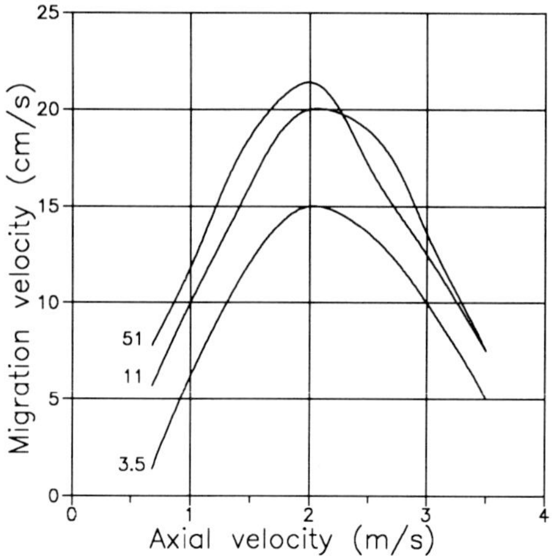

Figure 5.14 Migration velocity parameter *versus* axial velocity for 3.5-, 11- and 51-μm particles. Redrawn from [31].

migration velocity using low velocity is the result of the higher turbulence level.

Precipitators operating at elevated axial gas velocities must have a perfect gas distribution, as velocities above average might cause scouring and re-entrainment, thereby limiting the achievable precipitator efficiency. A plus for higher gas velocities is the relatively low level effect of turbulence improving efficiency.

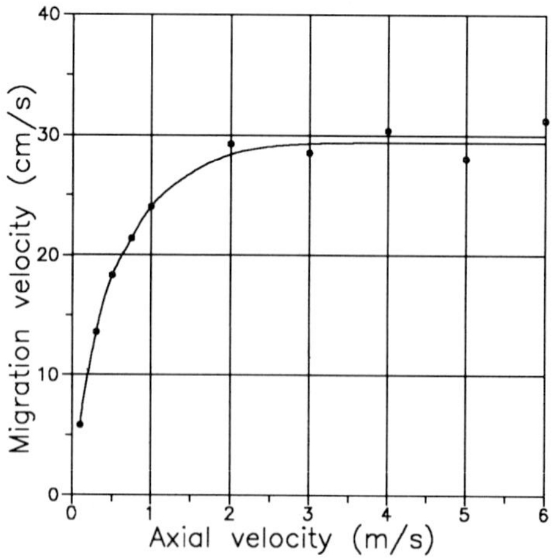

Figure 5.15 Migration velocity *versus* axial velocity. Calculated using the code of [30]. Re-entrainment not included.

5.4 Gas distribution

Dust-laden gases are led through ducts where the velocity is chosen according to various criteria. Dust concentration might vary between a couple of g/m^3 and several hundred g/m^3 and the conveying of the dust right up to the precipitator without particle deposition demands that the gas velocity in horizontal or almost horizontal ducts is not too low. Depending on particle size distribution, dust density, concentration and duct length, the gas velocity is normally chosen between 13 and 20 m/s at nominal plant output.

The need for an adequate conveying velocity is counteracted by the request for a reasonable pressure drop and restricted wear on walls and internal structures. Such considerations demand velocities below about 20 m/s. Some plants operate with a variable loading factor and if dust is deposited on the duct bottom at reduced production there is a risk that the bottom wall and the lower part of the side walls are cooled and will corrode. The cooling might also mean that the deposit becomes sticky and forms a more compact layer which cannot be redispersed when production rate increases. Loosely precipitated dust, deposited during periods of reduced production, will normally be re-entrained during run-up, which causes an increased dust loading to the precipitator. If the precipitator has not been sized to cope with such peaks in inlet concentration, the result might be increased emission, maybe even violation of the guarantee or legislative limit.

The optimal gas velocity in the space between electrodes varies from process to process, but for the sake of simplicity assume it is set at 1 m/s. The transition piece between raw gas duct and precipitator housing is, fluid dynamically, a diffuser, where kinetic energy is converted to potential energy or pressure. The transition piece between the precipitator housing and the clean gas duct is a contraction, which, fluid dynamically, is relatively unproblematic, having only a small effect on the upstream flow. The design of diffusers with an expansion ratio of up to 20, or more, is critical to provide an even velocity profile at the outlet, both in space and time. Normally, the demand for an even profile, means an apex angle not greater than 8° to 10° for a 'straight' diffuser. This is obviously an impossible approach for economic reasons; firstly there would be a risk that dust would be deposited on the bottom and, furthermore, small angle diffusers demand horizontal space, which is costly and rarely available, particularly if an existing precipitator is to be upgraded. (Most plant arrangements have close coupling with minimal duct lengths to optimize capital cost and space.)

This leads to the use of angle diffusers combined with the transition piece. Short length diffusers, or wide angle diffusers are fitted with perforated screens, a technique adopted from wind tunnel diffusers using nets or grids. If a diffuser has too wide an apex angle, the flow will separate from the walls and a centre jet with a relatively high centre velocity and recirculating flow close to the walls is developed. Separation is caused by the negative pressure gradient created by the diffuser itself. From time-averaged measurements it is known that the characteristic velocity profile has a peak at the centre line and reverse flow at the walls. The instantaneous profile for wide angle diffusers adheres to one wall at a time and separates completely from the other wall. In the next time interval the walls change their roles. In popular terms, screens create a pressure drop, depending on the screen porosity, which acts against the pressure rise of the diffuser. With multiscreens there is a tendency to have velocity peaks close to the diffuser walls instead of a at the centre axis. See reference [32] for information about methods for producing uniform gas flow.

It is obvious that the ability of the inlet transition piece to reduce velocity and create an even velocity profile depends on the velocity profile in the raw gas duct. This must not be too skew and its form at the inlet flange should at least be known to the precipitator designer. Not only the flange velocity distribution has to be known, but also the particle distribution, any temperature stratification, and the distribution of, say, O_2 and SO_2, all of which are important to the precipitator efficiency. The gas mixing is influenced by molecular and turbulent diffusion, and as for the temperature, by heat transfer from convection and radiation.

5.4.1 Standards

The US 'International Gas Cleaning Institute', IGCI, published EP-7 in November 1980. It deals with 'standards' for gas distribution and model testing [33]. In this publication, it is clearly pointed out that the gas distribution demands are subordinate to the emission guarantee and need not be corrected, if it does not comply with IGCI EP-7, as long as the emission guarantee is met. (The IGCI has since changed its name to ICAC, 'Institute of Clean Air Companies'.)

IGCI EP-7 lays down requirements as to the number of measuring points, their distribution and the subsequent evaluation and statistics, e.g. it demands that 85% of the velocities in a cross-section, 3 ft (0.9 m) downstream of the leading edge of the first field, are less than or equal to 1.15 times the average velocity. Furthermore 99% must be less than or equal to 1.40 times average. The same needs apply to a cross-section 3 ft (0.9 m) from the outlet of the last field. Specific demands with 1.10 instead of 1.15, and 90% instead of 85% are not motivated by precipitator physics, but merely in an attempt to demonstrate effective project management.

In the case of a presumed normal velocity distribution (which is not necessarily the case) the [1.15 | 85%] claim corresponds to a certain coefficient of variation, σ/v_m, σ being the standard deviation and v_m the mean velocity. See Figure 5.16, where the abscissa is normalized velocity, v/v_m, and the ordinate is the cumulative normal distribution. The tangent to the curve

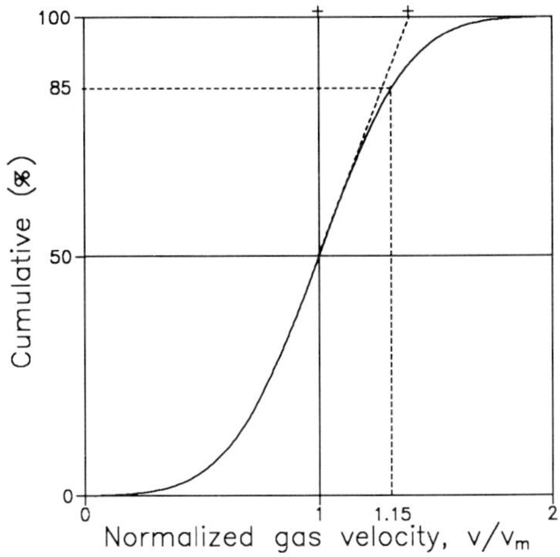

Figure 5.16 Cumulative normal distribution *versus* normalized velocity. Distance between crosses is $1.253\sigma/v_m$.

in point [1 | 50%] intersects the 100% horizontal line at $1.0 + 1.253\sigma/v_m$, and if the curve includes the point [1.15 | 85%], σ/v_m is approximately 0.145 or 14.5%.

Experience seems to indicate that σ is almost independent of mean velocity. In fact, the lower the average velocity, the more difficult it will be to obtain a good gas distribution, and σ/v_m increases with decreasing mean velocity.

5.4.2 Residence time

Provided that the precipitation occurs evenly along the collecting plates it can be concluded that the efficiency is dependent on the residence time. However, this is not the case for the bigger particles, which are precipitated after a relatively short axial length, while the full length is exploited for the smaller particles. This argument does not take re-entrainment, sneakage, sweepage or contra-emission into account.

The Deutsch equation, for the penetration P, had been derived with the implication that concentration profiles are even across the duct giving only axial changes in 'amplitude', presuming an infinite particle diffusivity or an infinite turbulence intensity level. In spite of such a strong presumption, the Deutsch equation (5.4) has given meaningful results and is still used worldwide. Implied in equation (5.1), for a given field strength, gas state and particle size distribution, is a migration velocity independent of axial gas velocity. With this constant migration velocity, White [6] used 'calculus of

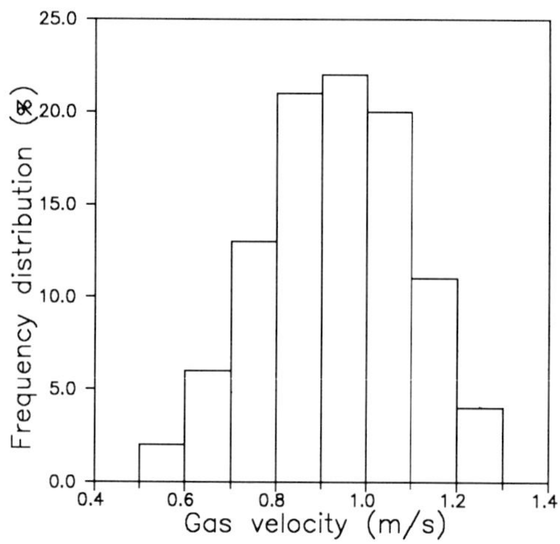

Figure 5.17 Frequency distribution of measured gas velocities. Readings are grouped in intervals of 0.1 m/s.

variations' to demonstrate that minimum penetration is achieved for a perfectly even velocity profile across the duct. In order not to confuse the matter with mathematical demonstration, P is calculated below for a precipitator with (1) constant gas velocity v_m, and with (2) $0.99v_m$ in one half and $1.01v_m$ in the other half. v_E/v_m is set at 0.15 and L/d at 50:

$$v = v_m \qquad P = 553.1 \text{ ppm}$$

$$v = 0.99v_m \text{ and } 1.01v_m \qquad P = 554.7 \text{ ppm}$$

the latter penetration being 0.3% higher than the penetration with constant gas velocity. Thus, it is made probable that the non-linear Deutsch expression gives minimum penetration for constant axial velocity.

If a velocity distribution function $f(v)$ is introduced, equation (5.4) can be modified. Figure 5.17 shows a measured discrete distribution, and Figure 5.18 a continuous approximation of the cumulative curve found from the summation of the step curve. Often $f(v)$ is assumed to be a Gaussian distribution with average velocity v_m and 'standard deviation' σ:

$$f(v) = (1/\sigma\sqrt{2\pi}) \cdot \exp[-(v - v_m)^2/2\sigma^2] \tag{5.14}$$

and hence:

$$P = \int_{v_{min}}^{v_{max}} P(v_m) f(v) dv \tag{5.15}$$

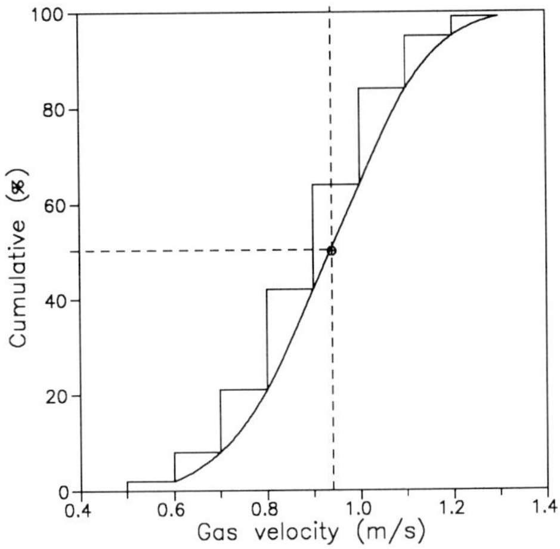

Figure 5.18 Cumulative distribution corresponding to the frequency distribution of Figure 5.17. Median velocity reads 0.94 m/s.

Idel'chik and Aleksandrov [34] evaluate $P = P(v)$ in a slightly different way:

$$P = \exp[(-v_E/v_m)(L/d)(1/M)] \tag{5.16}$$

where

$$M = 1/A_c \int_{A_c} (v/v_m)^2 \, dA_c \tag{5.17}$$

A_c being the cross-sectional area. See Figure 5.19 where penetration P is plotted against parameter M [34].

Gooch *et al.* [35] use a velocity weighted version of (5.15):

$$P = \int_{v_{min}}^{v_{max}} P(v_m) f(v) v/v_m \, dv \tag{5.18}$$

In Figure 5.20 the expressions (5.15), (5.16) and (5.18) are reproduced as graphs independent of σ/v_m. The actual gas distribution function $f(v)$ is considered as being a normal distribution. The expressions of White and Idel'chik and Aleksandrov correspond very well, giving a weaker dependence of the gas distribution than is found with (5.18).

Among precipitator vendors, theoretical or empirical curves based on σ/v_m, much like those of Figure 5.20, are used. In Figure 5.21 a correction curve based on measurements by Fläkt [36] in a pilot precipitator is given together with a measured curve from [37]. Points are calculated using the ω_b theory [38] and measured data from [39]. Maybe the truth lies somewhere between the full curve and the dotted curve.

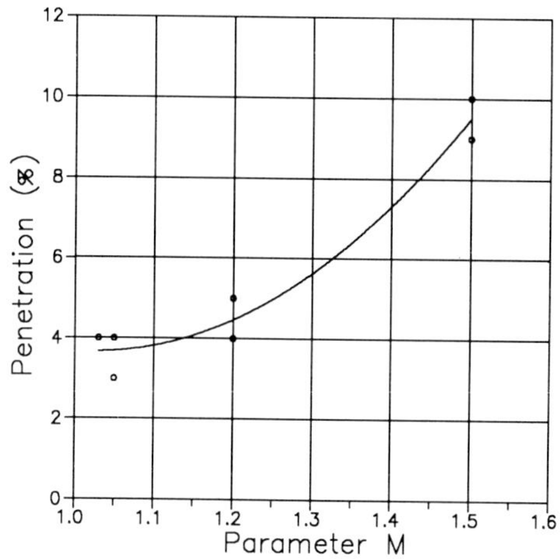

Figure 5.19 Penetration *versus* parameter M. Redrawn from [35].

5.4.3 *Space charge*

In the end it is the distribution of the dust particles, not that of the gases, that determines the precipitator efficiency. Especially the distribution of the fine particles, which are the hardest to catch, further down in the precipitator must be smooth. If the particle distribution is skew, then the space

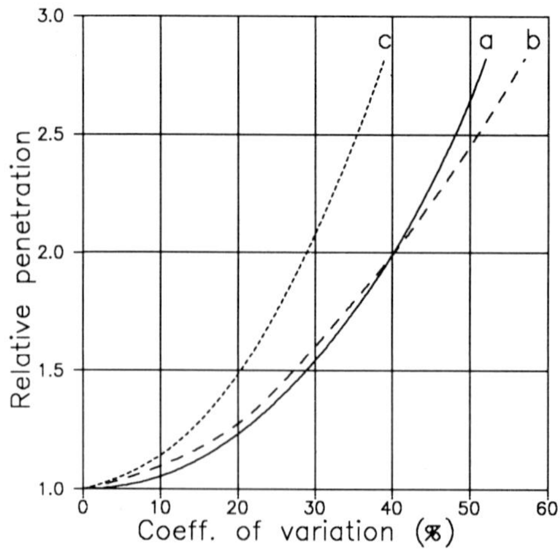

Figure 5.20 Relative penetration *versus* coefficient of variation. (a) equation (5.15), (b) equation (5.16) and (c) equation (5.18).

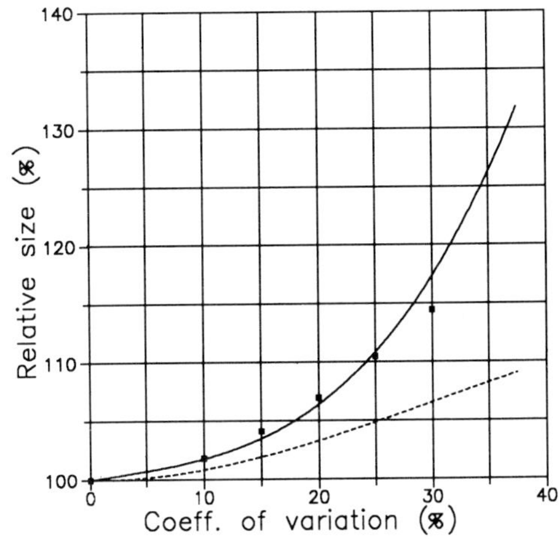

Figure 5.21 Relative precipitator size *versus* coefficient of variation. $\eta = 99\%$. Full line redrawn from [36], dotted line from [37] and squares from [38].

charge tends to disturb the current distribution. The term 'current capsizing' is often used for this phenomenon. Differences in dust mass concentration varying from top to bottom will always occur, but there is less difference in the number of small particles, which have the greatest surface compared with mass. The specific surface (m^2/kg) is, for a given dust mass density, proportional to $1/d_P$, d_P being the particle 'diameter' or characteristic transverse measurement. As charges adhere to the surface of the particles, the charge per unit mass will also be inversely proportional to d_P.

The difference in concentration between inlet and outlet planes is larger, e.g. if we consider a field efficiency of 90%, the outlet mass concentration is only one-tenth of that of the inlet. Yet, taking into account the fine particles at the outlet having the greatest surface area, the variation in space charge is less than a factor of 10. In principle, a balance will take place between inlet and outlet current densities, as long as current quenching, due to space charge saturation, does not occur: if the precipitation tends to decrease at the inlet, more particles are conveyed towards the outlet causing the current density there to decrease.

Crosswise the conditions are quite different. An individual duct hardly knows about the existence of the neighbouring ducts and not at all about the ducts on the far side of the precipitator. They are all exposed to the same voltage and a small dust concentration in a specific duct increases the local current density, due to a low space charge, thus creating a growing imbalance between ducts. For this reason it is better with a couple of ducts having high loading than a couple with low loading. This is why a gas distribution with high velocity in the outer duct, often seen in precipitators fitted with central inlet and low porosity screens, is less severe than one with much too low a velocity in the extreme ducts, which is likely to exist in similar configurations with high porosity screens.

Skew crosswise gas distribution was seen on earlier precipitators with transverse coupling of individual bus bar sections and a common rectifier. Recording the current of each section often revealed that one of the sections took the lion's share and left the other(s) with too small a share. Such arrangements result in reduced efficiency and, today, vendors normally provide each section with individual high voltage power supplies.

Current capsizing might, in cases with high resistivity dust, reduce the efficiency drastically, because high current density in a local area due to low space charge provokes contra-emission. The flow of electric current with reverse polarity disturbs the local performance and undersupplies other regions with current. As the total current is low, under high resistivity operation, the efficiency is expected to reduce rather significantly. Strange as it may seem, laboratory measurements with widely distributed contra-emission have demonstrated that there is less influence of gas distribution on efficiency for high resistivities. Yet, this is not the case for incipient or moderate contra-emission situations [40].

5.4.4 Re-entrainment

Dust once precipitated on the surface of the more permanent, 'old', dust layer is not necessarily caught once and for all. There has been some indication of re-entrainment taking place under normal operating conditions, even though the gas distribution is even, showing that precipitation efficiency is the net result of dust being deposited on the collecting surface and dust leaving the surface. Furthermore, it seems that the time constant involved can be in the order from a few seconds up to several days.

Unburned coal or particles, conductive for some other reason, present in the gas stream might be discharged upon reaching the collecting surface and re-entrained due to the lack of compressive or holding force from the electric field. They might also be only partly discharged making it easier for the gas to scour them off and redisperse them. Processes with conductive particles demand a correct gas distribution and a moderate axial velocity.

In cases where high resistive and conductive dust particles are present at the same time, it is important to increase the current density, above the contra-emission limit, in order to keep the conductive dust partially clamped to the collector. In such cases the demand for even gas distribution and moderate gas velocity is even more strict. During collecting plate rapping, lumps of dust are loosened and fall under the influence of gravity. Some are caught again and others fall into the hopper. Where the lumps slide along the dust layer on the plate, particles are loosened and redispersed. Likewise falling lumps will tend to disintegrate due to gas erosion or hitting obstructions, e.g. frame tubes, rapping bars, walkways and the hopper walls themselves. In all cases dust is re-entrained and redispersed reducing the collection efficiency.

Only a few collector plates are vibrated or rapped at one and same time in order to avoid too massive a re-entrainment factor. Some designs use systems where the neighbouring ducts are closed, or the gas velocity in the ducts is reduced, during rapping, using flaps or air curtains synchronized with the rappers; other designs close off the compartment flow during rapping. For the same reason synchronizing of rappers between serial fields is used in order to reduce the effect of rapping spikes, especially in cases demanding very low average emission.

One should be aware of the effect of falling lumps inducing strong downwash. Neglecting gas friction the terminal velocity can be calculated as $v_f = \sqrt{(2 \cdot g \cdot h)}$, g being the gravity acceleration and h the fall height. Putting $g = 9.8\,\mathrm{m/s^2}$ and $h = 10\,\mathrm{m}$, results in $v_f \approx 14\,\mathrm{m/s}$, a very high value, compared with the bulk velocity. The axial flow will be disturbed during rapping, and the gas distribution will be temporarily skew, and there is a risk of re-entrainment from the hopper. The phenomenon is predominant at the inlet end, where dust is most abundant.

5.4.5 *Erosion*

If the bulk velocity is increased above a certain level, say above 2 m/s, and the gas distribution is uneven, erosion will most likely occur; again an argument for ensuring a perfect even distribution and a not too high bulk velocity (see section 5.3).

5.4.6 *Sneakage and sweepage*

Sneakage is the term describing dust-laden gas not passing through the active electrode system, including areas where the discharge system has no corona points, e.g. at frame tubes, even though turbulence in the ducts tends to improve charging and thereby precipitation of the dust inside the active electrode system.

Gases flowing between the very outmost collector plates and the wall of the housing, and gases flowing above and below the electrode systems, carry dust and give rise to sneakage. If 0.1% of the dust passes untreated from the inlet to the outlet, it is easily seen that efficiency can never exceed 99.9%, however efficient the precipitating system might be. With $10\,\mathrm{g/m^3}$ dust loading the emission will never be less than $10\,\mathrm{mg/m^3}$.

In order to minimize sneakage, extra screens, so-called baffles, are put between the fields and below the collectors. These screens should be preferably perforated, porosity $\geqslant 10\%$, in order to prevent turbulence and recirculation behind the screen.

The greater the space above and below the electrode system and the smaller the collector height, the more the sneakage in terms of volume flow rate and hence the reason why low precipitators must be fitted with effective baffles.

Sneakage is estimated from velocity measurements in the regions above and below electrode systems. Some model laboratories judge the sneakage from the time it takes to empty the hoppers of smoke, injected before start-up of the fan.

5.4.7 Optimal distribution

As seen from the preceding sections the interaction between gas distribution and precipitator physics is complex, and the demand that the gas distribution should be very even is mostly based on the modified Deutsch expression such as equations (5.15), (5.16) and (5.18).

In the first field, where most coarse particles are found, there is no sense in trying to raise the heavy dust fraction to the upper region, as it finally has to be accumulated in the hoppers. This supports a distribution with velocities above average at the lower part of the field. From the second and further fields, it is the finer particles which are to be caught, suggesting the gas distribution should be fairly even until the exit from the last field. Particles re-entrained in the lower part of the last field might be swept out in the clean gas duct, and it is recommended that the bottom outlet velocities are below average. Outlet transitions fitted with gas distribution screens make it possible to adjust the vertical velocity profile, which is especially important in cases where the precipitator is designed with a 'bottom-type' outlet.

Figure 5.22 shows field-recorded vertical velocity profiles with an air load on a two field precipitator. The precipitator had a penetration about 0.07% and almost no rapping peaks for an inlet loading of some 40 g/m^3, so even though the distribution is far from even, it is believed that this distribution profile has a positive effect on the efficiency [40].

Arthur G. Hein [41] uses a two-dimensional model, dividing the length and height of a precipitator into rectangles, and looks at the net mass balance of each rectangle. Dust enters the rectangular cell from the upstream element, partly as re-entrained dust from upstream cells at a higher level, and leaves the cell, either because it has not been precipitated or because it has been re-entrained. The mass balance for all cells is collated and the system of equations solved using a computer. Hein states that 'controlled non-uniform gas distribution at the inlet and outlet faces can be used to improve the performance of the precipitator', and recommends that more than half the total gas flow should be supplied to the lower half of the treatment zone and should be exhausted from the upper half of the treatment zone.

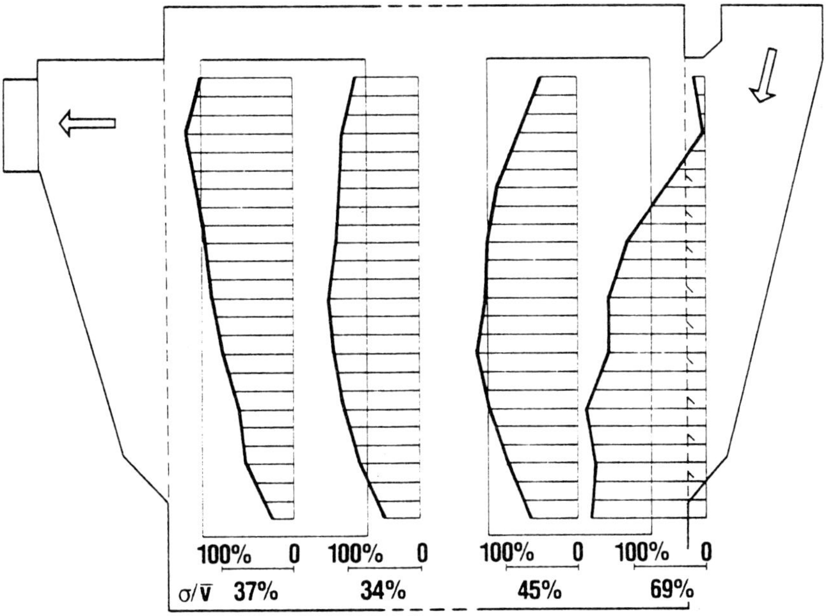

Figure 5.22 Vertical velocity profiles from air load field testing. From [40].

5.5 Model testing

The reason for the demonstration of a proper gas distribution in a scale model stems from the 1960s, where precipitators often did not fulfil their guarantee. Thus modelling became a demand, and stricter claims of gas distribution quality were introduced, e.g. IGCI EP-7.

Model studies are performed by vendors or by specialized laboratories, either when new transition pieces are to be designed or on request by the client or specification. In the latter case the whole duct system will normally be included. Such studies include first of all the gas distribution and system pressure drop, but also an assessment of the risk of dust drop-out in the raw gas ducting.

The model specification, measurement methods and evaluation of results are normally part of the client's tender document. Sometimes, the clients' consultant tightens the demands on the distribution beyond that reasonable, with respect to precipitator efficiency and cost, and sometimes beyond the physically possible. An extremely even distribution needs extended laboratory measurements and time plus a large number of internal flow baffle correcting devices.

The fact that it is possible to predetermine the gas distribution in a full-scale precipitator by performing a test on a model of scale 1:16 to 1:8 is explained by physical–mathematical dimensional analysis. Model testing demands geometrical, kinematic and dynamic similarity. In other words, the ratio between any linear dimension in the model and full-scale must be the same and the ratio between velocities in the model and full-scale at corresponding points must be constant. Finally the ratio between forces acting on the fluid in the model and full-scale must be identical; this is normally automatically fulfilled if there is geometrical and kinematic similarity. The physics of the flow is described by three Navier–Stokes equations, one per coordinate in space and the equation of continuity. Making the equations dimensionless, using characteristic mass, time and length and combinations thereof, it can be proven that the solution is universal, provided the dimensionless numbers appearing in the equations are equal in the model and full-scale. This includes the Reynolds number, Re, the Euler number, Eu, and the Mach number, Ma. The Euler number is automatically fulfilled as it depends on Re.

Re is the ratio between flow inertia and friction, while Ma is the ratio between the gas velocity and the speed of sound, i.e.

$$\mathrm{Re} = vd/\nu \quad \text{and} \quad \mathrm{Ma} = v/a = v/\sqrt{(\kappa RT)}$$

where v is velocity (m/s), d is characteristic length (m), ν is kinematic viscosity (m^2/s), a is speed of sound (m/s), κ is adiabatic power, R is gas constant (J/kg/K) and T is temperature (K).

If the model is ten times smaller than full-scale and $\nu_{FS} = 2\nu_M$, identical Re gives $v_M = 5v_{FS}$. For $v_{FS} = 1$ m/s, the model velocity becomes $v_M = 5$ m/s. In the duct, 20 m/s full-scale velocity gives 100 m/s model velocity, or Ma = 0.29, very close to the compressibility limit.

The net power, necessary to drive the air, is $W = \Delta p V \sim \psi^{1/2} \rho v^3 d^2$, where V is volume flow rate. Provided the pressure coefficient Ψ is the same in model and full-scale (fully turbulent flow) the powers are found to be:

$$W_{FS} \sim \rho_{FS} v_{FS}^3 d_{FS}^2 \tag{5.19}$$

$$W_M \sim \rho_M v_M^3 d_M^2 \tag{5.20}$$

The ratio between W_M and W_{FS} is:

$$W_M/W_{FS} = (\rho_M/\rho_{FS})(v_M/v_{FS})^3(d_M/d_{FS})^2 \tag{5.21}$$

With a density ratio of 1.4, a velocity ratio of 5 and scale factor of 0.10, the fan power needed in the model is 1.75 times full-scale power. This high power is prohibitive, hence praxis is to use the same axial velocity in the model as full-scale. The Reynolds number (Re) in the ducting of the model will still be in the fully turbulent range, while Re in the precipitator model will be in the low turbulent range, about 10^4. This reduces the Mach number to approx. 0.05 and the model fan power to 1.4% of that of full-scale.

IGCI EP-7 recommends to double the model duct width in order to double the Reynolds number, but this is hardly the proper solution at the inlet of the first field, because the straightening effect of the collector plates is of greater importance to the crosswise velocity distribution than is the correct Re based upon duct width.

The Reynolds number based upon lengths from the leading edge of internals, such as guide vanes, dividing walls and kicker plates in the model, are often in the laminar regime and their effect on the flow will therefore be underestimated.

Velocities in the model and full-scale are normally recorded with hot-wire, hot-film or vane-type anemometers. None of these probes, in their standard design, can indicate the flow direction. As the end result should be an even distribution, this is not of great importance but, where necessary, flow direction can be visualized using smoke, provided the smoke is injected with the same velocity as the local flow, or visualized using wool tufts. Knowledge of the velocity vector direction can help the troubleshooter find the best remedy for correcting the gas distribution.

The vane anemometer has the disadvantage that velocities below 0.3–0.4 m/s cannot be recorded, and a jet only partly filling the vane cross-section is registered with its full value. Apart from this, the vane probe is a reliable and rugged instrument.

Figure 5.23 Photograph of model in the laboratory of Lodge Sturtevant.

Some laboratories use one person to position the probe in the measurement cross-section and one to read the instrument. Other laboratories use slides with electric motors and electronic control and only one person to control the experiments. Data recording is done by pen or paper or by computerized data acquisition, the latter being capable of calculating and plotting the statistics and graphics simultaneously.

Figure 5.23 shows a photograph of a model in the Lodge Sturtevant laboratory. IGCI EP-7 requires the gas distribution to be mapped only at the inlet of the first field and the outlet of the last field, which does not reveal possible maldistributions in other internal cross-sections. Therefore, it is recommended to map at least one extra cross-section, e.g. the inlet of the second field, where the hopper tends to draw the gases downwards creating a bottom peak on the vertical velocity profile. This tendency is less pronounced, with precipitators having a trough-type hopper such as those used in lime reburning kiln precipitators.

Results are normally presented in a report which should include sketches of the inlet and outlet transitions, plus inlet and outlet screen geometries and guide vane arrangements. Velocity profiles, plane and isometric, iso-velocity plots, velocity vector plots and statistics are also normally included.

5.6 Computational fluid dynamics

With the appearance of powerful and fast computers, new possibilities for replacing time-consuming model testing and field testing have arisen. This involves solving the differential equations describing fluid motion, using either a finite volume or sometimes, but more rarely, a finite element method. The methods are named computational fluid dynamics or simply CFD.

Physically, the conservation of mass and Newton's second law are applied on the fluid, mathematically expressed as the equation of continuity and the Navier–Stokes equations in two or three dimensions. Even with today's fastest supercomputers, it would be impossible to divide the calculation domain in parts small enough to describe all the details in the flow field. Solving the equations directly is a discipline called direct numerical simulation, which is still restricted to very limited Reynolds numbers and small geometries. With the smallest eddies (of size, as per, the Kolmogorov length scale, $(\nu^3 \cdot \delta / U^3)^{0.25}$, ν being the kinematic viscosity $\approx 40 \cdot 10^{-6}\,\mathrm{m^2/s}$, δ a typical shear layer thickness $\approx 10^{-1}$ m and U the bulk velocity ≈ 1 m/s) the length scale is of order 0.3 mm. This means that a precipitator of dimensions $15 \times 15 \times 20\,\mathrm{m^3}$ should have a mesh number of order of one hundred million millions, far beyond the capacity of computers in the 1990s to tackle direct numerical simulation.

Instead the turbulent variables are taken as average values plus fluctuating parts, e.g. $u = U + u'$, u being the instantaneous x-velocity, U being the time average and u' the time-dependent fluctuation (time average of $u' \equiv 0$). Introducing these variables into the Navier–Stokes equation and time averaging leads to:

$$\rho D\boldsymbol{U}_{\text{av}}/Dt = \rho g - \nabla p_{\text{av}} + \nabla \cdot \tau_{ij} \tag{5.22}$$

where the stress tensor τ_{ij} is:

$$\tau_{ij} = \mu(\partial u_i/\partial x_j + \partial u_j/\partial x_i) - \rho(u_i'u_j')_{\text{av}} \tag{5.23}$$

The first term is the laminar stress, and the second term the turbulent stress. The turbulent part of the stress tensor is either found by solving the so-called Reynolds' stress equations or simply by expressing it using average flow values and the so-called Boussinesq assumption (J. Boussinesq, Paris, 1877): $\tau_t = \mu_t \partial U/\partial y$, where μ_t is the eddy (or turbulent) viscosity, U is an average velocity and y is a coordinate perpendicular to vector $\boldsymbol{U}$.

μ_t is determined using either Prandtl's mixing length theory, $\mu_t \equiv \rho l^2|\partial U/\partial y|$, l being the so-called mixing length, or using the k-ε model: $\mu_t \equiv C_\mu k^2/\varepsilon$, C_μ being a constant or coefficient, k being the turbulent kinetic energy, $k = \frac{1}{2}(u'^2 + v'^2 + w'^2)$, and ε being the turbulent dissipation, $\varepsilon = k^{3/2}/L_d$; where L_d is a length scale for the dissipating eddies.

The more complex the turbulence model is, the more differential equations must be introduced, which can only be solved by closing the system of equations by introducing algebraic expressions based upon empiricism.

After the break-through of IBM's computers and software in the 1950s, a great impetus was given to the development of physical and numerical methods for solving flow dynamic problems. At the end of the 1960s more research groups contributed to the development, among others researchers from the Mechanical Engineering Department at Imperial College in London [42,43] looked at the solution of internal flow problems, whereas the aerospace companies emphasized external flows for calculation of lift and drag.

The most common methods of today are based on transformation of the differential equations for conservation of energy, mass and momentum to difference equations, solved by integration after applying the given boundary conditions. Originally in two dimensions, the equations can be transformed and solved using streamfunction Ψ ($\rho u = \partial\Psi/\partial y$, $\rho v = -\partial\Psi/\partial x$) and vorticity ω ($\omega_z = \partial v/\partial x - \partial u/\partial x$), thus eliminating pressure as a variable. Later on, with more efficient computers, the equations were formulated and solved in the primitive variables, pressure and velocity, in three dimensions, thus avoiding the problem of defining proper boundary conditions for a vorticity having steep gradients close to the walls.

As velocity gradients are also steep at the walls, special attention is paid to the description of analytical near-wall variations, the so-called wall functions. The difference equations are often solved in Cartesian or orthogonal grids, equidistant or almost equidistant, using 'upwind differences' taking into consideration the local direction of flow. One problem with the grids is the appearance of 'numerical diffusion', which is maximal if the flow vector has a 45° angle with the grid axes. In recent years advanced grids have been developed, e.g. 'adaptive grids', where the mesh size changes according to the gradients of the variables, decreasing in size where gradients are steep. Another approach is the use of 'multigrids', where calculation shifts between a fine and a coarse mesh, thus smoothing out short- and long-wave variations.

The difference equations are solved indirectly (integrated) by iteration, sometimes several hundreds, and numerical stability is achieved by proper successive under- or overrelaxation. The criterion for having found the final solution normally is the mass continuity, and the calculation stops when the maximum residue is less than, say, 10^{-4}.

The 'solver procedure' is a chapter in itself using various numerical principles, continually updating the values in the domain as soon as new values have been found. A number of fast and efficient solvers are commercially available today. (See Patankar [44] for more details on the numerical principles.)

With more complicated geometries, such as precipitators including fancy transition pieces, guide vanes and screens, grid designing is difficult. A simple method is to use a parallelepipedal domain and block out all the outer elements until an approximate contour is achieved, leaving the geometric domain boundary as a step surface. This means that there is a limit to how precise the solution close to the walls can be, even though the internal flow is only slightly influenced. In fact, it is a grid generator, especially easy to use, which is the main issue for the operator. The easier the grid generation, the more different configurations can be calculated in a reasonable time, presuming, of course, that the solver is effective and fast. Less than 2 h per new contour and less than 1 h per modification would be ideal.

The real progress in mesh generation is a mesh that can be fitted without any consideration of the interfacing, at least as seen from the point of view of the user. Up to now mesh structures have had to fit where ducts and transition pieces, or transition pieces and precipitator housing meet, but new grid systems are appearing. Thus, it is possible to have an essentially polar mesh of a circular configuration corresponding with a rectangular mesh of a box, without any concern about mesh fitting at the plane where they meet. This block type is well suited for duct and precipitator analysis with respect to both gas distribution and pressure drop determination. Figure 5.24 shows this type of mesh from the code Star-CD.

The more advanced the mesh, the more variables to be treated and the stricter the convergence criteria, the more calculation time is needed to find

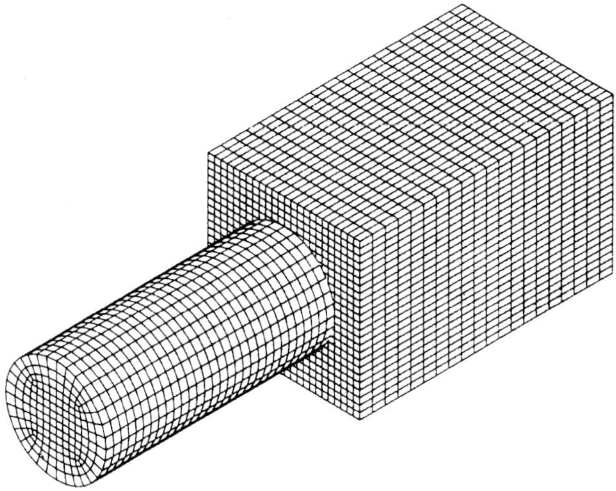

Figure 5.24 Example of arbitrary interfacing of the mesh of a cylinder and the mesh of a box. Star-CD.

a correct solution. The need for space in the memory and on the disc is another 'eye of a needle', the cell number increasing roughly with the product of the number of variables and the reciprocal of the mesh size to the third power. While details such as guide vanes, kicker plates and ladder vanes have to be modelled separately, there is a possibility to simulate perforated screens using some sort of porosity model such as a Darcy porosity. Screens with evenly distributed guide vanes might be modelled as a whole, but this calls for special routines, not normally commercially available. In the equations the so-called 'source' terms can be modified in order to reflect the influence of the gas distribution screens. (See Patankar [44].) Screens can also be modelled by 'blocking' out cells, but this procedure demands a very fine mesh with many nodes. The flow before a perforated screen is often recirculating, unless small screen guide vanes are used. Such flows can be identified using CFD, revealing velocity vectors parallel to the screen, making it impossible to improve the distribution by modifying the screen by increasing or decreasing the open area alone. Programs which cannot treat small thin oblique surfaces in an effective and easy way should therefore not be used. Like the use of porosity for simulating a perforated screen, it is possible to design a subroutine with a screen combined with distributed guide vanes and use it as a black box when the calculation is done in the screen domain. This would reduce the need for a complex mesh with a large number of grid points.

Next to the mesh generation, postprocessing, i.e. the presentation of the results, is the most important factor for the user. Commercial systems apply velocity vector plots and iso-curves for pressure and velocity, normally supported by colours. Isometrical presentations of results are often used, these being more fanciful than informative and less applicable than, e.g., 2-D velocity profiles. Figure 5.25 shows part of a precipitator with guide vanes in a manifold, and Figure 5.26 shows the streak lines of zero-mass particles. Unfortunately, it is not possible here to show vector plots or velocity profiles, because they are presented with relatively faint colours, difficult to reproduce in black and white. For further details see references [45] and [46].

A serious user of CFD must have access to the necessary hardware and software, invest in training and have at least one person engaged full time in calculations. Likewise, a company having its own CFD facilities is necessary in cases where external specialist firms are engaged to do some or all of the computational work, because it will make it more efficient to arrange test programs and to interpret the results.

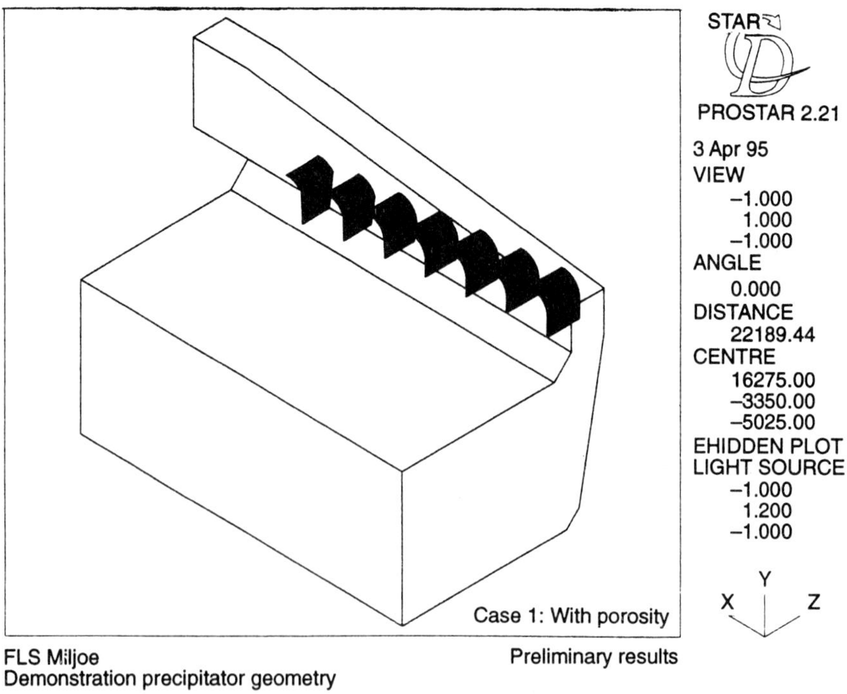

Figure 5.25 Modelling of guide vanes in the manifold above a precipitator inlet. Star-CD.

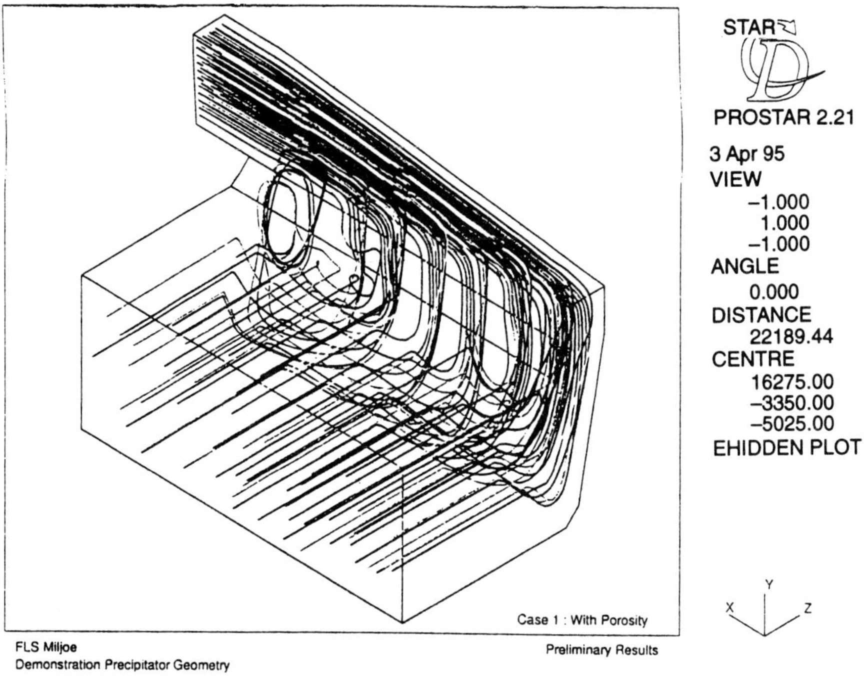

Figure 5.26 Streak lines of massless particles in manifold and inlet to a precipitator. Star-CD.

Suitable hardware work stations are IBM, HP, Digital, Vax, etc., whose prices depend on facilities. The system also requires a proper colour printer or plotter for graphical output. If the geometry is taken from a CAD system, it should be possible to transfer the geometry files to the mesh generator of the CFD system.

Software can be developed by the users themselves, using physical, mathematical and numerical backgrounds available from books. Most organizations, however, prefer buying a commercial package, because program development is extremely expensive and upgrading and introduction of new theories and facilities can hardly be maintained or justified by a small group of engineers occupied with other disciplines outside CFD analysis.

Commercial software packages are numerous and without favouring a particular code, names such as Flow3D, Fluent, Kameleon, Phoenics, Fidap, Star-CD and Viscous exist. Some systems can be supplied as user property, others can be hired on a yearly basis; some codes are open, i.e. there is access to the source code, others are 'closed' delivered as executable programs only. Upgrading, troubleshooting and hot line assistance are normally offered by the supplier.

Companies specializing in CFD offer their services, sometimes both CFD and model testing, and often institutes, e.g. aeronautical laboratories, technical universities and technological centres worldwide, perform calculations if required.

The advantages of numerical modelling are that (1) there is no need for violating the model laws, (2) the model will not have to be physically constructed and (3) the result is presented in a relatively short time. Yet, it must not be forgotten that the flair and know-how of the skillful fluid mechanics engineer cannot be replaced by a code, however fanciful it might be. While using the program for solving flow problems, the changing of the outer contour, changing design, position of guide vanes and screen porosity will give the operator the opportunity of gaining experience in a relatively short time. But practical experience and know-how can hardly be replaced by keyboard and screen alone. The solution from a CFD run must not be accepted uncritically: first of all because the solution to the difference equations is not necessarily the correct solution to the differential equations; secondly, because of possible limited applicability of the turbulence model and thirdly because of imprecise boundary conditions. Finally, the numerical solver has a limited precision and the calculation might not converge properly. Sometimes fluctuating residuals from the solution of the time-averaged equations indicate that there is no stationary solution to the problem. In order to gain confidence in a given CFD code, results should be field tested on existing installations.

5.7 Field testing

The full-scale demonstration of the velocity distribution, whether model tested, CFD, or not, can be performed at clients' or vendors' request before start-up. Clients' interest is confirmation of a proper distribution while vendors interest is in building-up and maintaining experience.

In most European designs of precipitators there is normally adequate access to the interspace between fields and before the first field and after the last field, so it is possible either to climb the emission frame with the velocity probe or to fit some sort of 'monkey' to the collecting plate edges to carry out the necessary velocity traverses.

In precipitators of the US type, access is normally more difficult and sometimes the probe is lowered into the precipitator from the top of the electrode system or from the roof.

As with model testing, hot-wire, hot-film or vane anemometers are used. In cases where the fan power is restricted to operating with cold air, the average velocity might be low. This speaks for using hot-wire or hot-film anemometers which have higher sensitivity at low velocity, but generally one should beware of low velocities as precision falls.

Two operators are needed inside the precipitator and at least one outside. In the duct a velocity probe, vane anemometer or pitot-static tube, is fitted connected to a recorder for verification of stable fan operation. This velocity signal can, in the case of fan or changing flow problems, to some extent be used for correcting the measured distribution.

Inside the precipitator sufficient light must be available and an intercom, for communication makes the job easier. Special inpection doors fitted with acrylic glass, because it is thus possible to find the way out in the event of a power failure, and to have radio communication with external personnel and the control room, adds to safety.

The time needed for traversing a cross-section is of the order of 1 h; however, large cross-sections, mapped by use of a climbing 'monkey', might need 2 h or more. To this, one must add the time for rigging up and dismantling. If time is scarce, simultaneous mapping in more than one cross-section is possible. If modifications are found necessary and the hardware has been prepared, e.g. different screen open areas and additional guide vanes, the measurement procedure after modification must be repeated until a satisfactory result is achieved.

In some cases it is necessary to return to the laboratory and resume the model testing or further numerical calculations in order to find a solution. This is a normal procedure if time in the field is restricted. Later on, if measurements have to be taken after the plant has been running, and as the precipitator will be dirty, special precautions should be taken to protect the personnel by using respiratory, eye and ear protection.

5.8 Dust build-up and wear

Internal gas distribution devices are subject to dust build-up and wear.

Guide vanes should be placed and designed correctly, so that build-up is minimized, and if the risk cannot be eliminated then rapping, vibration or acoustic horns for cleaning should be considered. This includes vanes on, or close to, screens. Horizontal vane surfaces should, in principle, be avoided; a better solution is oblique surfaces, if necessary double-angled, e.g. 45° upwards and 45° downwards. Furthermore, screen vanes should be integrated with the screen in order to avoid dust build-up on the vane, due to dust impacting on solid screen areas; such screens are often supplied with vibrators or hammers. Acoustic horns have also demonstrated their capability of keeping screens clean on many precipitators.

It must be emphasized that the designer should ensure that gas distribution demands are fulfilled, not only in the model or during field testing under air load, but also during normal operation.

Fly ash dust from cement clinker coolers and dust from some metallurgical processes consist normally of very sharp, abrasive grains, so if the gas velocity in the raw gas duct and at the entrance to the inlet transition piece is too high, there is a risk that wear of guide vanes and of the first distribution screen will occur. It must be remembered that coarse particles have much higher inertia than the gas; therefore, deceleration of the gas velocity right before an inlet flange does not necessarily reduce the velocity of the sharp particles. In some cases, wear is so severe that vanes and screens disintegrate so rapidly that extra plant outages, between regular overhauls, are needed in order to meet emission demands or in order to remove eroded pieces blocking the dust conveying system. If there are no alternative layout arrangements to reduce the risk of severe wear, then vanes and screens must be made out of thick or wear-resistant steel.

References

1. Deutsch, W. (1922) Bewegung und Ladung der Elektrizitätsträger im Zylinderkondensator. *Ann. Phys.* (*Leipzig*), **(4)68**, 335–44.
2. Cochet, R. (1961) Lois de charge des fines particules (submicroniques). Etudes théoriques — Contrôles récents spectre de particules. *Colloque International — La Physique des Forces Electrostatiques et Leurs Applications.* Centre National de la Recherche Scientifique, Paris.
3. McDonald, J.R., Smith W.B. and Spencer III, H.W. (1977) A mathematical model for calculating electrical conditions in wire-duct electrostatic precipitation devices. *J. Appl. Phys.*, **48(6)**, 2231–43.
4. Mayer-Schwinning, G. (1987) Walther Deutsch, a pioneer in electrostatic precipitation. *3rd ICESP*, Abano-Padova, Padova University, Italy, pp. 3–11.
5. Deutsch, W. (1931) *Ann. Phys.* (*Leipzig*), **5**, 251–63.
6. White, H.J. (1963) *Industrial Electrostatic Precipitation.* Addison-Wesley, MA, pp. 238–93.
7. Robinson, M. (1968) Turbulence in electrostatic precipitators... a review of the research literature. *Miner. Process.*, May, pp. 13–7.
8. Friedlander, S.K. (1959) Principles of gas–solid separation in dry systems. *Chem. Eng. Progr. Symp., Ser. 25*, **55**, 135–49.
9. Cooperman, P. (1965) Eddy diffusion and particle erosion in electrostatic precipitation. Toronto, APCA 58th Annual Meeting, Paper 65–132, APCA, Pittsburgh, USA.
10. Robinson, M. (1967) A modified Deutsch-equation for electrostatic precipitation. *Atmos. Environ.*, **1**, 193–204.
11. Crowe, C.T. and Stock, D.E. (1974) The effect of electrodynamic secondary flow on the performance of an electrostatic precipitator. *Heat Transfer Fluid Mech. Inst.*, 12–14 June 1974, 24th Proceedings, pp. 254–65.
12. Bernstein, S. and Crowe, C.T. (1979) Interaction between electrostatics and fluid dynamics in electrostatic precipitators. *2nd Symposium on the Transfer and Utilization of Particulate Control Technology*, July 23–27, Denver, Colorado, EPA 600/9-80-039b, Vol. III, pp. 125–45.
13. Gross, H. (1979) Zur wirkung der Turbulenz in Elektroabscheidern. *Staub* **39**, 197–202.
14. Leonard, G.L., Mitchner, M. and Self, S.A. (1980) Particle transport in electrostatic precipitators. *Atmos. Environ.*, **14**, 1289–99.
15. Leonard, G.L., Mitchner, M. and Self, S. (1981) Precipitation from turbulent flows. *1st ICESP*, Monterey, October, pp. 208–56, EPA, Pittsburgh, USA.
16. Yamamoto, Y., Nakamura, S. and Velkoff, H.R. (1980) Numerical study of secondary flow interaction in an electrostatic precipitator. *Innovative Numerical Analysis for Engineering Science*, University Press of Virginia, pp. 3–12.
17. Yamamoto, T. and Velkoff, H.R. (1981) Electrohydrodynamics in an electrostatic precipitator. *J. Fluid Mech.*, **108**, 1–18.

18. Thomsen, H.P., Larsen, P.S., Christensen, E.M. and Christiansen, J.V. (1982) Velocity and turbulence fields in a negative corona wire-plate precipitator. *4th Symposium on the Transfer and Utilization of Particulate Control Technology*, October 1982, Houston, Texas. (Vol. II, EPA-600/9-025b, Nov. 1984.)
19. Shaugnessy, E.J., Davidson, J.H. and Hay, J.C. (1984) The fluid dynamics of electrostatic precipitators: effects of electrode geometry. *5th Symposium on the Transfer and Utilization of Particulate Control Technology*, August 1984, Kansas City, Missouri, Vol. 2, paper No. 28 (EPRI CS 4404, Feb. 1986).
20. Larsen, P.S. and Sørensen, S.K. (1984) Effect of secondary flows and turbulence on electrostatic precipitator efficiency. *Atmos. Environ.*, **18(10)**, 1963–7.
21. Sørensen, J.N., Larsen, P.S. and Zamany, J. (1991) Experimental and numerical analysis for flows in negative corona precipitator. *8th Symposium on Turbulent Shear Flows*, Sept. 9–11, Techn. Univ. of Munich, Germany.
22. *Efficiency Measurements on a Laboratory Precipitator*. FLS miljø a/s, Internal Report in Danish.
23. Leonard, G.L., Mitchner, M. and Self, S. (1981) Precipitation from turbulent flows. *1st ICESP*, Monterey, October, EPA, Pittsburgh, USA.
24. Lind, L. and Bojsen, E. (1993) Experience with baffle-free collecting plates in an electrostatic precipitator. *10th EPRI Symposium on the Transfer and Utilization of Particulate Control Technology, and 5th ICESP*, Washington, April 3–8, pp. 36.1–14, EPRI TR 103048, Vol. 2.
25. Self, S.A., Kihm, K.D. and Mitchner, M. (1987) Precipitator performance improvement through turbulence control. *3rd ICESP*, Abano-Padova, Italy, pp. 443–80.
26. Self, S.A., Mitchner, M., Choi, D.-H., Kihm, K.-D. and Leach, R. (1986) Finite-diffusivity effects in single-stage precipitators–theory and experiments. *6th Symposium on the Transfer and Utilization of Particulate Control Technology*, February, New Orleans, Mississippi, EPRI CS 4918, Vol. 2, pp. 28.1–30.
27. Larsen, P.S., Christiansen, J.V. and Christensen, E.M. (1987) Secondary flows and turbulence in a pulsed, negative-corona, barbed-wire precipitator. *3rd ICESP*, Abano-Padova, Padova University, Italy, pp. 481–93.
28. Davidson, J.H. and McKinney, Peter J. (1989) EHD flow visualization in the wire-plate and barbed plate electrostatic precipitator. *Proc. IEEE*, pp. 2118–25.
29. Larsen, P.S. (1986) *Secondary Flows in Negative Corona Precipitator*. DCAMM Report No. 337, October, Technical University of Denmark.
30. Zamany, J. (1992) *Modeling of Particle Transport in Commercial Electrostatic Precipitators*. Ph.D. Thesis, ATV EF316, Copenhagen, September, ISBN 87-984457-0-7.
31. Dalmon, J. and Lowe, H.J. (1961) Experimental Investigations into the Performance of Electrostatic Precipitators for P.F. Power Stations. *Colloque International—La Physique des Forces Electrostatiques et Leurs Applications*. Centre National de la Recherche Scientifique, Paris.
32. Methods for producing uniform gas flow in processing equipment. *Br. Chem. Eng.*, July, 359–63.
33. *Gas Flow Model Studies* (1981) Publication No. EP-7, International Gas Cleaning Institute Inc., Revision 4, October or Institute of Clean Air Companies, Pub. No. EP-7, 1993.
34. Idel'chik, I.E. and Aleksandrov, V.P. (1974) The effect of the nonuniformity of the gas flow on the efficiency of electrostatic precipitators. *Teploenergetika*, **21(8)**, 60–2, 85–7.
35. Gooch, J.P., McDonald, J.R. and Oglesby Jr, S. (1975) *A Mathematical Model of Electrostatic Precipitation*. EPA-650/2-75-037, April.
36. Rosby, S.-O. (1974) *Precipitator Model Studies and Scaling-Up Experiences*. Fläkt, AB Svenska Fläktfabriken, Issue No. 2, Telub AB Teknikinformation, Växjö, 12 pp.
37. *Gas Distribution Influence on Precipitator Efficiency*. FLS miljø a/s, internal report in Danish.
38. Høegh Petersen, H. (1990) A precipitator sizing formula *4th ICESP,* Beijing, September.
39. H. Høegh Petersen and L. Lind, private communication, 1986.
40. Lind, L. (1986) Influence of gas distribution on precipitator performance. *6th Symposium on the Transfer and Utilization of Particulate Control Technology*, New Orleans, February, EPRI CS 4918, Nov. 1986, Vol. 2, pp. 33.1–15.
41. Hein, A.G. (1989) Dust reentrainment, gas distribution and electrostatic precipitator performance. *JAPCA*, **39(5)**.

42. Gosman, A.D., Pun, W.M., Runchal, A.K., Spalding, D.B. and Wolfshtein, M. (1969) *Heat and Mass Transfer in Recirculating Flows.* Academic Press, 338 pp.
43. Wolfhstein, M. (1969) *Convection Processes in Steady Two-Dimensional Separated Flows.* Imperial College of Science and Technology, Report EF/R/G/1, Thesis, January.
44. Patankar, S.V. (1980) *Numerical Heat Transfer and Fluid Flow.* Hemisher Publishing Corporation, New York.
45. Ettema, R. (1994) Electrostatic precipitator performance improvement through numerical simulation. World Cement, April, pp. 63–6.
46. Schwab, M.J. and Johnson, R.W. (1994) Numerical design method for improving gas distribution within electrostatic precipitators. *American Power Conference*, 56th Annual Meeting, Chicago, April 25–27, 7 pp.

6 The physical and chemical properties of particles and their effect on performance

K. PORLE AND K. R. PARKER

6.1 Particle size and shape

In the air pollution control industry, particles to be collected fall into one or more of the following general categories, i.e. dust, fume, ultrafine fume or mist.

Dust particles usually arise as a result of mechanical disintegration or communition of large lumps into smaller particles by grinding, crushing, etc. The particles are irregular in shape and their size refers to some average dimension and spans the range from above 200 μm down to about 1 μm (1 $\mu m = 1 \times 10^{-6}$ m). For inertial type collector technology (see chapter 1) the particle size is usually referred to the equivalent Stoke's diameter or free falling velocity. Any particles produced during communition having a size greater than 200 μm normally settle out so quickly that there is no real difficulty in their collection.

Pulverised fuel ash (PFA) is worth special mention, as worldwide, the largest application of electrostatic precipitators is for the collection of fly ash from coal fired utility type power generation facilities. The ash (PFA) carried over from the combustion zone comprises a complex mixture of material, predominantly silica/alumina based compounds, having a median particle size of around 15 μm. Using the above categories, this would be considered as 'dust', but during combustion, the coal particles are exposed to high temperatures e.g. 1200 °C and this results in some ash material being volatilised. This subsequently condenses in the cooler regions of the boiler to produce submicron particles, i.e. less than 1 μm, amounting to some 2% of the total ash. Hence PFA is a mixture of both dust and submicron size fume particles.

Fume is typically conferred to solid particles formed by either sublimation of solid phase material or condensation of a vapour phase condition, usually as a result of exposure to high temperature such as in combustion or smelting processes. As most of these particles have been, at one stage, in a liquid/molten phase, surface tension effects tend to produce spherical particles, usually in a small size band. A typical size range would be 0.1 to 1.0 μm. These particles do not have a measurable settling velocity, but exhibit strong Brownian motion.

Ultrafine fume is reserved for the smallest possible sizes, usually less than 0.1 μm; the particles are normally spherical in shape, e.g. tobacco smoke, or sometimes irregular, such as carbon black. With sizes less than 0.1 μm these particles are approaching molecular dimensions and exhibit very strong Brownian motion.

This Brownian motion is very necessary in the diffusion charging process for electrostatic precipitators, as discussed in chapter 3. Generally their spherical shape enables optical or laser-type measurements to be carried out to evaluate their size.

Mists or fogs are formed by the condensation of a vapour phase material on suitable nuclei to produce a suspension of small liquid droplets. Although atmospheric fogs can have fairly large droplet sizing, for the purpose of precipitation duties, only those mists having a particle size less than 1 μm will be considered. Being in a liquid phase, a mist particle is spherical and the smallest particles exhibit strong Brownian motion, which is essential to their effective charging.

In the pollution control field, particle suspensions are further characterised by their physical and chemical properties; for example, their size, structure, surface area, electrical resistivity, chemical reactivity and composition, cohesivity, ability to absorb electric charge and propensity for erosion, are some of the properties which have an effect on electrostatic precipitator design and/or performance. Those characteristics having the greatest effect on precipitation will be described in some detail, particularly as to how they impact on performance.

6.1.1 Particle sizing

A precipitator would see at its inlet a dispersoid of particulates being carried by a gas stream. This stream behaves like a fluid, splashes like a liquid, supports wave motion and can be pumped through conduits. One of the critical aspects of the dispersoid make-up is that of particle size. Particles of dust are generally made up of a myriad of different shapes and sizes and, unlike fume, are rarely spherical in shape, although some combustion processes can produce cenospheres, which are hollow and result from the outgassing of small pieces of carbon suspended in molten ash.

One of the difficulties facing the air pollution control engineer is to define what is meant by particle size and, then, how the size will impact on a specific method of collection. If the particle is truly spherical the size measurement is simply the diameter, but if the shape is irregular, then the definition and measurement becomes more complex.

The majority of dusts met in practice consist of a whole spectrum of sizes, rather than a single size, as are sometimes laboratory produced for a specific mono-dispersed aerosol. The spectrum comprises a continuous distribution of sizes, usually on a log/normal form of distribution. This enables the sizing

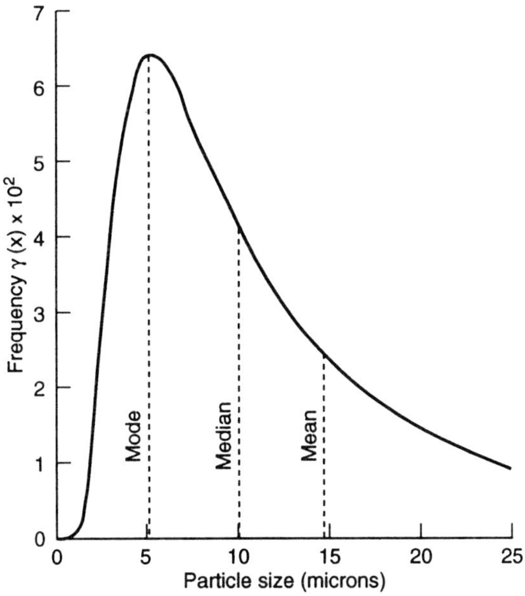

Figure 6.1 Typical particle size distribution of fly ash.

to be treated on a statistical analytical basis; for this, the size frequency distribution function $y_0(x)$ and the cumulative distribution function $y(x)$ are:

$$\gamma_0(x)\,\mathrm{d}x = \text{the proportion of particles in the size interval } x \text{ to } x + \mathrm{d}x.$$

and

$$y(x) = \int_0^x \gamma(x)\,\mathrm{d}x = \text{the proportion of particles in the size range 0 to } x.$$

A typical example of a frequency distribution function is illustrated in Figure 6.1. In this, three useful parameters are indicated, i.e. the mean, the median and mode points.

The mean particle size is defined as the size where 50% of the mass lies above and below that size, usually referred to as the d_{50} size.

The mode point or size is that with the highest frequency or concentration. This is useful in the case of electrostatic precipitation, where one considers whether or not corona suppression or space charge effects from submicron particles are likely to be significant.

The median size is where 50% of the number of particles lie above and below that value. This is important where one has to consider the fraction of particles below a certain value, such as in the proposed US PM10 regulations (PM10 particles $< 10\,\mu$m).

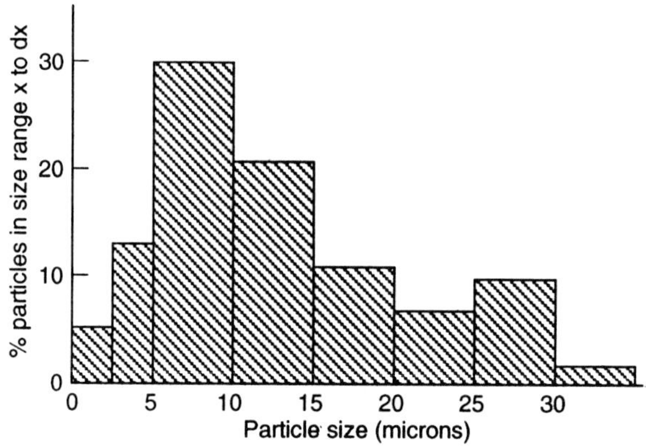

Figure 6.2 Typical particle size histogram.

Methods of determining the particle size of a particular dust do not produce a continuous distribution curve as indicated in Figure 6.1. Most evaluate the number or mass of particles within a specific size band or range, for example in a sedimentation analysis. This produces a histogram-type curve as shown in Figure 6.2. The mid-points of each specific size range are then taken and a cumulative curve is produced as in Figure 6.3. If this is plotted as a log probability curve, a straight line is usually derived (Figure 6.4); this however can deviate somewhat at both upper and lower ends. This deviation is not altogether surprising, since in practice a single large particle

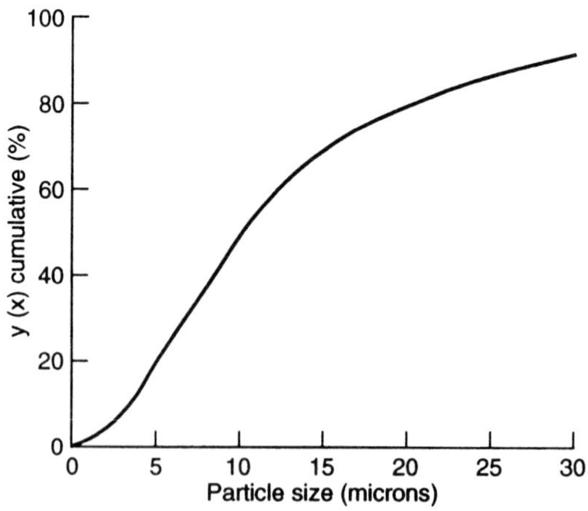

Figure 6.3 Cumulative particle sizing.

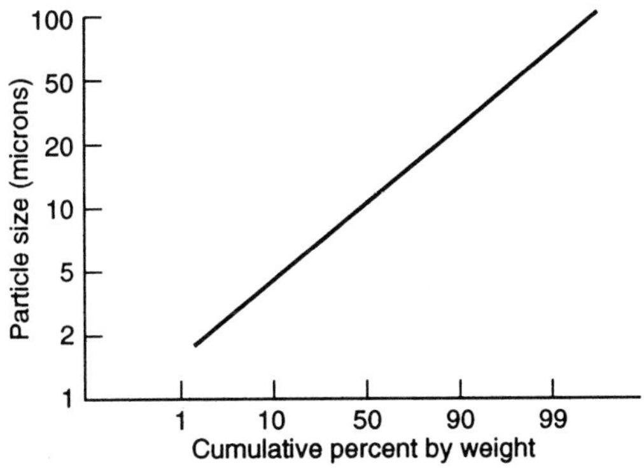

Figure 6.4 Log-probability curve for particle distribution.

has a big impact on the mass distribution at the coarse end and fine particles may arise not from communition of the feed material, but because of volatilisation/condensation, and so affect the fine end.

One of the problems with many analytical sizing methods is that of obtaining a truly representative sample for analysis, i.e. as the precipitator or downstream device really sees the dispersoid. Even if an isokinetic sample is obtained from sampling, most analytical methods redisperse the sample, so the sizing could measure agglomerates arising from the sampling which are not broken during redispersion, or true agglomerates in the flue which are. Materials which are volatilised and recondense in the system can be found adhering to larger particles upon which they have impacted or used as condensation nuclei; hence the question, does one consider the agglomerate mass as a whole or a big particle plus a number of separate small ones?. Figures 6.5 and 6.6 [1] show both a large particle with small fume-size particles attached and an agglomerate of ultrafine particles. For mass collection, the weight of the small fine particles has no significant effect on the overall efficiency, but, where a specific efficiency for a particular material or size is required, the need to know the true distribution is important.

Recent investigations [1, 2] carried out on high efficiency power plant precipitators, 99.85% +, using the latest sizing techniques, e.g. Berner low pressure impactors, differential mobility analysers (DMAs) and optical particle counters, have indicated that PFA, although predominantly monomodal at around 8 μm, has a smaller but distinct secondary mode at around 0.2 μm. This second mode is almost coincident with the minimum efficiency (maximum penetration) in the ESP fractional efficiency curve (see Figure 3.34 in chapter 3). Where specific emission requirements are to be met it is important to be aware of this bimodal distribution.

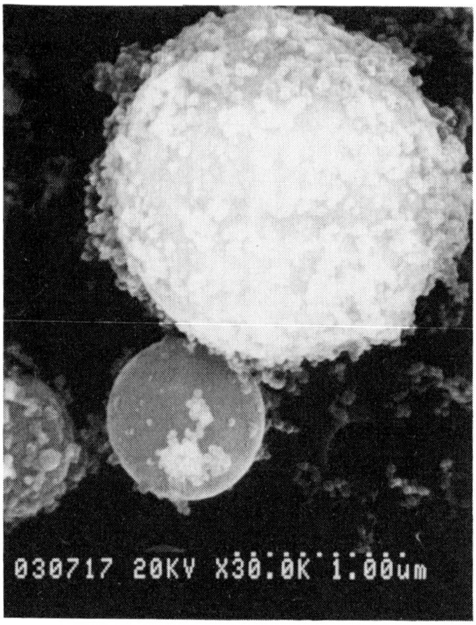

Figure 6.5 SEM micrograph of fly ash particles from the combustion of South African coal. Photograph of large particle with smaller attachments. Magnification 15 000 ×.

Another interesting fact arising during this work was that, with the lower combustion temperatures associated with fluid bed combustion (FBC), the number of particles below 0.1 μm was an order of magnitude lower than from conventional pulverised fuel (PC) fired units.

There are many different methods of measuring a parameter which can be assigned as a particle size. This parameter is defined in various ways, depending on the method or apparatus used; generally for irregularly shaped particles, the size is normally defined in terms of an equivalent diameter, which will depend on the physical or geometric properties of the particle.

The following lists the main types of equipment available and the principle associated with each method.

Sieving. This depends on passing the particles through a range of defined size apertures, usually in the form of a mesh, in descending stacked order. The particle retained by a specific mesh is normally allocated the mesh size, i.e. sizing is based on a linear dimension. The method is only acceptable for particles in excess of about 50 μm because of agglomeration difficulties. Some systems use wet sieving, which extends the range slightly downwards.

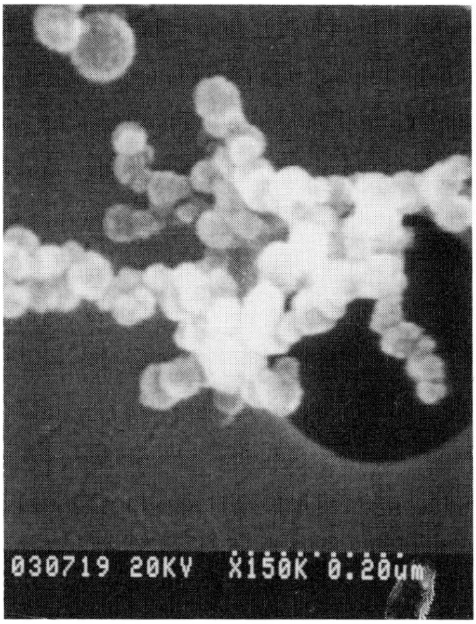

Figure 6.6 High-resolution SEM micrograph of fly ash particles from the combustion of South African coal. Photograph of agglomerated small particles. Magnification 15 000 ×.

Inertial systems. These depend on suspending the particles in a gas stream and modifying the gas flow such that the particle's momentum allows it to leave the stream and be captured. The parameter in this case is the equivalent Stokes' diameter, i.e. the diameter of a sphere having the same density and settling velocity as the particle.

Liquid sedimentation systems. These produce a measure of the Stokes' diameter, but instead of a moving air stream, a liquid column is used through which the particles fall. At specific times, samples are taken and the particulate concentration measured and back-calculated as the equivalent free falling diameter.

Volume measurement. In this approach, the particles are suspended in an electrolyte and passed through a small electrolytic cell. The change in cell current is then converted to a resistance effect and assuming only one particle is present in the cell, its volume can be calculated.

Area measurement. The particles suspended in a gas stream are irradiated by a laser beam and the change in beam intensity measured as a result of absorption/scattering by the particle.

Diameter measurement. For the smaller particles, such as those classified as fume, the electron microscope enables the diameter to be calculated. As these particles are usually spherical the equivalent size is that as measured. Larger irregular particles, using an optical microscope, are more difficult to assess, since they could be flat platelets.

From the foregoing, it is obvious that there is no single universal method of sizing and each differs in the parameter measured. For general gas cleaning duties, the determination of the equivalent Stoke's diameter offers advantages but, except for the cascade impactor, most methods rely on redispersing collected particulate samples, which because of agglomeration may introduce errors.

6.1.2 Particle shape and structure

While the above sizing methods are imporant in determining one of the particles' physical properties, they do not give a complete picture as far as electrostatic precipitation is concerned. Another parameter, the shape of the individual particles, needs to be assessed, since very few dusts are truly spherical and this means that their shape can play an important role in the overall precipitator efficiency.

Particles arising during the incineration of paper tend to retain their original platelet form, so have a large surface area in two planes but are very thin in the third. Another form for concern is that of partially combusted coal, which produces a voided coke-like particle having a large surface area but low mass. Both paper char and coke particles, comprised mainly of carbon, have very low electrical resistivities, which, as detailed in section 6.4, can be readily re-entrained by the gas flow after reaching the collectors.

The precipitation engineer is also concerned with processes giving rise to particles which are small in cross-section but are very long, e.g. hair or rod form. These particles tend to align themselves with the field lines and can join or 'chain-up' causing electrical breakdown.

In the case of fume and mists, because they usually arise through sublimation, volatilisation and condensation processes, surface tension effects during their transition phase tend to give rise to spherical particles usually 1 μm or less in diameter. Elutriation or sieving procedures are not practical in this size range and equivalent volume, area or actual measurement of diameter is normally carried out, typically giving results on a frequency basis before conversion to mass. If actual mass distribution is required, the cascade impactor, used as an *in-situ* device, will produce a true sizing and will determine any naturally agglomerated particles as an equivalent diameter sphere.

Although particles of carbon black have been found to 'chain-up', the major problem facing the precipitation engineer from small submicron particles is one of potential space charge and corona suppression effects on the electrical operation of the unit.

6.2 Optical properties

The optical properties of aerosols are of great importance to the air-pollution control engineer, because the degree of pollution is often judged by the appearance of the stack discharge. The colour of emitted smoke or particulates is determined for the smaller particles by their shape, size, and refractive index and for the larger particles, by their colour and surface area. Raleigh [3] carried out theoretical calculations on the scattering of light by spherical particles having diameters of 0.1 μm and less and proposed that light scattering is proportional to the sixth power of particle diameter and inversely proportional to the light's wavelength. This is mainly of interest as it explains the redness of the morning and night sky and the intense blue of outer space.

For particles greater than 0.1 μm Raleigh's Law does not hold and changes to a second power law of diameter. In the case of fresh tobacco smoke, the particles of tar are around 0.25 μm; this is below the wavelength of normal light, 0.4 to 0.8 μm, so the scattered light appears blue since this is the wavelength scattered most. When the smoke has been exhaled the particle size is much larger, because of condensed water vapour, and the smoke appears white, as the light is scattered more uniformly.

For even larger particles, scattering is mainly by reflection, i.e. in proportion to their surface area and is the basis of optical extinction type meters. The physics of the extinction meter is based on Lambert's Law, which states that, if a beam of light intensity I_0 is passed through a column of particles of length x, then the amount of light reaching the far side will diminish exponentially depending on the properties of the aerosol. This can be expressed mathematically as:

$$I = I_0 \cdot e^{-kx}$$

where k is a function of particle size and mass concentration, i.e. total surface area, or

$$I = I_0 \cdot e^{-Smx/4}$$

where S is the specific surface area of the particles (m^2/g), m is the mass concentration (g/Nm^3) and x is the path length (m).

In practice, the emission from a specific source, after a precipitator for instance, tends, even if the mass emission should vary, to maintain a roughly similar particle sizing and hence specific surface area. This means that any change in the opacity for a given path length is, as a first-order approximation, directly proportional to the mass concentration.

The following example illustrates how a shift of 20% in the legislative emission limit for a power station (utility plant), affects the opacity for a 10 m path length. If we assume the dust concentration to be 50 mg/Nm^3 and the dust has a specific surface area of 9000 cm^2/g, then from Lamberts' Law we have:

Emission. (mg/Nm^3)	$Smx/4$.	Transmission factor (I/I_0)
50	0.1125	0.894
40	0.09	0.914
60	0.135	0.874

For comparison, normal window glass reduces the amount of transmitted light by around 8%, so in the table, all conditions would give 'fairly clean' chimney conditions.

The above calculations indicate that, for a 20% shift in emission, the intensity of light received on the far side of the column changes by 2%, and although small, can be used for control purposes. In chapter 8, covering the electrics, this approach is in fact detailed for both precipitator performance optimisation and power saving by means of a feedback loop between the opacity meter and TR electrical control system.

6.3 Agglomeration

Particle agglomeration is attributable in the main to the collision and impaction of the smaller finer particles. These are in continuous motion, as the result of bombardment of the particles by gas molecules, to exhibit what is termed Brownian motion. The effect of the bombardment is maximised for the finer particles and, as they agglomerate, Brownian motion decreases, so the possibility of further collision is reduced.

In precipitation, the effect of agglomeration is to shift the particle sizing upwards; this will impact on their free-falling velocity as per the Stokes–Cunningham relationship and may reduce the possibility of space charge and corona suppression effects.

Another aspect, which is gaining prominence in the air-pollution field, is that of the need to efficiently collect toxic and heavy metals; these materials in most processes are very small in terms of mass concentration, but recent toxicology investigations have indicated their importance to health.

Heavy metals usually exist in the downstream area of the process as condensed submicron particles after passing through a volatilization stage in their life cycle. During condensation, some of the particles use larger particles as condensation nuclei, so it is not unusual to find small metal-rich fume adhering to larger inert particles, as indicated in Figure 6.5. This means that these fine heavy metal particles are removed along with the easier to collect coarser particles. This would help explain why solid phase toxic and heavy metal mass balance measurements, in spite of their

smaller particle size and potentially lower performance, compared with the coarser particles, produce almost the same order of efficiency as the bulk materials [4].

The effects of agglomeration and cohesion, although producing particle flocculation, are the result of different mechanisms and should not be confused. Agglomeration, as stated above, arises wholly in the gas phase because of Brownian motion and, except for impaction on some larger particles, predominantly applies to small submicron particles forming larger but still small units (see Figure 6.6). Cohesion, on the other hand, applies to the collected precipitated dust, as will be covered in section 6.4 below. Cohesion, is a measure of the binding mechanisms holding together all size particles, as a result of electrical and mechanical forces acting on the layer rather than individual particles.

6.4 Cohesivity

In electrostatic precipitation, particle cohesion plays an important role in the plant's performance, firstly in respect of how the particles are held to the collector and secondly how the particles hold together in their transference to the hopper, following rapping.

Particles after reaching the collector plate are held initially by electric forces and after losing their charge would, without the mechanical binding force, be readily released and possibly re-entrained by the gas stream. From Stokes' Law, with a precipitator gas velocity of 1.5 m/s, only bound particles greater than 500 μm equivalent diameter stand any chance of reaching the hopper. Without this particle cohesion the dust would be subject to massive re-entrainment, which would have a deleterious impact on efficiency.

The forces holding the particles to the collectors and themselves are a combination of electrical and mechanical Van-Der-Waal forces. Both are linked to the surface chemistry of the particle, the nature of which depends not only on the composition of the dust but also on the gas constituents to which the particle has been exposed. For particles having electrical resistivities above 10^{11} Ω-cm. the electrical holding force tends to predominate over the mechanical force, but for low resistivity dusts the mechanical force becomes more dominant, depending on the dust layer properties.

In general, light fluffy dusts adhere poorly, while dense or sticky dusts adhere well. Particles such as carbon and ionic salts, which can form 'snow flakes', have low packing densities and poor cohesion, so the particles are only loosely held together. An exception is where the ionic salt is deliquescent/hygroscopic and absorbs moisture from the flue gas which cements or binds the particles together. If the temperature should approach the gas dew-point the dust becomes sticky and may be difficult to remove from the internal components of the precipitator.

The importance of cohesivity and precipitator performance has been discussed in general terms by a number of investigators; for example, Lowe and Lucas [5] calculated the forces to which a dust particle would be subjected on the collector; Penney and Klinger [6] measured the cohesion of an electrostatically precipitated dust layer after the power had been switched off and compared the result with that of a mechanically formed layer with fairly good agreement.

Dalmon and Tidy [7], determined the relative importance of cohesion of both mechanical and electrical origin and then related the cohesive properties to precipitator performance. This investigation was carried out with two very different power station fly ashes, one relatively fine having high resistivity and another coarse containing 33% carbon and of low resistivity. Both dusts gave rise to poor precipitator efficiency.

The collected dust samples were initially washed to remove all surface conditioning effects/agents and injected into an oil-fired combustor rig fitted with a 250 mm hexagonal tube precipitator. Precipitator efficiency measurements were made with the dust being self-conditioned by the flue gas and then with various conditioning agents, known to reduce resistivity, injected into the rig, upstream of the precipitator with about a half-second exposure-time.

At the end of each run, a detachable section of the collector was removed and the bulk density and the minimum compressive load, i.e. the load to just observe further compression of the dust precipitated on the sheet, were determined. Dust from the collector plates at the end of each run was placed in a powder tensiometer [8] and the ultimate tensile strength (UTS) of the medium was determined for each sample over a range of bulk densities. (Although the cohesivity of the powder sample was not directly measured, Farley and Valentin [9] have shown that for powder beds the UTS is directly proportional to the cohesivity.)

The results from these measurements showed that each dust had a distinct but separate UTS vs. compressive load relationship, which was independent of conditioning agent, and was basically a function of particle size, shape and hardness. The force required to fracture the bed is proportional to the product of the mean strength of the individual forces and their number density in the fracture surface. For a specific ash, the tensile strength at a fixed bulk density can, therefore, be used as a measure of the particle/particle force, or cohesion, and thus indicates the relative ease of re-entrainment.

The precipitator performance, when treating the variously conditioned high resistivity fly ash, showed little dependence on the UTS value measured at the precipitated bulk density or the UTS at constant bulk density. There was, however, as expected, a distinct effect of resistivity on collection efficiency due to the conditioning agents and hence the conclusion, for high

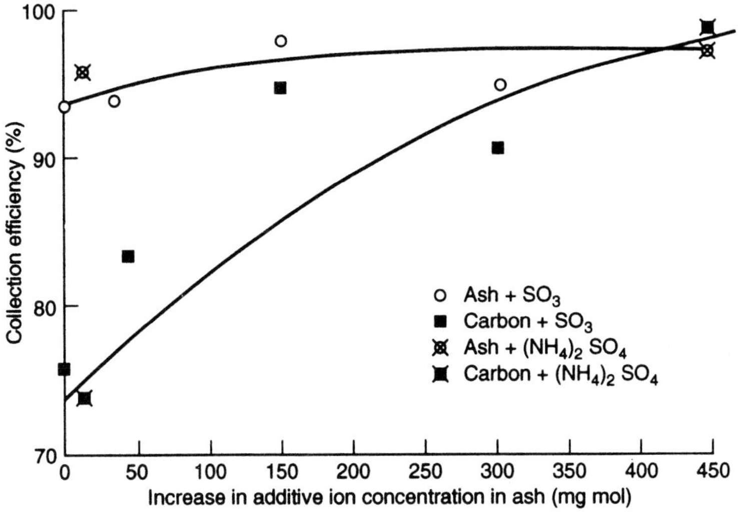

Figure 6.7 Effect of additives on precipitator efficiency for ash and carbon [7]. ○, ash + SO_3; ■, carbon + SO_3; ⊗, ash + $(NH_4)_2SO_4$; ⊠, carbon + $(NH_4)_2SO_4$.

resistivity dusts, that the improvement in performance is primarily by reduction in electrical resistivity.

With the coarser low resistivity ash (10^7 Ω-cm), the conditioning agents were not expected to produce any further reduction in resistivity; however, the addition of conditioning agents was found to significantly improve the performance. The collection efficiencies of the fly ash and carbon were separately evaluated for a range of injection rates for SO_3 and ammonium sulphate; these are reproduced in Figure 6.7. This clearly demonstrates that it is the carbon efficiency which is improved by the conditioning agent by preventing re-entrainment. When the UTS was measured at constant bulk density, the emv vs. UTS relationship was found to be linear (Figure 6.8), with the higher emvs being associated with the higher UTS. This increase in the individual particle/particle force is responsible for the improved collection efficiency of the carbon and the reduced re-entrainment potential.

Dalmon and Tidy [7] concluded that tensile strength measurements made on mechanically formed beds of highly resistive dusts gave results which closely align with those measured on a precipitated bed after the field was removed. Cohesivity will be present in any deposited layer and will be augmented by an equal force due to the electrical field when the ash resistivity is high. As the ash resistivity reduces, the electric field effect reduces and may even become repulsive, so particles are readily re-entrainable. Increasing the tensile strength of low resistivity particles by the

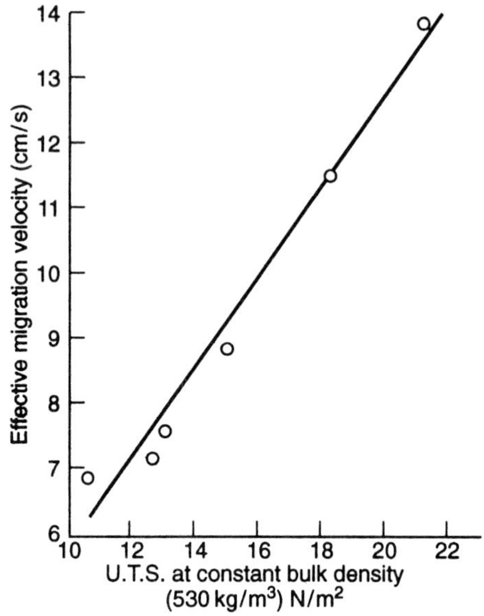

Figure 6.8 Relationship between effective migration velocity (EMV) and UTS for variously treated fly ash samples [7].

injection of conditioning agents produced increased precipitation efficiency, mainly from the capture of very low resistivity carbon particles, as a direct result of the higher cohesive forces prevailing, leading to reduced re-entrainment.

6.5 Particle electrical resistivity

Perhaps as regards precipitator engineering, the electrical resistivity of the particles is of paramount importance to performance and hence the size of the precipitator itself. As most particulates from power stations are complex compounds of basically silica and alumina, both having excellent insulating properties, one would anticipate resistivities in the 10^{14} Ω-cm and higher range as a bulk resistivity. The electrical resistivity of the particles can be modified, however, by the presence of impurities in the ash or where the surface of the particle is naturally conditioned by components present or added to the entraining gases. The nett resistance of a surface conditioned particle, even for a thin monatomic conductive coating, is significantly reduced. (Resistances in parallel.)

The electrical resistivity is determined by two mechanisms, the bulk volume conductivity, which is a function of the particle matrix constituents, and surface conduction. The latter is governed by the adsorbed surface layer, which is related to the surface reactivity and gas components.

The measurement of resistivity, because of this surface conditioning effect, makes laboratory determination difficult. There are, however, standard techniques and apparatus [10], where the sample is placed in a cell and the surrounding environment simulates the original gas condition in terms of temperature, water dew-point and, if necessary, acid gas concentration. To obtain comparative laboratory data, the accepted procedure uses a standard cell configuration and the dust is compacted to a set pressure. Once the apparatus has reached the correct 'equilibrium gas' condition, the current passing through the cell is determined for a range of applied voltages. In practice, resistivity evaluations are made for both increasing and decreasing temperatures to enable the peak and general resistivity profile to be found.

Laboratory measurements tend only to be comparative, since the packing density will not necessarily be the same as on the collector plates and the sample itself may have aged and so affected its surface reactivity. In addition, measurements by Goard and Potter [11], have shown that the resistivity, not unexpectedly because of the ionic nature of the conduction mechanism, is strongly field dependent.

For actual *in-situ* field determinations the US Southern Research Institute (SoRI) developed a point/plane apparatus [12], which is inserted into the flue. Dust is electrostatically precipitated onto the plate by energising the point electrode negatively, with respect to the plane or plate electrode. After a certain time, depending on the dust concentration, a circular plate attached to the point is carefully lowered onto the dust surface and the current, for a preset range of voltages, is measured. From this, and with known cell dimensions, the resistivity of the precipitated dust can be evaluated.

An alternative form of apparatus for both *in-situ* and laboratory measurements is a cell mounted in the base of a small 25 mm diameter sampling cyclone [13]. Gas is drawn through the cyclone, which collects all but the smaller submicron particles, and after rapping the cell to produce a constant packing density, the current through the cell is measured in the usual way, thereby enabling the resistivity to be calculated.

With either type of *in-situ* apparatus, while the environmental conditions are reproduced exactly, the actual ash sample is from a single point and may not be representative of all the ash and there is the possibility that the fines escape the apparatus completely.

A typical laboratory derived temperature/resistivity curve for a power station fly ash, is illustrated in Figure 6.9. This shows that at temperatures higher than 200 °C, where the surface conditioning effects have been destroyed, the resistivity follows classical theory of volume conduction, as a result of increased thermal motion of the molecules, producing a linear relationship with the inverse of absolute temperature. Below 200 °C, the

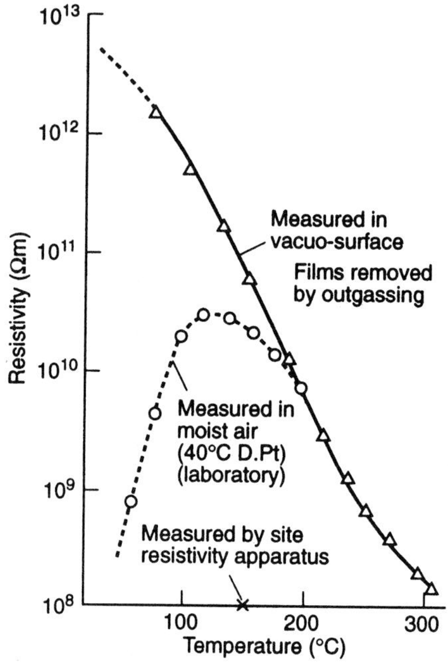

Figure 6.9 Effect of flue dust resistivity of surface film condition. Willington power station flue dust.

curve shows a decreasing resistivity as the surface conditioning, due in this case to moisture, becomes more significant as the temperature falls, because of the increasing vapour pressure. The result of vacuum outgassing the sample produces a continuation of the classical theory line confirming that the decreasing resistivity below 200 °C is the result of surface conditioning. Figure 6.10 has been developed for a dry process cement kiln dust which illustrates the effect of changing gas moisture content on particulate resistivity. From this, one can appreciate the need for water conditioning of this type of dust in order to avoid the problems of reverse ionisation and to limit the size of the precipitator.

While many engineers consider the electrical resistivity of the particulates to be all important, the problems of accurately determining the true value, as indicated above, really means that resistivity is only one of the tools used in the precipitation industry and one must not overemphasise its importance in the design and sizing of precipitators. The effect of resistivity on precipitator operation and performance can be summarised as follows.

For particles having resistivity in the range 10^{10} to 10^{11} Ω-cm the particle charging and discharging regime, when the particle arrives at the collector, proceeds normally and so has minimum impact on performance.

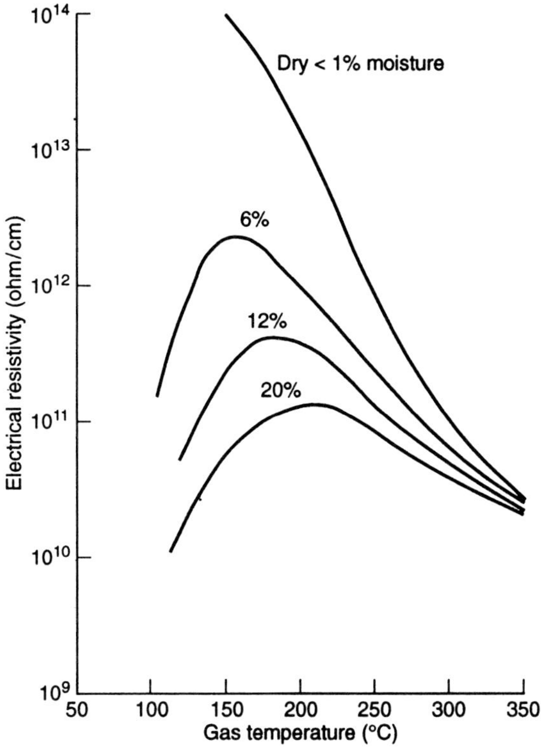

Figure 6.10 Effect of moisture content of entraining gas and temperature on cement dust resistivity.

As the resistivity increases, although the charging occurs normally, the particle on arriving at the collector only slowly loses its charge and a voltage begins to develop across the dust layer according to Ohm's Law. Dependent on the resistivity and layer thickness the voltage, for resistivities in the $10^{13}\,\Omega$-cm range, can reach a point where positive ions begin to be emitted from the surface of the dust. These positive ions cross the inter-electrode space and collide with and neutralise negative ions and charged particles, which significantly reduces precipitator performance. This condition is termed reverse ionisation, or back-corona, and the operating electrics exhibit a much reduced average voltage but an extremely high current. The average indicated voltage actually falls with increasing current, which is the opposite to a 'normal dust' operating condition.

Recent investigations of the applied voltage and current (AVC) waveforms have shown that, under reverse ionisation conditions, the peak applied voltage tends to be maintained, but with the high current flow, the ripple voltage significantly increases and the main feature is that the minimum ripple voltage level dramatically falls. To control/minimise the deleterious effects, modern AVC units monitor this minimum voltage condition and take appropriate corrective action to prevent the positive current 'runaway' (see chapter 8).

For slightly lower resistivities, or dusts having poor packing density, although a voltage develops across the dust layer, interstitial breakdown occurs through the layer and gives rise to a 'streamer' which results in field breakdown. The symptoms of this condition are a slightly reduced but normal average operating voltage, but a very low discharge current, any attempt to raise the current immediately resulting in increased flashover. Examination of the voltage waveform shows a minimum ripple voltage and any attempt to increase the current raises the threshold voltage resulting in breakdown. While this condition does not have such an impact on performance, as does the classical reverse ionisation or back-corona scenario, it nevertheless causes an efficiency limitation.

At the lower end of the resistivity range, for very conductive particles, such as metallics or carbon, the charge on arriving at the collector is lost so quickly that the particle sits on the surface as a neutral particle and can be entrained by the gas stream. While there are no specific characteristics identifying this phenomena, the potential re-entrainment results in poor collecting efficiency. Examination of the outlet dust samples shows an increasing quantity of these conductive particles (see section 6.6). The extent of the re-entrainment depends on a number of factors, such as particle cohesion, gas stream velocity and turbulence and the electrostatic forces acting on the particle.

Durham [14] working with low resistivity spray drier particles shows that, by theoretically examining the conditions existing at the boundary of the dust layer (Figure 6.11), it is possible for a repulsive electric force to develop which expels the particles back into the gas stream. From Figure 6.11 an electric field E_g exists at the surface of the layer as a result of the corona electrode voltage, which also creates the ion density N_i at the surface. The resultant current density J_g is in the positive x direction, although the ions move toward the collector.

We consider:

$$J_g = N_i \times e \times B_i \times E_g \text{ A/m}^2$$

where N_i is the ion density, e the charge on an electron and b_i the ion mobility.

Assuming homogeneous and steady-state conditions we have:

$$J_g = J_1 = J_p \text{ A/m}^2$$

where J_1 is the current density through the dust and J_p is the specific

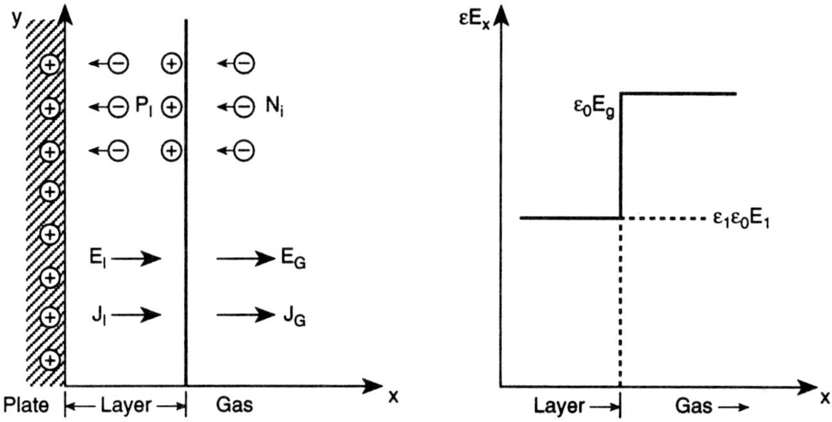

Figure 6.11 Boundary conditions for a layer of particles on the collector.

collector current. So

$$E_1 = J_1 \times \rho_1 \text{ V/m}$$

where E_1 is the electric field within the dust layer and ρ_1 the dust resistivity (Ω m).

The surface charge density is given by:

$$\sigma = \varepsilon_0(E_g - \varepsilon_1 \times \rho_1 \times J_p) \text{ c/m}$$

where ε_0 is the dielectric for free space and ε_1 the dust layer dielectric. From this, the surface charge density can be positive or negative, depending on the resistivity and layer conditions. The force acting on the surface charge density, f_x, is given by:

$$f_x = (E_g + E_1)/2 \text{ N/m}^2$$

If the surface charge density is positive, the force tends to pull the particles from the surface, and if negative, the force holds the particles to the surface.

Figure 6.12 has been taken from [14] showing the effect of these forces for a current density of 0.06 mA/m^2 and for different resistivities. For the higher resistivities, the attractive force holding the particles to the plate becomes predominant, whereas for the low resistivities there is an apparent electric repulsive force promoting re-entrainment.

While theory would suggest that low resistivity dusts, such as those arising from spray driers, should be subject to electrical re-entrainment, there are a number of full-scale plants fitted with precipitators, following spray drier desulphurisers, where re-entrainment does not appear to be a problem, [15]. This does not necessarily mean that the proposed theory is

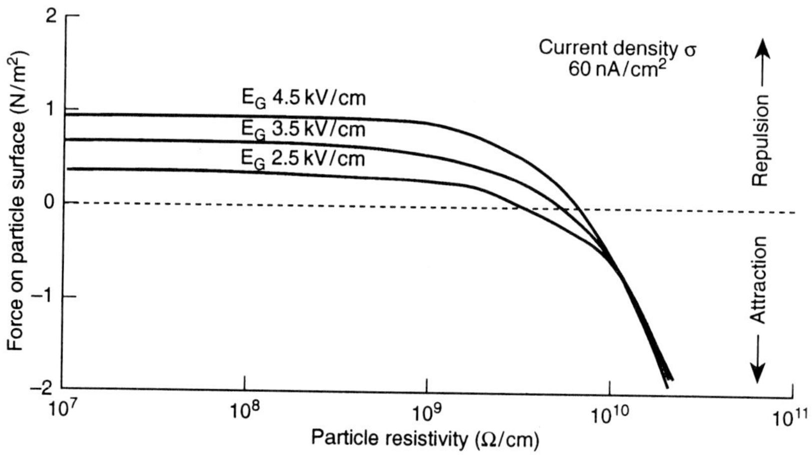

Figure 6.12 Force acting on particle surface for different electrical resistivities and field strengths [14].

incorrect but, as the properties of the particulates are controlled by a large number of factors, it is likely that a slight difference in the approach to dew-point temperature could make the dust more adhesive in spite of the lower resistivity.

Fortunately for the precipitation engineer, knowledge of the process or previous experience enables the correct choice/size of unit to be made to give a certain efficiency. If, however, an error of judgement occurs then there are various means of overcoming the worst effects, either by modifying the surface condition, by the injection of various chemicals, as described more fully in chapter 15, or by applying modern electrical energisation techniques, as covered in chapter 8.

6.6 Chemical composition and reactivity

Virtually all dusts met industrially consist of a multitude of chemical compounds in differing particle sizes, very few having identical composition because of the presence of impurities, either in the raw feed material or the fuel used in processing.

In the case of steam raising plants, the ash at the precipitator consists mainly of the fuel's ash components, mainly silica and alumina, which have been exposed to high temperature in an oxidising atmosphere. Some of the latest DeNOX technologies, however, employ an initial reducing atmosphere and a somewhat lower flame temperature which can affect ash chemistry. There is invariably a higher carbon carryover which has to be considered as it will effect precipitator performance.

In the power industry it is fairly usual to consider carbon carryover as approximately equal to the loss on ignition at 800 °C. In other industries the loss on ignition can be the result of the loss of water, the release of carbon dioxide from carbonates, the evaporation of alkali salts and some metallic components, so for industries outside power, the carbon needs to be determined in some other way.

At the precipitator inlet, the ash chemistry and reactivity is important in that it determines the particle resistivity. Raw coal and ash analyses in fact provide the precipitator design engineer with invaluable information on the probable fly ash resistivity and hence the potential size of precipitator for a specific efficiency. Chapter 15 provides a fuller explanation of Bickelhaupt's relationship [16] between coal and ash chemistry on fly ash resistivity.

In general, for steam raising units, the coal and ash components which produce high resistivity are silica and alumina, both excellent insulators in their own right, whereas those which reduce resistivity in the bulk media are coal sulphur and ash sodium.

Although during combustion most of the coal's pyritic sulphur forms sulphur dioxide, a small percentage converts to sulphur trioxide which is the important substance in significantly effecting resistivity. In a boiler flue gas there is always moisture present, either from the moisture or converted hydrogen in the coal or from atmospheric moisture; the gaseous SO_3 with this moisture reacts to produce condensed phase sulphuric acid. This then uses the surface of the particles as condensation nuclei, thereby reducing their resistivity, as conduction is now able to proceed through this absorbed or adsorbed layer of acid. The effect of surface conditioning is dependent on the amount of sulphur trioxide present, the gas temperature and hence vapour pressure as indicated in Figure 6.13 [17]. For coals containing in excess of 1.5% sulphur, there is sufficient natural surface conditioning at normal operating temperatures of 130 °C to give acceptable values of resistivity for effective precipitation.

Although the foregoing is generally true, there are incidents where the fuel, particularly the hydrocarbon content, is not totally combusted, but is volatilised off to coat the particles with an insulating-type material which can promote severe reverse ionisation-type operation. The phenomena/occurrence is probably more prevalent during plant start-up, when the furnace is cold and the oil start-up burners are not ideally set up, but can arise at other times, so one must be aware of this possible situation.

The presence of sodium in the ash behaves very differently, where, instead of surface conditioning, the sodium ions act as charge carriers, so high resistivity effects are negated to a certain extent. Measurements on a Japanese installation firing Australian low sulphur (0.5%) fuels, and therefore expected to give 'difficult' precipitation, showed that the precipitator

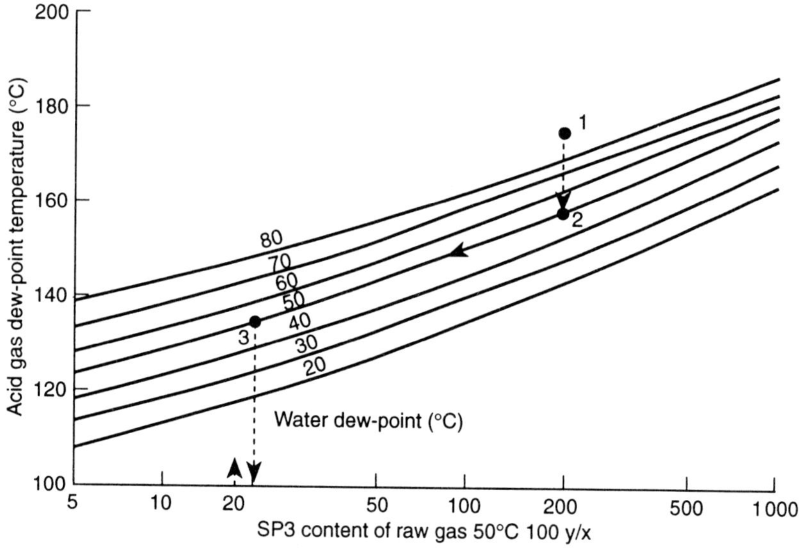

Figure 6.13 Relationship between SO_3 content of gas and dew-point temperature. (Source: *Chemical Engineering Progress*, August 1974 and April 1977). Example: SO_3 in raw gas, 200 ppm; gas temperature, 175 °C; moisture content, 11.6%; water dew-point, 50 °C; gas cooled to 135 °C; residual SO_3, 20 ppm; SO_3 condensed, 180 ppm.

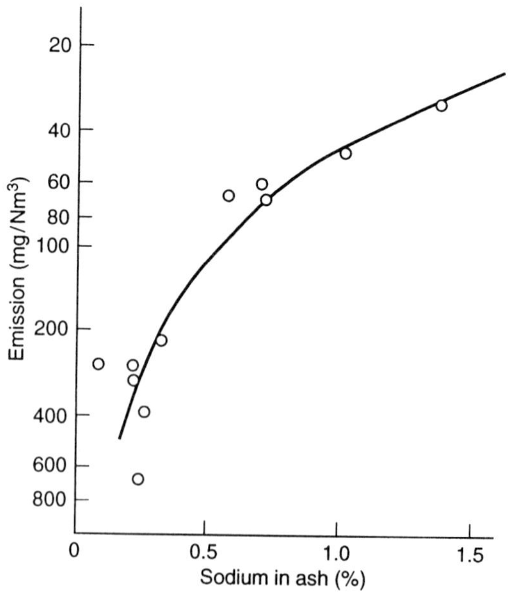

Figure 6.14 Effect of sodium in ash on emission for Australian coals, 0.4 to 0.54% sulphur.

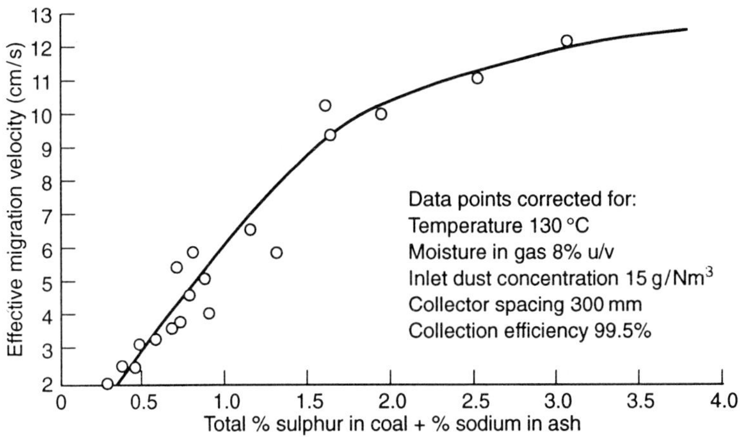

Figure 6.15 Relationship between precipitator performance and sulphur in coal and sodium in ash. Data points corrected for: temperature, 130 °C; moisture in gas, 8% v/v; inlet dust concentration, 15 g/Nm^3; collector spacing, 300 mm; collection efficiency, 99.5%.

emission was very dependent on the sodium present in the fly ash, as illustrated in Figure 6.14. Lithium is also anticipated to react similarly, but is only present in very small quantities, which are too small to have a significant impact on performance. Potassium, although having similar chemical properties to sodium, does not appear to react in the same manner.

Calcium and magnesium present in the ash tend to produce sulphates which are not effective conductors, so interfere with resistivity reduction by sulphur trioxide and thus must be considered as leading to increased resistivity effects.

Performance measurements on fly ash precipitators have been correlated with the coal and fly ash analyses and first-order precipitator sizing curves have been developed. Figure 6.15 plots the migration velocity necessary for a precipitator to give an efficiency of 99.5% [18] against the coal sulphur and ash sodium content. Other curves incorporating silica, alumina, calcium and magnesium, in addition to sulphur and sodium, have been developed but do not appear to significantly affect precipitator sizing requirements based on data from Figure 6.15.

Measurements of *in-situ* fly ash resistivity show good agreement with those calculated from the Bickelhaupt relation and Figure 6.16 has been plotted, showing how the resistivity varies predominantly with the sulphur and sodium components. It is interesting to note that, at around a resistivity of 10^{10} Ω-cm, the resistivity steeply increases for reducing values of sulphur plus sodium, whereas for higher values the resistivity only reduces slowly. This mirrors the performance curve as Figure 6.15; another interesting point

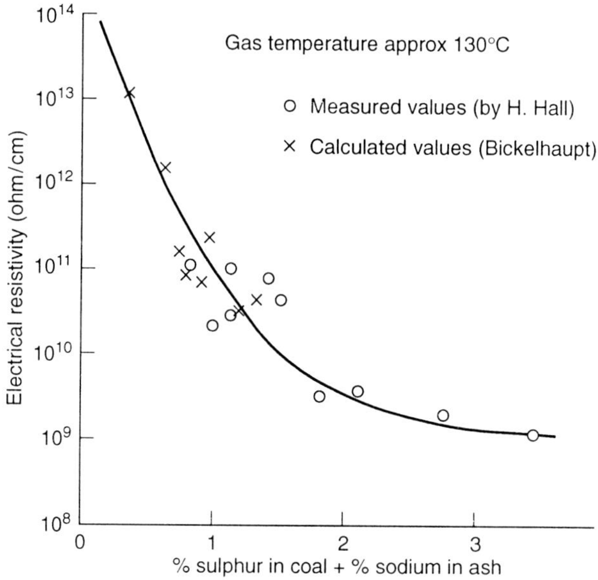

Figure 6.16 Relationship between fly ash resistivity and sulphur in coal and sodium in ash. Gas temperature approx. 130 °C. ○, measured values (by H. Hall); ×, calculated values (Bickelhaupt).

is that a resistivity of 2×10^{10} Ω-cm is the value which White [19], in 1963, indicated as being an acceptable resistivity for good precipitation.

For precipitators operating at high temperature, the coal, ash and surface chemistry is not so important, as volume conductivity is controlled more by thermal molecular motion. Some unexpected performance problems were however encountered on a US utility plant high temperature precipitator which were found to be the result of sodium ion migration within the collected dust layer itself (sodium depletion), the sodium concentrating on the surface. This produced the classic reverse ionisation phenomena which affected performance. This condition can be overcome by supplementing the sodium content of the fuel artificially by adding soda ash or other sodium products during combustion.

As an alternative to operating at high temperature to reduce particle resistivity, investigations, particularly with NSW Australian low sulphur coals, have shown that operating at a lower gas temperature, e.g. 100/110 °C, produces higher efficiencies by reducing the worst effects of back-ionisation.

For fairly inert dusts, e.g. from NSW, additional air heater surface can be installed to reduce temperature without the risk of serious air heater blockage. With other coals, a multitude of two fluid atomisers have been installed in the ductwork upstream of the precipitator to effect the temperature reduction. In practice the problems of maintaining the atomisers such that they don't result in duct blockage means that ideally a correctly sized cooling tower should be supplied. Reference has already been made to the application of spray driers ahead of precipitators; these effectively reduce temperature and increase the moisture level, both aiding the reduction is resistivity.

In the metallurgical field it is not unusual to find a high percentage of submicron fume arriving at the precipitator; this fume arises because of initial evaporation/volatilisation of metals and metalloids at the processing temperatures, followed by recondensation at the lower back end temperatures. The presence of these fume particles can give rise to severe space charge and corona suppression effects and the designer must be aware of them, so the correct type of discharge electrode and matching HT equipment can be supplied to overcome these possible problems. In practice it has been found that normally only the inlet field needs to have the high emission electrodes, since if this field operates normally, the subsequent fields see a very much reduced particulate loading and hence reduced space charge and corona suppression effect.

The carryover of alkali chlorides giving rise to 'snow flake' type formations, particularly in the cement industry using wet or semi-wet feed material, has a significant effect on precipitator performance, both in terms of efficiency reduction and chimney appearance. The formation of snow flakes is complex but is related to the quantity of soluble alkali material fed to the kiln. As the alkali material is derived from volatilisation and condensation, its particle size is very fine and consequently the highest concentrations are found in the outlet field hoppers. If this can be extracted separately and discarded, few precipitation problems are experienced; however, as in most plants the total precipitator catch is returned to the kiln, the amount of alkali material in the feed builds up to a point where the snow flake problems rapidly appear within the precipitator [20].

The term 'snow flake' was derived from scanning electron microscopy where the material was found to have a crystalline structure resembling snow. The material itself agglomerates into flocculant lumps of low bulk density, so is not only subject to significant gas re-entrainment, but also 'lifts' within the field area, as a result of ionic wind or thermal diffusion, and tends to collect across the collector top beams, which can lead to complete shorting of the HT in extreme cases.

Deliquescent/hygroscopic materials, or those close to their melting point, need special attention at the design stage, not that they cannot be satisfactorily precipitated as they are conductive and usually at equilibrium with the gas phase moisture. If temperatures or process conditions alter, e.g. during start-up and shut-down, then further moisture absorption can produce a sticky dust, which may impede maintaining the internals 'dust free' and so can affect operating conditions, or can give rise to hopper dedusting difficulties. Although special operating procedures and plant preheating can reduce the start-up and shut-down difficulties, it may be advantageous to consider wet, rather than dry, precipitation.

In the case of high carbon dusts, controlled conditioning is deliberately carried out to increase the cohesivity of the collected material to reduce re-entrainment losses.

Corrosion, as a result of the particulate composition, is normally a problem associated with wet or mist precipitators and its effect is fully covered in chapters 13 and 14. High temperature corrosion, however, has been experienced on older municipal incinerator precipitators operating at 300 °C where, as a result of chlorides being present in the feed, these when precipitated can transform from the ferrous to more stable ferric form by removing iron from the collector plates with their resultant thinning. Nowadays, a change in incineration practice, to give a low back end temperature, has largely resolved this particular problem, but the designer should be aware of this reaction when considering materials of construction for high temperature applications.

Various investigators have reported other elements in power plant fly ash which affect the resistivity and hence precipitator performance, e.g. iron and phosphorous. Iron is often an indication of the amount of pyrites in the coal and hence combustible sulphur, while phosphorous has been shown to be an effective conditioning agent in reducing the surface resistance of the fly ash particles [21]. Their exact mechanism in reducing the ash resistivity, however, needs further investigation.

It is likely that other elements present in the fly ash also influence resistivity/performance, but the foregoing are the most pertinent.

References

1. Kauppinen E.I. *et al.* (1995) Fly ash formation on PC boilers firing South African and Colombian coals. *EPRI/DOE International Conference on Managing Hazardous and Particulate Air Pollutants*, Toronto, Canada. Aug 15–17, in print.
2. Porle K. *et al.* (1995) Full-scale ESP performance after PC boilers firing South African and Colombian coals. *EPRI/DOE International Conference on Managing Hazardous and Particulate Air Pollutants*, Toronto, Canada, Aug 15–17, in print.
3. Lord Raleigh (1871) *Philosophical Magazine*, **41**, 107.
4. Parker K.R. and Novogaratz D. (1991) Electrostatic control of air toxics. *EPRI/EPA 9th Particulate Control Symposium*, Williamsburg, USA, Oct., Session EPRI TR 100471 **2**, Palo Alto, CA, USA, 1992.
5. Lowe H.J. and Lucas D.H. (1953) The physics of electrostatic precipitation. *British J. Applied Physics*, Supplement No. 2.
6. Penney G.W. and Klinger E.H. (1962) Contact potentials and the adhesion of dust. *Trans A.I.E.E.* **81** (1) p. 200–204.
7. Dalmon J. and Tidy D. (1972) The cohesive properties of fly ash in electrostatic precipitation. *Atmospheric Environment* **6**, p. 81–92.
8. Ashton M.D., Farley R. and Valentin F.H.H. (1964) An improved apparatus for measuring the tensile strength of powders. *J. Scientific Instruments*, **41**, p. 763–5.

9. Farley R. and Valentin F.H.H. (1968) Effect of particle size upon the strength of powders. *Powder Technology* **1**, p. 344–54.
10. IEEE Standard Criteria and Guidelines for the Laboratory Measurement and Reporting of Fly Ash Resistivity. IEEE Std 548–1984.
11. Goard P.R.C. and Potter E.C. (1978) Operational resistivity measurements on freshly generated fly ashes. *CSIRO Symposium on Electrostatic Precipitation*, Leura, pp. 3.1–8, CSIRO, Sydney, Australia.
12. Nichols G. and Spencer H. (1975) Test methods and apparatus for conducting resistivity measurements. Report prepared by the Southern Research Institute for the US. E.P.A. SoRI, Bieringham, AL, USA.
13. Cohen L. and Dickenson R. (1963) The measurement of the resistivity of power station flue dust. *J. Scientific Instruments*, **40**, p. 72–5.
14. Durham M.D. *et al.* (1991) Identification of low resistivity reentrainment in ESPs operating in dry scrubbing applications. *9th EPRI/EPA Particulate Control Symposium*, Williamsburg, USA, Oct., Session 5A EPRI 100471 **2**, Palo Alto, CA, USA, 1992.
15. Porle K. *et al.* (1991) ESP operation following spray dryers with low resistivity particulates. *Proc. 9th EPRI/EPA Particulate Control Symposium*, Williamsburg, USA, Oct., Session 5A EPRI 100471 **2**, Palo Alto, CA, USA, 1992.
16. Bickelhaupt, R.E. (1979) A technique for predicting ash resistivity. EPA 600/7-79-204 US.
17. Parker K.R. (1990) The wet ESP and its role in modern pollution control. *Proc. Xth Australian and New Zealand Air Pollution Control Conference*, Auckland, N.Z., Oct., pp. 23–30, Clean Air Soc., Auckland, N.Z.
18. Darby K. *et al.* (1991) The influence of sodium in fly ash on electrical resistivity and its impact on precipitator performance. *EPRI/EPA 9th Particulate Control Symposium*, Williamsburg, USA, Oct., Session 6A EPRI TR 100471 **2**, Palo Alto, CA, USA, 1992.
19. White H.J. (1963) *Industrial Electrostatic Precipitation*. Addison Wesley, USA.
20. Darby K. and Parker K.R. (1990) The use of electrostatic precipitators in the cement manufacturing industries for the control of dust emissions. *Proc. 4th International Conference on Electrostatic Precipitation*, Beijing, China. Oct., **3**, pp. 173–86, International Academic Publishers, Beijing, 1991.
21. Darby K. and Whitehead C. (1974) The use of electrostatic precipitators in current power station practice. *Proc. Inst. of Fuel Symposium on Changing Technology of Electrostatic Precipitators*, pp. 35–48, Inst. of Fuel, Adelaide, Australia.

7 Performance design considerations

C. COTTINGHAM

7.1 Introduction

In the design and sizing of any electrostatic precipitator, both the customer and vendor have certain responsibilities to ensure that the final plant will meet the full expectations of both. This chapter will review the requirements in detail in order to achieve the most competitive offer, with the minimum of technical exposure to the vendor, while still meeting all the customer's needs. The format to a large extent will be in the form of questions and answers, which will explain some of the methodology used in the ultimate sizing and configuration of the plant.

7.2 What are we trying to achieve?

The design and sizing of an electrostatic precipitator to be ultimately successful has to achieve the customer's requirements at the least cost to the customer and vendor alike.

To achieve this the customer has to be absolutely clear in his requirements:

(a) understand what has to be achieved, usually in terms of emission, i.e. to be aware of current and proposed future legislation;
(b) clearly define the design parameters, particularly at the inlet to the precipitator or outlet of the plant to which the precipitator is to be installed.

Although the above may be stating the obvious, there are many times when the supplier is asked to commit time and effort to satisfy customers who have little concept of what they really require.

The vendor has to assess the client's requirements and examine the following in detail.

(a) What process is being considered?
(b) Are all the relevant process parameters defined?
(c) What level of performance is required?
(d) Are there any site constraints?
(e) Does the client have any special technical requirements/preferences?

Only when all the above have been addressed can the vendor successfully begin to size and offer a plant that is technically acceptable.

7.3 Assessment of the process

The designer of electrostatic precipitators (ESPs), can be confronted by a bewildering array of processes and operating conditions (refer to chapters 12, 13 and 14). He has to be able to accurately assess the process conditions and configure the ESP to satisfy the customer and ensure that the ESP has minimum technical risk at the most competitive cost.

To illustrate the range of experience required, a list of a few of the processes are given below.

Boilers
Pulverised coal
Heavy fuel oil
Fluid bed (sand and/or lime beds)
Stoker (grate or spreader)
Biomass (wood, straw, bagasse, etc.)
Emulsified tar
CO recovery FCC
Black liquor recovery

Steel
Main sinter strand
Sinter deduster
Furnaces (blast, BOF, etc.)
Electric arc
Gas recovery
Secondary ventilation
Pelletiser

Non-Ferrous
Converters (copper lead)
Furnaces (copper lead)

Cement
Kiln (wet, dry, semi-wet)
Finishing mill
Clinker cooler
Alkali bypass
Raw mill

The above represent only the more commonly addressed processes. As can be seen, the designer has to be able to understand the wide and varied range of applications that the ESP operates on. The designer has to be able to assess the process parameters and determine how they relate to the ability of the ESP to meet the performance requirements.

For the designer to be successful, it is essential that he has a consistent method for analysing the process conditions. In trying to meet the client's requirements, the designer may be confronted by various changes to the design inlet parameters during the bidding process. If consistency is not used, then the designer risks the chance of losing his way and hence his credibility.

At all stages of the process analysis, the designer has to assess the impact of the following primary parameters:

(a) gas temperature
(b) dust analysis
(c) gas analysis
(d) dust particle size

(e) quantity of dust
(f) dust resistivity
(g) dust cohesivity
(h) gas flow rate

For every process, and each one of the above, the designer should have relationships that enable him to access the impact of the process on the performance of the ESP.

In the early days of electrostatic precipitation, the designer had a library of EMVs (Deutsch Effective Migration Velocities or Sizing Parameters) for every process. Corrections and variations to the 'library standards' were a matter of personal judgement and experience. This may have been acceptable 30 years ago, when both technical and commercial margins were greater, but in today's environment more precise techniques are essential. The experienced designer of today will have a background of reference plants and test data that will enable him to guarantee the plant performance, to much closer limits with much reduced technical margins. Coupled with this, there have been much published data on the precipitation process in the last 20 years, which can be judiciously used to expand/confirm his own in-house data.

7.3.1 Typical assessment

We can best illustrate the techniques required by examining the following worked example.

Let us assume we have an enquiry from a customer for a PC-fired boiler. The design parameters are given as:

Gas vol.	530 Am^3/s
Gas temp.	130 °C
H_2O in gas	7.0% v/v
Ash in coal	10% w/w
Ash resistivity	5.0×10^{11} Ω-cm
Ash size	50% < 20 μm

The designer will have his own database of tests that will enable a series of relationships to be determined. Typically a relationship between effective migration velocity (EMV) and the coal/ash characteristics (resistivity) is determined from test data.

These tests will have been conducted under various operating conditions, so the EMVs will have to be normalised to set levels of treatment time, temperature, H_2O, dust loading and collector plate spacing; otherwise any relationship will be meaningless. The normalised levels will be determined by the designer based on his experience.

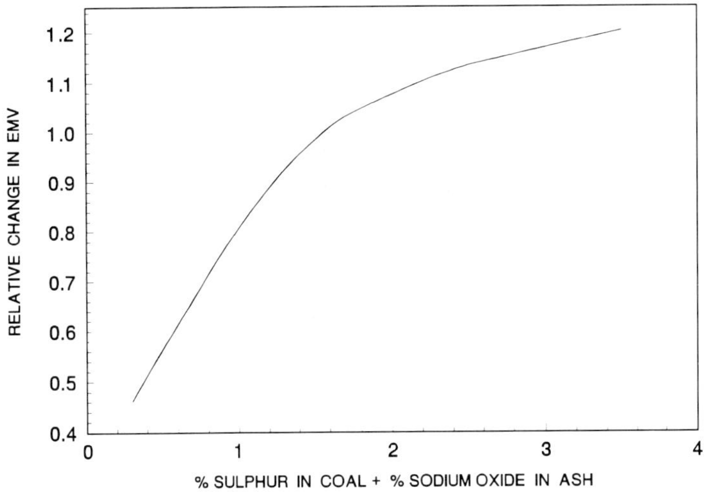

Figure 7.1 Relative change in EMV *vs.* coal ash characteristics.

Figure 7.1 presents a typical relationship between EMV and some coal/ash components [1]. For this the normalised levels are as given below:

Dust loading	$15\,G/Nm^3$
Temperature	130 °C
H_2O	8% v/v
Particle size	50% < 20 μm
Plate spacing	300 mm

Although the curve is plotted as the Deutsch EMV (w), the form would be similar for the modified Deutsch (w_k), or an EMV based on any other theory that the designer considers acceptable. If w is used, then the EMV has to be corrected to a constant treatment time, but if the relationship is based on w_k, then the EMV is automatically corrected for variations in treatment time.

The EMV determined from the coal/ash characteristics is then corrected to the customer's design parameters, by multiplying it by the relevant factors related to temperature, H_2O, particle size and dust loading. The designer then has a design EMV that can be used to determine the ESP plate area or treatment time.

A temperature relationship for a difficult fly ash is illustrated in Figure 7.2 [2]; this as explained in chapter 6 reflects the effect of temperature, primarily on the electrical resistivity of the fly ash, the efficiency (EMV) decreasing with increasing temperature to reach a minimum at around 160 °C, then rising as the resistance becomes more dependent on ionic rather than surface conduction.

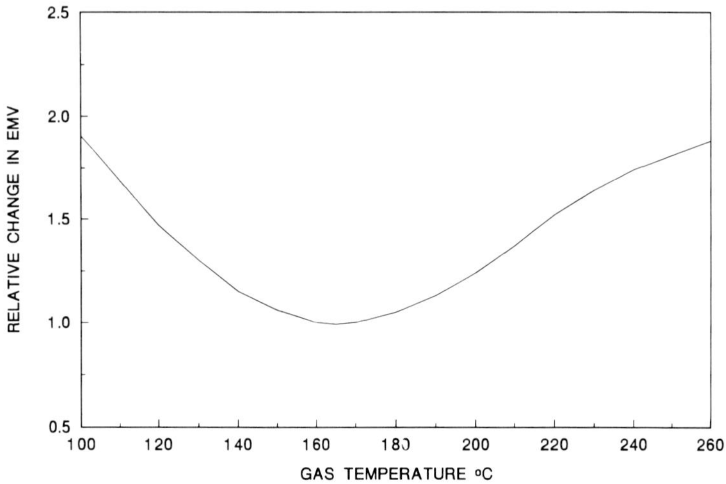

Figure 7.2 Relative change in EMV *vs.* gas temperature.

The main effect of moisture is to modify the electrical operating conditions; generally, the higher the moisture, the higher the operating voltage and hence performance.

The effect of particle size is complex in electrostatic precipitation, since, theoretically, the number of elementary charges that a particle receives is proportional to the surface area and hence its diameter. Particles below 0.2 μm diameter receive their charge by diffusion processes as a result of Brownian motion, while particles greater than 2 μm are charged by ionic collision means. This gives the typical dip in the fractional efficiency curve, as illustrated in Figure 7.3 [3] and in theory, chapters 3 and 8. This means that there is a specific relationship between particle size and performance which is typically based on the median or d_{50} size of the particles and hence specific surface.

Dust loading impacts on the precipitator size and configuration in the following manner. First it affects the design efficiency assuming a constant emission is required. Another aspect is the possible development of significant space charge or corona suppression, which requires special consideration in terms of discharge electrode design. Large dust catches, from high dust loadings, demand proportionately sized hopper and dust evacuation systems to prevent hopper overfilling, leading to mechanical damage to the precipitator internals.

A similar methodology, using a data bank relating to the different processes listed in section 7.3, would be used to establish the design parameters/contact time, etc., for each different application. In these instan-

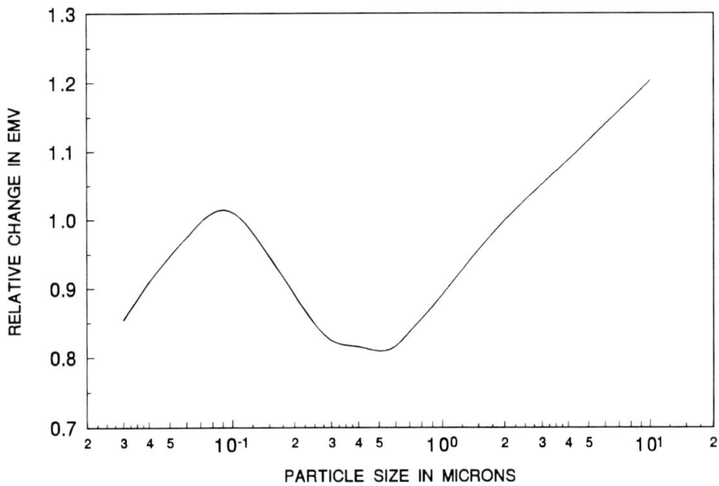

Figure 7.3 Relative change in EMV *vs.* particle size relationship.

ces, the variables would be related to functions other than the fuel and ash analyses.

7.4 Plate spacing

Before moving on to consider the configuration of the ESP, we should review the question of plate spacing.

Traditionally the ESP size has been evaluated using the EMV. The EMV is a precipitation industry design tool that is variable and is influenced by the process conditions given above and other factors. Before proceeding it is worth expanding upon the background of the Deutsch equation and EMV.

The basic equation of electrostatic precipitation was first developed by Walther Deutsch in 1922 [4]. Deutsch used a simple wire and tube type ESP to carry out his theoretical calculations, from which the collection efficiency was derived as a function of tube dimensions and electric field strength:

$$\exp - (2EwL/Rv)$$

where E is the field strength, w is the particle velocity per unit field strength, L is the tube length, R is the tube radius and v is the gas velocity.

As it was not possible to physically measure the factor 'w', practical precipitation engineers replaced it by the empirically determined EMV. At the same time, the tube dimensions and gas velocity were transposed into collecting plate area (A) and gas volume (V). This changed the original Deutsch relationship to the more well known:

$$\exp - (\mathrm{EMV} \cdot A/V)$$

In this relationship, the influence of plate spacing and gas treatment time on EMV is completely absent. Deutsch's original relationship contains both of these factors by considering the tube radius, gas velocity and tube length.

The above relationship between EMV and A/V (specific collecting area), became the accepted norm for many decades. The EMV was considered to be independent of collector plate spacing and the trend in the 1960s and 1970s specifying SCA forced ESP suppliers (against more enlightened suppliers' better judgement) to retain narrow spacing (220–300 mm) in order to be competitive.

There were some exceptions to this, e.g. on specialised applications where severe corrosion demanded expensive fabrication materials; in these, the plate spacing was successfully increased to reduce the cost of the installation. In fact, in the early days of commercial precipitation, investigations were successfully carried out using tube diameters of up to 48 inch (1.22 m), but power supplies limited their practical usage.

It was not until the 1980s, before 'wide spacing' (400 mm) was to be considered acceptable on a large scale and, with the exception of some unfortunate experiences in the United States, by the mid-90s, 400 mm collector spacing had been accepted throughout the world as a precipitator 'norm'.

7.5 Configuring the ESP

Once the required collecting area has been determined, the designer has to decide how to configure the ESP. There is almost as much skill and experience required to do this as is needed to determine the collecting area/contact time for a specific duty.

The designer is confronted by a large range of decisions:

(a) How many series fields?
(b) What size of collecting plate?
(c) How many transformer rectifier sets?
(d) How many electrical bus sections?
(e) Any special insulator considerations?
(f) Any special rapping considerations?
(g) What type of inlet mouthpiece?
(h) What aspect ratio is acceptable?
(i) What gas velocity is acceptable?

(j) Dust hopper considerations?
(k) Cohesivity/angle of repose of dust?
(l) Any corrosive element in the process?

The choice of collector size is generally dependent on the quantity of gas being treated but, as a general rule, the taller the collector plate the lower the cost. This basic consideration has to be balanced against the number of fields required, acceptable gas velocity range, minimum acceptable aspect ratio and the effect of wind loads on the support structure design.

All ESP suppliers have a range of collector plate sizes, from a minimum, of perhaps 1 m × 4 m, to a maximum size of say 5 m × 15 m. Over this entire range the collector plate should: (a) be manufactured to the required tolerances to ensure good electrical clearances; (b) be capable of being rapped with sufficient energy to adequately remove the dust; (c) maintain the electrical clearances within the life of the plant, i.e. remain mechanically stable; (e) be manufactured and installed cost effectively.

Coupled with the collector design, the discharge electrode system must also be capable of maintaining the electrical clearances in a cost-effective way, whether the discharge electrodes are mounted in a mast, bedstead frame, or are of the rigid 'unbreakable' type, as illustrated in Figure 7.4.

It will be appreciated from chapter 4 that the choice of discharge electrode format to meet a particular process is decided from the particulate

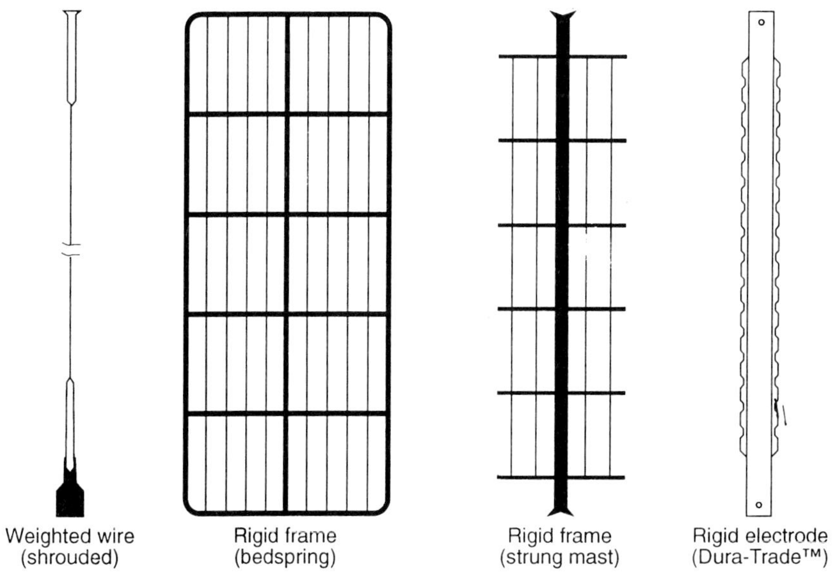

Figure 7.4 Discharge electrode formats and mountings.

loading and the size analysis of the dust being presented at the inlet. As most large precipitators comprise many kilometres of discharge electrode element, the cost of special profiles can prove expensive. The designer must consider, from the dust at the inlet, whether there is likely to be a significant space charge or corona suppression effect which will demand the use of controlled or high emission type electrodes. High emission electrodes usually have a complex configuration, and are hence not only more costly than a simple profile, but, if the corona suppression is not prevalent, then there will be a very high power consumption. This may mean that the selected transformer rectifier set is undersized to meet the required duty.

The ratio of the overall collector length to the collector height is called the 'Aspect Ratio'. Each designer will have an acceptable minimum Aspect Ratio depending upon the designer's experience, the design of the collector plate, the nature of the dust, the gas velocity in the electrical field and the quantity of dust being treated. For a given dust, the Aspect Ratio tends to increase as the inlet dust loading increases, and/or the emission requirements decrease. This rule can be modified if the dust is of a type that does not re-entrain easily.

The designer has to decide what is the minimum number of series fields that are required to achieve the required performance. For example:

Required efficiency $<98.0\%$, two fields would be used.
Required efficiency $\sim 99.0\%$, three fields would be considered.
Required efficiency $>99.7\%$, four or more fields would be the norm.

Choosing the number of fields is a compromise between minimum Aspect Ratio and cost. Generally the greater the number of series fields, the higher the overall cost, as each field is usually energised by its own transformer rectifier set or sets and has its own hopper. One exception to this is when precipitating a relatively easy dust in a small plant, where the cost of a transformer rectifier set is a large proportion of the total cost; in this case, a significant saving can be achieved by energising, for example, three mechanically independent fields with two transformer rectifier sets, one on the first field and one being shared by fields two and three. The plant readily absorbs the power to achieve the performance, but the three independent mechanical fields enable the rapping to be optimised.

The decision on the number of transformer rectifier (TR) sets is also an important consideration when designing large ESPs [6]. The decision is easy if the customer specifies the maximum plate area to be energised by each TR, however, more often than not, the designer has to make his own decision and he has the usual conflict between technical acceptability and economics.

The amount of subsectioning within each field depends upon the design of the ESP. As each subsection requires support insulators (generally two or four), the fewer the number of sections, the lower the cost. There is always a mechanical limit to the size of each subsection, which may limit the total number of subsections on a large power station plant, but the client may also specify a minimum number, so that there is more security should a subsection fail for any reason.

The gas velocity in the ESP treatment zone is an important design parameter. Most customers consider a 'rule of thumb' value of about 1 m/s, as an acceptable precipitator gas velocity; however, many vendors have experience of good performance with considerably higher gas velocities of 1.7 m/s and higher for some applications. Generally, the higher the Aspect Ratio and/or the greater the number of fields, the higher the gas velocity that can be tolerated. Obviously there is a maximum velocity that should never be exceeded, as rapping re-entrainment and possible gas scouring will have an increasingly deleterious effect on the emission [7].

The designer often has a considerable problem in reconciling the economics of the configuration and the gas velocity. For a given design treatment time and collector height, the only way the designer can alter the gas velocity is to vary the ESP field length. This is illustrated by the following table:

	Case 1	Case 2	Case 3
Collector size (m × m)	3 × 10	4 × 10	4.5 × 10
Number of fields/length (m)	3/9	3/12	3/13.5
Number of gas passages	22	17	15
Aspect ratio	0.9	1.2	1.35
Gas velocity (m/s)	1.14	1.47	1.67
Treatment time (s)	7.92	8.16	8.10

As can be seen, when the treatment time is virtually unaltered, the gas velocity increase is a function of the length of the field. Changing the collector height for the same treatment time will not alter the gas velocity.

There is always a possible conflict between acceptable Aspect Ratio, gas velocity and economics. The conflict is particularly apparent when designing very large ESPs and the designer is trying to utilise the largest collector plate in his range. With an application that requires a low gas velocity and a reasonable Aspect Ratio, the designer would be forced to reduce the height of the collector and therefore increase his costs. The responsible designer, however, will always put technical considerations before cost, so the plant would normally have the reduced collector height configuration.

Many common processes, such as PC boilers, MSW incinerators and cement kilns, present relatively minor process problems to the designer. The configuration of the dust hoppers, inlet mouthpieces, type of insulators, etc., all come within a standard range of designs and experience. However, the designer is also expected to produce designs for processes that may present special difficulties outside the pure precipitation duty.

Some processes require special consideration to be made of the inlet mouthpiece and gas distribution devices, to ensure freedom of dust deposition so that the gas distribution will remain unaffected. On non-ferrous roasting processes, the dust can be very high in sulphates and is therefore very sticky, so special provision is made to make sure the dust does not hang up within the internals and hoppers. In this respect, the form of rapping in terms of intensity needs consideration, particularly if the dust is very cohesive requiring very intense rapping and therefore the type of suspension and anvil may be of a special design to ensure maximum plant availability.

An area which needs careful consideration is that of high temperature applications, e.g. those over 300 °C [8], where differential thermal expansion between casing and substructure may mean that the width of the casing is somewhat limited, so that the foregoing optimum sizing considerations, e.g. bus section size, number of TRs and other factors, have to be reassessed to produce the lowest precipitator cost, while still minimising technical risk.

Other applications will demand special designs of insulators which would also have an impact on the final configuration and cost.

7.6 Conclusions

At the end of the design process, which will have considered all the above, the designer will have sized and configured the plant to meet the customer's requirements at the lowest cost. This complex process is inevitably a compromise but, as stated above, the last area to be compromised should be the technical credibility of both the designer and the plant.

Although each major precipitator supplier will have his own data bank, on which he will base the design and size of any new application, the methodology and approach will be similar to that outlined. It is not the intention of this chapter to contain sufficient information to enable the non-specialist engineer to design a precipitator, but to explain how the size was derived.

It is surprising that in spite of the many different precipitator designs that have been used industrially, the size of any competitively sized precipitator, for a given duty, is roughly similar, regardless of the supplier. While all suppliers would consider their knowledge superior, in actual practice the data must be comparable; otherwise there would be considerable differences in the plant sizings.

References

1. Darby, K., Cottingham, C. R. and Radai, J. (1991) The influence of sodium in fly ash on electrical resistivity and its impact on precipitator performance. *EPRI/EPA 9th Particulate Control Symposium.* Williamsburg, USA October 1991, Session 6A EPRI TR 100471 **2**, Palo Alto, CA, USA.
2. Electricity Commission of New South Wales Research Note No. 59. *Liddell Power Station Investigation of Requirements of Electrostatic Precipitators*, ECNSW, Sydney, Australia.
3. Parker, K.R. (1990) The wet ESP and its role in modern pollution control. *Proc. 10th Clean Air Society for Australia & New Zealand Conference*, Auckland, March, pp. 23–30, Clean Air Soc., Auckland, NZ.
4. Deutsch, W. (1922) Bewegung und Ledung der Electricilatstrager in Zylinder Kondensator. *Ann. Phys.*, **68**, 335–44.
5. Darby, K. and Novogoratz, D. (1990) Increased plate spacing in electrostatic precipitators. *EPRI/EPA 8th Particulate Control Symposium*, San Diego, CA, Session 5A EPRI, Palo Alto, CA, USA.
6. Ramsdell, R.G. (1973) Practical design parameters for hot and cold ESPs. *American Power Conference*, Chicago, May.
7. Dalmon, J. and Lowe, H.J. (1961) Experimental investigations into the performance of ESPs for PF power stations. Colloque International–La Physique des Forces Electrostatiques et leurs Applications. Centre National de la Recherche Scientifique, Paris 1961.
8. Darby, K. and Parker, K.R. (1975) The electrostatic precipitator at high temperature. *National Society for Clean Air Annual Conference*, Newcastle-on-Tyne, March, Clean Air Soc., London, UK.

8 Electrical operation of precipitators

V. REYES

8.1 Introduction

The operation of any precipitator is closely related to its electrical energization, i.e. the way the power is delivered to the precipitator. Therefore, this aspect will be covered in detail. The ancillary electrical equipment used in the precipitator, such as rapping systems, insulator and hopper heaters and purge air system, will not be discussed.

The electrical energization is the key factor for a satisfactory precipitator operation. One of the pioneers and main contributor to this field [1] once expressed it very clearly: 'a precipitator functions by the Grace of God and electrical energization; if electrical energization is good, one's state of Grace in other areas can perhaps be somewhat less than perfect'.

The technology and equipment used in the electrical operation of precipitators have undergone a considerable improvement after the second World War. This is mainly a result of considerable efforts in research and development in Japan, USA and Europe. In the following, only the technology available at the present will be presented. This includes traditional DC, intermittent and pulse energization. A fourth method is emerging, based on the switch mode power supply technology. This is a very promising technology but at the moment its use is limited to low current levels. For this reason and for lack of space, this technology will not be covered. Moreover, for the same reason, an historical review of the development of the electrical equipment used in precipitators will not be included.

8.2 Precipitator performance and electrical energization

This dependence can be derived from the fundamental equation for precipitator efficiency [2]:

$$\eta = 1 - \exp\left(-\frac{A}{Q}\omega\right) \tag{8.1}$$

where A is the collection surface, Q is the gas flow rate and ω is the particle-migration velocity.

The migration velocity, ω, is a direct measure of the rate of collection of the particles. From equation (8.1) the inverse proportionality between ω and the collection surface A indicates that an improvement in ω means that A

may be correspondingly decreased, provided the collection efficiency remains at the same value. Therefore, ω is considered as a generalized performance parameter and it is the natural link between electrical energization and collection efficiency.

From laboratory measurements on pilot precipitators, and from theoretical analysis, it has been found that ω is proportional to the product of the mean and the peak precipitator voltage [2]. Collected data from different plants have shown that ω is also proportional to the mean precipitator current. Hence, ω may be expressed by

$$\omega = k_1 \frac{P_c}{A} \tag{8.2}$$

where P_c is the corona power and k_1 is a constant which depends on the gas state, particle composition, precipitator geometry and size, etc. P_c/A *is* the corona power density. If ω is expressed in ft/s (1 ft/s = 30.5 cm/s) and the power density in W/ft^2 (1 W/ft^2 = 10.75 W/m^2), as seen in Figure 8.1, the proportionality constant k_1 has been estimated to be 0.67 [2].

P_c takes into account the ω dependence on precipitator current and voltage. As will be seen in section 8.5, the precipitator current and voltage delivered by commercial power supplies are not pure DC magnitudes, so a simple calculation is not possible. By using the actual waveforms and computer calculations, a useful expression for P_c can be determined [3]:

$$P_c = \frac{1}{2}(V_{mean} + V_{peak})I_{mean} \tag{8.3}$$

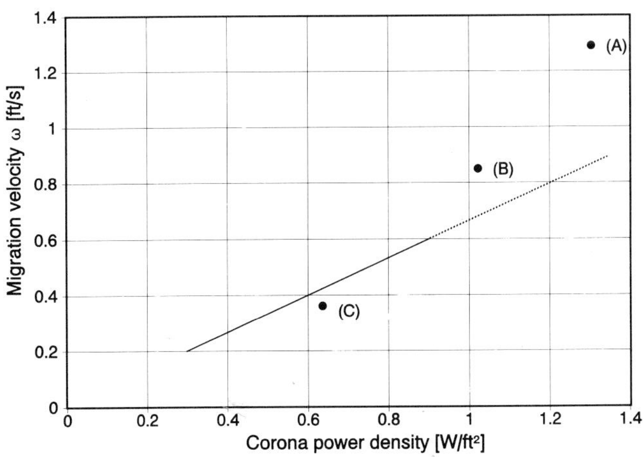

Figure 8.1 Migration velocity as a function of corona power based on empirical data. Reference line redrawn from White [2].

Equation (8.3) indicates that the collection efficiency depends directly on the mean voltage, the peak voltage and the mean current of the precipitator.

8.2.1 Examples

To illustrate the dependence of the migration velocity, ω, on corona power, three practical examples with traditional DC energization have been chosen and represented in Figure 8.2.

(A) A fly ash precipitator (one chamber, two bus sections) for a boiler burning an English/Polish coal blend ($S = 1\%$). In Figure 8.2, ω is plotted as a function of the corona power density, for this case, where the dust resistivity is low. This curve shows a steady increasing ω for an increasing power density. The corona power density is limited by sparking in both sections at around 60–70% of rated current, but at this level the curve becomes flat. This means that optimum precipitator performance is obtained with approximately $14\,W/m^2$ of collection surface. The migration velocity used here is that modified by Petersen, i.e. ω_B [4].

(B) A fly ash precipitator (two chambers, eight bus sections) for a boiler burning high S (2.6%) American (Appalachian) coal. In this case, the migration velocity ω is lower than in the previous case, but the corresponding curve in Figure 8.2 also shows an increasing ω for an increasing power density, which is typical for low resistivity conditions.

(C) A fly ash precipitator (two chambers, five bus sections) for a boiler burning low S (0.26%) South African (Middelburg) coal, causing

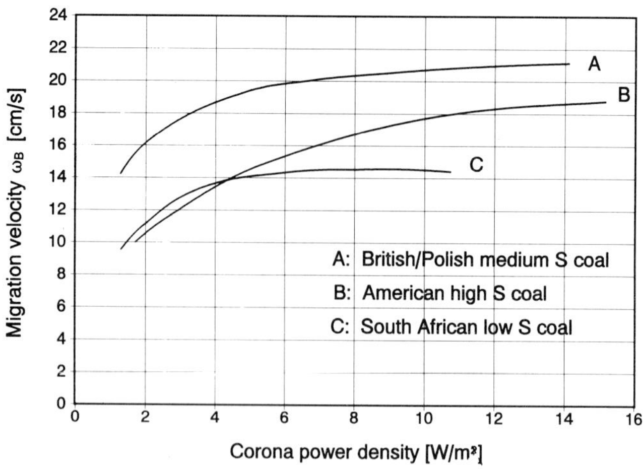

Figure 8.2 Migration velocity as a function of the corona power for three different coal blends.

resistivity problems. In this case, the optimal migration velocity ω is lower than in the previous case and it is clearly seen that above a power density of 6 W/m^2, the performance of the precipitator is almost constant, with a slight decreasing tendency. This suggests that 6 W/m^2 is the limit for the useful corona power and the surplus power is wasted in back-corona generation.

In the case of severe back-corona, occurring over most of the collection surface in each bus section, the limit for the useful corona power is more definite as the precipitator efficiency begins to fall, as suggested by Figure 8.14.

The curves in Figures 8.2 and 8.14 indicate the dependence of ω on the resistivity of the fly ash, which also determines the useful corona input power. This must not be confused with the installed power, i.e. the rated power that can be delivered by the high voltage power supplies. The useful corona power can be defined as the power level giving the optimal ω or performance.

In Figure 8.1 the results of the three examples are also plotted. These show a large divergence in relation to the reference line.

To summarize

- The electrical energization is, in practice, one of the most important factors in obtaining a high precipitator efficiency.
- The particle-migration velocity ω is the basic link between precipitator efficiency and electrical energization.
- The corona power can be expressed as a function of the mean current, the mean voltage and the peak voltage.
- The particle-migration velocity ω may be related to different electrical quantities, the corona power seeming to be the most appropriate.
- The proportionality constant k_1 can be used as a reference, but the ESP manufacturers base the sizing of their precipitators and the useful corona power for a particular process from their own collected data. (Both factors are closely related to the resistivity of the collected dust.)

8.3 Corona suppression and space charge effects

These effects are better understood by analysing the current–voltage relationship for a particular precipitator under different operating conditions. The determination of the current–voltage (i–v) characteristics has been thoroughly covered in previous publications [2, 5, 8], so this matter will not be reviewed in detail here.

The theoretical determination of the i–v characteristics for a duct electrode (wire-plate) geometry is a very difficult task. This requires the solutions of the quasi-static electrodynamic Maxwell equations, relating the electric field, the potential and the space charge with the resulting current density field. The problem arises from two particularities of the electrical conduction in gases:

- the electric field that transports the charge also creates it, i.e. when the precipitator voltage exceeds the corona onset level, the ion charges move more rapidly and new ions are generated at a higher rate.
- the moving ionic space charge is comparable with the electrostatic charge on the surface of the electrodes, i.e. the ions distort the field which both generates and transports them.

These characteristics of gas conduction produce a strong interaction between electric field and space charge density, which makes it impossible to calculate these two quantities independently of each other. Furthermore, the differential equations governing the phenomena are highly non-linear.

8.3.1 Electrical characteristics with air load

The *i–v* characteristics have been determined by analytical methods [7], using an approximation in relation to the ion space charge effect. The resulting equation is shown in Appendix 8.A. Here can be seen the influence of the electrode geometry, temperature, pressure, ion mobility, etc., when the precipitator is clean and no gases are passed through it (air load).

The *i–v* characteristics can also be determined by using numerical analysis, normally by using computer calculations, which is more and more extended, as explained in chapter 9. In one of proposed methods [9] the current density is determined for different positions in the interelectrode space for various values of applied voltage (see Figure 8.5). The average current density is then determined for each voltage level, and these values are plotted in the traditional way.

It is possible to find examples of good agreement between calculated and measured *i–v* characteristics obtained from both methods, but due to limited space, only one example will be given. The *i–v* characteristic shown in Figure 8.3 corresponds to an industrial precipitator for a waste incineration plant having rigid electrodes and 400 mm duct spacing, and indicates a good agreement between the measured and the calculated *i–v* characteristic.

8.3.2 Characteristics with dust load

For the same precipitator, *i–v* characteristics have been obtained with the precipitator in operation. The characteristics shown in Figure 8.4 correspond to the first field and the last field. These have been measured for 100% boiler load, a dust load at the ESP inlet equal to 5 g/Nm^3 and a gas

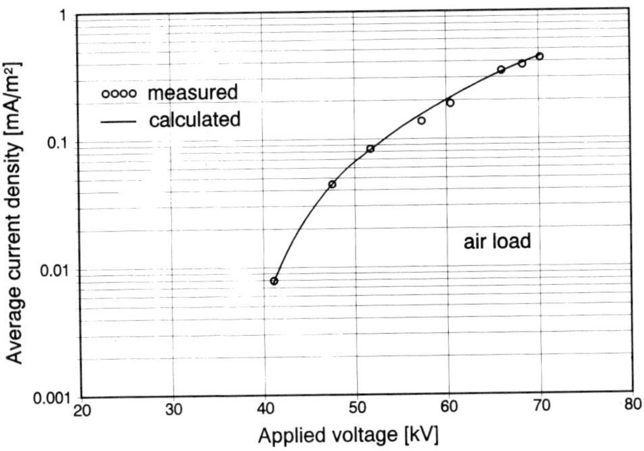

Figure 8.3 Average current density as a function of applied voltage in the case of air load and clean precipitator.

temperature of 165 °C; again, good agreement between the measured and the calculated characteristic for the first field has been obtained [9].

These i–v characteristics illustrate the effect of the particle space charge. As the particle concentration falls along the precipitator, so does the particle space charge density. The charged particles in the interelectrode space contribute to the total space charge density, i.e.

$$\rho_T = \rho_i + \rho_p \tag{8.4}$$

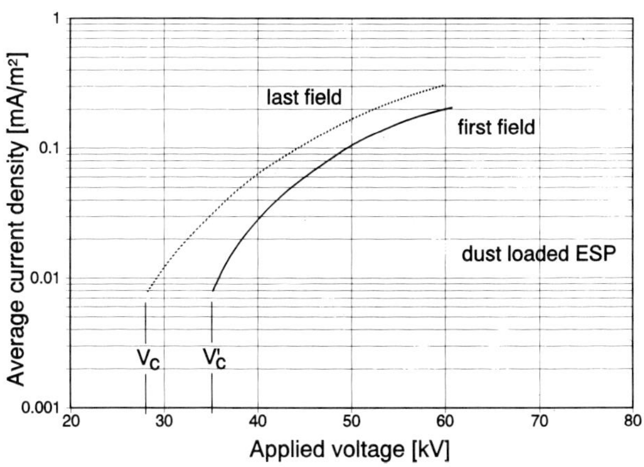

Figure 8.4 i–v characteristics for first and last section with dust-laden gas and normal operating conditions.

where ρ_i is the ion space charge density and ρ_p the particle space charge density expressed normally in C/m^3.

Owing to the low drift velocity of the charged particles, in comparison with the gas ions, the space charge effect of the particles is greater than that of the ions, especially in the region of high particle concentration. In the following it will be assumed that the particle space charge is negligible in the last field. Then the corona onset potential V_c will have the value shown in Figure 8.4.

The particle space charge in the first field partly shields the discharge electrodes and the effect is to weaken the electric field near the discharge electrode surface. In order to attain the critical corona onset field, it is necessary to apply a higher voltage to the discharge electrodes. This is equivalent to an apparent increase in the corona onset potential and the result is that the i–v curve is displaced towards higher voltages with respect to the i–v curve of the last field.

Assuming the particle charge density is constant in the first field, the apparent corona onset potential V_c' can be expressed [5] as follows:

$$V_c' = V_c + \frac{\rho_p}{2\varepsilon_0} s^2 \tag{8.5}$$

where ε_0 is the permittivity in the free space and s is one-half duct spacing and ρ_p is expressed by:

$$\rho_p = \varepsilon_0 p E S_p \tag{8.6}$$

where p is a dimensionless constant related to the relative dielectric constant of the particles, E is the electric field and S_p is the particle surface area per unit volume of gas (m^2/m^3).

The effect of the particle space charge is a partial suppression of the corona current. This is seen by comparing the two curves in Figure 8.4 for the same applied voltage. It is seen, for the same voltage, that the current in the first field is lower than that in the last field. Under extreme conditions, with a very large particle concentration (or a large S_p) at the inlet, the corona can be suppressed or quenched totally.

The corona suppression can also be observed, especially at the inlet field of the precipitator. By using a computer model it has been possible to determine the current density along the field in the gas direction for various voltages [6, 9]. This is illustrated in Figure 8.5 for the same precipitator. It will be seen that the current is suppressed at the inlet of the field and then starts increasing as the particle space charge decreases along the field because of the precipitation of the particles. This allows the ion space charge to increase and in turn the current density increases. When the ion density increases, the particles receive a larger amount of charge and become

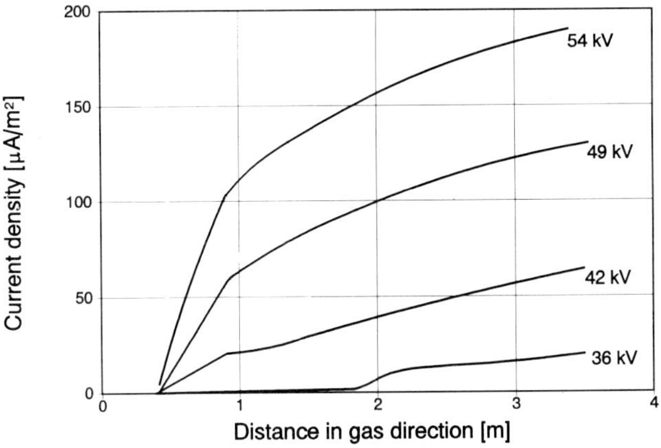

Figure 8.5 Average current density along the inlet field as a function of the distance for various voltage levels.

precipitated faster. This interaction results in a sudden increase in the calculated current density which can be interpreted as a threshold between the low and high corona regions. For high dust concentration at the inlet, or low applied voltages, the threshold moves towards the outlet of the field.

8.4 High tension sectionalization

One of the problems encountered by the designer of electrostatic precipitators, in order to achieve the required collection efficiency at minimum risk, is whether to use a few large bus sections or more smaller bus sections, i.e. to find the optimal degree of sectionalization.

From the previous section, 8.3, it becomes clear that a large electrostatic precipitator cannot be designed as a single bus section, with all the discharge electrodes energized by one common high voltage power supply, because of the uneven current density along the precipitator resulting from the particle space charge, which falls as the dust laden gas becomes cleaner and cleaner. With large particle concentrations, the space charge effect causes partial corona suppression. This effect is larger for the discharge electrodes near the inlet and may be negligible for those near the outlet; therefore, it is impossible to match inlet and outlet discharge electrodes electrically. If all the electrodes are energized from one source, the outlet electrodes will consume most of the current and the converse will happen with the inlet electrodes.

Obviously, the discharge electrode arrangement must be sectioned in order to obtain good precipitator performance. The maximum precipitator efficiency will be theoretically obtained by energizing each discharge electrode separately [2], i.e. optimum energization.

There are other reasons for not using few large bus sections; probably the most important is the reduction in the sparking voltage level, and in consequence the operating voltage and corona power, resulting in a loss of precipitator efficiency. Other reasons are that the construction details are never perfect after the precipitator has been in operation for a long period of time. These may be:

- misalignment of the discharge electrodes
- sharp points or corners in the collecting plates
- vibration of the discharge electrodes
- excessive dust build-up
- electrode warping

Any one of these factors can produce a localized reduction of the sparking level.

Another reason for a poorer efficiency, when using very wide bus sections, is bad gas distribution, or put in another words: it is more difficult to obtain a good gas distribution when very wide and/or high sections are used.

Other authors have mentioned, as a problem, the mismatch between the capacitance of a large bus section (C is proportional with the collection area) and the reduced internal impedance when large TR sets are used. The result may be very large current surges when sparking occurs, which is considered as poor electrical stability. This, however, is no longer a problem, as this can be solved as explained in section 8.5 and [10]. A difficulty that still exists, in the case of localized sparks, is a potential reduction in the life of the electrodes, because of spark erosion due to the larger energy dissipated during the spark. The stored energy in the system immediately before the spark is proportional to the capacitance of the bus section and the square of the applied voltage ($Q = \frac{1}{2}CV^2$).

Finally, there are a number of aspects which also have to be mentioned. One is the higher rapping losses where large sections are used. The other is the case of back-corona, where smaller electrically optimized sections could improve the precipitator performance [9]. This latter aspect is based on theoretical analysis, but its benefits have to be demonstrated by full-scale tests.

All these factors reflect a need for a larger degree of sectionalization. Ramsdell (see [2] and [6]) has derived an exponential relationship between the collection efficiency and the number of bus sections needed per 300 000 m^3/h of gas treated. This is based on performance data for more than 30 fly ash precipitators. The relationship is plotted in Figure 8.6; the curve shown is valid for the utility system analysed by Ramsdell, but for other coal types and operating conditions it is displaced. In spite of the curve being only roughly correct, it serves as a valuable guide in the design

Figure 8.6 Necessary number of sections per 300 000 m^3/h gas recommended in order to obtain a required collection efficiency. Redrawn from [6].

of fly ash precipitators. Those precipitators which fail to meet the requirements expressed by Figure 8.6 may be expected to perform poorly or marginally on the average [2].

There are many examples of improved performance when two bus sections, previously connected in parallel and energized by one TR set, have been separated and energized by its own TR set. The reason for not using a larger degree of sectionalization is normally economic, as the cost of power supplies and high voltage insulation is lower for large sections. This puts economic pressures on the design engineer who tends to minimize the number of sections. A more profound analysis, however, may indicate that a smaller precipitator with more sections and TR sets could be more efficient and cheaper. Finally, it should be mentioned that smaller and cheaper power supplies could economically allow an increase in the degree of sectionalization, so that more efforts in this area are necessary.

8.5 Traditional DC energization

The traditional DC energization used in the last two decades is obtained with the power supply depicted in Figure 8.7. In order to control the corona power, the line voltage is regulated by a thyristor controller (phase control by a pair of antiparallel thyristors) before it is applied to the primary of the high voltage transformer. By choosing a suitable turns ratio (n) the primary voltage is raised to the desired secondary level and then rectified by a high voltage silicon bridge rectifier. The rectified secondary voltage is applied directly to a precipitator section without additional filtering.

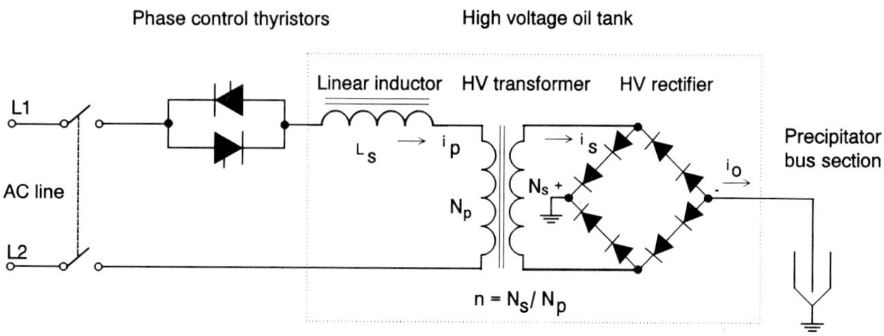

Figure 8.7 Main circuit of a traditional power supply for electrostatic precipitators.

The high voltage rectifier is connected in such a way that the output voltage has a negative polarity, i.e. negative corona is generated in the precipitator.

Normally a current limiting inductor (L_s) is included in series with the primary winding of the transformer in order to increase the short-circuit impedance of the power supply to limit the current during sparking, arcing or short-circuit in the ESP.

8.5.1 Basic principles

The output voltage of the power supply, applied to a precipitator bus section, is controlled by varying the firing angle of the thyristors, i.e. by delaying or advancing the firing instant in relation to the zero crossing (cross-over) of the line voltage.

The principle is shown in Figure 8.8 for a 50 Hz power supply. This shows the waveforms obtained when the thyristors are fired at a time equal to 3 ms after the line voltage zero crossing, i.e. a firing angle $\alpha_o = 180 \times 3/10 = 54$ (°el). When a thyristor is fired, it changes from a state of high resistance to low resistance and the primary current starts increasing and becomes as shown in Figure 8.8a. Its magnitude and duration is determined by the current limiting inductor (L_s), the leakage inductance of the HV transformer and the precipitator load.

The output current (i_o), the so-called precipitator current, has a waveform corresponding to the rectified secondary current. The amplitude difference between the primary and the secondary current is given by the turns ratio ($i_s = i_p/n$). The precipitator current and the resulting output voltage, the so-called precipitator voltage, are as depicted in Figure 8.8b. These are negative because of the negative corona, but in this chapter, for the sake of simplicity, their absolute values will be used. These waveforms are typical for a medium size power supply energizing a normal precipitator load.

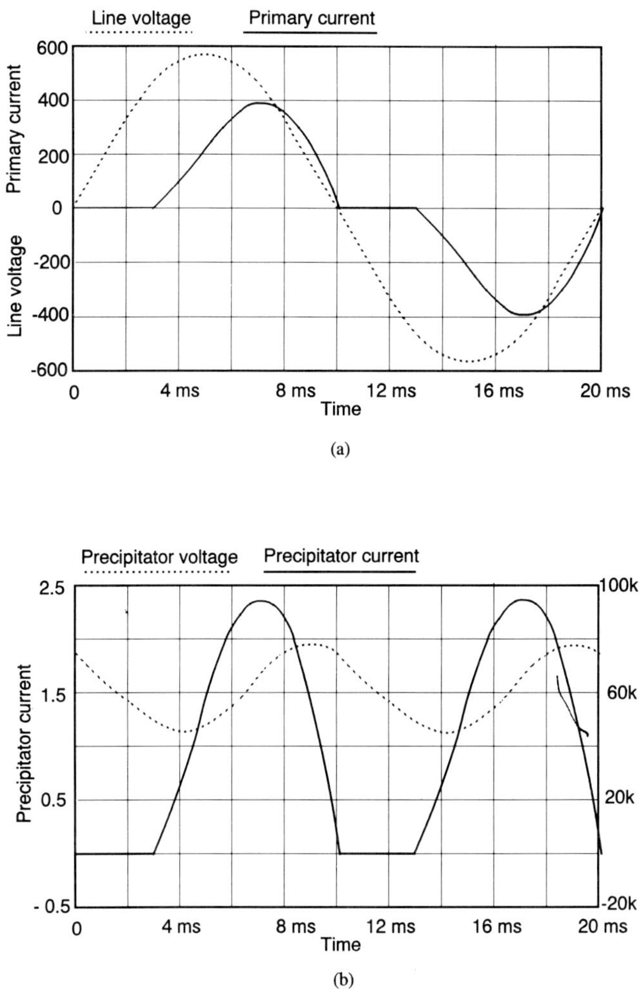

Figure 8.8 Voltage and current waveforms with DC energization and a firing angle of 54° (3 ms).

The precipitator voltage has a considerable ripple, because of the inherent capacitance of the precipitator section and the fact that no filtering is used. The precipitator voltage is characterized by its:

- peak value ($V_{o\,peak}$)
- mean value ($V_{o\,mean}$)
- minimum value ($V_{o\,min}$)

The peak value depends on the charge delivered in one current pulse (Q_p), i.e. the area under the current pulse. Because the precipitator

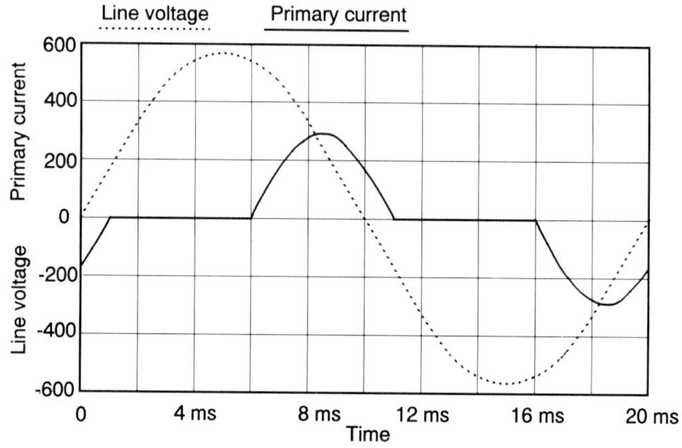

(a)

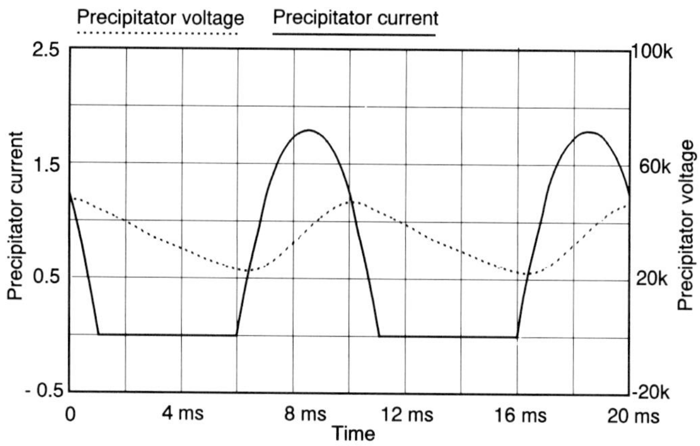

(b)

Figure 8.9 Voltage and current waveforms with DC energization and a firing angle of 108° (6 ms).

capacitance is discharged in the time interval between current pulses the voltage decreases. The voltage level can also be decreased by delaying the firing angle α_o, resulting in a lower precipitator voltage as seen in Figure 8.9b. These waveforms illustrate the situation existing when the firing instant occurs 6 ms after the zero crossing. Table 8.1 shows the most relevant quantities and their values for both firing angles. These values were obtained by computer-aided design evaluation.

It is clearly seen that both the precipitator voltage and the precipitator current are reduced when a delayed firing is used. The extreme case

Table 8.1 Current and voltages with two different firing angles

Firing angle	t_o(ms)	3	6
Primary current	$I_{\text{p rms}}$(A)	223	145
Precipitator current	$I_{\text{o rms}}$(mA)	1400	920
	$I_{\text{o peak}}$(mA)	2350	1800
	$I_{\text{o mean}}$(mA)	1030	576
Precipitator voltage	$V_{\text{o peak}}$(kV)	78	47
	$V_{\text{o mean}}$(kV)	61	35
	$V_{\text{o min}}$(kV)	46	23

corresponds to $t_o = 10\,\text{ms}$ ($\alpha_o = 180°$) giving a stationary current and voltage equal to 0.

The earliest firing of the thyristor is influenced by the value of the precipitator voltage at the firing instant which is equal to $V_{\text{o min}}$. This is greater than the corona onset voltage and is influenced by the precipitator geometry and process conditions. With precipitator loads and a 50 Hz power supply, the earliest firing is in the order of 2 to 3 ms, corresponding to 36 to 54°.

Another important quantity is the form factor (FF) of the precipitator current, defined as

$$\text{FF} = \frac{I_{\text{orms}}}{I_{\text{omean}}} \tag{8.7}$$

where

$$I_{\text{orms}} = \sqrt{\frac{1}{T}\int_0^T i_o^2(t)\,\text{d}t} \tag{8.8}$$

and

$$I_{\text{omean}} = \frac{1}{T}\int_0^T |i_o(t)|\,\text{d}t \tag{8.9}$$

The output current $i_o(t)$ is periodic and $|i_o(t)|$ is its absolute value.

A calculation of the form factor of the precipitator current for $t_o = 3\,\text{ms}$ gives a value of 1.4. This is a typical design value used in power supplies.

Since the precipitator current is not a pure sinusoidal wave, it is not easy to calculate its mean and rms value. This can be done, however, by computer simulation or by using approximations, as indicated in Appendix 8B.

8.5.2 *High voltage power supply ratings*

The high voltage power supply for electrostatic precipitators is characterized by the turns ratio (n) and the short-circuit impedance present in the main circuit (X_{Lsc}). It is predominantly inductive and is composed of the leakage reactance of the transformer and the linear inductor L_s, as illustrated in Figure 8.7. This inductor is added in order to increase X_{Lsc} above 5 to 10%, which is a typical value of the leakage reactance of the HV transformer. The aim is to limit the current surges that might occur during sparking in the precipitator and thus:

- increase the lifetime of the precipitator internals
- protect the electrical equipment, and
- obtain a more stable electrical operation.

The value used by most of the manufacturers is in the order of 30–40%, which reduces the current surges to 3.3 and 2.5 times the rated value, respectively.

By knowing the short-circuit reactance, X_{Lsc}, and the turns ratio, n, all the important quantities of the power supply can be determined for different loads. The ratings can be expressed in different ways, but in this chapter the most usual practice in Europe will be used. The following quantities are normally indicated on the rating plate:

• Precipitator mean current ($I_{o\,nom}$)	mA
• Precipitator peak voltage @no load ($V_{o\,nom}$)	kV_{pk}
• Primary rms current ($I_{p\,nom}$)	A
• Line voltage ($V_{L\,nom}$) and frequency (f)	V and Hz
• Apparent input power (S)	kVA

8.5.2.1 Precipitator mean current. The precipitator current has the waveform depicted in Figures 8.8b and 8.9b. Its mean value is given by

$$I_{o\,nom} = \frac{1}{T}\int_0^T i_o(t)\mathrm{d}t \tag{8.10}$$

where $i_o(t)$ is the precipitator current and T is the period of the line frequency.

The rated precipitator mean current is the maximum mean current the power supply is able to deliver to a load, without exceeding the design value of the current form factor. Some manufacturers carry out tests on the power supply with a RC-load, which simulates a precipitator load. Others use a pure resistive load, which gives a lower current form factor. This means that, at rated mean load current, the rms value of the secondary and the primary currents have a lower value compared with the corresponding values of a precipitator load.

8.5.2.2 Primary rms current. The waveform of the primary current is shown in Figures 8.9a and 8.9b. Its rms value is defined as

$$I_{\mathrm{Pnom}} = \sqrt{\frac{1}{T}\int_0^T i_{\mathrm{p}}^2(t)\mathrm{d}t} \tag{8.11}$$

This can also be expressed by equation (8.12).

$$I_{\mathrm{Pnom}} = n\,\mathrm{FF}\,I_{\mathrm{onom}} \tag{8.12}$$

where n is the turns ratio, FF is the form factor used by the manufacturer (typical value 1.35 to 1.4) and $I_{\mathrm{o\,nom}}$ is the rated precipitator mean current.

8.5.2.3 Precipitator peak voltage at no-load. At no-load, the output current of the power supply $i_{\mathrm{o}} = 0$ and the primary current is equal to the magnetizing current of the HV transformer. As this is negligible compared with the rated primary current, the peak voltage at no-load is equal to

$$V_{\mathrm{onom}} = \sqrt{2}\,nV_{\mathrm{Lnom}} \tag{8.13}$$

where $V_{\mathrm{L\,nom}}$ is the rated rms value of the line voltage.

8.5.2.4 Apparent input power. This quantity is especially important in the sizing of the electrical installation. The apparent input power is defined by equation (8.14).

$$S = I_{\mathrm{Pnom}} V_{\mathrm{Lnom}} \tag{8.14}$$

i.e. the product of the line voltage and the primary current.

The active input power is given by

$$P = S\cos\varphi_1 \tag{8.15}$$

where φ_1 is the phase angle between the line voltage and the fundamental frequency component (50 Hz) of the primary current ($\cos\varphi_1$ is normally known as the power factor).

The active power cannot be expressed as a rated value, because it varies with the precipitator load for the same rated precipitator mean current. The active power is normally measured with a wattmeter or calculated by means of computer simulation.

The power factor is normally better than 0.8 at rated current, assuming the form factor is approximately 1.4.

8.5.2.5 Example. The waveforms depicted in Figure 8.8 are obtained with a power supply having the following data:

- Rated precipitator mean current: 1000 mA
- Rated precipitator peak voltage: 90 kV
- Line voltage: 400 $\mathrm{V_{rms}}$

- Short-circuit reactance: 35%
- Design form factor: 1.4

The precipitator capacitance used is 130 nF.
The turns ratio is

$$n = \frac{1}{\sqrt{2}} \frac{V_{o_{mean}}}{V_{L_{nom}}} = \frac{1}{\sqrt{2}} \frac{90\,000}{400} = 159$$

The precipitator rated rms current is

$$I_{o_{nrms}} = I_{o_{nom}} \mathrm{FF} = 1000 \times 1.4 = 1400\,\mathrm{mA}$$

The rated primary rms current is

$$I_{p_{rms}} = nI_{o_{rms}} = 159 \times 1.4 = 223\mathrm{A}$$

The rated apparent power is

$$S = I_{p_{nom}} V_{L_{nom}} = 223 \times 400 = 89.2\,\mathrm{kW}$$

8.5.3 *Influence of the linear inductor*

Because this aspect has not been illustrated sufficiently in the past, a brief explanation will be given in relation to the importance of this component.

8.5.3.1 Advantages and disadvantages. The linear inductor's main function is to limit the current surges during sparking in the precipitator, but, at the same time, it has a number of advantages related to the electrical operation of the precipitator. To illustrate these advantages, Figure 8.10 shows the waveforms of the current and precipitator voltage when the linear inductor is not used, i.e. the current is only limited by the leakage reactance of the high voltage transformer. In Table 8.2 are shown the values obtained, when the short-circuit reactance of the transformer is ≈9%, for the same rated mean current and voltage as obtained with a linear inductor. The figures in parentheses correspond to the values obtained with a normal short-circuit reactance of 35%.

Comparing these results with those obtained with a normal short-circuit reactance, valid for the same precipitator mean current and precipitator load, the following disadvantages can be pointed out:

- The peak value of the precipitator current is higher and its duration is shorter, giving a higher form factor (36% higher).
- The higher form factor causes a higher primary current and apparent input power for the same precipitator mean current.
- The time occurrence of the precipitator voltage peak is closer to that of the line voltage peak. Since sparking occurs around the peak of the

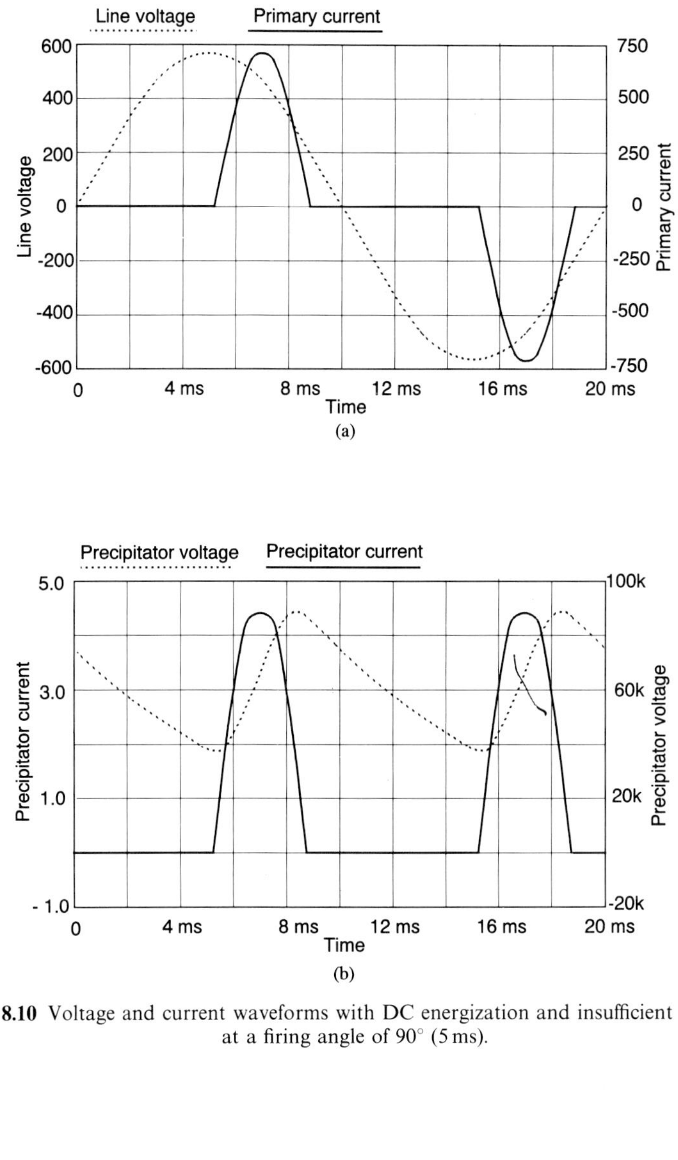

Figure 8.10 Voltage and current waveforms with DC energization and insufficient inductance at a firing angle of 90° (5 ms).

precipitator voltage, the current surges will have a higher amplitude and a longer duration. These characteristics are detrimental for stable operation of the precipitator, especially in relation to voltage recovery after spark.

- The precipitator current is obtained with a delayed firing angle, giving a larger phase angle between the line voltage and the fundamental component of the primary current. This results in a lower power factor which is not desirable by the power utilities.

Table 8.2 Current and voltages with insufficient shortcircuit impedance (comparison with values obtained with normal impedance level)

Firing angle	t_o(ms)	5	(3)
Primary current	$I_{p\,rms}$(A)	302	(223)
Precipitator current	$I_{o\,rms}$(mA)	1900	(1400)
	$I_{o\,peak}$(mA)	4400	(2350)
	$I_{o\,mean}$(mA)	1000	(1030)
Precipitator voltage	$V_{o\,peak}$(kV)	90	(78)
	$V_{o\,mean}$(kV)	61	(61)
	$V_{o\,min}$(kV)	38	(46)

As theoretical advantages, the following can be maintained:

- The precipitator voltage waveform is more pulsating because the peak value is higher. This could have a positive effect in the case of high resistivity particles. (At the present this problem is solved by using intermittent energization or pulse energization. See later sections.)
- In the case of precipitator loads requiring a higher voltage, the power supply is better suited to deliver its rated current because of the lower voltage drop in the short-circuit reactance.

8.5.3.2 Physical implementation. The linear inductor may consist of an air coil inductor or an iron core inductor with a suitable air gap giving a linear characteristic. This is normally placed inside the high voltage tank and its inductance cannot be changed. This is a good economic solution, provided a good match exists between the size of the power supply and the bus section.

Some manufacturers place the inductor inside the control cabinet, which can then be provided with tappings. These allow one to change the inductance value to overcome any mismatch between the power supply and the bus section. This solution has the inconvenience of occupying space in the control cabinet and causing acoustic noise.

In recent years, a variable inductor has been introduced, whose inductance value changes inversely with the value of the primary current inside a certain range. An automatic control loop increases the inductance value when the primary current decreases in order to keep the form factor constant.

8.6 Intermittent energization

Intermittent energization (IE) is a recent method introduced in the early 1980s, with the purpose of saving energy and improving the collection efficiency with high resistivity dusts. This energization form is also known under other trade names like energy control, semi-pulse, variopulse, etc.

IE operation emerged as a cheaper alternative to pulse energization which was already developed and used commercially in the solution of high resistivity dust problems.

8.6.1 Basic principles

Intermittent energization is obtained with the same electrical equipment employed in traditional HV power supplies. The difference resides in the

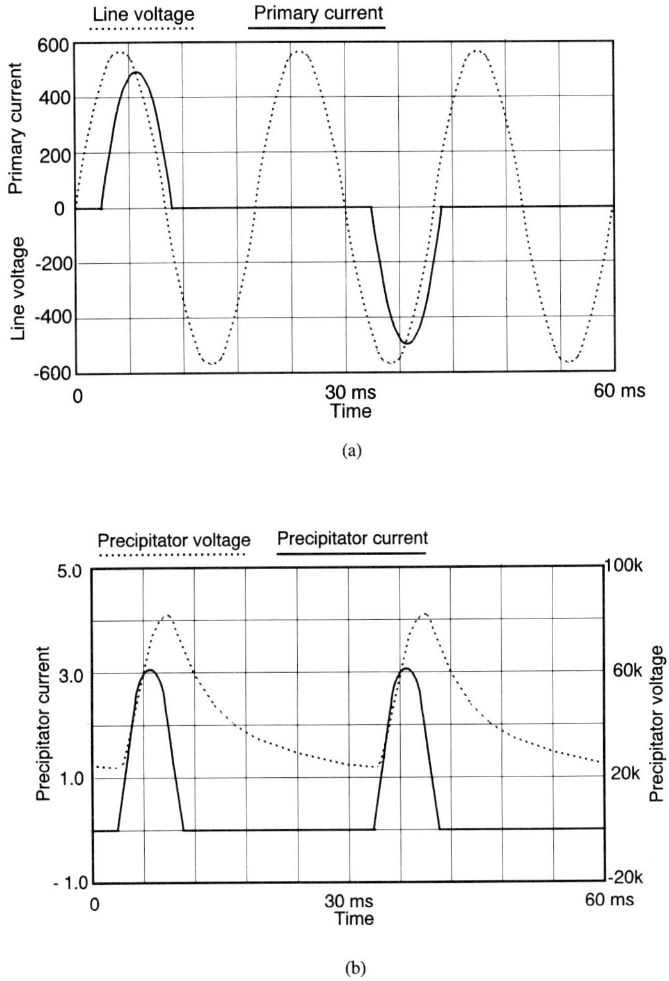

Figure 8.11 Voltage and current waveforms with intermittent energization and a firing angle of 54° (3 ms). Two out of three pulses suppressed ($D = 3$).

automatic voltage control equipment, which allows the suppression of a certain number of half-cycles of the primary current delivered to the transformer by the AC line.

This suppression is obtained by not firing the phase control thyristors in the respective half-cycles or alternatively, by using a firing angle $\alpha_o = 180°$. The principle is illustrated by the waveforms in Figure 8.11, for a 50 Hz line. This example shows six half-cycles of the line frequency and illustrates the case where two out of three current pulses are suppressed.

Figure 8.11a shows the primary current in relation to the line voltage. The corresponding waveforms of the precipitator voltage and current are shown in Figure 8.11b. The firing angle corresponds to 3 ms after the zero crossing of the line voltage.

8.6.2 Comparison with traditional DC energization

The waveforms in Figure 8.11 show the following differences compared with DC energization for the same firing angle in Figure 8.8:

- The peak value of the precipitator voltage is higher.
- The minimum (trough) value of the precipitator voltage is lower.
- Due to the suppression of two current pulses, the mean and the rms values of the precipitator current are reduced.

As a consequence, the following can be established:

- The mean value of the precipitator voltage is lower.
- The corona power delivered to a particular bus section of a precipitator is lower.
- The power consumption of the precipitator is reduced.

Table 8.3 Current and voltages with intermittent energization (comparison with values obtained with DC energization)

Energization form		IE	(DC)
Primary current	$I_{p\,rms}$ (A)	172	(223)
Precipitator current	$I_{o\,rms}$ (mA)	1080	(1400)
	$I_{o\,peak}$ (mA)	3100	(2350)
	$I_{o\,mean}$ (mA)	476	(1030)
Precipitator voltage	$V_{o\,peak}$ (kV)	82	(78)
	$V_{o\,mean}$ (kV)	41	(61)
	$V_{o\,min}$ (kV)	24	(46)

These results are summarized in Table 8.3 and discussed in the following paragraphs:

8.6.2.1 Peak voltage. The peak voltage is higher because the area under the current pulse is greater. As this area corresponds to the electric charge Q_p delivered to the precipitator section, the larger the value of Q_p, the higher the peak voltage because the precipitator load has an inherent capacitive component.

8.6.2.2 Minimum voltage. Because of the longer time interval between current pulses, i.e. without receiving electrical charge, the precipitator is discharged towards the corona onset voltage. This causes a lower minimum value.

8.6.2.3 Mean voltage. This is proportional to the area under the precipitator voltage in one energization duty cycle (in this example: three half-cycles of the line frequency). Because of the lower minimum value the mean value also becomes lower.

8.6.2.4 Mean current and degree of intermittence. The mean value of the precipitator current is reduced due to the suppression of a number of current pulses. This suppression of current pulses is expressed by the so-called 'degree of intermittence' D.

D is defined as the number of half-cycles included in one energization duty cycle divided by the number of current pulses in this time interval. In the example shown in Figure 8.11 the energization cycle is three, so the degree of intermittence $D = 3$. Another example: if the thyristors are fired once and then kept blocked for the next ten half-cycles, then $D = 11$.

The degree of intermittence is also expressed by other names, e.g. the so-called 'charge ratio' δ. This is defined as the number of current pulses in one energization duty cycle divided by the number of half-cycles included in the energization cycle. For instance, the charge ratio in Figure 8.11 is:

$$\delta = 1{:}3$$

expressing that the IE operation, in this particular case, consists of one current pulse out of three. The charge ratio is the reciprocal value of the degree of intermittence.

Assuming the area under the current pulse is the same for IE and DC energization, and if the mean current obtained with DC energization is I_{DC}, the mean current obtained with intermittent energization I_{IE} can be expressed by

$$I_{IE} = \frac{I_{DC}}{D} \tag{8.16}$$

As shown in Figure 8.11 the area under the current pulses may sometimes be higher with intermittent energization by a factor k. Then, the mean current is

$$I_{IE} = k\frac{I_{DC}}{D} \tag{8.17}$$

where k may typically vary between 1 and 1.5. The factor k in the example shown in Table 8.3 is 1.39.

The mean current is then reduced by a factor k/D. Assuming that the primary current pulse has a similar waveform in both cases, the rms value is reduced by a factor equal to k divided by the square root of D.

There are two aspects where IE is inferior compared with DC energization:

- the form factor of the primary current, and
- the saturation of the transformer magnetic core

From Table 8.3, it is clearly seen that the relationship between the rms value and the mean value of the precipitator current, i.e. the form factor, is higher than the one obtained with DC energization. This results in a corresponding higher form factor for the primary current, which is equivalent to a higher harmonic content.

Because of the pause interval introduced by IE, the value of the magnetic induction variation (ΔB), which is possible to use without saturating the core of the HV transformer, is lower than in the case with DC energization. This problem can be avoided in different ways, e.g.

- by using a larger transformer core
- by using an auxiliary current pulse before the main one
- by using a larger linear series reactor
- by using a core with high remanence and low eddy losses

8.6.3 *Collection efficiency*

Intermittent energization (IE) reduces the corona power and in consequence produces energy savings. The question arises, in which way does this reduction in the corona power influence the collection efficiency of the precipitator? The normal way to evaluate the collection efficiency obtained with intermittent energization is to compare it with the efficiency obtained with DC energization. A practical way is to compare the two migration velocities and then to find the enhancement factor H defined as

$$H = \frac{\omega_{IE}}{\omega_{DC}} \tag{8.18}$$

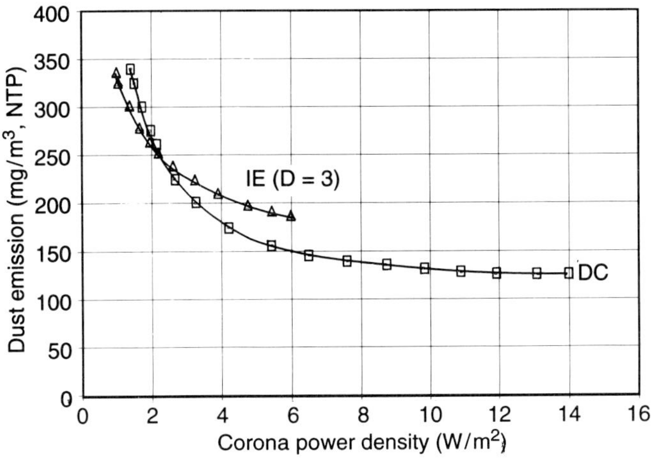

Figure 8.12 Stack emission as a function of the corona power with DC and IE for low resistivity conditions.

where ω_{IE} is the migration velocity obtained with intermittent energization and ω_{DC} is the migration velocity obtained with DC energization.

The required collection surface is inversely proportional to ω for a given collection efficiency, so if $H > 1$, then H is the factor by which the collection surface of a DC energized precipitator should be increased in order to obtain the same collection efficiency provided by intermittent energization.

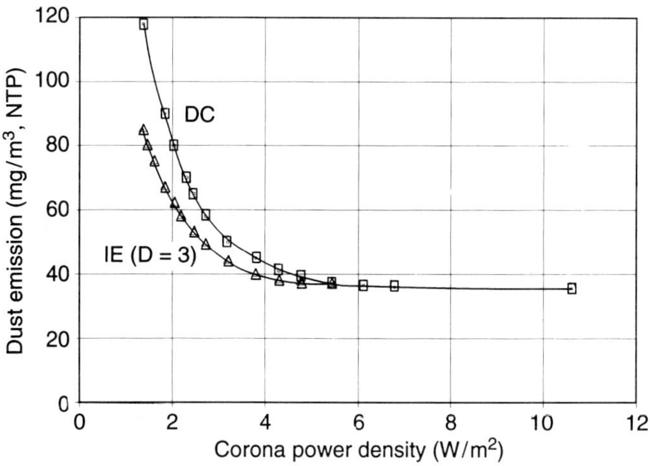

Figure 8.13 Stack emission as a function of the corona power with DC and IE for medium resistivity conditions.

Intermittent energization results in a lower precipitator mean voltage and a lower precipitator mean current. Then, for the case of low resistivity dusts (see Figure 8.2), it can be assumed that this reduction in corona power will result in a lower migration velocity and therefore a lower collection efficiency. This is illustrated in Figure 8.12, where dust emission is plotted as a function of the corona power density P_c/A for DC and intermittent energization (intermittence degree $D = 3$). The precipitator collected fly ash from a boiler fired with an English/Polish coal blend causing no resistivity problems. In this situation IE results in a lower performance and therefore it is not advantageous for this application ($H < 1$).

In Figure 8.13, the same curves are plotted for a precipitator collecting fly ash from a boiler fired with South African coal, which results in some degree of back-corona. It is seen that both energization forms produce the same minimum dust emission ($H = 1$), but, if a higher dust emission is allowable (>40 mg/m^3 NTP), then IE is better for the same power density.

In Figure 8.14 the dust emission is plotted as a function of the power density for DC and intermittent energization ($D = 7$) for fly ash arising from a boiler fired with Russian coal, which caused severe back-corona in the precipitator. It is seen that IE is clearly better than DC energization ($H > 1$), since it results in a lower dust emission at a lower power consumption.

These three practical examples show that:

- Intermittent energization always results in a lower power consumption, but it does not always give the highest collection efficiency.
- The performance with intermittent energization is closely related to the resistivity of the collected dust:

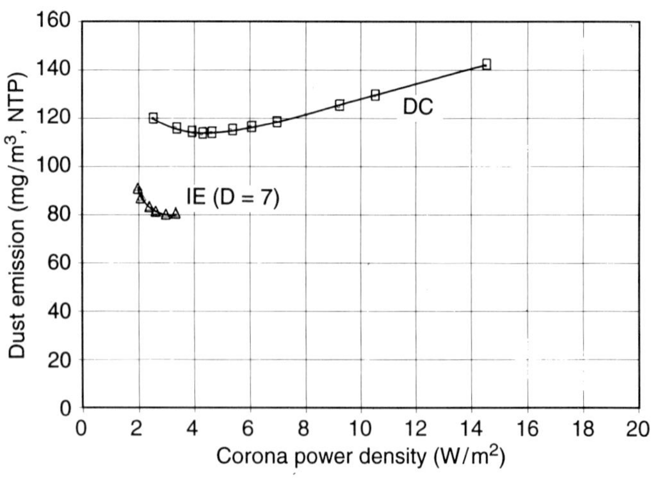

Figure 8.14 Stack emission as a function of the corona power with DC and IE for high resistivity conditions.

in the case of high resistivity, IE is superior to DC energization.
in the case of medium resistivity IE is as good as DC energization and sometimes better.
in the case of low resistivity, IE is inferior to DC energization, since it results in a lower precipitator performance.

The above-mentioned statements are generally valid for a great number of applications. The ESP manufacturers, however, need to consider other factors when giving guarantees, regarding the enhancement factor H that can be obtained by using IE. For instance, one difficult case is when the precipitator has to collect high resistivity fly ash and at the same time the unburnt carbon content of the dust is high.

8.7 Automatic voltage control and instrumentation

8.7.1 Introduction

Other authors and contributors to this field [1, 2] have in the past emphasized the importance of optimizing electrical energization in order to obtain the maximum collection efficiency. To accomplish this objective, two important aspects have to be taken into account:

- a good matching between the size of the TR set and the energized bus section.
- a good automatic voltage control (AVC) unit.

The first aspect has been covered in section 8.5. As explained, the key parameter governing the corona power delivered by the TR set to a particular bus section is the firing angle of the thyristors. This angle is determined by the control unit for every half-cycle of the line frequency and must have the correct value according to the existing operating conditions within the precipitator.

Automatic control is closely related to the instrumentation used. Therefore, in the following, a brief review of the signals used in the AVC units will be given.

8.7.2 Instrumentation

The signals used in the AVC units by different manufacturers are not always the same. The Europeans have a long tradition of using the precipitator current and voltage, the so-called 'secondary values', whereas the Americans have preferentially used the 'primary values', but in recent years the tendency has been to incorporate the secondary values in their AVC units.

Furthermore, the installation of opacity (or extinction) meters in the stacks is more and more common, and in some countries they are compulsory, especially in connection with new plant. The signal delivered by these

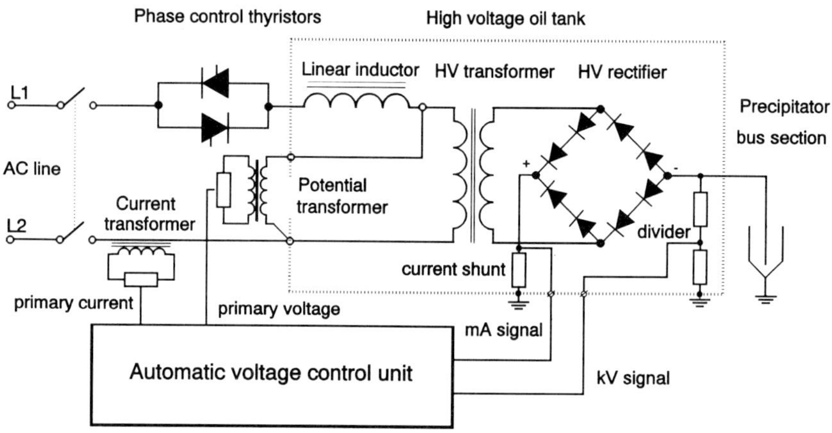

Figure 8.15 Signals used in the instrumentation and automatic control of a high voltage power supply.

meters is used for continuous monitoring of the stack dust emission, but sometimes it is also used by the control units. The purpose of the opacity meter in conjunction with the AVCs is:

- optimization of the operation of the precipitator, and
- energy savings under easy operating conditions.

The signals normally used by the control units are depicted in Figure 8.15.

8.7.2.1 Secondary values. These are the precipitator voltage and the precipitator current. The voltage is measured by means of a voltage divider and the current by means of a measuring resistor or current shunt.

In a modern control unit the following quantities are normally measured and displayed:

- precipitator mean current ($I_{o\,mean}$)
- precipitator mean voltage ($V_{o\,mean}$)
- precipitator peak voltage ($V_{o\,peak}$)
- precipitator minimum (trough) voltage ($V_{o\,min}$)

In the past, normally the mean values were used for the control tasks but, in the last decade, the importance of measuring, for instance, the minimum value has become more significant [10].

This value is vitally important in evaluating the operation of a precipitator collecting high resistivity dust. Other important tasks are the automatic detection of back-corona and the automatic control of the degree of intermittence [10, 11].

The peak value is important in determining the sparking level in the precipitator and in optimizing the electrical operation, e.g. during voltage recovery after a spark, and in computing the corona power in an approximated way (see equation 8.3).

In the past, these values of minimum and peak voltage were measured with an oscilloscope using a voltage divider connected to the discharge frame if one was not built into the TR set. This measurement was cumbersome for less experienced plant personnel, but today it is a standard feature in many modern AVC units using solid state circuitry (ICs).

8.7.2.2 Primary values. These signals are also shown in Figure 8.15. Some control units use them in the automatic voltage control and/or monitoring tasks. Nowadays, it is recognized that the use of the secondary values is superior for the automatic control of the high voltage and in the evaluation of the precipitator operation, e.g. spark detection, voltage recovery after spark and back-corona detection.

The primary values, however, can be used in important monitoring tasks, e.g. in the determination of:

- rms value of the primary current
- rms value of the primary voltage
- active power delivered to the TR set
- apparent power delivered to the TR set

The primary voltage and current are measured by means of a potential and current transformer, respectively. In this way these signals can readily be connected to the AVC unit, as the transformers provide isolation and adequate signal levels.

The active power delivered to the TR set has two components:

- the corona power delivered to the bus section, and
- the losses in the TR set (transformer iron and copper losses, silicon diode conduction losses, etc.).

The active power delivered by the AC line has two main components:

- the active power delivered to the TR set, and
- the conclusion losses in the thyristors.

Because the last is negligible, compared with the corona power, the active power measured by means of the primary values is approximately equal to the active power delivered to the TR set.

The primary values are also important in monitoring tasks like:

- state of the phase control thyristors
- line overcurrent

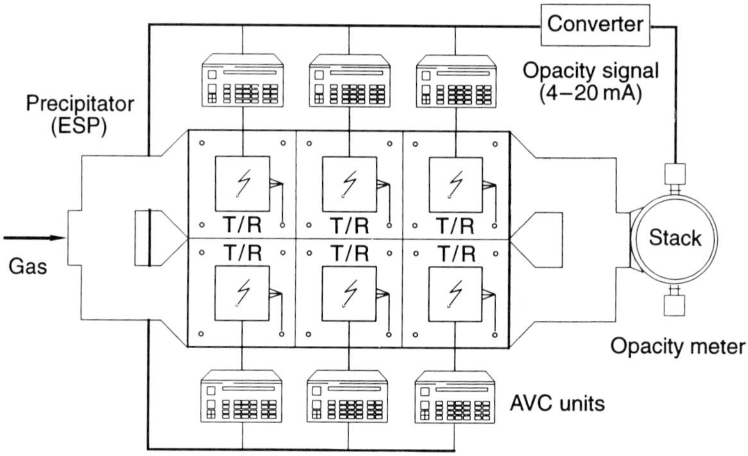

Figure 8.16 Simple energy management system (EMS) for an ESP.

- line overvoltage
- saturation of the HV transformer
- too high a form factor, etc.

8.7.2.3 Opacity signal. The principle used is depicted in Figure 8.16. This shows a three-field precipitator with two bus sections per field. Each section is energized by a separate TR set, which in turn is controlled by an AVC unit. The opacity (or extinction) meter is mounted in the stack and delivers a 4–20 mA signal to a converter, where the opacity signal can be filtered, converted to a digital signal, etc., before it is connected to the individual control units. This signal is a measure of the dust emission in the stack based on a calibration performed by previous gravimetric measurements. The measured dust emission is compared with a setpoint in each control unit, resulting in a control action on the corona power in order to accomplish a particular objective.

This is a simple and economic solution and must not be confused with the more expensive approach, where the control units are connected to a common communication bus, which is connected to a 'master' computer, or to a plant computer, via a 'gateway' unit. This computerized central control of precipitators is covered later in section 8.9.

8.7.3 Basic control principles

Under ideal operating conditions, the firing angle of the thyristors could be controlled manually, but in practice this is impossible. Most of the processes

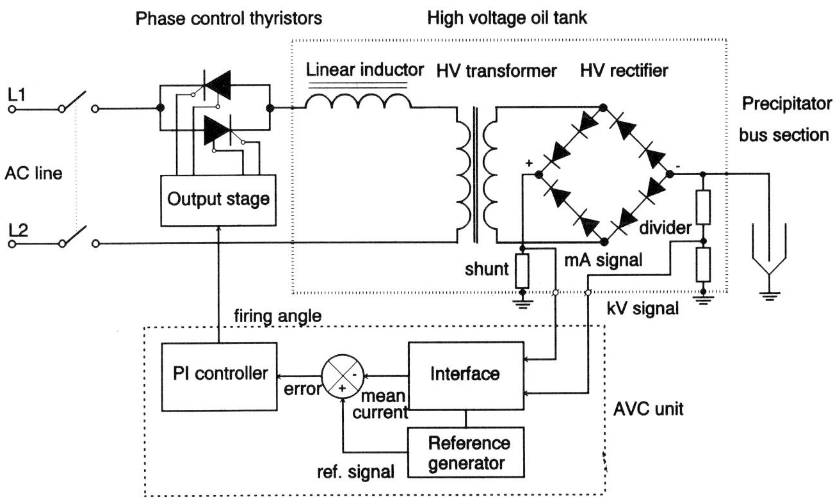

Figure 8.17 Principle of the closed loop automatic control of the precipitator current.

which use electrostatic precipitators are subject to both slow and fast changes in the operating conditions. The gas flow, gas temperature, gas humidity, fuel, raw material, etc., can change frequently. In order to keep the collection efficiency as high as possible under difficult conditions, more powerful and sophisticated control units are appearing all the time.

One basic architecture is illustrated by the block diagram in Figure 8.17. Here it is supposed that the mA signal is used as the feedback signal, i.e. the precipitator mean current is the controlled parameter in a closed loop. In other words, the firing angle of the thyristors is varied by a proportional-integral (PI) controller in such a way that the mean current follows a reference signal (or a time varying setpoint) as closely as possible.

The firing pulses to the thyristors are delivered by an output stage, providing an adequate signal level and isolation from the AC line. The kV signal is also shown connected to the control unit, but is mainly used in connection with spark detection and voltage recovery as explained in the next section.

Nowadays, most of the control units are based on microprocessors and peripheral circuits, which offer very powerful performance because of their inherent memory and computing capabilities. The reference signal varies as a function of time, according to a programmed control strategy.

The basic control principle is illustrated in Figure 8.18. In this example, the mean current is increased linearly at a rate of rise R, until a spark occurs or an upper limit is reached. R is normally expressed in %/min where 100% corresponds to the rated current of the TR set.

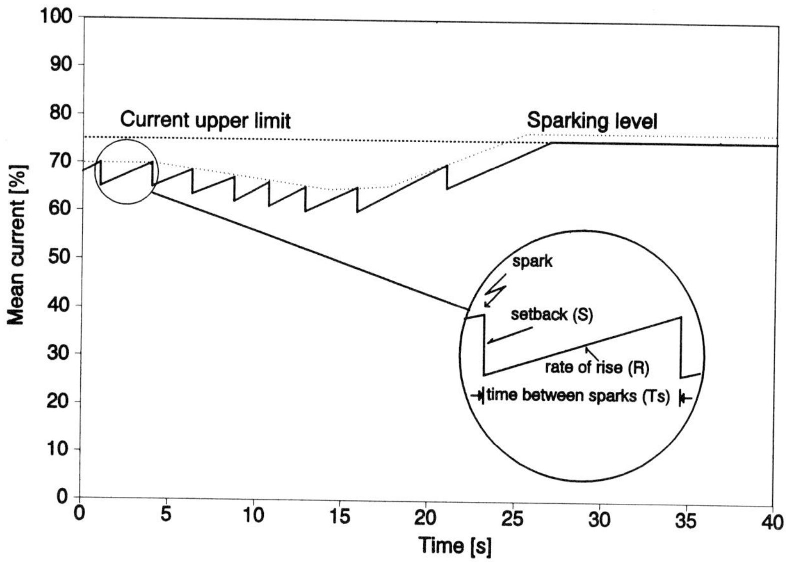

Figure 8.18 Basic control strategy for the precipitator mean current.

The sparking level changes in the way shown. This is fairly constant at the beginning, and then falls; it remains low during a short period and then increases again. When a spark occurs, the current is automatically reduced by a constant setback value S. In this example S has an absolute value expressed as a percentage of the rated current (5%). Assuming a constant sparking level, the spark rate SPR can be expressed as the reciprocal value of the time interval between two sparks T_s.

From the zoomed area in Figure 8.18 it can be seen that the rate of rise is determined by

$$R = \frac{S}{T_s} (\%/\text{min})$$

Then the spark rate can be expressed by

$$SPR = \frac{1}{T_s} = \frac{R}{S} (\text{sparks/min}) \tag{8.19}$$

Equation (8.19) indicates that a high spark rate can be obtained with a high rate of rise R and a small setback S of the controlled variable. Conversely, a low rate of rise and a large setback give a low spark rate. In the example shown in Figure 8.18, the rate of rise $R = 100\%/\text{min}$ and the setback $S = 5\%$; then the spark rate, at a stable sparking level, will be 20 sparks/min.

When the sparking level is decreasing, the time between sparks is shorter and the spark rate is higher, but if the sparking level increases, the spark rate becomes lower. When the controlled variable reaches the upper limit the spark rate becomes zero.

The parameters *R* and *S* or *S* and SPR are normally found as settings in all control units. *S* is used as an absolute or relative value. The way these parameters are used and set by the specialists of the precitator suppliers is different, each advocating having the best control strategy. Often this is based on tradition, for the processes where the control units are normally used, or in particular characteristics of their precipitator design.

Because of the variety of existing control strategies, e.g. current/spark rate, voltage hill climbing, etc., only the general principles will be reviewed. In assessing a particular control unit it is important that this has been proven for difficult processes, e.g. those with fast varying operating conditions, like metallurgical plants, cement kilns, etc. It is also important that the control unit has been developed by the precipitator manufacturer, i.e. by people with experience in precipitator theory and operation.

With respect to the basic control strategy as shown in Figure 8.18, and in order to maintain a high corona power level, during varying conditions:

- the rate of rise *R* has to be high
- the setback *S* has to be as small as possible
- the spark rate *SPR* has to be high, but an upper limit must exist.

A limit for the spark rate exists, beyond which the collection efficiency starts falling because of 'precipitation time' lost in voltage recovery. Moreover, too high a spark rate may be detrimental to the life of the internal parts of the precipitator and the high voltage equipment.

8.7.4 Spark detection and voltage recovery

One of the important objectives in a modern control unit is to obtain a fast recovery of the precipitator voltage after a spark; in this way it is possible to maximize the voltage-time integral and maintain a high collection efficiency. The fast voltage recovery is obtained if:

- unnecessary turn-off time intervals (this turn-off time is also called 'deionization time', 'quench time', etc.) of the control thyristors are avoided, and
- the voltage is raised to the highest possible level within a few half-cycles of the line frequency.

It is also important to perform this voltage recovery without a new spark arising, i.e. 'multiple sparking' should be avoided. Voltage recovery is closely related to spark detection, so this aspect will be briefly covered below.

8.7.4.1 Spark detection. Most of the control units classify the sparks in two types, according to their intensity:

- a light spark (or spitting), where the precipitator instantaneous voltage rises to a certain level after the spark within a very short period of time;
- a severe spark (or arcing), where the precipitator instantaneous voltage remains low after the spark for a certain period of time.

Figure 8.19 illustrates the two types of sparks and the voltage recovery performed by a good modern control unit. Figure 8.19b shows that, even in case of a severe spark, the voltage can be raised to a high level, without the utilization of turn-off time and without the occurrence of multiple sparking.

The question whether or not to utilize a turn-off time is one of the less understood problems in automatic voltage control techniques. There are manufacturers of modern control units who recommend in their user manuals to use a turn-off time in order to avoid the occurrence of arcs in the precipitator. The alternative method to avoid such a problem will be explained in the next section.

A fast voltage recovery is also closely related to the detection method used. In the past, the primary values and the precipitator current have been used, but the use of the instantaneous precipitator voltage has proven to be superior, as discussed in reference [10].

8.7.4.2 Voltage recovery. The problem of recovering the precipitator voltage within a few half-cycles of the line frequency without introducing turn-off time does not have a simple solution. One of the difficulties is to know how much the instantaneous voltage can be raised without the occurrence of a new spark, i.e. the determination of the 'aimed level'. Then, the next problem is to find the value of the firing angle which will provide this aimed level.

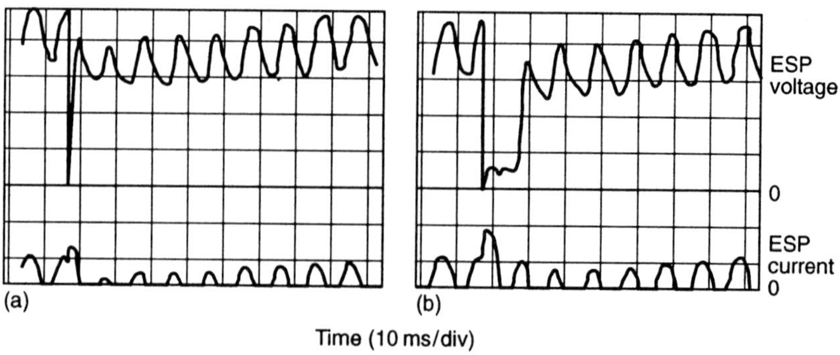

Figure 8.19 Classification of sparks according to their intensity. Light spark (a) and severe spark (b).

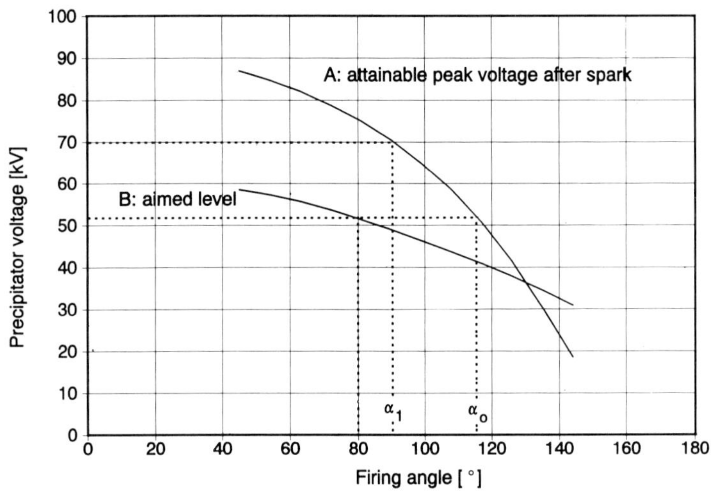

Figure 8.20 Control principle for a fast voltage recovery after spark.

This problem and its solution are illustrated by means of the curves in Figure 8.20. Curve B shows a typical variation of the mean voltage as a function of the firing angle during DC normal operation, while curve A shows the attainable peak voltage in the first half-cycle after the spark. Experience has shown that the aimed level can be represented by the curve B, without causing multiple sparking, and at the same time giving an acceptable precipitator voltage level. Sometimes a higher aimed level might be used, but the probability of sparking in the recovery period is quite high.

8.7.4.3 Example. The problem will be illustrated in the following example. Let us assume that the control unit is firing the thyristors at 80°. The aimed level for voltage recovery (52 kV) is shown by the dotted line. After a spark, the preset setback gives an increased firing angle α_1, this will give a voltage level determined by the intersection of curve A and firing angle α_1 (70 kV). It is clearly seen that this level is too high compared with the aimed level and this will undoubtedly cause multiple sparking. The correct firing angle is α_0, determined by the intersection of the dotted line representing the aimed level and the curve A.

At low voltage levels beyond the crossing of curves A and B, the problem is reversed. If the closed loop control is not opened, then the firing angle will be too high and results in too a low voltage level and a slow voltage recovery. The recommended solution is:

- to store the curve A in the memory of the control unit;
- to open the control loop in case of spark and find the right firing angle according to the aimed level and curve A;

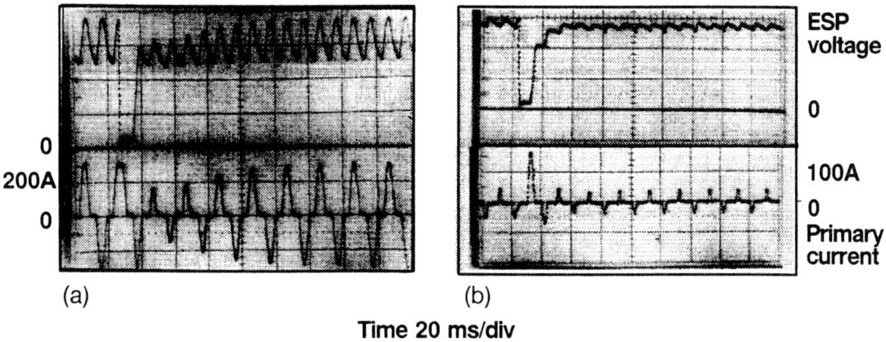

Figure 8.21 Reaction of a modern control unit to sparks at high (a) and low (b) current levels (courtesy FLS miljø a/s).

- to close the control loop and perform the required setback;
- to continue with the normal control strategy after the setback is performed.

To illustrate that a satisfactory voltage recovery can be obtained automatically, both at high and low current operation, and without using turn-off times, the oscillograms shown in Figure 8.21 are included. They speak for themselves and no further explanations need to be given.

Another important feature obtained with this method can be seen in Figure 8.21:

(a) At higher current operation, the first pulse current used to raise the voltage to the aimed level, and the immediate following ones, are lower than the current pulses at normal operation.
(b) At low current operation, the current pulse used to raise the voltage is higher than at normal operation.

It can be concluded that the electrical equipment is not subject to overload, in connection with a spark or arc, if the above-mentioned method is used. In this respect, it is necessary to remember that the condition for obtaining this result is the use of a suitable high short-circuit reactance, as mentioned in section 8.5.

8.7.5 Back-corona detection and corona power control

The occurrence of back-corona in one or more precipitator sections has been normally determined by examining the corresponding i–v curve. In the past, this curve was measured as a plot of the precipitator mean current *vs.* mean voltage. The criterion used for determination of back-corona was the

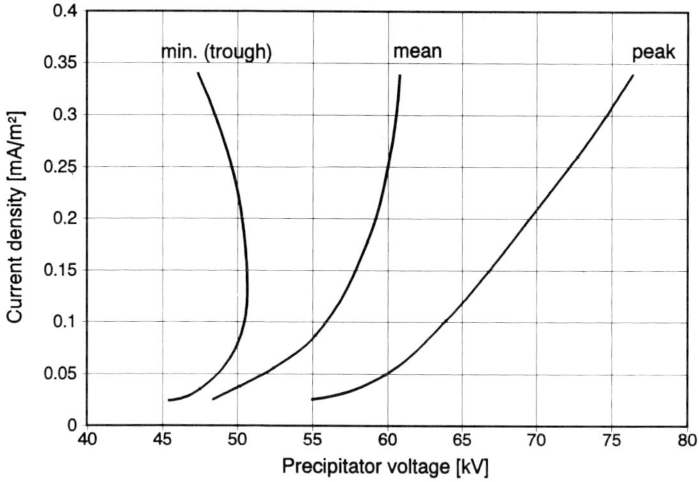

Figure 8.22 *i–v* characteristics in the case of back-corona conditions.

slope of the *i–v* curve; if the slope was ∞ or negative, it indicated back-corona. But in the early 1980s it was considered that this method was not sensitive enough [10], and a better indication was obtained by using the *i–v* curve, where the mean current is plotted against the minimum value of the precipitator voltage, as illustrated in Figure 8.22.

This shows that the curve taken as a function of the minimum voltage changes slope at a relatively low current level, while the curve based on the mean voltage still has a positive slope. The curve based on peak voltage shows a positive slope, irrespective of the presence of back-corona.

8.7.5.1 Automatic detection. Not all voltage control units include a back-corona detector. The units which include back-corona detection invariably use different principles:

Slope of i–v curve. This corresponds to the method described above, where the minimum voltage of the precipitator is used. This method, however, has the disadvantage that the power levels have to be reduced in order to find the inflexion point.

Voltage waveform at spark. This method is also based in the minimum value of the precipitator voltage. By comparing its value before and after a spark the occurrence of back-corona can be determined. Back-corona exists if the minimum value after the spark is higher than the value before the spark. This situation is illustrated by the oscillogram shown in Figure 8.23.

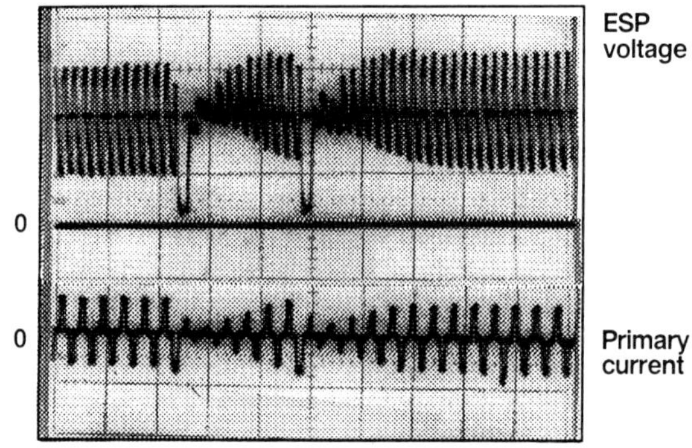

Figure 8.23 Precipitator voltage and primary current during spark in the case of back-corona.

By observing the oscillograms shown in Figure 8.19 and Figure 8.21, it can be concluded that back-corona does not occur in these cases.

This method has the advantage that it does not need to reduce power levels, because sparks are used. In the case of 'no sparking' conditions, a blocking period, where the thyristors are not fired, is used instead.

8.7.5.2 Corona power control. The corona power has to be reduced in the case of back-corona in order to extinguish it. As seen in the previous section there are two means to accomplish this:

- to change to intermittent energization and find the optimal degree of intermittance, and
- to reduce the electric charge (Q_p) delivered to the precipitator, i.e. the area under a current pulse, by delaying the firing angle of the thyristors.

This task has traditionally been performed by plant personnel, but over the last decade the tendency is to do it automatically. One approach has been to include this function into the AVC units; this is a cheap solution, but it is difficult to obtain the optimal settings when only the typical instrumentation signals, like those shown in Figure 8.15, are used by the AVC unit.

The optimization of the degree of intermittence (D) can be performed in combination with the automatic detection of back-corona. If during the detection back-corona is found, then D is increased, but if no back-corona is detected, D is reduced.

The optimization of the electric charge delivered by each current pulse, Q_p, is difficult. One method employed is to maximize the minimum value of

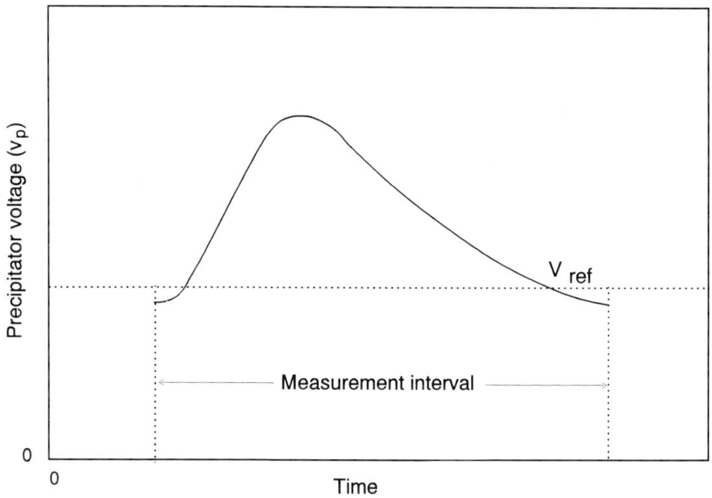

Figure 8.24 Determination of the figure of merit based on a comparison of the precipitator voltage with a reference voltage.

the precipitator voltage while remaining at the optimal degree of intermittence.

A combined method for the optimization of D and Q_p also exists [12]. Here D and Q_p are varied, so a variety of combinations of these two quantities is obtained. For each combination, a figure of merit is determined and the combination giving the best figure of merit is selected for control. Another method of determining the figure of merit is to compare the instantaneous voltage waveform with a reference voltage during the time interval where the corona discharge takes place, the reference voltage being set to about the corona onset voltage. The method is depicted in Figure 8.24. The figure of merit can be determined as the integral $I = \int v_p(t) \cdot (v_p(t) - V_{ref})dt$. Other ways of determining a figure of merit exist, but these will not be covered here.

In order to improve the above-mentioned tasks, some units incorporate the opacity signal in the optimization of the corona power, as shown in Figure 8.16. But this is not 100% effective, as a particular unit does not know if there has been a change in the settings of another control unit, or a change in the operating conditions of the precipitator. Therefore, the present tendency is to place this task in a supervisory computer, which has all the relevant process and stack emission signals connected to it. With these data, the computer can optimize the settings of the individual control units accordingly. This 'supervisory computer control' of the precipitator is covered briefly in section 8.9.

8.8 Pulse energization

8.8.1 Introduction

The advent of commercial pulse energization systems in the early 1980s was one of the major technical developments in the energization of electrostatic precipitators. In fact, pulse energization was the first really new method since Cottrell's development of the traditional power supply based on a high voltage transformer/rectifier [1].

This method was developed to improve the collection of difficult high resistivity particles, one of the main shortcomings of electrostatic precipitators. Pulse energization consists of short duration high voltage pulses superimposed on a 'base voltage'. The pulse systems developed for a single stage precipitator operate normally in one of the two following pulse widths: a microsecond range or a 100 μs range. In the following only the latter type will be covered.

The high voltage pulses are repeated at a certain frequency in the range of 1 to 400 pulses/s (pps). A typical waveform of the applied voltage is depicted in Figure 8.25 for a frequency of 100 pps, and for comparison purposes the voltage waveform with traditional DC energization is included. The differences are quite clear:

- the narrow high voltage pulses have a high amplitude
- the base voltage is kept close to the corona onset voltage

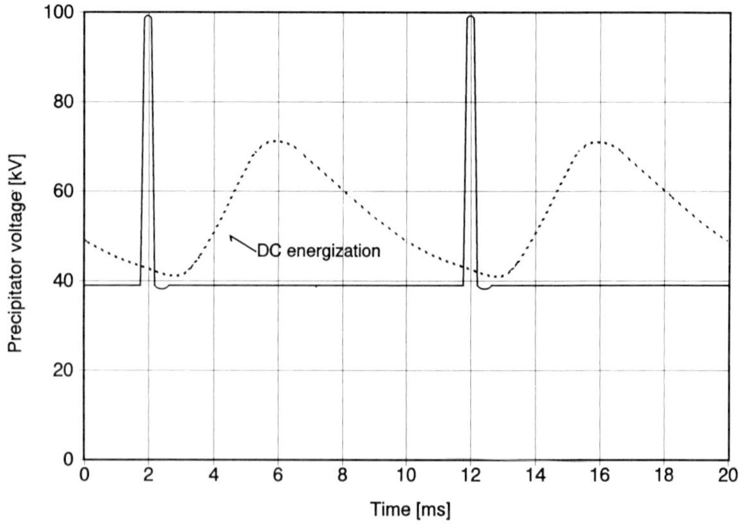

Figure 8.25 Typical voltage waveform obtained with pulse energization and comparison with DC energization.

- the peak value of the precipitator voltage is equal to the base voltage plus the pulse amplitude and its value is higher than for DC energization.

Historically, the first full-scale pulsing tests on precipitators were performed by White and Hall in the late 1940s [1,2]. The commercial application of their pulse system was hampered mainly by the lack of a reliable high voltage switch and by high power consumption. With the advent of high frequency switching thyristors in the 1970s and refined systems, in order to reduce power consumption, companies in USA, Japan and Europe developed their own pulse systems and were able to demonstrate full-scale tests in the late 1970s [14].

Nowadays, 15 years later and with hundreds of commercially operating pulse energization systems worldwide, there are a number of American, Japanese and European companies supplying pulsers as standard products. They are used in special applications where pulse energization is the best economical/technical solution to a problem dust.

8.8.2 Electrical configuration

As depicted in Figure 8.25, a pulse system has to deliver a narrow high voltage pulse superimposed on a base voltage, so a pulse generating circuit and a base voltage power supply are required. Moreover, the pulse amplitude, the base voltage and the pulse repetition frequency have to be varied according to a certain control strategy and this function is performed by a special control unit.

A block diagram is shown in Figure 8.26, where the basic elements of the system can be seen. In the construction of their systems, the various manufacturers have used different approaches, mainly as a result of patent

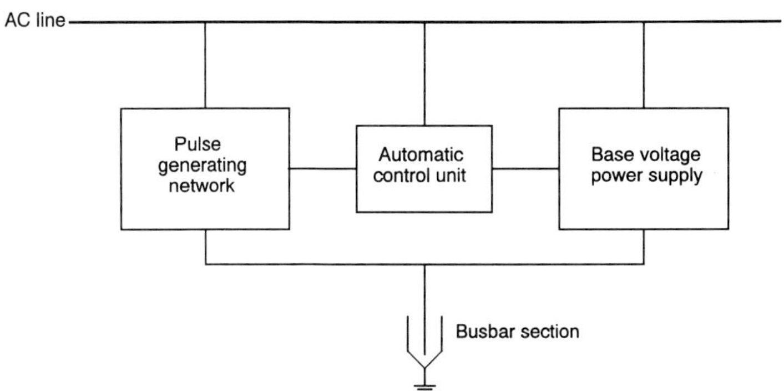

Figure 8.26 Simplified block diagram of a pulse energization system.

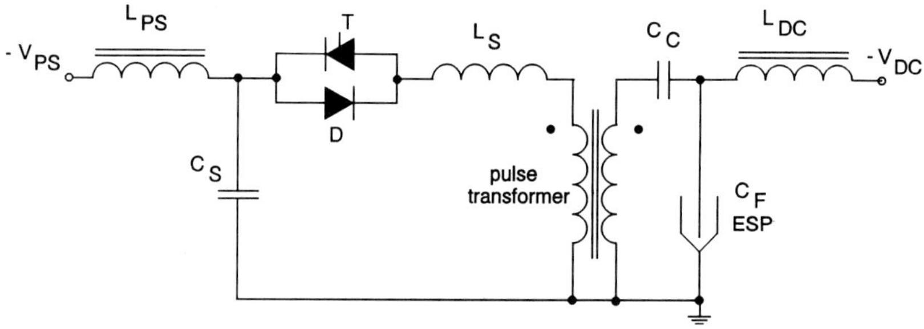

Figure 8.27 Main circuit of a pulse energization system with switching at low potential.

protection. But in principle, there are two main architectures; one based on switching at low potential as seen in Figure 8.27 and one based on switching at high potential [15] as seen in Figure 8.28. The first type normally uses two high voltage tanks (one for the pulse generator and one for the base voltage) and separate cabinets for the automatic control unit and power control devices. The second type normally uses one control cabinet and one high voltage tank resembling a traditional power supply. Both use an energy recovery principle based on a series LC resonant circuit, where the precipitator, represented by its capacitance, is one of the components of the circuit. The high voltage switch consists of a number of thyristors in series, each having an antiparallel diode.

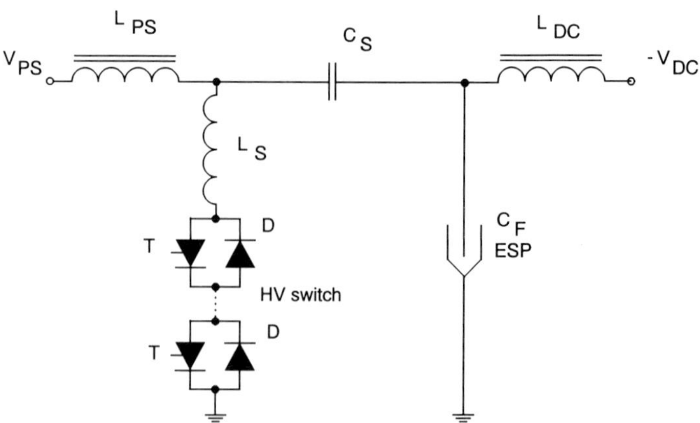

Figure 8.28 Main circuit of a pulse energization system with switching at high potential.

8.8.2.1 Pulse system with pulse transformer. This is depicted in Figure 8.27. The base voltage is delivered by a separate power supply represented by V_{DC}. The pulse generating circuit includes a power supply V_{PS} and the series oscillating circuit consisting of the storage capacitor C_S, the coupling capacitor C_C, the inductance L_S and the precipitator capacitance C_F.

Before the generation of a pulse, C_S is charged to the voltage $-V_{PS}$ and the precipitator is charged to $-V_{DC}$. When the thyristor T is fired, oscillation is initiated and the current through the main circuit has the waveform shown in Figure 8.29. During the positive half-cycle the current circulates through the thyristor T and this is turned off around the zero-crossing when the current falls below the holding level. Because of the energy stored in the inductance L_S, the diode D is forced to conduct and the current circulates in the opposite direction until it becomes zero. The current remains at zero until the thyristor is fired again after a period of time, corresponding to the pulse repetition frequency. The resultant precipitator voltage has the waveform shown in Figure 8.29.

The energy recovery, which is fundamental for the commercial utilization of pulse energization, occurs during the negative half-cycle of the pulse current. Here, the energy delivered to C_F, and not used in the corona generation, is returned to C_S and stored there during the time interval between pulses and used in the generation of the next pulse.

The coupling capacitor C_C avoids a short-circuit of the power supply V_{DC} by the secondary winding of the pulse transformer. The pulse transformer approach allows switching at a lower potential, but has the disadvantage of a higher price, weight and volume. Assuming an ideal pulse transformer, no

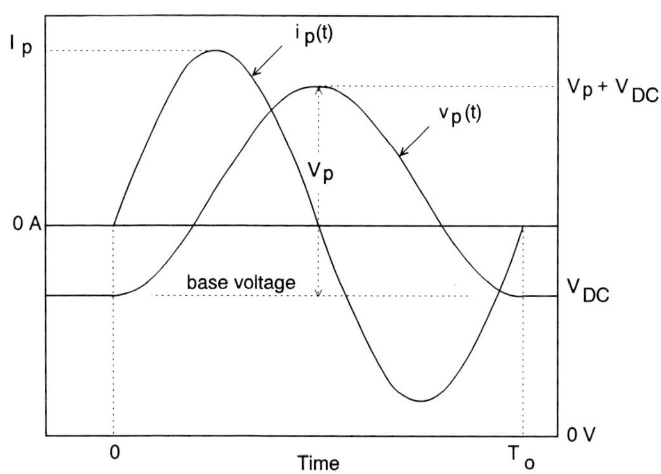

Figure 8.29 Idealized pulse voltage and current waveforms.

losses in the circuit and a very large storage capacitor, the pulse current can be expressed by:

$$i_p(t) = I_p \sin(\omega_o t) \tag{8.20}$$

where I_p is the peak value of the pulse current and ω_o is the angular frequency of the oscillation. The pulse width T_o can be expressed as a function of ω_o.

The pulse precipitator voltage is then:

$$v_p(t) = \frac{1}{C_F} \int i_p(t)\mathrm{d}t = \frac{V_p}{2}(1 - \cos \omega_o t) \tag{8.21}$$

where V_p is the amplitude of the pulse voltage.

The quantities I_p, V_p, ω_o and T_o can be determined as follows:

$$I_p = \frac{V'_{PS}}{\sqrt{\dfrac{L'_S}{C_{eq}}}} \tag{8.22}$$

$$V_p = 2V'_{PS}\frac{C_{eq}}{C_F} \tag{8.23}$$

$$\omega_o = \frac{1}{\sqrt{L'_S C_{eq}}} \Rightarrow T_o = 2\pi\sqrt{L'_S C_{eq}} \tag{8.24}$$

where

$$V'_{PS} = nV_{PS} \tag{8.25}$$

$$L'_S = n^2 L_S \tag{8.26}$$

$$C_{eq} = \frac{C_F C_C}{C_F + C_C} \tag{8.27}$$

8.8.2.2 Example. For the typical values given below, the amplitude of the pulse current, the amplitude of the pulse voltage and the pulse width are calculated as follows:

$$V_{PS} = 4\,\text{kV} \qquad C_C = 500\,\text{nF}$$

$$L_S = 25\,\mu\text{H} \qquad n = 10$$

$$C_F = 100\,\text{nF}$$

From equations (8.25), (8.26) and (8.27):

$$V'_{PS} = 10 \times 4 = 40\,\text{kV}$$

$$L'_S = 10^2 \times 25 \times 10^{-6} = 2.5\,\text{mH}$$

$$C_{eq} = 100 \times 500/(100 + 500) = 83.3\,\text{nF}$$

From equations (8.22), (8.23) and (8.24):

$I_P = 40 \times 10^3/\sqrt{(2.5 \times 10^{-3}/83.3 \times 10^{-9})} = 231\,\text{A}$

$V_p = 2 \times 40 \times 83.3/100 = 66.6\,\text{kV}$

$T_o = 2\pi\sqrt{(2.5 \times 10^{-3} \times 83.3 \times 10^{-9})} = 91\,\mu\text{s}$

8.8.2.3 Pulse system without pulse transformer. This is depicted in Figure 8.28. The base voltage is delivered by a power supply represented by V_{DC}. The pulse generating circuit includes a power supply delivering V_{PS} and the series oscillating circuit consisting of the storage capacitor C_S, the inductance L_S and the precipitator capacitance C_F. In this configuration, the semi-conductor switch is placed on the high voltage side and consists of quite a large number of thyristors in series [16].

Before the generation of a pulse, C_S is charged to the voltage $(V_{PS} + V_{DC})$ and the precipitator is charged to $-V_{DC}$. When the thyristors T are fired the oscillation is initiated and the pulse current and the pulse voltage have the waveforms as shown in Figure 8.29.

The expressions for determining the pulse current and the pulse voltage are the same as before, i.e. equations (8.20) and (8.21). In this case, however, the quantities I_p, V_p, ω_o and T_o are expressed by:

$$I_p = \frac{V_{PS}}{\sqrt{\dfrac{L_S}{C_{eq}}}} \tag{8.28}$$

$$V_p = 2V_{PS}\frac{C_{eq}}{C_F} \tag{8.29}$$

$$\omega_o = \frac{1}{\sqrt{L_S C_{eq}}} \Rightarrow T_o = 2\pi\sqrt{L_S C_{eq}} \tag{8.30}$$

where

$$C_{eq} = \frac{C_F C_S}{C_F + C_S}$$

8.8.3 *Main features of pulse energization*

During normal operation, the base voltage V_{DC} is kept about the corona onset level, and the pulse amplitude and the pulse frequency are varied according to a certain strategy. The results of using pulse energization in a precipitator can be better expressed by the i–v characteristics of the respective bus section, as shown in Figure 8.30. Here, the current density is

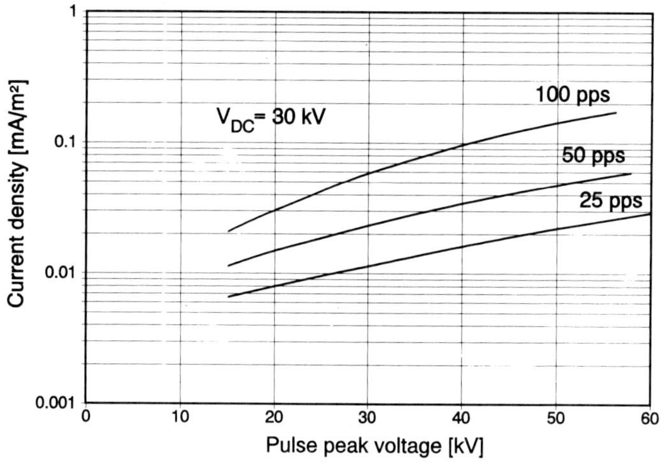

Figure 8.30 Typical *i–v* characteristics obtained with pulse energization (courtesy FLS miljø a/s).

plotted as a function of the pulse amplitude for a constant base voltage (V_{DC}) and three different pulse repetition frequencies. The current density is defined as the precipitator mean current divided by the collection area. The precipitator mean current is also called the emission current (I_F).

This typical family of curves shows the following interesting features of pulse energization [17]:

- the precipitator current can be varied by changing the pulse frequency in spite of the precipitator peak voltage ($V_{DC} + V_P$) being kept constant;
- the precipitator peak voltage is high due to the short duration of the pulses;
- the slope of the curves is rather flat (this is also the case with high resistivity dust).

These features of pulse energization give the following advantages:

8.8.3.1 Current control capabilities. The *i–v* characteristic (Figure 8.30) shows that the precipitator current (I_E) can be controlled independently of the precipitator voltage by varying the pulse repetition frequency. This allows the current to be reduced to around the onset of back-corona, without reducing the precipitator voltage, i.e. the precipitator can operate at low current and high precipitator voltage. This means a more suitable electrical energization for high resistivity dust compared with traditional DC energization, where current control cannot be performed without reducing precipitator voltage.

The dense ionic space charge produced by a pulse shields the discharge electrode and reduces the electrical field strength at its surface. This causes the suppression or limitation of the corona discharge during the rest of the pulse period. The consequence is a flat *i–v* characteristic.

8.8.3.2 Current distribution. With DC energization, the corona discharge tends to be localized at discrete spots on the discharge electrode. With the application of narrow pulses of high amplitude superimposed on a base voltage around the corona onset voltage, the pulse peak voltage significantly exceeds the corona onset level. This produces an intense corona discharge during a pulse and a correspondingly dense ionic space charge. A discharge electrode with very spotty corona under traditional DC energization can literally be made to glow with pulse energization [18]. This results in a better current distribution along the electrode, and this can be extended to the whole pulse energized bus section. This improved current distribution has been confirmed by measurements on a laboratory duct precipitator [17] and also measurements on pilot precipitators [20].

These results show that with DC energization, the current density at the beginning of the precipitator section is very low and increases along the section in the direction of the gas (see section 8.3). With pulse energization, the current density along the precipitator section is considerably more uniform.

A good current distribution on the collecting plates is important in order to avoid the initiation of back-corona due to localized spots of high current density.

8.8.3.3 Electrical field strength in the interelectrode space. With DC energization, free electrons are constantly generated producing an ionic space charge density and a field strength that, in principle, does not vary with time.

With pulse energization, where the base voltage is kept just below the corona onset level, free electron and negative ions are generated only during the pulse period. During this time, the ionic space charge crosses a part of the interelectrode space and during the time interval between pulses, the space charge moves towards the collecting electrode impulsed only by the base voltage field. As a consequence, the space charge and the field strength vary with time at each point of the interelectrode space.

Measurements on a pipe laboratory precipitator gave the results depicted in Figure 8.31. The field strength in relation to the values obtained with DC energization are plotted as a function of time. After a pulse, the electrical field is determined by the base voltage and the moving space charge, its strength increasing until the front reaches the collecting electrode. Thereafter the field strength decreases until all the ions have reached the collecting electrode.

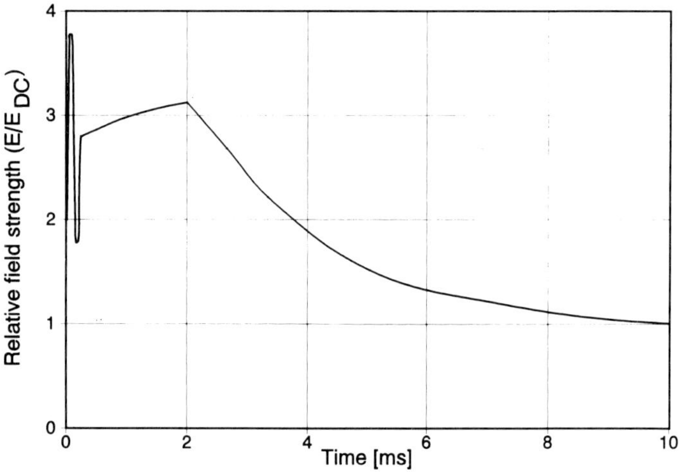

Figure 8.31 Electric field as a function of time after a high voltage pulse has been fired.

8.8.3.4 Particle charging. For particles of 1 μm and larger, electric field charging is the predominant mechanism, and the saturation charge is determined by the maximum field strength created by the ionic space charge. With DC energization, a particle at a certain position is surrounded by the ionic space charge, and its saturation charge depends on the constant electric field at that position. With pulse energization the particle charging occurs when the space charge passes the particle and its saturation charge is determined by the maximum field strength during the passage of the space charge (see Figure 8.31). Because the maximum field strength with pulse energization is much higher than with DC energization, this method provides enhanced particle charging.

Measurements have shown that the best results are obtained with a high pulse amplitude, because this causes a higher ionic space charge density.

8.8.4 Power consumption

A precipitator section can be represented by a capacitance (C_F) in parallel with a current generator accounting for the electronic, ionic and dust space charge current. Each pulse has to raise the voltage across the capacitance, from the base voltage (V_{DC}) to the peak voltage ($V_{DC} + V_p$). This means that a considerable amount of energy has to be used. Supposing that in the circuit of Figure 8.27 the coupling capacitor C_C is much larger than C_F, the energy supplied by the pulse system only for charging C_F is:

$$W_c = \frac{1}{2} C_F V_p^2 \tag{8.31}$$

For typical values ($V_p = 60\,\text{kV}$, $C_F = 100\,\text{nF}$), the energy required is $W_c = 180\,\text{J}$. If this has to be repeated 200 times/s, a large amount of power (36 kW) has to be used. As the energy necessary for the corona discharge is small compared with the energy needed to charge C_F, the power consumption becomes excessive if the pulse system does not include means for energy recovery. The system shown in Figures 8.27 and 8.28 includes this feature in the series oscillating circuit. Here, when the voltage across C_F is at its maximum, the pulse current is zero. Then it reverses and in the negative half-cycle the surplus energy is stored in the storage capacity C_S.

The power consumed by a precipitator section energized by a pulse system with energy recovery can be expressed by [17]:

$$P_r = I_E V_{DC} + c I_E V_p \tag{8.32}$$

where the constant c has been found experimentally to be approximately equal to 0.5.

8.7.4.1 Example. A pulse system is operating at $0.1\,\text{mA/m}^2$, $V_p = 60\,\text{kV}$, $V_{DC} = 40\,\text{kV}$ and energizes a $3000\,\text{m}^2$ bus section. Determine the power consumption of the bus section. The precipitator mean current is: $I_E = 0.1 \times 3000 = 300\,\text{mA} = 0.3\,\text{A}$. Applying equation (8.32)

$$P = 0.3 \times 40 + 0.5 \times 0.3 \times 60 = 12 + 9 = 21\,\text{kW}$$

This consumption corresponds to a power density of $7\,\text{W/m}^2$, which is a typical value for medium resistivity dusts. For high resistivity dusts, the required power density is quite low, hence with a low pulse repetition frequency (2 to 20 pps) the precipitator mean current is correspondingly reduced.

8.8.5 Collection efficiency

Similar to the case with intermittent energization, the improvement in the precipitator performance is closely related to the resistivity of the collected dust. This improvement is normally expressed by the enhancement factor $H = \omega_p/\omega_{DC}$, where ω_p is the migration velocity obtained with pulse energization and ω_{DC} is the migration velocity obtained with DC energization.

The enhancement factor H obtained with very high resistivity dust, reported by American [18], Japanese [20] and European [14] companies is about 2. The enhancement factor H, as a function of the dust resistivity, can be expressed by the curve shown in Figure 8.32. This comparison with DC energization indicates that at low resistivity levels, both energization forms produce the same result, and at high resistivity levels, pulse energization is much better [21].

Another way to express these results is by saying that the precipitator performance with pulse energization does not fall as much as with DC energization for an increasing dust resistivity.

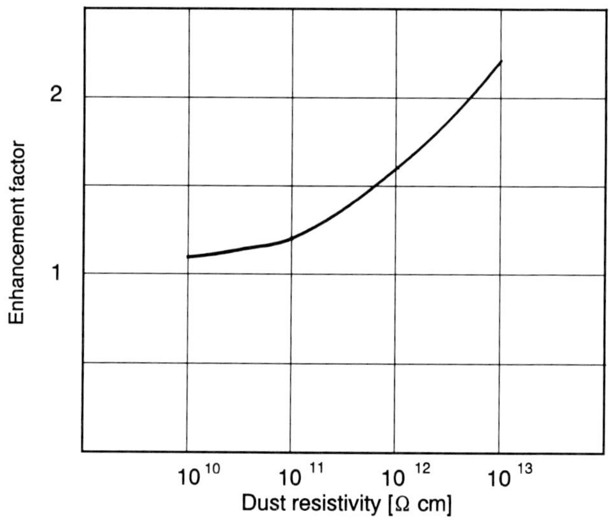

Figure 8.32 Enhancement factor obtained with pulse energization in relation to DC energization as a function of the dust resistivity.

A comparison between pulse and intermittent energization at high resistivity levels, expressed by the enhancement factor, gives a typical figure of 1.2 to 1.5 favourable to pulse energization. The precise value used by the precipitator manufacturer is assessed for each particular application, and is based on experience for the particular process, process conditions, dust composition, etc.

8.8.6 Applications

As previously mentioned pulse energization is used in the collection of very high resistivity dust. In this case, the enhancement factor is high and can compensate for the higher price of the pulse systems.

Typical applications are precipitators for:

- Four-stage preheater cement kilns (without or with insufficient water conditioning of the kiln gases).
- Coal fired power and steam generating boilers.
- Limestone, dolomite and magnesite kilns.
- Sinter strands for iron ore agglomeration.

The application of pulse energization is not restricted to new precipitators. It can also be a very effective solution for improving the performance of existing precipitators having resistivity problems. In this case, it is important to bear in mind that the mechanical condition of the precipitator has to be good. This means effective rapping of the discharge and collecting electrodes, good gas distribution, good electrode alignment, etc. Otherwise, the expected enhancement factor will not be obtained, because of low pulse voltage and/or precipitator current limitations.

For example, when the precipitator is collecting high resistivity fly ash, the current density is kept low by using a low pulse frequency in order to eliminate back-corona. But where the carbon content of the dust, carried by the flue gas, is high, the precipitation requires a high current density and this conflicts with the need for low current density to eliminate back-corona; hence the result is a higher stack emission.

8.8.7 Summary

Pulse energization is normally ideal for precipitators collecting high resistivity dusts. The improved precipitator performance in the collection of medium and high resistivity dust, compared with traditional DC energization and IE, is due to the combined effect of the following features:

- better particle charging
- higher collecting field strength
- better current distribution
- better current control capability

The enhancement factor obtained with pulse energization is mainly due to the application of a high pulse amplitude. Therefore, the mechanical condition of the precipitator has to be good. In the collection of fly ash, the concentration of low resistivity particles, like unburnt coal, has to be low; otherwise the expected performance will not be achieved.

8.9 Supervisory computer control

In the last decade, supervisory computer control for precipitators has become more common. A traditional solution has been a stand-alone computer, implemented with minicomputers in the past, and now with powerful PCs. The use of PCs has resulted in lower prices, which has meant that these systems are becoming more affordable to customers.

With increasing plant automation and more advanced graphics operator interfaces available in the control room, the integration of the precipitator control in the plant computer system is an increasing demand. Most of the plants built recently have a system similar to that depicted in Figure 8.33. The workstation shown represents the main computer providing the major services and where the application programs are run. Normally, a number of operator view stations with display, keyboard and control facilities are connected to the main computer *via* a common bus. This main computer

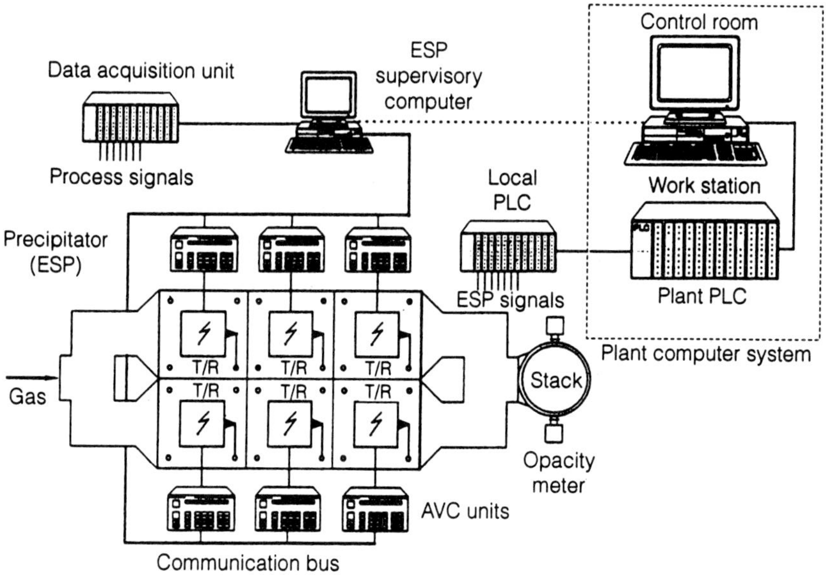

Figure 8.33 Supervisory computer control of an ESP with stand-alone computer (courtesy FLS miljø a/s).

communicates *via* another bus with the plant PLC (Programmable Logic Controller) system, which in turn exchanges data with the local PLCs in charge of the control of local areas or main equipment of the plant. In case of communication problems, the local PLCs are designed to continue with their alloted control tasks. This is the principle of the so-called decentralized control system (DCS).

This principle is combined with application programs included in the main computer providing the so-called Supervision, Control and Data Acquisition system (SCADA). The result is a system with powerful graphics operator interface and control functions.

It is understandable, therefore, that for the plant management and personnel interested in operating the whole plant from their operator view stations, a stand-alone precipitator control computer in the control room is not readily accepted.

A typical approach to this problem is the connection of the precipitator control equipment to a common communication bus. This bus is then connected to the plant computer *via* a so-called gateway unit, containing the two communication drivers, necessary for communication and data exchange, between the plant computer and the precipitator bus. In this way, the advanced control functions for the precipitator may reside in the plant computer system, or if needed, in a precipitator computer connected to the precipitator bus.

8.9.1 *Stand-alone computer*

A typical architecture is illustrated in Figure 8.33, which shows a six-bus section precipitator, each controlled by an AVC unit. In the European precipitators with rigid frame design, the AVC unit normally controls the high voltage power supply plus the rapping systems of the respective bus section. The six AVC units communicate with the precipitator supervisory computer *via* a common communication bus. This supervisory computer also receives data from the process and the stack opacity through a data acquisition unit (normally a PLC). The ESP is operated from the control room, through a local PLC, which takes care of the start/stop functions, alarm indication, voltage and current display, etc., and is connected to the plant PLC system *via* the system communication bus.

In this configuration all the advanced control functions are placed in the supervisory computer. Among them, the following features can be found:

- Menu based graphic operator interface, providing:
  screen displays of the precipitator operating conditions;
  screen displays of the operation of the HV power supplies and rapping systems (voltages, currents, corona power, timers, etc.);
  display and setting of parameters;
  remote start/stop;
  trend analysis, etc.
- Optimization of the corona power (by means of D and Q_p), according to the existing operating conditions and dust resistivity.
- Energy management system for costs savings whenever possible.
- Optimization of the rapping sequences (and reduced rapping losses).
- Automatic measurement and display of i–v characteristics.
- Automatic precipitator start-up and shut-down routines.
- Alarm handling and fault diagnosis.

One example of a commercial system for power plant precipitators, including all these features, is fully described elsewhere [22]. In this system the optimization functions are performed by means of Fuzzy Logic, and the aim of this so-called expert system is to act as a specialist being at the plant 24 hours a day.

A variant for this system is the direct communication between the precipitator supervisory computer and the plant computer, shown by the dotted line. This requires the existence of a communication driver in each computer. This approach makes the data adquisition unit superfluous, as the process data can be exchanged directly. This solution is attractive when the communication driver, residing in each computer, has already been developed. If this is not the case, the respective development can be a time-consuming and expensive affair.

8.9.2 Supervisory computer control via *a gateway unit*

A typical and simple solution to the above situation is the one depicted in Figure 8.34. The common communication bus for the control units is normally an accepted industrial standard. If this is not the case, this feature has to be included in the AVCs or a suitable converter has to be used. When this requirement is met, the gateway unit makes communication possible between the standard bus and various PLC systems with the most common trade marks. This communication problem is similar for other types of equipment used in industrial plants, which has made it necessary to develop the required communication drivers. This development has resulted in a considerably lower price, because these gateway units can now be obtained as stock components.

When the connection of the AVCs *via* a standard bus and a gateway has been established, the supplier normally has other modules which can be connected to the standard bus to exchange data with each other as shown in Figure 8.35.

The standard bus runs through the whole plant and the different modules can be placed physically at the point where they are required. As an example, a PC acting as supervisory control of the precipitator can be used, which includes the various functions mentioned in section 8.9.1. In the same way, a remote terminal for operation of the AVCs can be connected to the standard bus if neither local nor control room operation is required. Furthermore, intelligent I/O units or small PLCs can be connected. Where a more powerful PLC is required, this can be connected to the standard bus

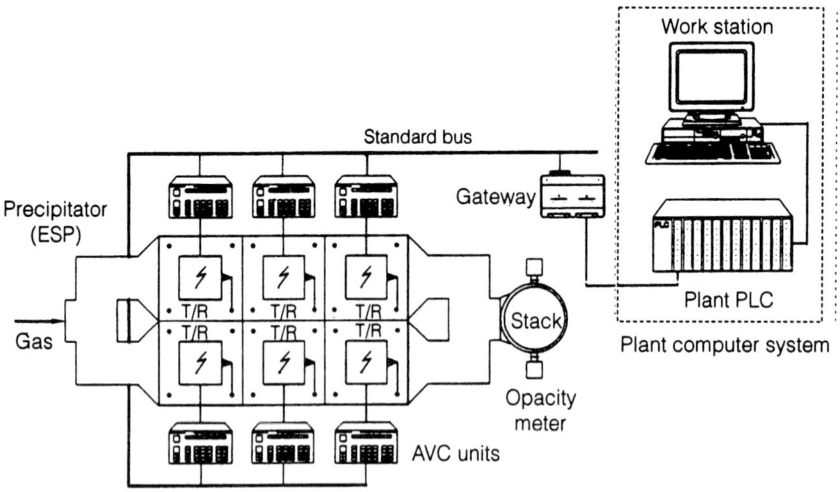

Figure 8.34 Supervisory computer control from the plant computer system *via* a gateway unit (courtesy FLS miljø a/s).

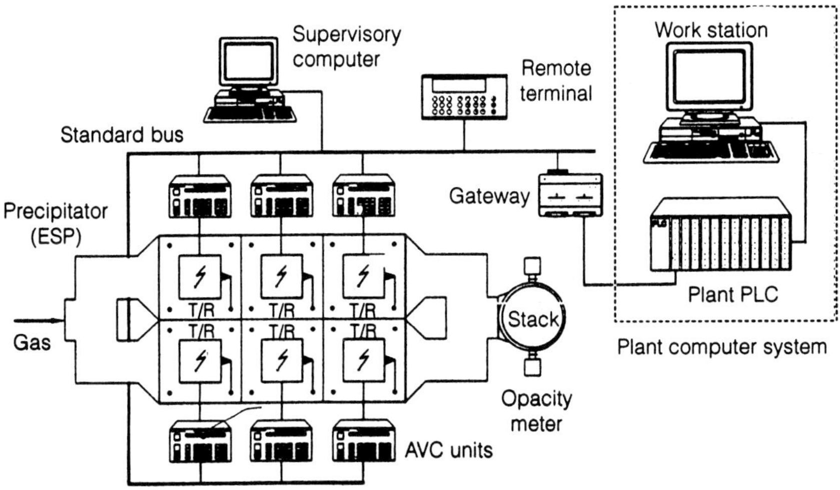

Figure 8.35 Supervisory computer control with dedicated ESP computer integrated in the plant computer system by means of a gateway unit (courtesy FLS miljø a/s).

via a gateway unit. This architecture looks very attractive at the moment especially because of its flexibility and modularity.

8.9.3 Advanced control functions

Among the features offered by a precipitator supervisory computer, one of the most important is an advanced control strategy including functions like:

- Optimization of the corona power, including
  optimization of the degree of intermittance (*D*)
  optimization of the charge delivered per current pulse (Q_p)
  optimization of the spark rate and current setback, etc.
- Optimization of the rapping sequences, including
  optimization of the off-time between rapping of the collecting plates
  synchronization between rapping of adjacent sections
  programmable power-off rapping sequences.
- Energy management system.

A supervisory computer is able to fulfil these requirements in a satisfactory way because it is constantly receiving information about:

- Process condition (gas temperature, flow, feed rate, O_2, SO_2, etc.).
- Stack opacity (dust emission).
- Currents and voltages in the individual bus sections.
- Status of TR sets, rappers, timers, parameters, etc.

Based on these data the supervisory computer can optimize the relevant settings in each automatic control unit (AVC) like:

- Degree of intermittence.
- Upper current limit (Q_p).
- Spark rate and current setback.
- Rapping off-time, etc.

These settings are changed at regular intervals, or when required, and their positive effect assessed by analysing the opacity signal, while simultaneously checking that the process conditions have not changed, by means of a trend analysis of their respective signals.

These relevant settings cannot be optimized in the same degree by the AVCs operating as stand-alone units, i.e. only relying on the electrical feedback signals (mA and kV).

For all these reasons, it is believed that the architecture depicted in Figure 8.35 will become more and more accepted. The control room personnel will operate the precipitator from their view stations in the normal way *via* a gateway communication, but, at the same time, the precipitator supervisory computer running in the background, will overtake and perform more and more advanced control and monitoring functions, like precipitator event and alarm indication and handling, fault diagnosis, etc. These features, complemented with precipitator start-up and shut-down automatic routines, will result in an intelligent and powerful supervisory computer control. This will relieve the plant personnel from tedious work routines and will provide cost advantages because of power savings, easier and better maintainance, a higher plant availability and lower average dust emission.

Appendix 8.A

I–V Relationship for a wide-plate geometry with air load

From the three Maxwell equations governing the electric field Poisson's equation can be derived:

$$\nabla \cdot E = \frac{\rho}{\varepsilon_o} \tag{8.A.1}$$

where ρ is the charge density (C/m^3) and ε_o is the permittivity of free space ($8.85 \cdot 10^{-12}\,F/\text{m}$), which is valid for gases under normal precipitator conditions.

The solution of Poisson's equation for a wire-plate geometry shown in Figure 8.A.1 is a formidable task.

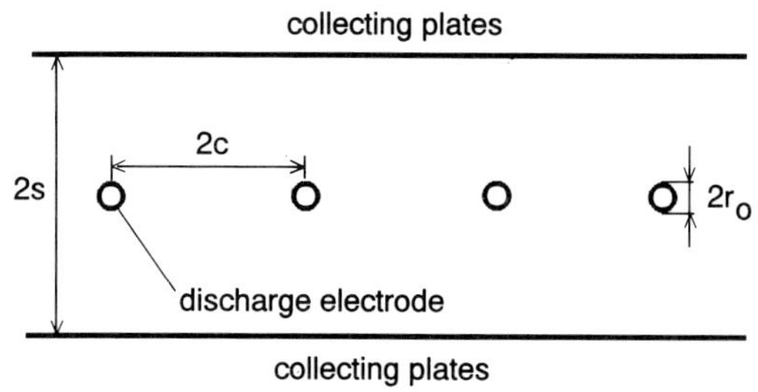

Figure 8.A.1 Electrode geometry.

The solution is simplified if it is assumed that the current is small and the alteration of the potential by the ion space charge can be represented by the same value found for the more simple wire-pipe geometry.

The average current density, as a function of the potential at the discharge electrode (see [2, 5]), can be expressed by:

$$j_s = \frac{\pi \varepsilon_o b}{cs^2 \ln(d/r_o)} V(V - V_c) \quad (\mathrm{A/m^2}) \tag{8.A.2}$$

where: b is the ion mobility ($2.1 \times 10^{-4}\,\mathrm{m^2/Vs}$ for negative corona in air), d is an equivalent cylindrical radius ($d = 4s/\pi$ for $s/c \leqslant 0.6$ (for other s/c values see the cited references)), and V_c is the corona onset potential expressed by:

$$V_c = r_o E_c \ln(d/r_o) \tag{8.A.3}$$

The corona onset field E_c has been found empirically [2, 5] and for negative corona in air it is expressed by:

$$E_c = \delta' \left(32.2 + \frac{0.864 \cdot 10^5}{\sqrt{r_o \delta'}} \right) \quad (\mathrm{V/m}) \tag{8.A.4}$$

(if r_0 is expressed in m).

The relative gas density δ' is conventionally expressed in relation to 1 atm. and 25 °C, i.e.

$$\delta' = \frac{298}{(298 + T)} P_a \tag{8.A.5}$$

Appendix 8.B

Approximated calculation of the mean and rms values of the ESP current

For the current waveform shown in Figure 8.B.1, the mean value, the rms-value and the form factor can be expressed by the following approxi-

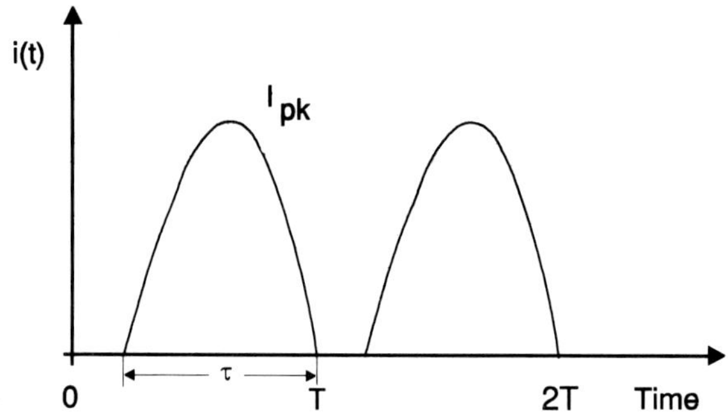

Figure 8.B.1 Precipitator current waveform.

mated equations:

$$I_{\text{mean}} = \frac{2I_{\text{pk}}}{\pi}\left(\frac{\tau}{T}\right)$$

$$I_{\text{rms}} = I_{\text{pk}}\sqrt{\frac{\tau}{2T}}$$

$$\text{FF} = \frac{\pi}{2}\sqrt{\frac{T}{2\tau}}$$

Note: It is left to the ardent reader to check the results shown in Table 8.1 using the waveform from Figure 8.8b.

References

1. Hall, H.J. (1971) Trends in electrical energization of electrostatic precipitators. *Proceedings of the Electrostatic Precipitator Symposium*, February 1971, Birmingham, AL, USA, pp. 177–89, SoRI Publication, Birmingham, AL, USA.
2. White, H.J. (1963) *Industrial Electrostatic Precipitation.* Addison Wesley Publishing Company, Reading, MA, pp. 198–226.
3. Reyes, V. (1991) *Determination of corona power in a precipitator section.* Unpublished FLS Miljø Company Report.
4. Petersen, H.H. (1990) A precipitator sizing formula. *4th International Conference on Electrostatic Precipitation*, Beijing, China. September, International Academic, Beijing, 1993, pp. 330–8.
5. Robinson, M. (1971) in *Air Pollution Control.* (W. Strauss ed.). Wiley-Interscience, New York, pp. 241–52.
6. Oglesby, S. and Nichols, G. (1978) *Electrostatic Precipitation.* M. Decker, New York, pp. 39–54.
7. Cooperman, P. (1960) A theory of space charge limited currents with application to electrostatic precipitation. *Trans. AIEE*, **79 I**, 47.
8. Cooperman, G. (1979) A new current–voltage relation for a duct precipitator valid for low and high current densities. *Trans. IAS* **79**, *IEEE, pp.* 146–7.

9. Zamany, J. (1995) Numerical modelling of electrodynamic conditions influenced by particle space charge and resistivity in ESPs of complex geometry for industrial applications. *Conference Electrostatics' 95, IOP,* York, England, April, Poster Session. Inst. of Physics, London, UK.
10. Reyes, V. (1987) Comparison between traditional and modern automatic controllers on full scale precipitators. *Proceedings of the EPRI/EPA Symposium*, March, Nashville, USA, Session 2A, EPRI, Palo Alto, CA, USA.
11. Reyes, V. (1990) Methods and apparatus for detecting back corona in an ESP with ordinary or intermittent energization. *US Patent 4,936,876*. June 26.
12. Jacobsson, H. and Porle, K. (1994) Method of controlling the supply of conditioning agent to an electrostatic precipitator. *PCT Patent Application WO 94/20218*, September 15.
13. Hall, H.J. (1990) History of pulse energization in electrostatic precipitation. *J. Electrostatics*, **25**, 1–22.
14. Petersen, H.H. and Lausen, P. (1979) Precipitator energization utilizing and energy conserving pulse generator. *2nd Symposium on the Transfer and Utilization of Particulate Control Technology*, Denver, USA, July, EPA Vol. II, pp. 352–68.
15. Kide, L. (1977) Electrostatic precipitator arrangements. *US Patent 4,052,177,* October 4.
16. Reyes, V. and Taarning, C. (1990) A semiconductor high voltage switch for pulse generation. *Proceedings of the Power Conversion Conference*, Munich, Germany, June, pp. 348–61.
17. Lausen, P., Petersen, H.H. and Jørgensen, H.J. (1981) Theory and application of pulse energization. *1st International Conference on Electrostatic Precipitation.* Monterey, USA, October, pp. 531–53, APCA Publishing Co., Pittsburgh, USA.
18. Feldman, P. and Aa, P. (1981) Operating results from the first commercial pulse energization system to enhance electrostatic precipitator performance. *Proceedings of the American Power Conference*, Chicago, USA, April.
19. Petersen, H.H. and Lausen, P. (1981) Application of energy conserving pulse energization for precipitators–practical and economic aspects. *3rd Symposium on the Transfer and Utilization of Particulate Control Technology*, Orlando, USA, March, pp. 291–302, Vol. 1, EPA Publication.
20. Fujishima, H. and Tomimatsu, K. (1990) Applications of an electrostatic precipitator with pulse energization system. *4th International Conference on Electrostatic Precipitation*, Beijing, China, September, pp. 419–30, International Academic Publishers, Beijing, 1991.
21. Schioeth, M. (1987) Five years' experience with pulse energized precipitators on power plants burning a wide range of coal. *Proceedings of the 3rd International Conference on Electrostatic Precipitation*, Abano, Padova University, Italy, October, pp. 197–207.
22. Reyes, V and Lausen, P. (1993) Utilization expert computer systems for control and operation of electrostatic precipitators. *10th Particulate Control Symposium and 5th International Conference on Electrostatic Precipitation*, Washington, USA, April, Session C4, EPRI TR 103048 Vol. 2, Palo Alto, CA, USA.

9 Precipitator sizing methods and models of electrostatic precipitators

C. PAULSON AND M. REA

Editor's note

The method of sizing precipitators has traditionally been the preserve of the precipitation using sizing factors (effective migration velocity), derived from efficiency measurements using precipitators operating on similar process plant. Originally the approach was based on the Deutsch equation, but as higher efficiency/lower emissions were demanded to satisfy legislative levels, sizing based on the Deutsch equation was found unsatisfactory and over the past 20 years or so, suppliers have found it necessary to use a modified equation, such as those derived by Matts-Ohnfeld of ABB Flakt or Petersen of FLS Miljo.

While the use of this type of modified equation has resulted in significantly improved prediction levels, the sizing is still based on data derived from units operating on similar processes and inlet conditions, etc. Most major suppliers have, over the years, built up detailed data banks, collating precipitator performance against specific process/inlet conditions/dust characteristics, and the use of these data banks has enabled precipitator installations to be sized so as to minimise technical/contractual exposure risks for the supplier.

In addition to determining performance characteristics from installed and operating plant, many suppliers have operated pilot sized precipitators in the field and laboratory, handling a few $m^3 s^{-1}$, not only to derive the precipitation factor or EMV, but also to evaluate the effect of process variables, e.g. temperature, gas velocity, contact time, moisture, etc. The use of the pilot precipitator has, over the years, produced significant amounts of data, but to be fully representative for plant sizing purposes, the inlet conditions to the pilot should be identical to those the full-scale precipitator will meet in practice.

The following chapter, after reviewing the modified equations of Matts-Ohnfeld and Petersen, describes the work carried out by CSIRO in Australia in developing a further modified sizing equation for coal-fired power station precipitators. This work, under the direction of Colin Paulson, was carried out in the laboratory using a small pulverised coal fired combustor fitted with precipitators to produce a range of inlet conditions similar to those met in practice. Although the modified equation gives good results it is still based on measured data.

The rapid development of computers over the past two decades has led a number of investigators to approach sizing from a theoretical numerical standpoint. A number of programs have been developed, but to date, although the investigators and various research organisations consider the derived data to be reliable, they are not so readily accepted by the precipitation industry for sizing purposes. Prof. Massimo Rea of Padova University reviews some of the earlier programs and concludes with the work being carried out by the University in conjunction with ENEL (Italy's Electricity Authority).

In addition to this reported work, other similar programs are being developed by some of the major suppliers and research organisations, for example the Electric Power Research Institute (EPRI) of the US. It is suggested that, as this type of program is rapidly developing, the reader should maintain contact with the various organisations to keep informed of the latest status of programming.

9A Precipitator sizing methods

C. PAULSON

9A.1 Theoretical considerations

9A.1.1 Basic dust-collection equation for gas in a duct

Consider a uniform duct along which passes a suspension of fine particulate matter in a turbulent gas. It is assumed that the walls of the duct collect any dust particles that arrive there, and it is further assumed that the turbulence, whilst sufficient to prevent the appearance of a zone with severely reduced particle concentration next to the walls, is not so vigorous that particles already collected are continually being re-entrained back into the gas stream.

Under these conditions it may be expected that the rate of arrival of particles at a selected small area of wall will be directly proportional to the volume concentration of particles in the immediate vicinity. Since those particles impinging on the wall collect there, the volume concentration of those passing by is correspondingly reduced. This loss of dust burden in the gas is progressive along the duct and the rate of dust collection at successive areas of wall correspondingly reduces.

At a selected cross-sectional plane distance l into the duct the above rate process may be expressed by:

$$\mathrm{d}n_1/\mathrm{d}t = kn_1 \tag{9A.1}$$

where n_1 is the number of particles in unit volume at plane l, t is time and k is a constant.

If the full length of the duct is L, then the total loss of particle concentration is obtained by integration of equation (9A.1) from $l = 0$ (where the inlet particle concentration is n) to $l = L$ (where the outlet particle concentration is n_0). The result is

$$\ln(n_L/n_0) = kt = \ln(1 - \varepsilon) \tag{9A.2}$$

where t is the average time a particle would take to traverse the duct, ε is the fractional collection efficiency of particles, i.e. $(n_0 - n_L)/n_0$ and $(1 - \varepsilon)$ is the fraction of particles not collected. This is termed the slip.

Given a constant throughput of dusty gas, t is directly proportional to L, so that

$$\ln(1 - \varepsilon) = \text{constant} \times L \tag{9A.3}$$

Equation (9A.3) shows that, under these conditions, complete particle collection ($\varepsilon = I$) can only be attained in a duct of infinite length.

Now, if we assume that the duct is a cylinder of radius r, so that the internal surface area, $A = 2\pi rL$, it follows that t, the treatment time, is $\pi r^2 L/V$, where V is the volumetric gas flow. It is also assumed that the dust particles are small enough not to suffer significant slip relative to the turbulent gas. From equation (9A.2)

$$\ln(1 - \varepsilon) = k\pi r^2 L/V = kAr/2V = k\alpha r/2 \tag{9A.4}$$

where $\alpha = A/V$, otherwise known as the specific collecting area (SCA) of the duct.

From equation (9A.2) we note that k has the dimensions $(\text{time})^{-1}$, so that $kr/2$ in equation (9A.4) has the dimensions of a velocity. We can therefore rewrite equation (9A.4) as:

$$\log(1 - \varepsilon) = -\alpha w/2.303 \tag{9A.5}$$

where w is the mean drift velocity of the particles to the cylinder wall and the negative sign appears because k is negative from the outset in equation (9A.1) because the dust burden is decreasing.

Taking a practical example, it is possible to achieve $\varepsilon = 0.5$ using an ordinary steel pipe about twenty-five times longer than its diameter with α set at $100\,\text{m}^2$ for every $\text{m}^3\,\text{s}^{-1}$ of dusty gas throughput. Equation (9A.5) then shows that w is $0.007\,\text{m}\,\text{s}^{-1}$. If it were desired to achieve $\varepsilon = 0.99$ (a 99% collection efficiency) then the pipe would need to be lengthened to almost 170 times the diameter. Ten times the original pipe length would theoretically catch 99.9% of the incoming particles.

The impingement dust separator described above is impractical and primitive, but is included here because it is a valuable introduction to better devices. Thus, if a suitable transverse force is imposed on the particles suspended in the gas stream, then clearly w is considerably increased and the dust separator is much more efficient for the same size.

The following are examples of transverse forces which could be used in gas cleaning devices.

(a) Magnetic forces. These are applicable only to particulates that respond well to such forces and thus have restricted scope. Magnetic coatings on non-magnetic particles are helpful.
(b) Centrifugal forces. These are applicable to the dusty gas stream as a whole, and are the basis of the cyclone separator.
(c) Differential pressure forces. These are applicable when the wall of the duct is permeable to gas but hardly at all to the particulate matter, as, for example, in the vacuum cleaner and the industrial fabric filter.
(d) Elcctrical forces. These are applicable to charged particles in an applied electric field and are the basis of the electrostatic precipitator.

9A.1.2 *Electrostatic precipitation*

9A.1.2.1 Basic principles. An electrostatic precipitator must provide three essential functions:

(a) the suspended particles must be given an electric charge;
(b) the particles must be subjected to an electric field to enable them to migrate from the gas stream to a suitable collecting electrode;
(c) the collected material must be removed from the collecting electrode in an efficient manner and deposited in a receptacle with the minimum amount of loss.

These operations are usually achieved by applying a high DC voltage, usually negative polarity, to a wire placed adjacent to an earthed plate and passing the dirty gas between the wire and the plate. The high voltage forms a corona around the wire causing the gas to ionise. The ions thus produced attach themselves to the particles suspended in the gas and the particles move in the electric field between the wire and the collector toward the collector. Typical values for this collecting field are 2–4 kV cm^{-1}, and the electric force acting on the particle can be many hundreds, or even thousands, of times gravity.

The layers of collected particles are removed from the collecting electrode by rapping the electrode which causes the material to fall into a hopper below the collector. This rapping process is critical to precipitator performance because a balance must be achieved between keeping the electrode clean and rapping too hard. Over-rapping causes particle re-entrainment into the gas stream with a consequent loss of performance.

9A.1.2.2 The fundamental efficiency equation. Considering equation (9A.5) in the context of the electrostatic precipitator the mean particle drift velocity consists of two components: (i) the mechanical component (w_m) caused by unassisted impingement (0.007 ms^{-1}) shown above, and (ii) the electrostatic component (w_{es}) caused by the movement of the charged particles in the electric field. So

$$\begin{aligned}\log(1-\varepsilon_p) &= \alpha(w_m + w_{es})/2.303\\ &= \log(1-\varepsilon_m) - \alpha w_{es}/2.303 \qquad (9A.6)\end{aligned}$$

where ε_p is the fractional efficiency of the precipitator and ε_m is the mechanical efficiency of the electrically dead precipitator. It should be noted that both ε_p and ε_m can be measured so that if α is known then w_{es} can be estimated from equation (9A.6).

If ε_m is small (or zero), which is very rare, then $\log(1-\varepsilon_m)$ is zero and equation (9A.6) becomes

$$\log(1-\varepsilon_p) = -\alpha w_{es}/2.303 \qquad (9A.7)$$

This is one expression of the well known Deutsch equation which was first published in 1922 [1] and has been commonly used to estimate electrostatic precipitator performance. Although the defect of omitting ε_{m} has been recognised for many years this equation is usually quoted as

$$\log(1 - \varepsilon) = -\alpha w_{\mathrm{e}}/2.303 \tag{9A.8}$$

where ε is the fractional efficiency of the precipitator and w_{e} is the effective migration velocity.

9A.1.3 Improvement of the Deutsch equation

The Deutsch equation is based on a number of assumptions. These include spherical particles, even distribution of the dust in the gas, the gas velocity and the electric current on the collecting plate, together with no particle re-entrainment into the gas stream on rapping. These assumptions make the equation unusable for practical purposes except for interpolating between measured values.

It has therefore been necessary to develop improvements of this equation to allow for the actual operating conditions prevailing in operating precipitators. A number of these improvements to the Deutsch equation are discussed below.

9A.1.3.1 Matts-Öhnfeld equation. This improvement, which is widely used to assess the performance of electrostatic precipitators, was developed by A.B. Svenska Fläktfabriken [2]. In this method the Deutsch equation (9A.8) is modified to

$$\log(1 - \varepsilon) = -(\alpha \cdot w_k)^{0.5}/2.303 \tag{9A.9}$$

where w_k is the modified migration velocity.

This technique can only be used when the electrical conditions of the precipitator are maximised and the temperature is constant. In theory, at these conditions, w_k will be constant for all values of α (SCA). Therefore by measuring the efficiency and SCA for a precipitator w_k can be calculated and then using this w_k value the size of precipitator required can be calculated for a different efficiency, or, of course, a new efficiency can be calculated if a different SCA is assumed.

9A.1.3.2 F.L. Smidth equation. This equation has been used by F.L. Smidth for sizing electrostatic precipitators since 1967 [3]. In this case the Deutsch equation has been modified in the form

$$1/(1 - \varepsilon) = (1 + b w_{\mathrm{B}} \alpha)^{1/\mathrm{b}} \tag{9A.10}$$

where w_{B} is the effective migration velocity in the first infinitesimal part of the ESP and b is an empirically determined exponent found to be 0.22 in most cases.

Like the Matts-Öhnfeld equation, because the exponent is empirically determined, this equation gives a much better estimate of the precipitator performance than the original Deutsch equation. This is despite the fact that it requires a number of simplifying assumptions including disregarding mechanical collection, re-entrainment, non-uniform gas distribution and back-ionisation.

9A.1.3.3 Extended Deutsch equation (CSIRO) It is known [4] that

$$q_B = (d^2 E_c/4)(3k/(k + 2))$$

where q_B is the saturation charge, d is the particle diameter, E_c is the charging field and k is the dielectric constant, and from electrostatic theory

$$F_1 = q_B E_p$$

where F_1 is the force on the charged particle and E_p is the precipitating electric field. So

$$F_1 = pE_c E_p d^2/4$$

where $p = 3k/(k + 2)$.

From Stokes' law the viscous drag F_2 on a spherical particle is given by

$$F_2 = 3\pi\eta dw$$

where η is the viscosity of the fluid and w is the particle velocity for steady conditions $F_1 = F_2$. So

$$w = (dE_c E_p/12\pi\eta)p \tag{9A.11}$$

so we see that

(a) as E_c and E_p are directly proportional to the applied voltage V, w is raised to the square of the applied voltage. For example, a 25% increase in voltage from say 32 kV to 40 kV gives an increase in w of over 56%;
(b) w is directly proportional to particle size, so that if the particle size can be increased by 50%, w is increased by the same proportion;
(c) k and η are constant at a fixed temperature, but they vary with temperature. An increase in temperature from 120 °C to 330 °C causes an increase in w of about 12% due to k and a decrease of 25% due to η, so even over this large temperature variation the composite effect will be small.

If (9A.11) is substituted in (9A.8), the Deutsch equation becomes

$$\log(1 - \varepsilon) = -C_2 \alpha dpE_c E_p$$

and if E_c and E_p are proportional to V,

$$\log(1 - \varepsilon) = -C_3 \alpha dpV^2$$

where C_3 is a constant.

Therefore, for a given feed material at constant temperature

$$\log(1 - \varepsilon) = C_4 \alpha d V^2 \quad (9A.12)$$

where C_4 is a new constant. This equation shows that the collection efficiency is directly proportional to the specific collecting area (α), the particle size (d) and the square of the applied voltage (V^2).

It has been determined that a precipitator which is electrically dead may still collect particles at a relatively high efficiency. Previously it was believed that the mechanical efficiency was low and mainly arose from gravitational settlement. The advent of pilot plants and the modifications of the Deutsch equation made it relevant to measure this efficiency. It has been found that, on occasions, figures as high as 40–50% are evident and that this mechanical efficiency is nearly always significant. This mechanical effect, which is mainly due to the turbulent motion of the gas causing particles to impinge on dusty surfaces, is additional to the electrostatic effect.

To allow for this effect a term must be added to equation (9A.12). Thus

$$\log(1 - \varepsilon) = \log(1 - \varepsilon_0) + C\alpha d V^2 \quad (9A.13)$$

where ε_0 is the mechanical efficiency.

However, the charging of particles and electrostatic collection are not possible until a corona is present so the term V^2 is not effective below V_g, the corona starting voltage. As the voltage is increased from 0 to V_g the collection efficiency exceeds ε_0 because some particles carry an electric charge already and, as a precipitating field already exists, a small increase in efficiency is found. Thus ε_0 increases to ε_g at the corona starting voltage V_g, and then above V_g

$$\log(I - \varepsilon) = \log(1 - \varepsilon_g) + C\alpha d V^2 \quad (9A.14)$$

This equation is called the extended Deutsch equation. The migration velocity term in the original Deutsch equation (9A.8) has been replaced by particle size and voltage components which can be measured. The collection efficiency (ε), the efficiency at $V_g(\varepsilon_g)$ and the SCA (α) can also be measured and so, using the extended Deutsch equation, precipitator performance over a range of operating conditions can be plotted with the log of the slip [$\log(1 - \varepsilon)$] as ordinate and αV^2 as the abscissa. The ordinate may be marked off as percentage slip or collection efficiency. This plot is called the performance line [5–7]. The general shape of the performance line with percentage slip as the ordinate is shown in Figure 9A.1.

The point A is the mechanical efficiency of the system and occurs when $V = 0$. Point B is the onset of corona and occurs at V_g. The line AB is not parallel to the abscissa because, as discussed above, some particles have a small natural charge which allows some precipitation to take place. The line BC is the effective performance line and if suitable values of α and V are

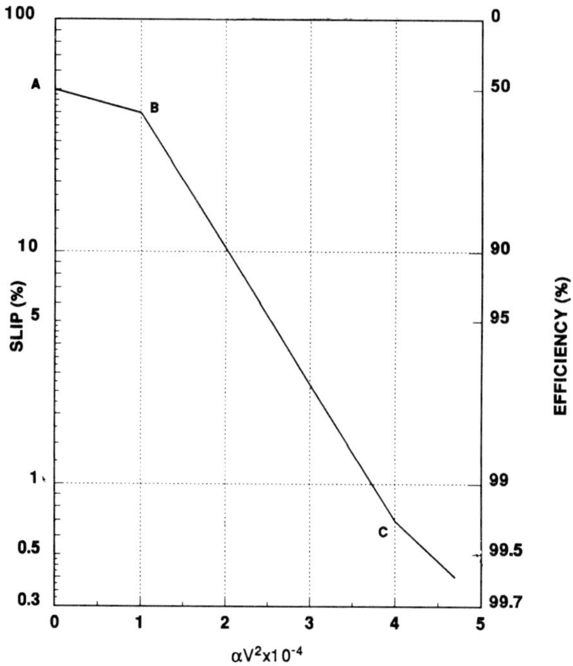

Figure 9A.1 Precipitator performance line.

selected then theoretically the line could be extended to very high efficiencies. This does not happen in practice because re-entrainment due to rapping puts an upper limit on the efficiency that can be obtained and the slope of the line decreases, as shown in Figure 9A.1, beyond C. Because the extended Deutsch equation has the form $y = a + bx$, where $b = Cd$, the slope of the performance line is proportional to the particle diameter at constant temperature. A practical example of a performance line is given in Figure 9A.2. In this graph each point represents a separate precipitator efficiency test on a common dust.

9A.1.4 Factors affecting electrostatic precipitation

9A.1.4.1 Particle size. By substituting in the expression for w derived from Stokes' law which is given in equation (9A.11) it can be shown that $w = 0.095dE_cE_p/\eta$ for an average non-conducting particle. Therefore, at constant temperature and electrical conditions the migration velocity (w) is proportional to the particle size (d).

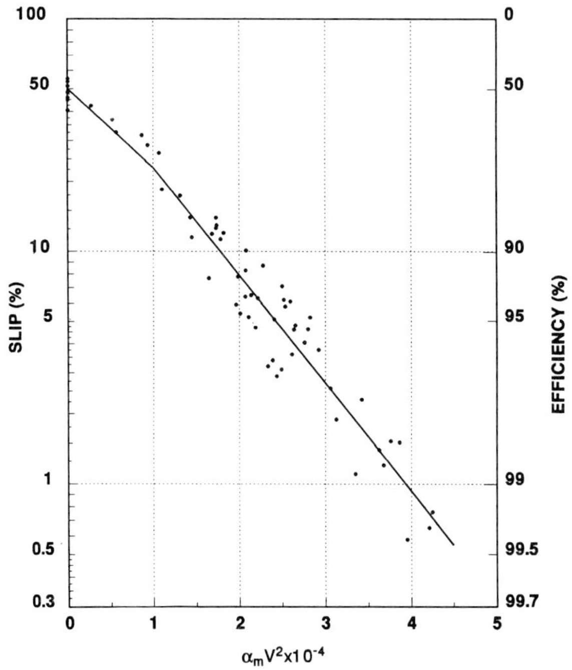

Figure 9A.2 Precipitator performance.

The various equations derived up to now strictly apply only to particles of the same size. However, in most cases, the dusts fed to precipitators consist of particles with a range of different sizes with a distribution characteristic of the material from which the particles originate. Assuming that the particles collect independently according to equations (9A.13) and (9A.14), then each size will collect at a different rate, the larger ones more efficiently than the smaller ones [8]. Obviously, some single particle size is required to represent the size distribution and it is usual to assume that this is the mass median diameter (MMD), which is the diameter above and below which 50% of the weight of the particles lie. Diagrams of the effects of changing MMD on the efficiency of a precipitator [9] show that at an SCA of $60\,m^2\,m^{-3}\,s^{-1}$ a dust with an MMD of 5 μm will exhibit a 4% slip, whilst at an MMD of 10 μm the slip is reduced to 1.5% and at 25 μm it is 0.6%.

This relationship is also demonstrated by the slope of the performance line It has been shown that, for a range of dusts, the slope of the performance line increases as the MMD increases [10]. This effect of particle

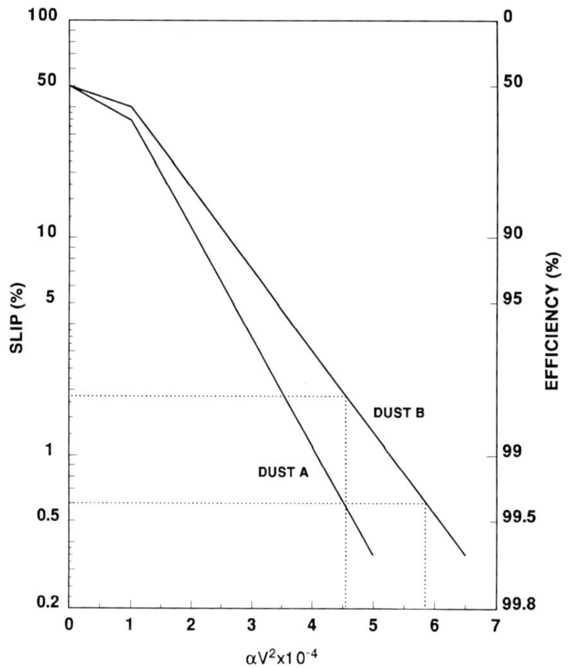

Figure 9A.3 Effect of particle size on the slope of the performance line.

size on the slope of the performance line is shown in Figure 9A.3. The MMD of the Dust A is 10 μm whilst the MMD for Dust B is 6.5 μm.

The precipitator alters the particle size distribution of the dust, so that as the material passes through the precipitator the dust becomes finer. Therefore, the latter sections of a precipitator collect material at a much lower efficiency than the earlier parts.

9A.1.4.2 Temperature. In some industrial plants it is possible to select the position of the electrostatic precipitator in the system. For instance in pulverised-coal-fired power stations the positioning of the precipitator may be either after the air heater, where the gas temperature is in the range 120–180 °C, or before the air heater, where the temperature is in the range 320–400 °C. Earlier it was claimed [11, 12] that precipitation proceeds less erratically at the higher temperature ('hot-side' precipitation) than at the more common lower temperature ('cold-side' precipitation).

The argument for this was as follows. The 'cold-side' precipitation of fly ash relies on surface electrical conduction for the successful passage of current, because the glassy interior of the ash particles has a prohibitively high resistivity at this temperature; so, if the surface conducts poorly, the precipitator will suffer electrical congestion (such as back-corona) [4, 13] with a consequent loss of efficiency. However, 'hot-side' precipitation should not depend on surface conduction because at the higher temperatures the electrical resistivity of the particle has been reduced to allow 'volume' conduction through the bulk of the particle. Thus 'hot-side' precipitators should not suffer from resistivity impediments and so should not exhibit the variability of 'cold-side' precipitators in collecting different coals and ashes.

In fact, full-scale experience with 'hot-side' precipitators showed that they were not, on the whole, more effective than 'cold-side' precipitators. When this finding was coupled with the fact that the 'hot-side' precipitator had operational difficulties due to the higher temperature, their use was discontinued except for special circumstances. A comparison of the performance of precipitators operating at different temperatures has been made by Darby and Whitehead [14]. They confirmed that in many cases the 'hot-side' unit was no smaller than the 'cold-side' unit and at the same time the engineering was more demanding.

The effect of temperature has also been discussed elsewhere [12, 15–17], but these publications do not discuss the effect of temperature on all the parameters affecting electrostatic precipitation. To examine the effect of these factors the extended Deutsch equation should be stated in full. So expanding equation (9A.14)

$$\log(1-\varepsilon) = \log(1-\varepsilon_g) + C_2 p\alpha d V^2/\eta \tag{9A.15}$$

whilst below the corona starting voltage, V_g

$$\log(1-\varepsilon) = \log(1-\varepsilon_0) + C_2 p\alpha d V/\eta \tag{9A.16}$$

These factors will be considered separately under gaseous, particulate and electrical classifications.

(a) Gas-related temperature effects. Viscosity (η) rises slowly as temperature increases giving greater drag forces on the particles and hence reducing migration velocity. The increase in gas viscosity over the temperature range of, say, 120 °C to 330 °C will decrease the slope of the performance line by about 25% which, if no other factors were affecting the precipitator, could reduce the efficiency from 99% to 97%.

Temperature affects SCA (α) if gas volume units are used but not if gas mass units are used. For example, in a process generating a fixed mass rate of gas and particles the SCA in units of $m^2/m^3\,s^{-1}$ (α_v) decreases with rising temperature according to Charles' law. So, in a given precipitator, α_v at 120 °C is greater than that at 330 °C by a factor of $(330 + 273)/(120 + 273) = 1.53$, which is sufficient to reduce an efficiency of 99.5% to 95%. However, if the SCA is calculated in units of $m^2/kg\,s^{-1}$ (α_m) there is no change with temperature, but the gas velocity increases in a fixed installation, probably causing greater dust re-entrainment.

(*b*) *Particulate-related temperature effects.* The mechanical efficiency (ε_0) may be affected by the increase in velocity as temperature is increased and also by a change in the surface characteristics of the particles, making them either more or less sticky depending on their specific sources and thermal history. Thus, it is difficult to estimate *a priori* the effect of temperature on mechanical efficiency.

The dielectric constant (k) which is involved in the factor p ($p = 3k/(k+2)$) increases with temperature. The effect on p of raising the temperature from 120 °C to 330 °C will be an increase in the slope of the performance line of about 12%.

The particle size (d) will not be affected by temperature, but as discussed, the particle stickiness may increase or decrease, thus affecting the clustering of particles into agglomerates and this, in turn, will influence the effective particle size and hence the slope of the performance line.

(*c*) *Electrically related temperature effects.* The fractional collection efficiency (ε_s) at the corona starting voltage (V_g) is greater than the mechanical efficiency (ε_0) because any natural charge possessed by some particles allows them to collect electrostatically at voltages below V_g. The difference between ε_0 and ε_g is small, however, and so the temperature effect on ε_g is likely to follow that of ε_0.

The corona starting voltage (V_g) can be calculated from the equation [18]:

$$E_g = 30\delta + 9(\delta/a)^{1/2} \tag{9A.17}$$

where E_g is the electric field in kV cm^{-1}, a is the corona wire radius in cm and δ is the gas density relative to that at 25 °C.

As the temperature increases the gas density decreases and so E_g and hence V_g are reduced. Furthermore, as the temperature increases the kinetic energy of the gas molecules increases resulting in a higher current flow but a reduced maximum sustainable voltage. This reduction in V_g causes the performance line to move towards the ordinate, causing a reduction in the value of the slip (or an increase in efficiency) at any given voltage. This is offset if there is a fall in the maximum operating voltage (V_{max}) [12, 19].

The effect of temperature on resistive impediments to precipitation is difficult to assess because it is difficult to measure the resistivity of a dust in the presence of an electric field in an operating precipitator. However, considering the change from surface to volume conductivity discussed earlier, the operational resistivity of highly insulating dust layers in precipitators is expected to rise to a maximum at some intermediate temperature and then fall as 'hot-side' temperatures are approached. The sharpness of this maximum will depend on the readiness of the dust surfaces to change as they cool, and the temperature at which the maximum occurs will mainly be determined by the dust composition.

These contrasting temperature effects make it very difficult to predict the overall effect of increasing temperature on precipitator performance. For instance pilot plant results for seven different bituminous coals [20] show that there appears to be little advantage in precipitating the fly ash at 330 °C rather than 120 °C. If current density is limited to the normal working maximum at 120 °C of $2 \times 10^{-4}\,\mathrm{A\,m^{-2}}$ then the size of the precipitator required for operation at 330 °C is larger in five of the seven cases. This situation can only be reversed in favour of 'hot-side' precipitators if considerably higher currents (up to $5 \times 10^{-4}\,\mathrm{A\,m^{-2}}$) are accepted. Furthermore, information published on hot precipitators built in the USA [21] shows that at 330 °C precipitators designed to operate at 99% efficiency have SCA values up to $60\,\mathrm{m^2/m^3\,s^{-1}}$. This is equivalent to $100\,\mathrm{m^2/kgs^{-1}}$ which is the same size as a precipitator with an SCA of $90\,\mathrm{m^2/m^3\,s^{-1}}$ at 120 °C. This is not a small precipitator by international standards.

(*d*) *Dust resistivity.* If the electrical resistivity of the dust is high enough there will be a large voltage gradient across the dust on the collecting electrode causing electrical breakdown in the interstices between the particles. This initiates a 'back'-corona which will impede precipitation by partly discharging particles during their approach to the affected area of the collecting plate. Therefore, the electrical resistivity of the collected layer can be critical to proper precipitation and so resistivity measurements have long been popular in this method of particulate control.

In an operating precipitator the collected dust layer cannot impede the precipitating action unless its resistivity (ρ_d) exceeds that of the gas displaced during the collecting process (ρ_g).

For example, a typical gas at 125 °C in a dust-free full-scale precipitator will ordinarily conduct a current density of $2.5 \times 10^{-4}\,\mathrm{A\,m^{-1}}$ on the grounded collected plate with the applied field close to electrical breakdown at $4 \times 10^5\,\mathrm{V\,m^{-1}}$ ($0.4\,\mathrm{kV\,mm^{-1}}$). By definition the resistivity of the gas (ρ_g) is the applied field across it per unit current density, so that under the conditions cited the value of ρ_g is $4 \times 10^5/2.5 \times 10^{-4} = 1.6 \times 10^9\,\Omega$-m. Thus, the dust layer must have a resistivity (ρ_d) exceeding $1.6 \times 10^9\,\Omega$-m at a field of $0.4\,\mathrm{kV\,mm^{-1}}$ to interfere with its own collection in the precipitator as described.

The importance of the volt/amp (or field/current density) characteristic of a dust-free precipitator is that it enables ρ_g to be found, this being the maximum dust resistivity that the precipitator can handle at a given operating condition without the possibility of collection impediment.

The question then arises what impediment is incurred when ρ_d exceeds ρ_g to a selected degree. An answer in principle to this question has been

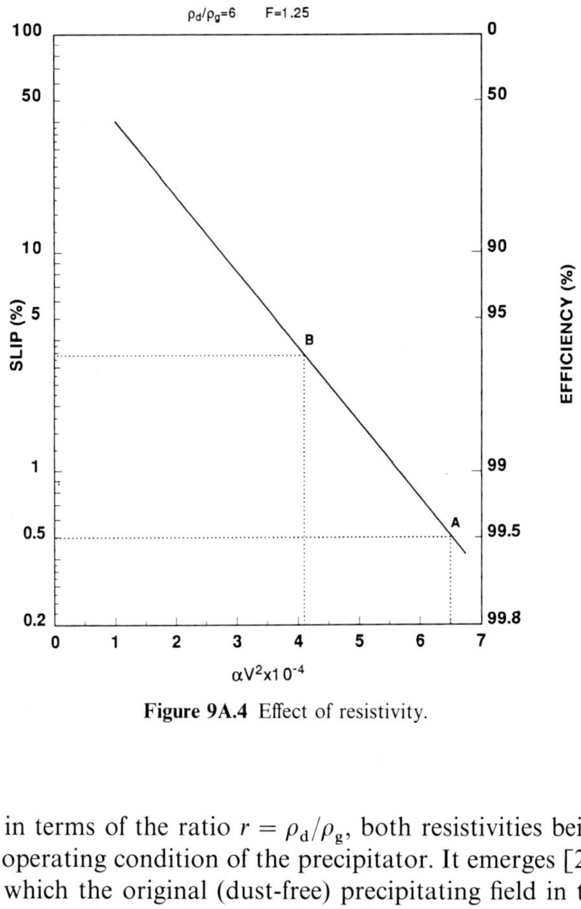

Figure 9A.4 Effect of resistivity.

worked out in terms of the ratio $r = \rho_d/\rho_g$, both resistivities being referred to the same operating condition of the precipitator. It emerges [22] that the factor F by which the original (dust-free) precipitating field in the precipitator is reduced following the collection of an excessively resistant dust depends on the thickness of the layer according to:

$$F = 1 + (t/d)(r - 1) \tag{9A.18}$$

where t is the dust layer thickness and d is the wire-to-plate distance and assuming the original applied voltage can be maintained.

From (9A.18) if t/d is at its reasonable maximum of 0.05 and r is 6, then $F = 1.25$, and an original precipitating field of 0.4 kV mm^{-1} is reduced to 0.32 kV mm^{-1}, just as if an applied voltage of 40 kV had fallen to 32 kV. In a typical case, this means that a collecting efficiency of 99.5% is reduced to 96.6% (an increase in slip by a factor of six) as shown in Figure 9A.4. Note that this assumes that no back-corona takes place.

At the instant when the dust layer is dislodged by rapping, the former full collecting efficiency is almost restored, only to start falling immediately as the dust layer accumulates again. It follows that high-resistivity dust requires special optimisation of the rapping cycle to achieve the least impediment to collection efficiency.

The electrical effect of a rising dust resistivity on the precipitator function is that the applied voltage is increasingly transferred from the gas (where it is effective for precipitation) to the dust layer (where it is not). This effect reaches a limit when the voltage across the dust is high enough to cause electrical breakdown of the dust, and the phenomenon of back-corona sets in.

Back-corona adds considerably to the power demand of the precipitator and causes instability. To avoid it the applied voltage must be reduced, and the efficiency then falls at a faster rate with rising r values. Eventually the voltage is so diminished that the normal corona cannot be sustained and the precipitator ceases to function because particle charging has ceased.

The measurement in principle of the resistivity of a poorly conductive powder layer is a comparatively simple matter, but to make that measurement relevant to the performance of a precipitator in which the powder has been or will be collected is often extremely difficult. The difficulties lie in the great sensitivity of high resistivity material to surface alteration caused by adsorption or desorption of water vapour and ionogenic gases. The surface may be subject to comparatively rapid and irreversible changes in air and at elevated temperatures, and the resistivity will also be electric-field dependent [23]. These features render most of the resistivity measurements reported in the literature for such particles impracticable to interpret usefully. As things stand the volt/amp behaviour of a precipitator both dusty and dust-free is a good diagnostic guide to a resistivity problem in a precipitator.

9A.2 Practical considerations

9A.2.1 Interpretation of test results

9A.2.1.1 Test results. The information required to apply either the w_k method, the w_B method or the CSIRO extended Deutsch equation can be obtained from tests carried out on either full-scale or pilot-scale plants. In each case great care must be taken to replicate the conditions to be expected on the plant being investigated. On many occasions this is easier on pilot plants than on full-scale plant but one major problem with pilot plants is to be completely sure that the dust being collected is exactly the same in physical and chemical condition as that to be collected on the full-scale. It must be remembered that the surface condition of the dust may have a very large effect on the performance of the precipitator. Any dust removed from a process may undergo considerable surface change when placed in a different atmosphere or even just allowed to change temperature.

9A.2.1.2 Application of equations. The precipitator's collection efficiency or slip (percentage escaping from the precipitator) can be measured by extracting isokinetic samples from the pipes entering and leaving the precipitator. Commonly each sample passes through a preweighed filter where the dust is separated from the gas, then through a cooler where the gas is cooled to ambient temperature, part of the water in the gas is condensed and the condensate collected. The saturated gas then passes through a dry gas meter and is vented to atmosphere *via* a pump. The weight of dust is determined by reweighing the thimble and the volume of the gas is determined from the volume passing through the gas meter and the volume of the condensate. Thus, the dust loadings into and out of the precipitator are determined and hence the collection efficiency or slip can be found.

The results have two major uses: (a) to supply information to precipitator users and/or manufacturers on the precipitation characteristics of previously uncollected dusts; (b) to investigate the electrostatic precipitation process.

To interpret these measurements the improved Deutsch equations described above are used. The interpretation of precipitator test results using two of these methods is given below.

(a) w_k method. Equation (9A.9) states that

$$\log(1 - \varepsilon) = -(\alpha \cdot w_k)^{0.5}/2.303$$

where for maximised electrical conditions ε is the fractional collecting efficiency; α is the specific collecting area ($m^2/m^3\,s^{-1}$) and w_k is the modified migration velocity ($m\,s^{-1}$).

Full-scale electrostatic precipitators follow the w_k relationship and, therefore, knowing the w_k value for a dust, estimates can be made of the collection efficiencies of other installations once the original collection efficiency is known. However, there is always some uncertainty in this because there is no guarantee that all installations will have exactly the same w_k value even when collecting the same dust.

With these limitations in mind, to demonstrate the use of w_k a pilot-scale investigation procedure was set up to determine efficiencies of collection, and hence w_k values, over a wide range of SCAs for the collection of five different fly ashes generated from a small pulverised coal combustor. The testing was performed by varying the number of stages used in the electrostatic precipitator at the combustion test facility and so varying the specific collecting area.

The w_k values calculated from these tests were then used to make predictions of the emissions from a given plant, assuming that 90% of the fly ash reached the inlet of the electrostatic precipitator. Using this procedure results of the type shown in Figure 9A.5 may be obtained.

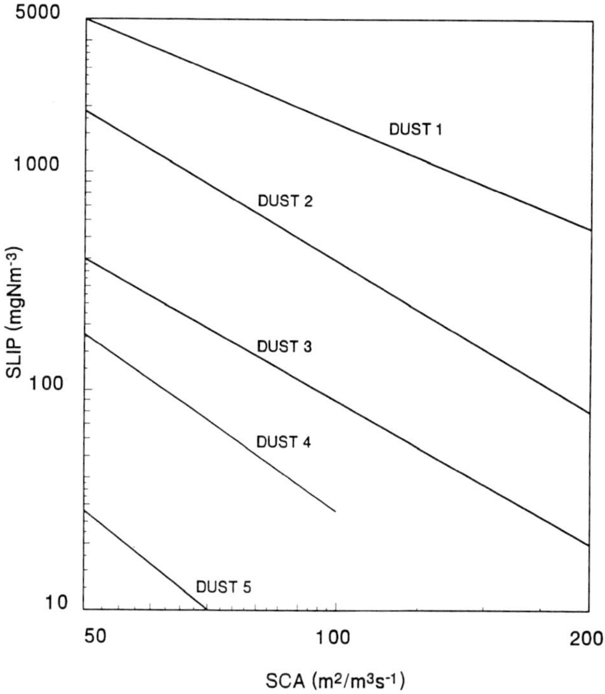

Figure 9A.5 Predictions of plant emissions using w_k values.

From the tests it was found that the w_k values did vary from test to test for the same coal. For instance, for one coal the results showed a variation of w_k between 4 and 8 cm s^{-1} whilst for another coal the results showed a variation between 40 and 70 cm s^{-1}. This variation is sometimes due to tests being carried out over a range of voltages according to the maximum available at the time of the test. Therefore, it is important to carry out a number of determinations of w_k in order to obtain the most accurate estimate of the value.

(*b*) *Performance line method.* Once corona voltage has been achieved equation (9A.14) applies.

$$\log(1 - \varepsilon) = \log(1 - \varepsilon_g) + C_1 \alpha d V^2$$

where ε is the fractional collection efficiency, ε_g is the fractional collection efficiency at the corona starting voltage, α is the SCA, d is the mass mean particle size, V is the applied voltage and C_1 is a constant. Considering constant mean particle size

$$\log(1 - \varepsilon) = \log(1 - \varepsilon_g) + C_2 \alpha V^2$$

where C_2 is a different constant.

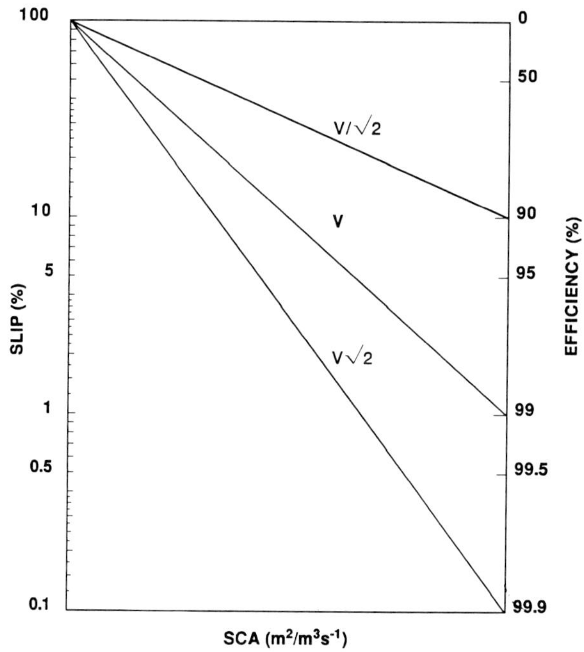

Figure 9A.6 Effect of voltage change.

This final relationship can be used to plot a performance line [5, 6, 10]. The general shape of the performance line is shown in Figure 9A.1 and is explained above.

The performance line is useful for demonstrating the dependence of precipitator efficiency on applied voltage and particle size. These effects are shown graphically for ideal cases in Figures 9A.6 and 9A.7, in which $\log(1 - \varepsilon)$ is shown as precipitator electrostatic collection efficiency. In Figure 9A.6 the three lines correspond to three different applied voltages covering the ratio of 2:1. If a vertical section is taken at the extreme right of the figure it is seen that a precipitator slip of 1% at constant α and applied voltage V is reduced to 0.1% by increasing the voltage by a factor $\sqrt{2}$ but is increased to 10% upon decreasing the voltage by the same factor. It is clear that the voltage is extremely important to maximise and that only one or two kilovolts at 40 kV produce a significant increase in collection efficiency.

In Figure 9A.7 the voltage variable is included in the abscissa, and the three lines correspond to three different particle sizes covering the ratio 4:1. Once again taking the vertical section at the extreme right of the figure, it is seen that at constant α and V the finest dust (represented by the uppermost line) corresponds to a slip of 10%, whereas the dust four times

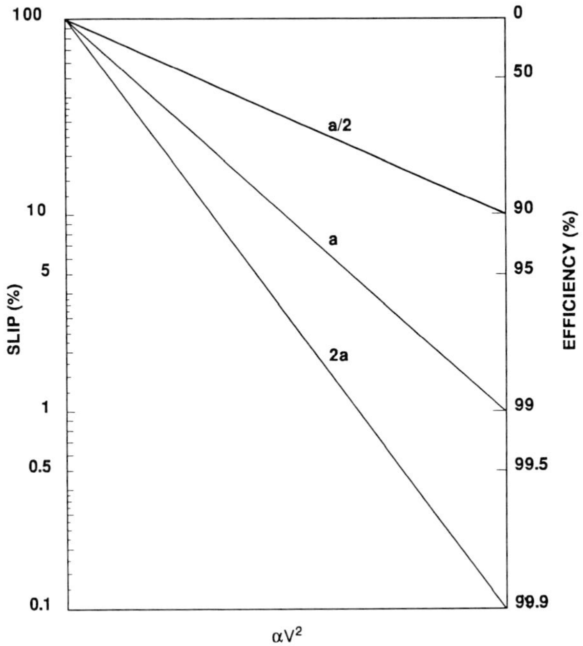

Figure 9A.7 Effect of particle size.

coarser (the lowest line) is precipitating with 0.1% slip under the same operating conditions. Bearing in mind that the average particle size of, say, fly ashes from different coals can readily vary over a ratio of 2:1, Figure 9A.7 demonstrates that a doubling of particle size alone is capable of explaining a change in electrostatic slip from 10 to 1% for example.

The minimum number of test required to properly establish a performance line is six. The sequence of these tests is as follows:

Test	SCA	Voltage	Point on Figure 9A.8
1	Value of most relevant α	V_m (maximum voltage	A
2	α	V_s (corona starting voltage)	B
3	Value to give $\varepsilon > 99\%$	V_m	C
4	α	$V_s + (V_m - V_s)/3$	D
5	α	$V_s + 2(V_m - V_s)/3$	E
6	α	0	F

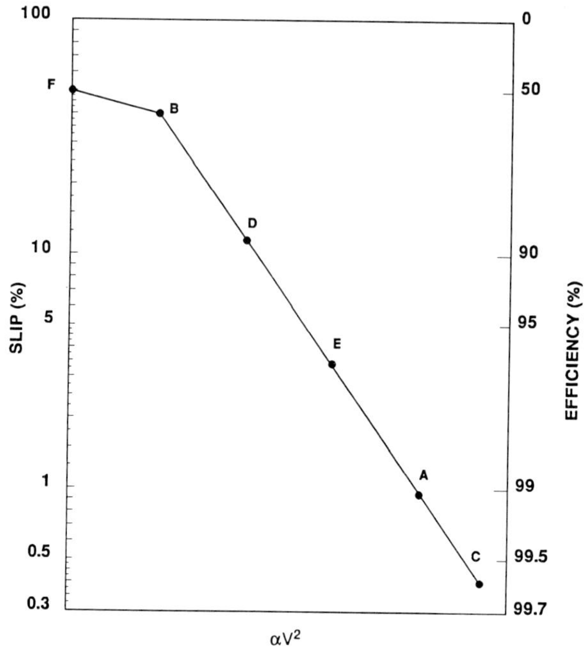

Figure 9A.8 Tests for performance line.

The position of these points is shown in Figure 9A.8. This is the minimum number of points required to establish a meaningful performance line, and in practice ten and often more are required.

9A.2.1.3 Usefulness of equations

(*a*) *Investigation of an unknown dust.* When an unknown dust is investigated the first activity is to establish a performance line such as Figure 9A.2. This information can be used as a guide to the precipitator manufacturer provided there is a good relationship between this pilot plant and the full-scale [24]. Quite often the performance of an unknown dust can be compared with that of a well known one, and hence the precipitator manufacturer can estimate from his experience the size and type of precipitator required for a new dust. When the performance lines for two dusts are compared the variation in performance at constant conditions can easily be seen (Figure 9A.3). For instance, from this figure we see that at a constant operating condition of $\alpha V^2 = 4.6 \times 10^4$, the slip for dust A is 0.6%, whilst for dust B it is 1.9%. So that for these two dusts more than three times as much dust would be emitted for dust B as for dust A. On the other hand if

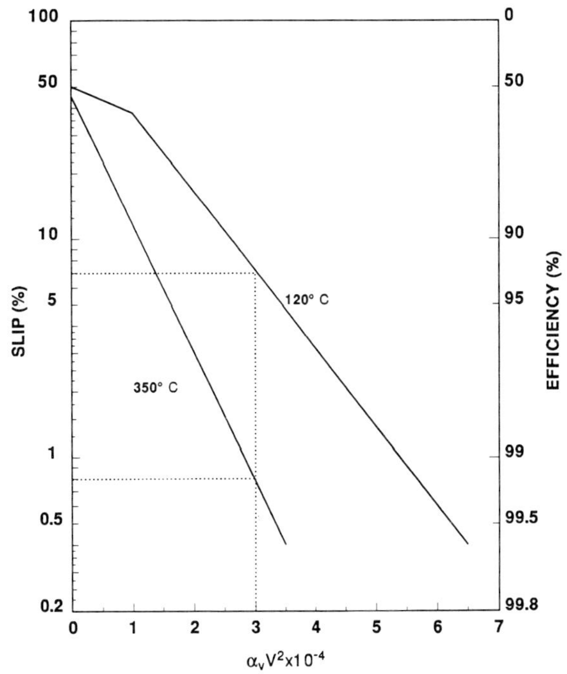

Figure 9A.9 Precipitator results for the same dust obtained at two different temperatures.

a slip of 0.6% was required for both dusts then the value of αV^2 for the two dusts would be 5.8×10^4 and 4.6×10^4, respectively. Therefore, at the same operating voltage the precipitator for dust B is more than 25% larger than that for dust A.

(*b*) *Investigation of the effect of temperature.* The effect of temperature on the performance of a precipitator has been investigated extensively using the pilot plant and the performance line technique for interpreting the results.

Figure 9A.9 shows some precipitator results for the same dust obtained at two different temperatures plotted on the usual $\alpha_v V^2$ basis. From this plot it would be easy to interpret the results as showing a considerable improvement at 370 °C over those at 120 °C. For instance at $\alpha_v V^2 = 3 \times 10^4$ the slip at 120 °C is 7% and at 370 °C is 0.8%. However it is not possible to compare these results on this basis because as the temperature is increased the volume of gas also increases and therefore α_v measured in $m^2\,m^{-3}\,s^{-1}$ is temperature dependent.

When the temperature is varied the specific collecting area of the precipitator must be based on the weight of gas being treated which is

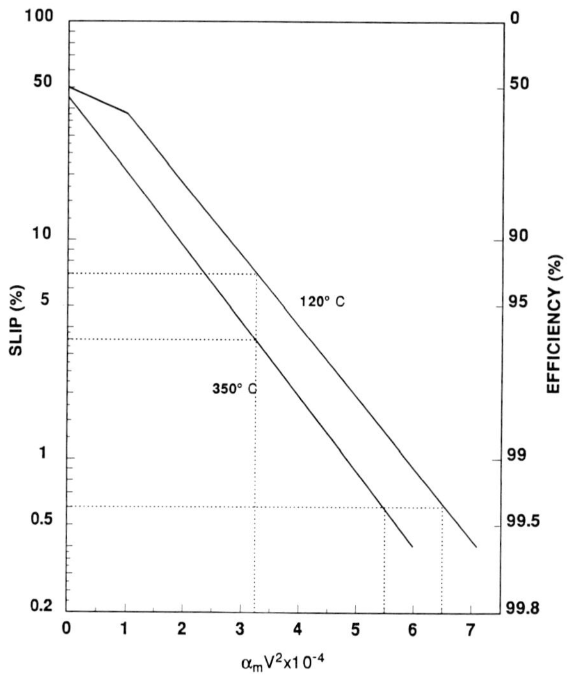

Figure 9A.10 Same results plotted independently of temperature.

independent of temperature. Thus, a new specific collecting area, which we call α_m, must be used with the units $m^2/kg\,s^{-1}$. Figure 9A.10 shows the same results plotted independently of temperature and now it can be seen that, in this case, the effect of temperature is not so great. At $\alpha_m V^2 = 3 \times 10^4$ the slip at 120 °C is still 7% but at 350 °C it is now 3.5% and if a slip of 0.6% was required the $\alpha_m V^2$ value increases from 5.5×10^4 to 6.5×10^4 as the temperature is increased from 120 °C to 350 °C, an increase in precipitator size of 18% at constant operating voltage.

The effect of temperature on the required precipitator size is, however, not linear. There would be an increase in the required size and a subsequent decrease in size as the temperature was changed from 120 °C through 180 °C and on to 370 °C. The severity of this effect could only be determined by pilot-scale or full-scale testing.

(*c*) *Investigation of the effect of dust composition.* Electrostatic precipitation data at 120 °C for 29 Eastern Australian bituminous coals burned in the CSIRO pilot-scale facility have been correlated with readily measured characteristics of the coals and their laboratory ashes [25]. These data are shown as crosses in Figure 9A.11.

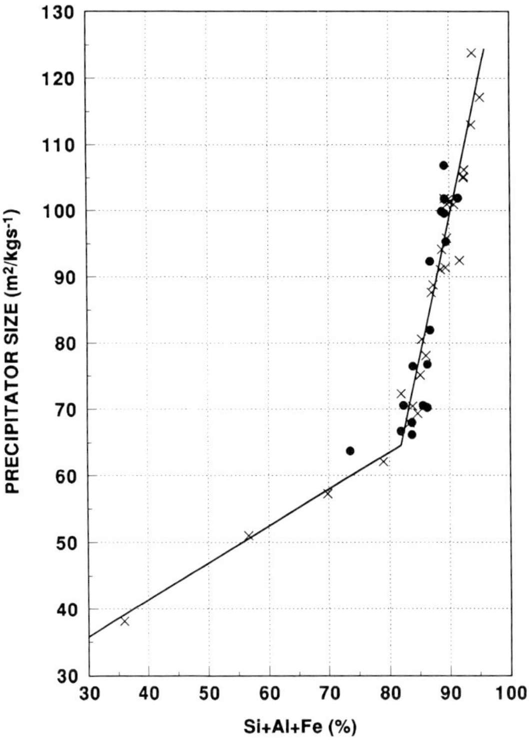

Figure 9A.11 Electrostatic precipitation data correlated with readily measured characteristics of Eastern Australian bituminous coals and their laboratory ashes.

It may be hypothesised that, if a fly ash is refractory, the particles will firstly have less opportunity to enlarge by coalescence while molten in the flame and secondly the collected ash is likely to be less electrically conducting. Thus silicon, through its refractory oxide quartz, and silicon and aluminium through the formation of aluminosilicates may adversely affect electrostatic precipitation. Furthermore, iron under oxidising conditions may have the same effect.

Figure 9A.11 shows the correlation between α_m, the specific collecting area, corrected to 15% mineral matter in the coal, in $m^2\,kg^{-1}\,s^{-1}$ to achieve an emission of 0.1 g Nm^{-3} and the sum of the elements silicon, aluminium and iron expressed on the ash content such that $Si + Al + Fe + Ti + Mn + Ca + Mg + Na + K + P + S = 100\%$. The figure shows that there are two adjoining straight lines which make up the correlation. Since its publication this correlation has been tested with other coals on the CSIRO pilot plant and the results for the next 17 coals tested show excellent agreement with the original correlation. These can be seen in Figure 9A.11 as the results denoted by circles.

The conversion factor f to convert α_m (corresponding to 15% ash) into the precipitator size at the actual ash percentage, A, may be obtained from

$$f = 1.364 - 0.488 \log_{10}[(100/A) - 1] \qquad (9A.19)$$

Thus, it is now possible, knowing the basic coal and ash analyses, to obtain a reasonable prediction of the size of precipitator required for a specific coal. However, pilot-scale and/or full-scale precipitator tests may still be required to confirm this size.

9A.3 Precipitator modelling

Mathematical modelling and practical testing are the two ways of simulating the performance of a full-scale electrostatic precipitator.

9A.3.1 Mathematical modelling

9A.3.1.1 Computer models. A number of computer models for electrostatic precipitators exist and typical examples can be found in the literature [9, 26–29].

The practical application of these models is difficult. In their 1975 paper, Gooch and Francis [9] said: 'Calculation of overall collection efficiency of polydispersed particulate in an electrostatic precipitator from theoretical relationships gives results considerably higher than those obtained from measurements on full-scale units for coal-fired power boilers.' Corrections to the idealised or theoretical collection efficiency to estimate the effects of non-uniform gas flow, re-entrainment and gas by-passing the electrified sections reduce the overall values of calculated efficiency to the range of values obtained from field measurements. These calculations suggest that the theoretical model may be used as a basis for quantifying performance under field conditions if the major non-idealities were to be quantified.

9A.3.1.2 Regression equation models. Several regression equation models have been published. Two examples are outlined below.

Frisch and Coy [11] proposed an equation which had the form

$$\eta = 1 - e^{[-k(P_c/A)^a (A/Q)^b (V)^c (x)^d]}$$

where η is the fractional efficiency, P_c/A is the power density ($W\,m^{-2}$), A/Q is SCA ($m^2/m^3\,s^{-1}$), V is the average treatment velocity ($m\,s^{-1}$), x is the mass median particle size (μm), k is the regression constant and a, b, c, and d are regression coefficients. This equation was combined with: corona characteristics, charging mechanism, resistivity, fuel specifications, temperature, etc.

Tassicker [30] proposed the equation

$$w = a_0 + a_1 \ln\left(\frac{A}{Q}\right) + a_2 \log \rho + a_3 \rho + a_4 \left(\frac{T}{100}\right) + a_5 \left(\frac{T}{100}\right)^2 + \frac{a_6}{d} + a_7 z + a_8 \log_{10} z$$

where w is the effective migration velocity ($\mathrm{m\,s^{-1}}$), A/Q is the SCA ($\mathrm{m^2/m^3\,s^{-1}}$), ρ is the resistivity ($\Omega\,\mathrm{m}^{-1}$), T is temperature (°C), d is mass median particle size (m), z is ash content (%), a_n are the regression coefficients.

It can be seen from these two equations that these mathematical models are complicated and require a considerable amount of information which on many occasions will not be readily available. It should furthermore be noted that these two equations have completely different forms, with the first one being a product whilst the second is a sum.

9A.3.2 Practical testing

Many organisations have pilot-scale rigs and laboratory-scale equipment designed to assess the electrostatic precipitation characteristics of dusts. In some cases, such as fly ash generated in the combustion of pulverised coal, this dust must first be produced in the laboratory.

9A.3.2.1 Laboratory scale equipment. Coal ash may be produced in a small laboratory furnace, designed for small amounts of coal which consist of

- a gas mixture of air, oxygen and propane
- a pulverised coal feeder
- a burner for the gas/coal mixture
- an electrically heated vertical tube furnace
- a heated ash collector.

The ash is generated by dropping the coal particles through a propane flame and collecting the resulting ash. In this manner an ash similar to that produced in a power station is generated [31–33].

The characteristics of the dust are then measured. Key measurements include the resistivity of the dust, the effect of the dust on corona characteristics in a laboratory-scale electrostatic precipitator, determination of the dielectric constant of the dust which affects its ability to be charged and so collected, measurement of its particle size distribution, and other parameters which affect re-entrainment.

These are then used to assess the precipitability of the dust [30, 31, 34–36]. With or without these results it is still important, whenever possible, to carry out pilot plant tests to obtain operating information on the performance of the dust in an electrostatic precipitator before building a full-scale plant.

9A.3.2.2 Pilot-scale equipment. A pilot-scale electrostatic precipitator may be either a transportable piece of equipment which is taken to a full-scale plant and then fed with dirty gas by a side-stream of the main gas flow, or may be a fixed installation in a laboratory attached to a dust feeding system.

For some processes the dust must be generated as part of the electrostatic precipitator assessment process. For instance, fly ash from the combustion of pulverised coal can only be properly tested if freshly generated fly ash in its original flue gas is considered. In this case a test furnace would be used. The time/temperature history of this test furnace must be carefully designed to be similar to the time/temperature history through power-station boilers.

The flue gases exit from the test furnace into the heat exchanger followed by a splitter. The purpose of this splitter is to allow varying quantities of flue gas to be passed to the test electrostatic precipitator so that the specific collecting area can be altered without changing the number of stages on-line.

Electrostatic precipitators with a tubular configuration are commonly used for these tests. These have number of advantages, which are itemised below:

no sneakage
minimum edge effects
minimum corona wind problems
minimum temperature control problems
less mechanical dropout with upward facing entries to each zone.

The main difficulty which has been foreseen with either a wire cylinder or wire plate configuration has been mechanical drop-out, which is common in pilot-scale installations. It is felt that this could be minimised with a tubular configuration by directing the gas carrying the fly ash upwards at the bottom to each stage by fitting turning vanes. Commonly the installation comprises a number of vertical tubular stages to allow for variations in the specific collecting area. The operating temperature of the precipitator should be variable up to about 400 °C allowing simulation of cold-side or hot-side precipitators.

Each precipitator stage should be energised by a single transformer rectifier set. Intermittent energisation is a low cost control strategy which can be easily fitted to the electrostatic precipitator and is extremely beneficial in minimising the difficulties associated with the collection of high resistivity fly ashes. Using this energisation one or two half-waves are imposed from the transformer rectifier set for the test electrostatic precipitator followed by an adjustable period of no energisation. During this period the voltage at the emitting electrode terminal decays to a minimum value which is controlled to the maximum level that can be sustained before currents run away due to back ionisation.

The major factors taken into consideration in the design of pilot-scale electrostatic precipitators are:

- SCA/velocity relationship
- rapping
- gas distribution
- electric field
- particle size and distribution.

All these factors cannot be taken into account at the same time, but the two most important – the electric field and the SCA/velocity relationship – must be given priority. A description of a typical pilot-scale electrostatic precipitator and its mode of operation has been given by Darby [37]. This paper also details the corrrections which may need to be applied to the pilot-scale results to translate them into full-scale design data.

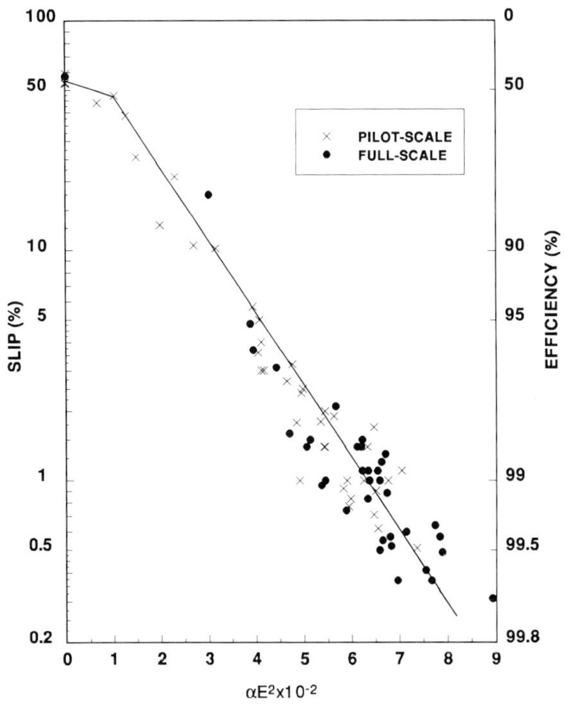

Figure 9A.12 Good relationship between pilot-scale and full-scale plants.

If these pilot-scale test results are to be applied to full-scale plant it is important that there is a good relationship between the pilot and full-scale plants. An example of such a relationship is shown in Figure 9A.12 from which it can be seen that, in this case, the relationship between the two plants is very good.

References

1. Deutsch, W. (1922) Bewegung and Ladung Der Elekrizitätsträger Im Zylingerkondensator. *Ann. Phys.*, **68**, 335.
2. Matts, S. and Ohnfeldt, P.O. (1963) Efficient gas cleaning with SF electrostatic precipitation. *SF Rev. 1963–1964*, **6, 7**. 105–22.
3. Petersen, H.H. (1990) A precipitator sizing formula. *4th International Conference on Electrostatic Precipitation*, Beijing, China, September, International Academic Publishers, Beijing, 1992, pp. 330–8.
4. White, H.J. (1963) *Industrial Electrostatic Precipitation*. Addison-Wesley Publishing Co., Reading, MA, p. 297.
5. Paulson, C.A.J. and Potter, E.C. (1974) Reduction of particulate emissions to air by improved assessment of electrostatic precipitators. *2nd National Chemical Engineering Conference*, Queensland, Australia, July, pp. 404–11, Inst. Chem. Engs, Australia.
6. Paulson, C.A.J., Potter, E.C. and Kahane, R. (1974) New ideas on precipitation technology from the CSIRO combustion rig. *Institute of Fuel Symposium on the Changing Technology of Electrostatic Precipitation*, Adelaide, Australia, November, pp. 26, Inst. Fuel, Adelaide, Australia.
7. Potter, E.C. and Paulson, C.A.J. (1974) Improvement of electrostatic precipitator performance by carrier-gas additives and its graphical assessment using an extended Deutsch equation. *Chem. Ind.*, pp. 532–33.
8. White, H.J. (1975) Role of electrostatic precipitators in particulate control – a retrospective and prospective view. *J. Air Pollut. Control Assoc.*, **25(2)**, 102.
9. Gooch, J. and Francis, N. (1975) A theoretical-based mathematical model for calculation of electrostatic precipitator performance. *Symposium on Electrostatic Precipitation for the Control of Fine Particles*, Pensacola Beach, FL, USA. EPA-650/2-75-016, September.
10. Paulson, C.A.J., Kahane, R. and Potter, E.C. (1976) Electrostatic precipitation of flyash from a range of Australian coals. *Institute of Fuel Conference on Energy Management*, Sydney, Australia, November, Inst. of Fuel, pp. 20.1–12, Sydney, Australia.
11. Frisch, N.W. and Coy, D.W. (1974) Sizing electrostatic precipitators for high temperature collection of flyash. *Institute of Fuel (Australian Membership) Symposium on the Changing Technology of Electrostatic Precipitation*. Adelaide, Australia, November, 28 pp., Inst. of Fuel, Adelaide, Australia.
12. Walker, A.B. (1974) Hot-side precipitators. *Symposium on Electrostatic Precipitators for the Control of Fine Particles*, Pensacola Beach (Florida). *EPA-65012-75-016*, September.
13. Schmidt, W.A. (1949) Electrical precipitation and mechanical dust collection. *Ind. Eng. Chem.*, **41**, 2428.
14. Darby, K. and Whitehead, C. (1974) The use of electrostatic precipitators in current power station practice. *Institute of Fuel (Australian Membership) Symposium on the Changing Technology of Electrostatic Precipitation*, Adelaide, Australia, November, 35 pp., Inst. of Fuel, Adelaide, Australia.
15. Matts, S. (1975) 'Cold-side' precipitators. *J. Air Pollut. Control Assoc.*, **25**. 146.
16. Hall, H.J. (1975) Design and application of high voltage power supplies in electrostatic precipitation. *J. Air Pollut. Control Assoc.*, **25**, 132.
17. Tassicker, O.J. (1975) Some aspects of electrostatic precipitator research in Australia. *J. Air Pollut. Control Assoc.*, **25**, 122.
18. Cooperman, P. (1960) A theory for space-charge limited currents with application to electrical precipitation. *Trans. IEE, Pt. 1*, **79**, 47–50.
19. Shale, C.C. (1967) New concept of electron detachment for air in negative corona at high temperature. *US Bureau of Mines Information Circular 8353*.

20. Paulson, C.A.J., Potter, E.C. and Kahane, R. (1978) The influence of temperature on electrostatic precipitation performance. *CSIRO Conference on Electrostatic Precipitation*, Leura, Australia, August, pp. 12.1–15, CSIRO, Sydney, Australia.
21. Kiff, J.W. (1976) Performance data for Western Precipitation's hot precipitators. *Western Precipitation Seminar on High Resistivity Flyash Collection*, Sydney, Australia, March, 35 pp. Joy Manufacturing Co., Sydney, Australia.
22. Potter, E.C. (1988) Principles of practical gas cleaning. *3rd CSIRO Conference on Gas Cleaning*, Medlow Bath, NSW, Australia, August, pp. 2.1–17, CSIRO, Sydney, Australia.
23. Goard, P.R.C. and Potter, E.C. (1978) Operational resistivity measurements on freshly generated flyashes. *CSIRO Symposium on Electrostatic Precipitation*, Leura, Australia, August, pp. 3.1–8, CSIRO, Sydney, Australia.
24. Potter, E.C. and Paulson, C.A.J. (1975) What size precipitator? A new basis for designing plant to specified dust-extraction performance. *Proceedings International Clean Air Conference*, Rotorua, New Zealand, February, pp. 159–75, Clean Air Soc., Sydney, Australia.
25. Paulson, C.A.J., Potter, E.C. and Vale, J.W. (1986). Correlation of some readily-measured parameters of coal and flyash with electrostatic precipitator performance. *The World Clean Air Congress*, Sydney, Australia, August, pp. 420–7, Clean Air Soc., Vol. 3, Sydney, Australia.
26. Theodore, L. and Pardini, J. (1971) Application of modelling and simulation techniques to the design of electrostatic precipitation. *64th Annual Meeting of the Air Pollution Control Association*, Atlantic City, NJ, USA, June, APCA, Pittsburgh, USA.
27. Theodore, L. and Eastmont, T. (1972) Simulation of an electrostatic precipitator – effect of velocity, particle size, particle mass flow rate and electrostatic force distribution on collection efficiency. *65th Annual Meeting of the Air Pollution Control Association*, Miami Beach, FL, USA, June, APCA, Pittsburgh, USA.
28. Theodore, L., Reynolds, I. and Navarette, R. (1973) Results of a new technique for calculating collection efficiencies of electrostatic precipitators. *66th Annual Meeting of the Air Pollution Control Association*, Chicago, IL, USA, June, APCA, Pittsburgh, USA.
29. Reynolds, J., Mercando, A. and Theodore, L. (1976) The effect of voltage in two-stage electrostatic precipitator efficiency: comparison between model and experiment. *69th Annual Meeting of the Air Pollution Control Association*, Portland, OR, USA, June, APCA, Pittsburgh, USA.
30. Tassicker, O.J. (1974) Performance of cold-side and hot-side electrostatic precipitators treating high resistivity flyash. *Institute of Fuel (Australian Membership) Symposium on the Changing Technology of Electrostatic Precipitation*, Adelaide, Australia, November, p. 19, Inst. of Fuel, Adelaide, Australia.
31. Sullivan, K.M. (1975) *A comparative study of laboratory flyash and power station flyash.* Australian Coal Industry Research Laboratories Ltd., P.R. 75–10, ACIRL, Sydney, Australia.
32. Sullivan, K.M. (1975) *A comparative study of laboratory flyash and power station flyash. Part 2.* Australian Coal Industry Research Laboratories Ltd., P.R. 76–12, ACIRL, Sydney, Australia.
33. Baker, J.W., Sullivan, K.M. and Tassicker, O.J. (1977) Assessment of a laboratory technique for predicting the precipitability of flyash derived from a coal bore core. *Proceedings of the Fourth International Clean Air Congress*, Tokyo, Japan, May, Clean Air Soc., Tokyo, Japan.
34. Baker, J.W. and Sullivan, K.M. (1976) The examination of the electrostatic precipitability of coal. *Institute of Fuel (Australian Membership) Conference on Energy Management*, Sydney, Australia, November, pp. 19.1–15, Inst. of Fuel, Adelaide, Australia.
35. Tassicker, O.J. and Sullivan, K.M. (1973) Estimation of precipitator performance for collection of flyash by examination of low sulphur bore cores. *66th Annual Meeting of the Air Pollution Control Association*, Chicago, IL, USA, June, APCA, Pittsburgh, USA.
36. Baker, J.W. and Sullivan, K.M. (1978) Developments in electrical testing of fly ash in relation to electrostatic precipitation. *Proceedings of the Sixth International Clean Air Conference*, Brisbane, Australia, May, Ann Arbor Science Publishers, Michigan, USA, pp. 223–38.
37. Darby, K. (1981) The use of pilot testing in field and laboratory. *1st International Conference on Electrostatic Precipitation*. Monterey, CA, USA, October, APCA, Pittsburgh, USA.

9B Models of electrostatic precipitators

M. REA

9B.1 Basic concept

The performance of an ESP, like that of any other industrial process, generally depends on some input or process variables and on the design and operating parameters (Figure 9B.1).

In the case of an ESP the main input or process variables are:

- the flow rate, the temperature, the pressure, the chemical composition of the gas;
- the size distribution, the chemical composition of the solid particles which determine several physical parameters like the resistivity, the cohesivity and adhesivity, etc.

The main design parameters are:

- the cross-section which determines the gas velocity,
- the duct width and height,
- the applied voltage and its function in time (its control),
- the shape of electrodes (both the emitting and the collecting electrodes),
- the type and frequency of rapping,
- the design of the hopper and the uniformity of the air flow.

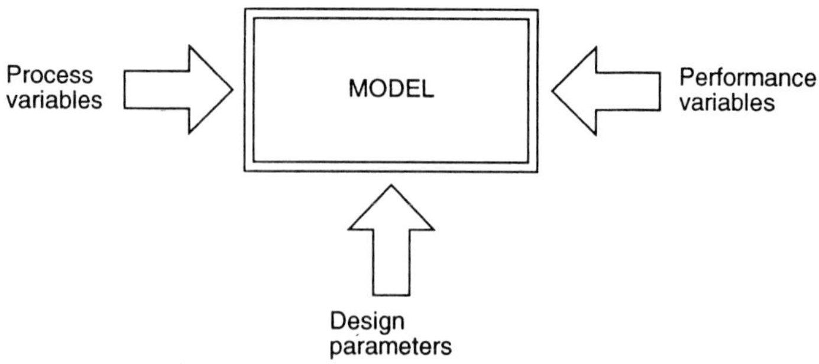

Figure 9B.1 General representation of a model.

Finally the main performance variables are:

- the collection efficiency
- the derivative of the collection efficiency in respect of the input or process variables, commonly called 'flexibility'.

In this respect any model simply represents the set of equations which relates the performance variables to the input variables and the design parameters. An appropriate model is needed to define the design parameters in order to attain the required performance with given input variables. It is quite difficult to define the equations relating the output variables to the input variables and the design parameters, as even an exhaustive identification of the input variables represents a difficult task. It is obvious that simplified models can be profitably used which take into account efficiency as the output variable and consider only those input variables and design parameters whose variability mostly influences this output variable. The difficulty of solving the set of equtions is today reduced by the availability of powerful numerical methods and computers; it is worth mentioning that as the computing power in solving any set of equations has increased, scientists are prone to take into consideration more input variables and are less stimulated to identify those input variables and those physical phenomena which have most influence on the efficiency (output variable). Several input variables are multidimensional, i.e. they are distributed in space and are frequently variable in time; this brings a great increase in complexity when determining and solving the set of equations relating to any model. Sometimes it can be a reasonable approximation to neglect the distribution in space and/or the distribution in time; in these cases unidimensional or two-dimensional models are proposed.

The phenomena playing a role in electrostatic precipitation may be-characterised by time constants dissimilar even by several orders of magnitude; in these cases it is advisable to initially solve the equations of the fastest developing phenomena and then place the results as constant parameters in the equations which describe the slow varying phenomena. For instance, the equations describing the corona discharge are solved in order to determine the time-averaged value of the space charge and of the electric field distributions; these values are taken into account in the equations describing the charging of the solid particles and their drift/migration to the collecting electrodes.

9B.1.1 The Deutsch equation

The oldest and simplest model still widely used is the Deutsch equation (Figure 9B.2). Its proposer, Mr. Walther Deutsch (1885–1957) was a great German scientist and technician [1]. He proposed that the particles are driven to the collecting plates by the electrical field with a constant velocity

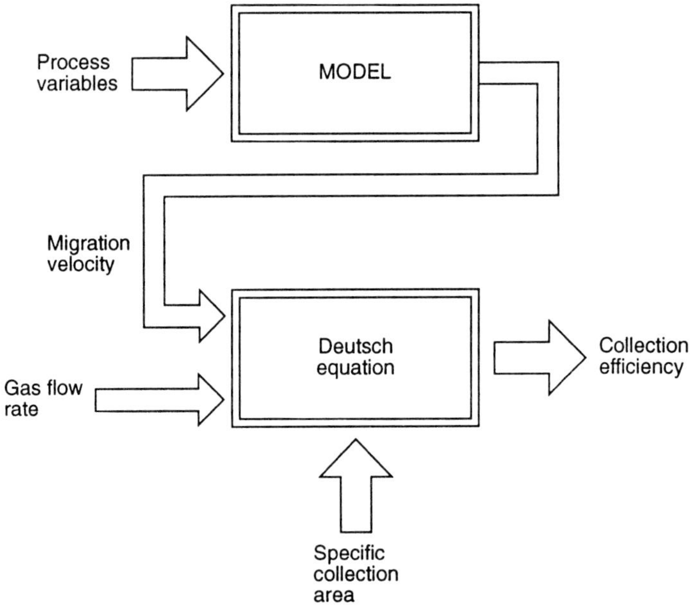

Figure 9B.2 The model based on the Deutsch equation.

(migration velocity). Of course this velocity will depend on several other process variables, but it can be experimentally determined or evaluated by comparison with other similar installations. In these hypotheses the collection efficiency can be evaluated using the following 'Deutsch' equation:

$$\eta = 1 - \exp(-\omega \cdot \mathrm{SCA}) = 1 - \exp\left(-\frac{\omega \cdot S}{v \cdot A}\right) = 1 - \exp\left(-\frac{\omega \cdot S}{G}\right)$$

where η is efficiency, ω is migration velocity (velocity of the solid particle toward the collecting plate), v is mean gas velocity, S is the surface of the collecting plates, A is overall cross-section of the electrofilter, G is gas flow rate $= v \cdot A$ and SCA is specific collection area $= S/G$.

The Deutsch equation may be used in two different ways:

- once the efficiency has been measured, the migration velocity (which is then called *effective*) can be computed;
- once the effective migration velocity has been estimated, for instance because of the similarity with another precipitator, the collecting surface (S) can be evaluated in order to achieve the desired efficiency with the given gas flow rate (G).

It is interesting to mention how the equation has been derived because it points out the influence of some process variables and it explains some suggested later modifications by other workers [2]. Let us assume that:

- the particles are completely charged;
- because of gas turbulence, the particles are uniformly distributed at any cross-section;
- the gas velocity does not influence the particle velocity toward the plates;
- the particle always moves at its electrical terminal velocity;
- there is no interaction between particles;
- the collision of ions and neutral gas molecules has no effect;
- at a short distance from the collecting plate the gas flow is laminar and the particles are driven to the plate with velocity, ω.

With reference to Figure 9B.3, let H and W be the height and width of the gas duct respectively, L be the length in the direction of the gas flows, and v the velocity of the gas stream. After a time increment t, during which the gas stream has moved a distance $L = vt$, all the particles N present in the boundary layer, at a small distance d from the plate, will be driven to the collection plate. If ω is the drift velocity of the particles to the plate, it follows that $d = \omega \cdot t$ and $N = H \cdot W \cdot L$. The incremental equation for particle removal will be:

$$\frac{\Delta N}{N} = -\frac{\omega \cdot \Delta L}{v \cdot W}$$

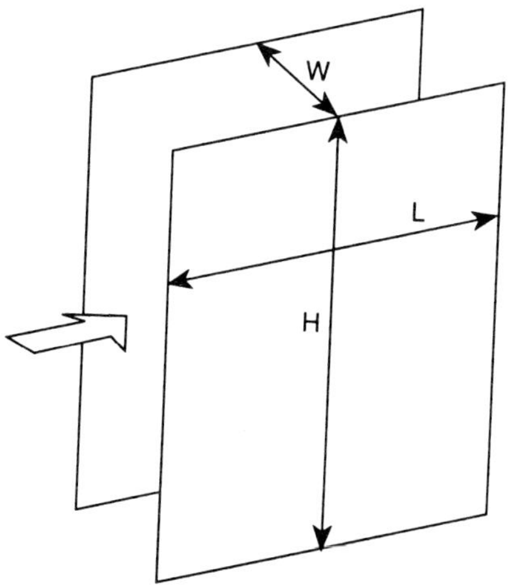

Figure 9B.3 Single duct of an electrostatic precipitator.

which, after integration, gives:

$$N = N_0 \cdot \exp\left(-\frac{\omega \cdot S}{v \cdot A}\right)$$

where N_0 represents the particle concentration at the inlet and N the particle concentration after a length L of gas path. The efficiency, expressed as the ratio of the collected particles to the inlet particles, in the above after rearrangement, becomes the Deutsch equation.

9B.1.2 Charging of particles and the modified Deutsch equation

Several improvements of the Deutsch equation were proposed in order to take into account other process variables which were recognized to play an important role while still taking advantage of the simplicity and efficiency of the Deutsch equation.

The most important modification of the Deutsch equation is that suggested by Matts and Ohnfeldt [3]; its purpose is to take into consideration the influence of the particle size distribution on the collection efficiency. In order to understand the reasons for this it must be remembered that the corona discharge developed by industrial precipitators produces negative ions which impact with the solid particles flowing between the collectors and transfer to these some of their electric charge. Collision of the negative ions with the solid particles is generally due to random motion, but the electric field distorted by the solid particles may either favour or hinder the collisions depending on the charge on the particle and on its size.

In addition to the motion of the particle towards the collecting plate, there is the resultant effect of several forces (see Figure 9B.3) acting on the particles, some of which depend on the size of the particle, itself.

$F_e = q \cdot E$	electrical force
$F_\eta = 6 \cdot \pi \cdot a \cdot \eta \cdot \omega$	viscous drag force
$F_g = m \cdot g$	gravitational force
$F_i = m \dfrac{d\omega}{dt}$	acceleration force

where E is the applied electric field, q is the charge on the particle, η is the viscosity of medium, a is the radius of the spherical particle, ω is the velocity of the particle, m is the mass of the particle and g is the acceleration due to gravity.

If the variable $q(a)$, evaluated following the model of charging and the resultant of the previous listed forces acting on the particle, is solved, the migration velocity of the particle ω becomes

$$\omega = \frac{E_p}{6 \cdot \pi \cdot \eta} \frac{k \cdot T}{e} \cdot \ln\left[\frac{(1 + \pi \cdot a \cdot v \cdot N_0' \cdot e^2 \cdot t)}{k \cdot T}\right]$$

where E_p is the electric field close to the plate, k is the Boltzmann constant, T is the temperature, e is the charge on the electron, v is the gas velocity and N_0' is the negative ion density.

In particular this implies that migration velocity is a function of particle size a and that the Deutsch equation should be evaluated for each class of particle size. If a log-normal distribution of particle size is assumed, the Deutsch equation becomes:

$$\eta = 1 - \exp(-\omega_k \cdot \mathrm{SCA})^k$$

where k represents a factor taking into account several process variables.

Another important modification to the Deutsch equaton is that proposed by Cooperman [4] which takes into consideration a diffusional force acting on the particles due to the different concentration along the precipitator, which reduces the effective migration velocity. The suggested equation takes the following form:

$$\eta = \left(1 - \frac{\beta}{1 - \alpha}\right) - \exp[-\omega \cdot (1 - \alpha) \cdot \chi \cdot \mathrm{SCA}]$$

where α is the coefficient of variable erosion, i.e. re-entrainment due to rapping, β is the coefficient of constant erosion, i.e. re-entrainment due to gas scouring and χ is the ratio between dust concentration close to the plate and average value.

9B.2 The modern approach to computer modelling

9B.2.1 Early models

The development of computers during the 1970s gave rise to the development of computer-based models for electrostatic precipitation. Several scientists of the Southern Research Institute in the US have been involved in this task guided by Oglesby and Nichols [2]. The basic approach was to subdivide the precipitator into small incremental lengths L and to subdivide the particles into a number of particle size increments a so that the Deutsch equation becomes:

$$\eta_{i,j} = 1 - \exp(-\omega_{i,j} \cdot \mathrm{SCA}_j)$$

where $\omega_{i,j}$ is the migration velocity of the jth particle size and SCA_j is the specific collection area for the jth incremental length. The collection efficiency *i* for the *i*th particle size becomes:

$$\eta_i = \frac{\sum_j \eta_{i,j} \cdot N_{i,j}}{N_{i,j}}$$

where $N_{i,j}$ are the number of particles of the *i*th particle size per cubic meter of gas entering the *j*th increment of length.

The overall collection efficiency becomes:

$$\eta = \sum_i \eta_i \cdot P_i$$

where P_i is the percentage by mass of the *j*th particle size. In fact this model takes into account the space distribution of flow and the size distribution of dust.

In these models a number of important phenomena like rapping re-entrainement, flow pattern were represented by correction factors. Some other phenomena had to be neglected such as:

- the influence of the dust on the space charge density and electric field strength;
- the influence of the dust layer thickness and resistivity on the electric field including the back-corona;
- the formation of the dust layer, i.e. the probability of the charged particles being collected by the dust layer or causing the re-entrainement of a part of the dust layer;
- the erosion of the dust layer produced by the gas flow;
- the re-entrainement of dust produced during rapping of the collecting plates.

The approach which considers the basic physical equations can be applied to the elementary particles (solid and gaseous) present in a small gas volume, or on a small surface of the collecting plate and numerically, then integrates this equation all over the electrostatic precipitator, is becoming very popular with a number of investigators.

The main problem with the models produced by this approach is the dimensions of the cells into which the precipitator has to be divided; in order to respect the physics of many important phenomena, the mesh should be so fine that even large computers would be unable to perform the computations. The technique being adopted is to use a coarse mesh for the precipitator and a finer mesh for each cell.

9B.2.2 Model by Cañadas et al. [5]

Figure 9B.4 represents the flow chart of a recently proposed computer model [5]. Referring to the original paper for details the fundamental variables considered as substantially independent are: the applied voltage,

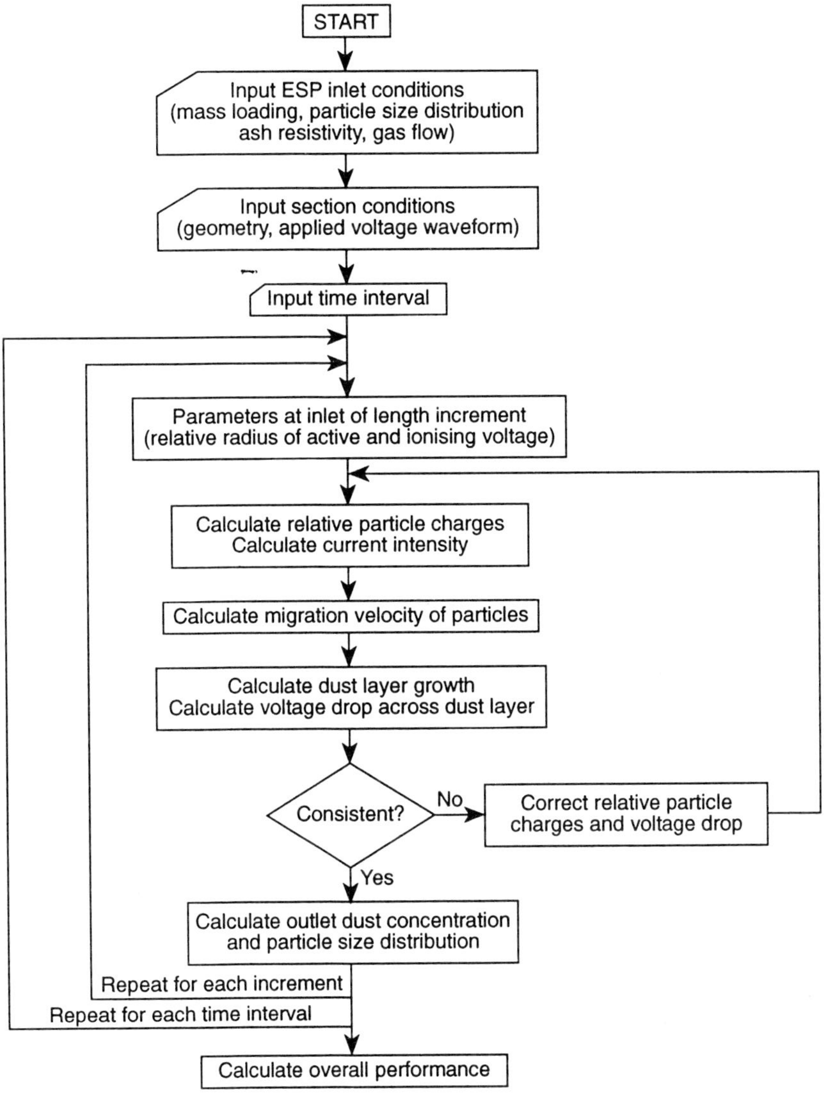

Figure 9B.4 Flow chart of the model proposed by Cañadas *et al.* [5].

the ionising voltage, the radius of the active zone, the current intensity arriving at the plates, the voltage induced by the particle space charge, the relative particle charge of each particle size, the migration velocity of each particle size, the dust layer thickness, the particle layer resistivity and the voltage drop across the deposited layer.

The basic equations considered by the model are as follows.

1. the electric field is computed as a function of the applied voltage, the distance between the collecting plates and discharge electrodes, the electrode system confuguration factor and the shape factor of the electrostatic field;
2. the relative particle charge is computed as a function of the ion mobility, the distance between the collecting plates and discharge electrodes, the voltage waveform, the electrode system configuration factor, the charging coefficient, the ionising voltage, the effective velocity of the gas molecules and the particle diameter;
3. the current/voltage characteristic is computed as a function of the ion mobility, distance between the collecting plates and discharge electrodes, current attenuation factor, net available voltage, voltage induced by particle space charge, collecting capacity of plates, electrode system configuration factor;
4. the migration velocity is computed as a function of the permittivity of the gas, the particle charging coefficient, the peak voltage value, the maximum value of the field shape factor, the mean voltage value, the shape factor of the electrostatic field in the deposit layer, the distance between plates and discharge electrodes, the electrode system configuration factor, the relative particle charge and the particle diameter;
5. the dust layer thickness is computed as a function of the applied voltage, the particle layer resistivity, the current density in the deposit layer, the specific gravity of the dust layer, the distance between the collecting plates and discharge electrodes, the collecting capacity of the plates, the electrode system configuration factor, the gas viscosity, the particle concentration, the shape factor of the electrostatic field in the deposit layer, the maximum particle charge attainable by ion bombardement, the relative particle charge and the particle diameter;
6. the voltage induced by particle space charge is computed as a function of the density of particle space charge, the distance between the collecting plates and discharge electrodes, the permittivity of the gas.

It is evident that a number of empirical coefficients are used, such as the shape factor of the electrostatic field, the electrode system configuration factor and the current attenuation coefficient. This means that the model is suitable for predicting, for a given precipitator, the variation of the collection efficiency from the actual value due to some variation of the operating conditions or of the structural dimensions.

9B.2.3 Modelling at Padova university [6]

An even more detailed model which claims to be able to predict the collection efficiency by simply referring to the design characteristics of the

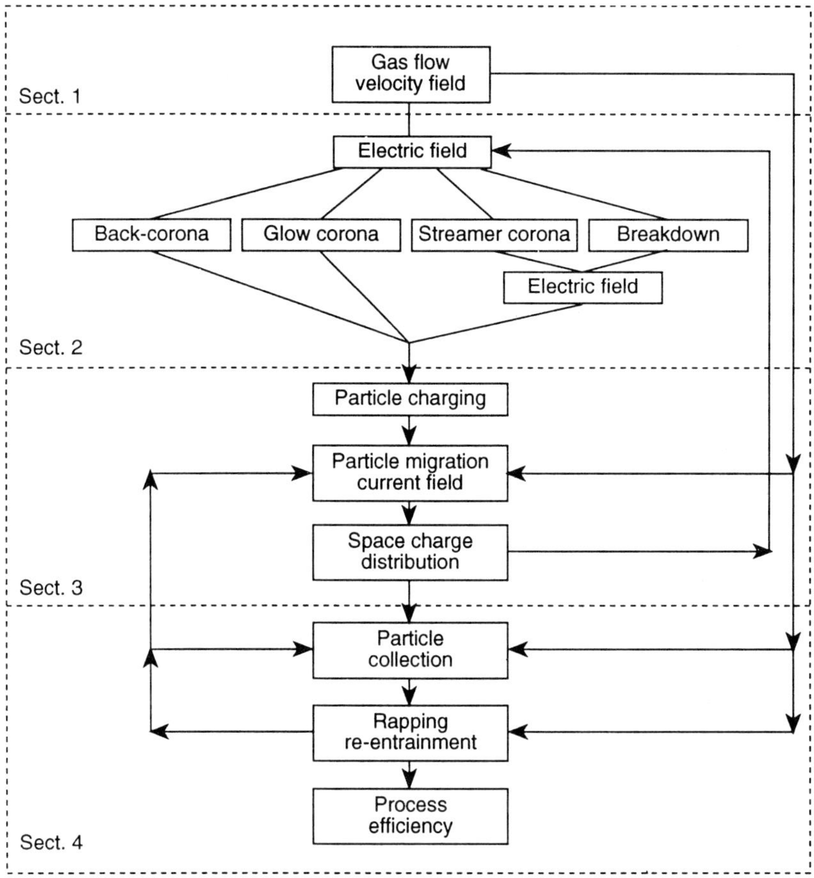

Figure 9B.5 Model developed by Prof. Gallimberti, University of Padova [6].

ESP and to planned operating characteristics is being developed at the University of Padova under a research contract from ENEL [6].

The flow chart represented in Figure 9B.5 shows that the model is organised into the following sections and subsections:

1. gas and particle motion: it evaluates the actual fluid-dynamic conditions of the gas motion and the velocity field which retards the dust particles;
2. electric field and discharge processes, subdivided into the modules:
   2.1 electric field which solves the Poisson equation to determine the electric field;

 2.2 back-corona which takes into consideration the emission of positive charges from the collecting plates;
 2.3 glow-corona which simulates the continuous flow of positive and negative charges in DC operation;
 2.4 streamer corona which simulates the production of charge due to impulse streamer discharge;
 2.5 breakdown which simulates the critical conditions resulting in electrical breakdown.
3. particle charging and migration, subdivided into the modules:
 3.1 particle charging which models the progressive charging of the solid particles;
 3.2 particle migration and current field which takes into consideration the forces applied to the particles, the charge exchanges and the turbulent diffusion;
 3.3 space charge distribution evaluates the charge distribution needed for the updating of the electric field;
4. particle collection, subdivided into the modules:
 4.1 particle collection simulates the laminar flow close to the collecting plates;
 4.2 rapping and re-entrainement simulates the sudden injection of particles from the dust layer due to rapping or the continuous one due to scouring;
 4.3 process efficiency, determines the overall collection efficiency of the precipitator.

This model, recently completed, is presently under test. Up to now it has been able to predict with high accuracy the voltage/current characteristics, when the precipitator is loaded with different monodisperse particle concentrations and differently energised: conventional AC rectified voltage or DC voltage with short voltage pulses superimposed. It has also successfully verified the prediction of the back-corona onset as a function of the applied voltage and the dust layer resistance and the prediction of the breakdown voltage at a given voltage and dust layer thickness.

Any model cannot avoid a number of numerical constants which combine the effects of several minor physical process; the main feature of this latter model is that it requires very few 'adjusting parameters', namely the following:

- the roughness of the emitting wire, which also takes into account the possible dust layer formation of the wire;
- when the corona discharge produces streamers (when voltage pulses are superimposed) the number of streamers per unit length of emitting wire;
- the turbulent diffusion which depends on the gas composition and temperature;
- the re-entrainement due to the erosion of the dust layer;

- the dust re-entrainement due to the rapping; this parameter has to be computed in advance following a model of the dust layer.

The input quantities required by the model may be grouped into the following items:

- the data characterising the geometry;
- the data characterising the gas (composition and velocity distribution at the inlet cross-section) and the particulate (composition and size distribution at the inlet cross-section);
- the data characterising the applied voltage.

References

1. Mayer-Schwinning, G. (1987) Opening Lecture at the *Third International Conference on Electrostatic Precipitation*, Abano, University of Padova, Italy.
2. Oglesby, S. and Nichols, G. (1978) *Electrostatic Precipitation*, Marcel Dekker, New York.
3. Matts, S. and Ohnfeldt, P.O. (1963) *Efficient gas cleaning with SF electrostatic precipitators*, A.B. Svenska Fläktfabriken Report, **6(7)**, 105–22.
4. Cooperman, P. (1969) A new theory of precipitator efficiency. Paper no. 69–4, Fourth *APCA Meeting*, APCA, Pittsburgh, USA.
5. Cañadas, L., Navarrete, B., Salvador, L. and Rodriguez-Aragones A. (1993) PRELEC: a mathematical model of electrostatic precipitation, *10th EPRI Particulate Control Symposium and the Fifth International Conference on Electrostatic Precipitation*, Washington, USA, pp. 21.1–15, EPRI TR **2**, 103048, Palo Alto, CA, USA.
6. Bellagamba, B., Lami, E., Mattachini, F., Gallimberti, I., Turri, R., Gazzani, A., Tromboni, U., Pasinetti, A. and Sala, R. (1993) A mathematical model for simulation of large scale electrostatic precipitators" *10th EPRI Particulate Control Symposium and the Fifth International Conference on Electrostatic Precipitation*, Washington, USA, pp. 25.1–14, EPRI TR **2**, 103048, Palo Alto, CA, USA.

10 Sampling and analysis for particles and heavy metals in gas streams

G.B. NICHOLS and E.B. DISMUKES

10.1 Sampling and analysis

The determination of the performance of air pollution control devices, as well as the emission rates to the atmosphere, requires the sampling and analysis of the materials carried by the gas stream exiting from the process plant. There are a number of specific standards that govern the measurement methods established by the appropriate government agencies responsible for the control of emissions, as well as standards developed and specified to be followed to determine the performance level of control devices, either for establishing contractual guarantees or for information required for other reasons. The specific standards that apply depend upon the particular country where the information is to be acquired. For example, in the United States, the sampling and analysis methods are described in the Code of the Federal Regulations (40 CFR), while in Europe the DIN standards generally apply. It is not the intent of this chapter to delineate all the specific codes that apply, but rather to provide a discussion that applies to sampling and analysis in general that provides insight into the methodology. Regardless of the purpose for the measurement and the specific standard that applies, it is necessary to obtain a representative sample of the materials contained in the gas stream. The standards that apply in various locales are given in references 1–4.

There are a number of different measurements needed to completely characterize the emissions from a particular source. The first is a simple measure of the mass of the condensed liquid or solid particles emitted. These consist of particles and droplets suspended in and carried by the gas stream. The stream velocities in ductwork are usually sufficiently high that particles, other than the large ones, do not settle out significantly, but there is always a tendency to develop some stratification in the ductwork. This stratification is related to the differential gravitational settling velocities of the large heavier particles, compared with the lower settling velocities of the smaller, less massive ones. Bends, obstructions, internal supports, leaks and fans also contribute to the formation of non-uniform distributions of particles in the cross-section of the ductwork. Thus, it becomes necessary to establish a protocol that samples the entire cross-section of ductwork in a statistically representative manner.

Items of interest in a gas stream emanating from any process may include solids, liquids and vapors. A total characterization of the emission requires that these items be sampled representatively. The appropriate method to use for sampling either particles or droplets requires sampling the entire duct cross-section with enough sampling points to be statistically acceptable and to extract these samples isokinetically. This means that the gas velocity inside the sampling nozzle be identical with that just prior to entering the nozzle. Individual sampling standards will identify the minimum number of sampling points to adequately sample a given gas stream. In general, the greater the distance from turns or other obstructions, the smaller the number of sampling points; there is however, a minimum acceptable distance, upstream and down, from a flow upsetting obstacle for sampling. This should be addressed in the applicable sampling standard.

The necessity for isokinetic sampling is associated with the behavior of the constituents in the gas stream. If the item to be sampled is a gas, logic would suggest that isokinetic sampling is not required, except for the case where in-leakage or other factors may lead to a non-uniform distribution of the gases. The mass and inertia force of the gas molecules of interest are the same as the other gas molecules in the stream and will be sampled at the same rate as the carrier gas. However, both solid and liquid materials suspended in the gas stream will have significantly greater gravitational and inertia forces acting on them than the gas molecules. Figure 10.1 illustrates the three conditions of sampling discussed. Figure 10.1a represents isokinetic sampling, i.e. the velocity in the sampling nozzle matches that in the ductwork just prior to sampling. Next, consider the case where the gas stream in the sampling nozzle has a velocity greater than the velocity in the duct as illustrated in Figure 10.1b. At the inlet to the nozzle, the gas flow lines will converge into the nozzle. The inertia force acting on the particles, which increases with size and mass, will cause the larger particles to flow through the gas in the converging flow into the nozzle and some portion of the larger particles will be missed, leading to an underrepresentation of the large particles. This will lead to a collected mass sample less than that in the main gas stream and a bias in the particle size distribution of the sample collected.

Conversely, when the velocity in the sampling nozzle is less than that in the main gas stream, the sample collected will also be unrepresentative of that in the main flow stream. Figure 10.1c illustrates this undersampled case. In this example, the gas stream from the main duct will be diverted around the nozzle. The inertia force on the particles will now cause the larger particles to flow through the diverging gas stream and into the nozzle. This sampling error leads to oversampling the particles, with a sample collected that has more mass per unit volume than the main gas stream. Typically, isokinetic sampling within 5–10% of the local velocity is adequate to collect an acceptable sample.

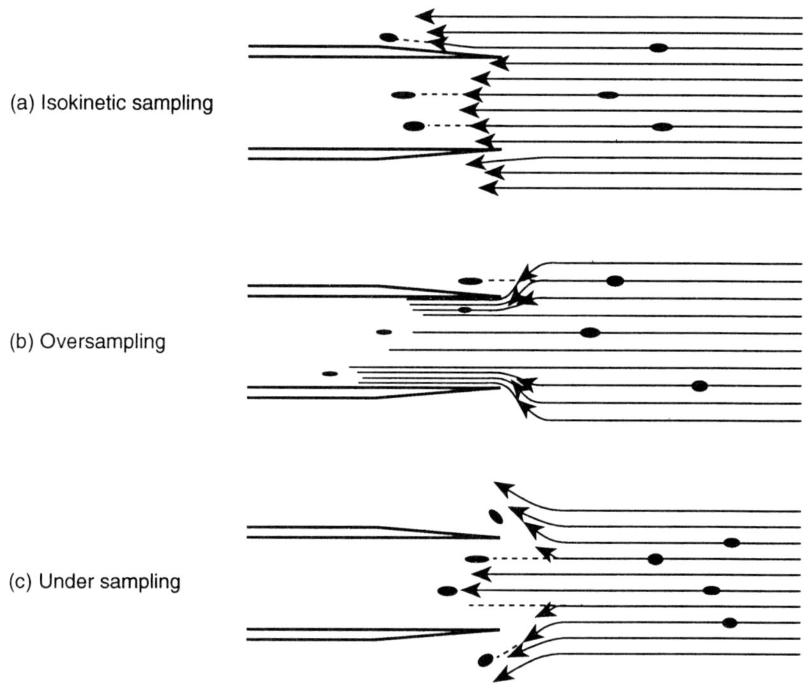

Figure 10.1 Illustrations of isokinetic, over and undersampling.

The test methods used throughout the world for determining the mass loadings and/or emissions are similar in that the sampling trains used consist of some type of probe equipped with a streamlined nozzle, a filter, a gas flow measuring device and a pump or other means for pulling the gas stream through the sampling system. The probes typically are equipped with thermocouples and pitot tubes to measure the gas temperature and velocity. The sampling system is referred to as *in-situ* if the filter is located in the gas stream proper and extractive if the filter is external to the flow stream. Bubblers and liquid filter traps may also be added for specialized testing needs. An example of an extractive sampling system is given in Figure 10.2 [5]. If the filter holder was mounted just behind the sampling nozzle the example would represent an *in-situ* probe.

Special care must be exercised if the gas stream contains condensable materials and the sampling method selected is extractive. Some methods specify the temperature of operation for the external filter depending upon the purpose for the measurement. If the purpose is to evaluate the performance of a particle control device, the filter should operate at or slightly above the process gas stream temperature. If the purpose is associated with the

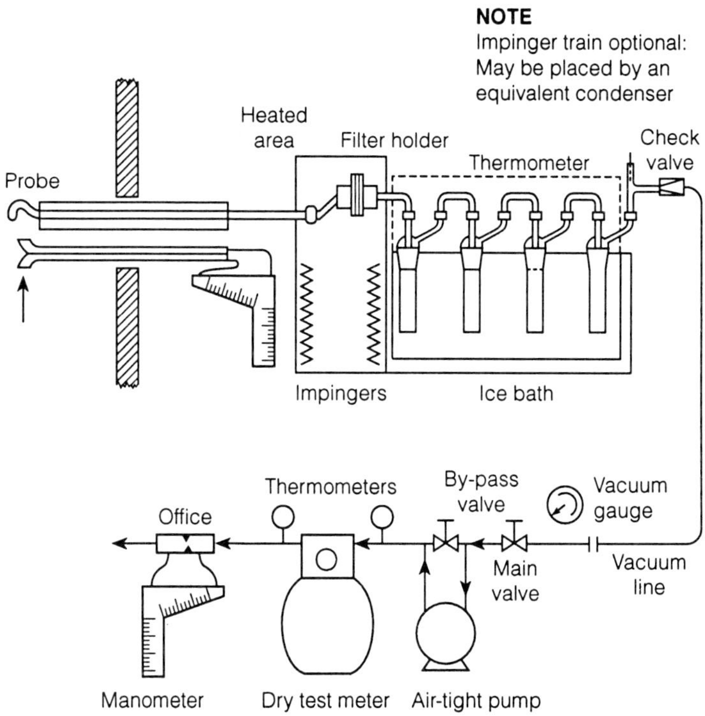

Figure 10.2 EPA Method 5 particulate matter sampling train.

determination of compliance with regulations, the applicable standard must be consulted. If the sampling system is operated at a temperature below the process stream temperature, the condensables may be collected on the filter medium and be reported as particulates. In both extractive and *in-situ* cases, the particulate material that adheres to the inside surfaces of the probe nozzle and probe liner upstream from the filter must be retrieved and added to the filter catch for a representative sample to be collected.

There are several other types of mass measurement systems that are available for specialized purposes. Some of these include Beta beam absorption measurements through a previously collected sample, light scattering transmissometers and absorbers, and multiple light frequency instruments. These instruments are typically used for monitoring performance rather than for detailed performance or mass measurements. As such they are only mentioned for completeness. Their operation will not be described. References 6–10 identify some instruments of these types.

A second characteristic of the particles or droplets suspended in the gas stream is the particle size distribution of the particles as they exist in the gas.

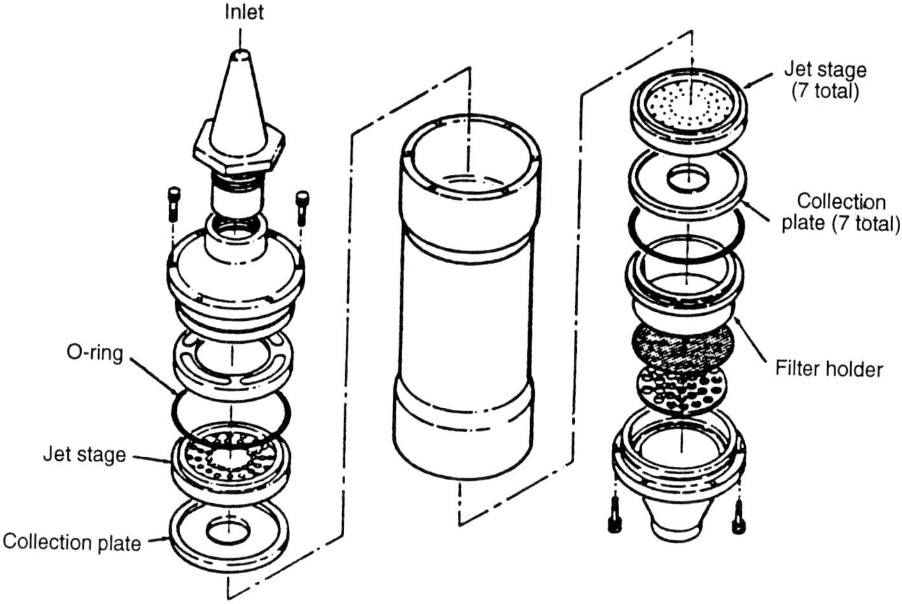

Figure 10.3 University of Washington Mark III inertial impactor.

Both laboratory and *in-situ* methods are used to determine the particle size distribution, but the laboratory methods will not be discussed in this chapter.

Inertial impactors and multiple cyclones are useful for determining the particle size distributions of particles suspended in a gas stream. Both systems operate under the principle of inertial separation. As mentioned above, isokinetic sampling is required to obtain a representative sample. The flow path of the particle-laden gas stream is caused to make abrupt turns such that the inertial forces acting on the particles cause them to move across the gas flow stream lines to deposit the particles on an impaction surface. Inertial impactors have a number of stages designed with increasing stream velocities in each section. These streams flow onto a flat plate where the gas stream is forced to make a right angle turn with the appropriate size of particle being retained on the impaction surface and the smaller particles carried on by the stream. In the next stage, where the velocity is higher, smaller particles will now impinge on the impaction plate. Finally, a back-up filter serves to collect the remaining particles to complete the process. Figures 10.3 and 10.4 illustrate an inertial impactor and a multiple cyclone sampling system, respectively [5].

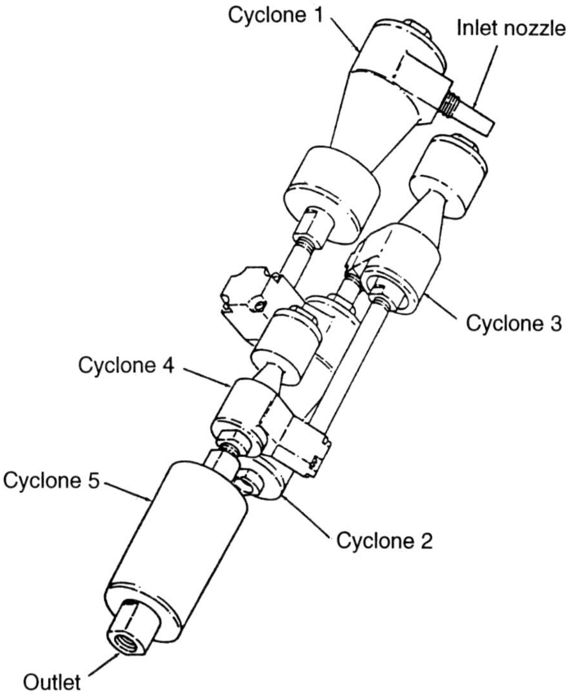

Figure 10.4 EPA/Southern Research Institute five-stage series cyclone.

The cyclone system operates in a similar manner. Each stage of the cyclone system is designed to operate at increasing velocities to collect reducing smaller particle fractions. The larger particles are collected in the inlet cyclone and the smaller ones downstream. Back-up filters are also used with cyclones when appropriate. Cyclone systems are typically used in high mass loading situations or where large particle samples are desired for analytical or other purposes.

There are conflicting requirements for conducting particle size distribution measurements in a duct system with a non-uniform gas velocity distribution. First, there is the requirement for isokinetic sampling to collect a representative sample of the particles. This leads to the requirement for adjusting the sampling velocity in the sampling probe to match the local gas velocity at the sampling point. There is also the requirement to be able to determine the particle size distribution of the collected sample. Since the cutpoint for each stage on either an inertial impactor or series cyclone is related to the gas velocity through the stage, it is necessary to maintain a constant velocity through the system while sampling. This precludes adjusting the flow at each point to maintain isokinetic velocities at each sampling point. In practice, the inertial impactor or cyclone is operated at the average gas velocity for the cross-section of ductwork being measured, thereby keeping the cutpoints for each stage constant for the measurement.

Currently, there is an interest in determining the respirable fraction of the particles suspended in a gas. In some instances, particles smaller than 10 μm diameter (PM-10) are considered to be appropriate, while in others, those smaller than 2.5 μm (PM 2.5) are selected. The actual decision about which is appropriate to regulate is still pending. The measurement may be required to be made either in the ambient air or in a confined gas stream. Either inertial impactors or cyclones could be appropriate for this determination.

In addition to the need for sampling for mass emissions and particle size distribution, there is a current interest in sampling for specific substances contained in the exiting gas stream. These items are referred to as air toxics, volatile organic substances and heavy metals. The heavy metals subset of the items of interest require sampling and analysis techniques significantly different from other aspects of sampling. These techniques will be discussed in the following section.

10.2 Heavy metals

10.2.1 General considerations

At temperatures pertinent to particulate control, heavy metals occur in flue gas both as particulate matter and in the vapor state. There is, of course, a significant range of temperatures where particulate-control devices operate. The range is from temperatures as high as 371 °C (700 °F), for hot-side ESPs, to temperatures as low as 52 °C (125 °F), which are reached in wet units. This range of temperatures permits some metals to exist primarily as vapors and others primarily as particulate matter, depending upon circumstances.

Even at the highest temperatures of concern, certain metals are insufficiently volatile to occur other than as solids. Barium is certainly one such example; even at the lowest temperatures of concern, on the other hand, one specific metal — mercury — is volatile enough in the elemental state to exist wholly in the vapor state. The factors concerning mercury that determine the primary state of actual occurrence include both thermodynamics and kinetics. The prevailing evidence is that, regardless of the interplay of these two factors, mercury occurs in the flue gas mainly in the vapor state. The volatilities of three of the more volatile metals, their oxides and the dichloride of one (mercury) are depicted in Figure 10.5. The sources of these data are References 11–15.

The analysis of heavy metals in flue gas must be based upon some of the same fundamental sampling strategies that are critical for particulate matter.

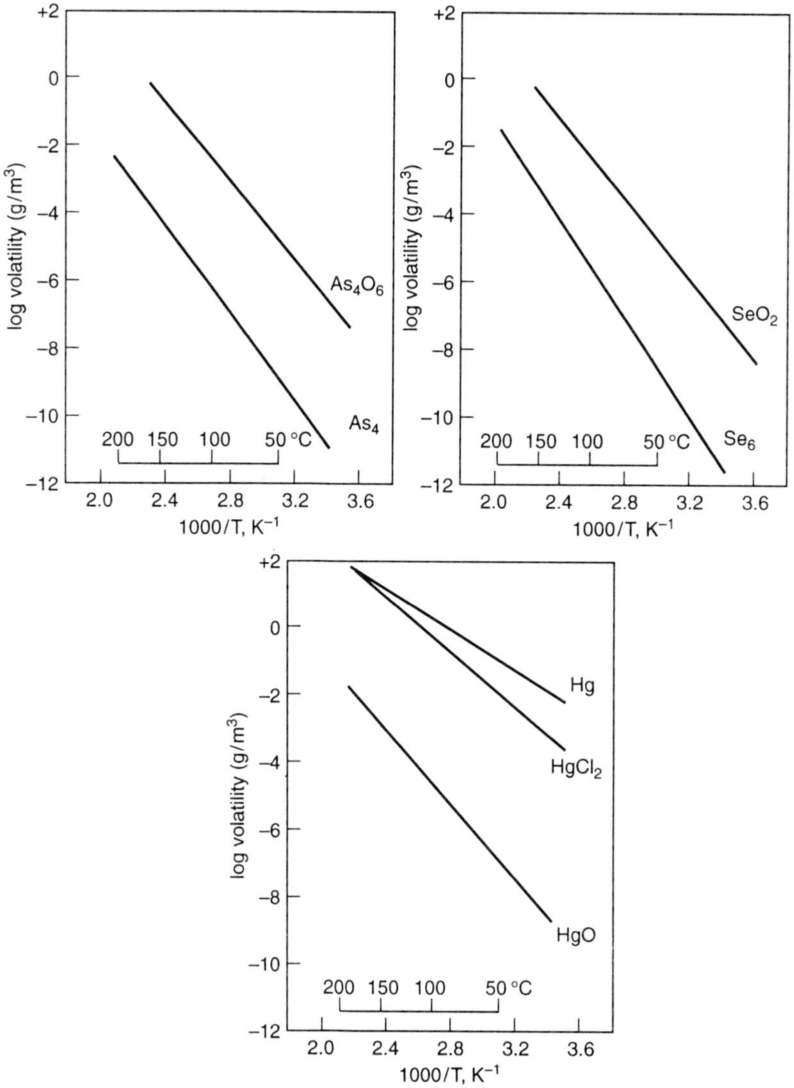

Figure 10.5 Volatilities of metals and compounds with relatively high vapor pressures. Reprinted from *Fuel Processing Technology*, **39**, E.B. Dismukes, Trace element control in electrostatic precipitators and fabric filters, 403–16, (1994) with kind permission of Elsevier Science–NL, Sara Burgerhartstraat 25, 1055 KV Amsterdam, The Netherlands

In general, sampling for metals occurring either as particulates or vapors must be done by traversing the entire duct in question, to ensure that non-uniform concentration distributions are taken into account. There may be the temptation with vapors to compromise this principle, based on the argument that vapor concentrations are far less likely to be stratified than particulate concentrations. Still, the sampling strategy must allow for the possibility that temperature gradients will change vapors to particulates, or *vice versa*, and must allow for the possibility that air in-leakage will create concentration gradients even in vapors.

Testing analytical data for material balance should be regarded as an important, if not essential, aspect of analyzing heavy metals in flue gas. The data currently obtained for heavy metals are often in doubt as to accuracy, and the testing of data for material balance is a useful technique for determining which metals have been determined successfully and which have not. The material balance exercise may embrace an entire plant, beginning with the fuel and extending to all waste streams. On other occasions, just incoming and outgoing streams at a control device may provide an adequate basis for material balance considerations [16].

10.2.2 Sampling methods for multiple types of heavy metals

The state of the art in the United States for multi-element sampling of heavy metals is incorporated in the so-called Method 29 of the US Environmental Protection Ageny. This method has not yet achieved status as an officially sanctioned method; that is, it has not yet been formally included in the Code of Federal Regulations [1]. It should ultimately appear in Title 40, Part 60, but it will not appear earlier than the Fall of 1996. Nevertheless, the method has long been unofficially known as Method 29, and has been used for several years for sampling programs under official US Governmental auspices. It remains in use today as the primary method for sampling flue gas for multiple metals.

Method 29 is fundamentally just a modification of Method 5 for measurement of total particulate concentrations (Method 5 has appeared in 40 CFR 60 for more than two decades). Method 5 extracts flue gas isokinetically from a gas duct through a heated probe into externally mounted devices that, first, collect particulate matter on a heated filter and, then, collect any condensable vapors in ice-chilled impingers.

40 CFR 60 actually describes the basic Method 5 and a total of eight variations designated as Methods 5A through 5H. The basic method stipulates that the filter be maintained at 121 °C (250 °F); two of the variations permit higher temperatures in an effort to mininize the presence of sulfate salts or sulfuric acid as a component of the filter catch. Method 29 retains the use of the 121 °C (250 °F) temperature; thus, in principle, it distinguishes operationally between particulate and vaporous metals on the basis of an arbitrary filtration temperature.

Water is the only collection medium employed in Method 5 and its variations. Other media containing acids and oxidizing agents, on the other hand, are used in Method 29. There are essentially two different sampling media in the impingers of Method 29. The first impinger traversed by the

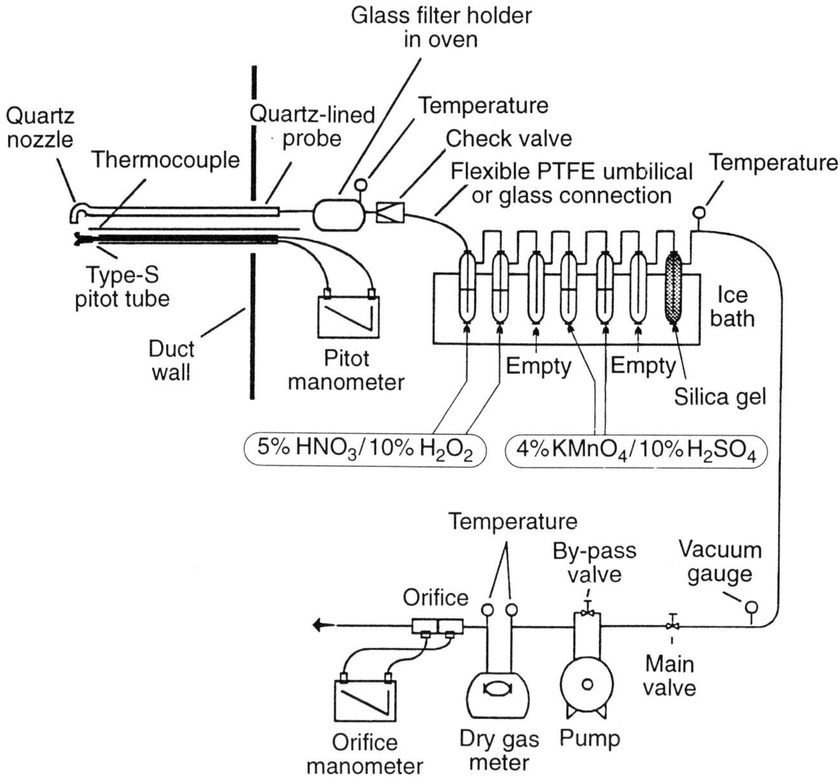

Figure 10.6 EPA Method 29 multiple metals sampling train.

sample gas stream contains a mixture of hydrogen peroxide and nitric acid; the second contains a mixture of potassium permanganate and sulfuric acid. The sample collected in the peroxide–nitric acid mixture is analyzed for all metals of concern; the permanganate–sulfuric acid mixture is analyzed only for mercury. See Figure 10.6 for a sketch of the Method 29 sampling train [17].

One of the alternative methods for sampling of heavy metals is the so-called HEST method of John Cooper [18]. In comparison with Method 29, the HEST method is much simpler to use and provides samples that are far more easily analyzed. In essence, it employs a Teflon or quartz particulate filter followed by charcoal impregnated filters to retain any metals that occur in the vapor state rather than in particulate matter. The method produces analytical results with non-destructive instrumental analysis of the filters; it avoids the complex sample digestion and dissolution scheme required for samples from Method 29.

10.2.3 Sampling methods for mercury alone

Method 29 can, by choice, be used for sampling mercury when this is the only metal of interest. Some time after Method 29 had been in use for total mercury, there came the belief that the two liquid media in the impingers can provide data on different vaporous species of mercury. The peroxide impingers seem to collect principally the vapors of oxidized mercury (believed to be mainly $HgCl_2$), whereas the permanganate will definitely collect the elemental form of the vapor that remains in the gas stream after the oxidized vapor has been removed. In summary, Method 29 offers the prospect of determining particulate mercury (oxidized mercury in chemical forms such as the oxide), vaporous mercury in the oxidized state (likely $HgCl_2$), and vaporous mercury that is in the free elemental state.

The so-called Bloom method [19] was the first widely used method for sampling mercury for the specific purpose of speciating the vapors. This method depends upon the use of solid rather than liquid collecting media. A cartridge containing soda lime is believed to be selective for collecting oxidized vapor; a back-up cartridge packed with iodated carbon collects the elemental vapor that remains (as well as any oxidized vapor that is not collected by the soda lime).

A more recently developed method for mercury, described by Keith Curtis of Ontario Hydro [20], may be regarded either as a modification of Method 29 or a modification of Method 101A [1]. As a modification of Method 29, the change is the use of an aqueous solution of KCl in place of the peroxide–nitric acid solution. The rationale for aqueous KCl is that $HgCl_2$ is soluble in water and stabilized in that medium as the $HgCl_4^{-2}$ complex ion with excess chloride ions. As a modification of Method 101A, the change is the introduction of the KCl solution as a sampling medium ahead of the permanganate. In either event, this method provides samples of particulate mercury on a filter, oxidized mercury ($HgCl_2$) in the KCl solution, and elemental mercury in the permanganate solution.

10.2.4 Metal analysis in the laboratory

Method 29 has been used to determine some 15 to 20 trace metals. Antimony, arsenic, barium, beryllium, boron, cadmium, chromium, cobalt, copper, lead, manganese, mercury, molybdenum, nickel, selenium, and vanadium comprise the list of analyses of interest in recent investigations of utility plants. Method 29 recommends that most of the metals of interest be determined by inductively coupled plasma atomic emission spectroscopy (ICAPES), which provides the advantages of simultaneous measurement of numerous target metals and, as a rule, acceptable levels of sensitivity. In certain instances, however, because of limited sensitivity on the part of

ICAPES, other methods are selected instead. For mercury, the choice is cold-vapor atomic fluorescence spectroscopy. For arsenic and selenium, atomic absorption spectroscopy with a graphite furnace is preferred.

10.2.5 Prospects for real-time monitoring

Costs of metals analysis in terms of money and time make it highly desirable to produce instruments capable of real-time, continuous emission monitoring. The particular concerns about mercury have led to emphasis on continuous emission monitors for this metal. ADA Technologies, Inc., is an American firm that claims to have CEM for mercury [21] that is soon to be commercially available. The analytical principle for mercury calls for conversion on all compounds to the elemental vapor and measurement of the vapor by ultraviolet absorption.

References

1. Title 40 Code of Federal Regulations Part 60, *Reference Methods* US Government Printing Office, Washington, revised annually and dated July 1 each year.
2. *BSI Standard ISO 9096.*
3. *VDI Standard 2066.*
4. J15 Z 8803–1970, Japan.
5. McDonald, J.R. and Dean, A. (1980) A manual for the use of electrostatic precipitators to collect fly ash particles. EPA publication *EPA-8-600/8-80-025*, May.
6. Opacity Meter, Monitor Labs Inc. 74 Inverness Dr. East, Englewood, CO 80112, USA.
7. Forward Scattering, Insitec, 2110 Omega Rd. Suite D, San Ramon, CA 94583, USA.
8. Back Scatter, Environmental Systems Corp., 200 Tech Center Dr. Knoxville, TN 37912, USA.
9. Beta Gauge, Graseby-Anderson, 4801 Fulton Ind. Blvd., Atlanta, GA 30336, USA.
10. Triboelectric, Auburn International, Inc., 8 Electronics Ave., P.O. Box 2008, Danvers, MA 01923, USA.
11. Smith, J.D. (1973) Arsenic, antimony and bismuth, In: J.C. Bailar, Jr *et al.* (Eds.) *Comprehensive Inorganic Chemistry*, Vol. 2, Pergamon Press, Oxford, pp. 547–683.
12. Behrens, R.G. and Rosenbkat, G.M. (1972) Vapor pressure and thermodynamics of octahedral arsenic trioxide (arsenolite). *J. Chem. Thermodyn.*, **4**: 175.
13. Neumann, K. and Lichtenberger, E. (1939) Molecular-weight determination and vapor pressure of selenium. *Z. Phys. Chem.*, **A184**: 89.
14. Pupp, C. *et al.* (1974) Equilibrium vapor concentrations of some polycyclic aromatic hydrocarbons, As_4O_6 and SeO_2, and the collection efficiencies of these air pollutants. *Atmos. Environ.*, **8**: 915.
15. Chase, M.W., Jr *et al.* (1985) *JANAF Thermochemical Tables*, 3rd edition J. Phys. Chem. Ref. Data 14: Supplement No. 1.
16. Vann Bush, P. *et al.* (1995) Sampling and analytical challenges for air toxics assessments. *EPRI/DOE International Conference on Managing Hazardous and Particulate Air Pollutants*, Toronto, 15–17 August, in print.
17. Methods Manual for Compliance with the BIF Regulations, *EPA/530-SW-91-010*, December 1990.
18. Cooper, J.A. (1994) Recent advances in sampling and analysis of coal-fired power plant emissions for air toxic compounds. *Fuel Processing Technol.*, **39**, 251.

19. Bloom, N. (1991) Mercury speciation in flue gases: overcoming the analytical difficulties. *Conference on Managing Hazardous Air Pollutants – State of the Art*, Washington, DC, 4–6 November, EPRI, Palo Alto, CA, USA, pp. 148–60.
20. Curtis, K. (1994) Ontario Hydro Technologies, private communication, October.
21. Schlager, R.J. *et al.* (1995) Continuous monitors for measuring emissions of mercury and particulate matter. *EPRI/DOE International Conference on Managing Hazardous and Particulate Air Pollutants*, Toronto, 15–17 August, in print.

11 The commissioning of electrostatic precipitators

D.A. STYLER and J.C. WESTBURY

11.1 Introduction

As the marketplace has become more and more competitive over the last several decades, so has the need to design an electrostatic precipitator that will meet its design duty with a margin in which both supplier and purchaser are confident.

As a prerequisite to this need, the designer must be able to size the precipitator with confidence. To do this he needs to be happy in the knowledge that his base data are formulated on results obtained from plant that is in extremely good condition, both mechanically and electrically.

The starting point in ensuring that this state exists, and has a better chance of remaining so, is the initial commissioning of the equipment. A dictionary defines the word 'commissioning' as getting something 'ready for active service, to assign to perform a task or function'.

A plant that has been commissioned properly and is seen to be performing well in its early life induces a sense of well-being in both the vendor and his client. In the case of the former, it strengthens and enhances the validity of his technical database. For the latter, it promotes the idea that the precipitator is well worth maintaining properly by using and keeping to a managed maintenance programme.

Taken to an extreme, commissioning can be considered as a function which is not only necessary but should commence from the point at which the first piece of steel or concrete is laid down on the site. If this part has not been manufactured or installed correctly, then the remainder of the construction and commissioning exercises can only be a compromise, to a lesser or major degree.

Even though there are a substantial number of manufacturers producing their own type of electrofilter, all have the same basic elements contained within them. These fundamentals would be discharge electrodes, collecting electrodes, electrode rapping equipment, high voltage insulators, gas distribution equipment and a casing to house them all in. It is because these principal parts are common to all manufacturers that the fundamental procedure for commissioning will remain more or less the same for each.

When considering the procedure undertaken from initial commissioning to hand-over as a fully operational plant to the client, the process can be split into three major parts. Each of these parts, at various stages, will be running either in isolation to, or in parallel with, each other.

These three parts can be identified as follows:

(a) mechanical commissioning
(b) electrical commissioning
(c) process commissioning.

The timing of these functions within the commissioning programme and whether or not they overlap in any way may well vary, not only from contract to contract but also between two ostensibly identical contracts, as a result of prevailing site conditions or client-imposed criteria.

11.2 Mechanical commissioning

The action of mechanical commissioning can be resolved into four phases: construction, post construction, cold and then hot commissioning. These phases can also be split further into a number of component parts. As with the commissioning programme, they too can be carried out as discrete actions or be involved in combinations of actions from different components.

As has previously been mentioned, it is important to ensure that the construction of the precipitator is monitored in as much that the criteria that affect the performance of the unit are checked and approved as the programme progresses. In order to follow the construction phase activities through, Figures 11.1 and 11.2 show a sectionalised, large, steel cased electrofilter and an exploded diagram of the main features of the casing, respectively. They both identify and name the various component parts for easy reference.

11.2.1 Construction stage

11.2.1.1 Substructure and casing. Once all the civil engineering work which is directly associated with the filter structure has been passed as acceptable, the first stage of this construction procedure can begin. Table 11.1 indicates the typical tolerances which must be achieved for this civil engineering stage, and a guide to the tools and instruments to obtain them. The first stage of construction will be the placing of the understructure onto the foundations. As the support columns are positioned, their verticality and diagonal relationships must be verified. The diagonal measurements are taken to ensure that the casing will not end up as a parallelogram. The degree of accuracy required for verticality is that which can be obtained by the use of a theodolite. Provided that the substructure fabrication was to tight tolerances and any shimming, required to obtain the correct height, has been inserted, the pads at the uppermost extremities of the columns will now

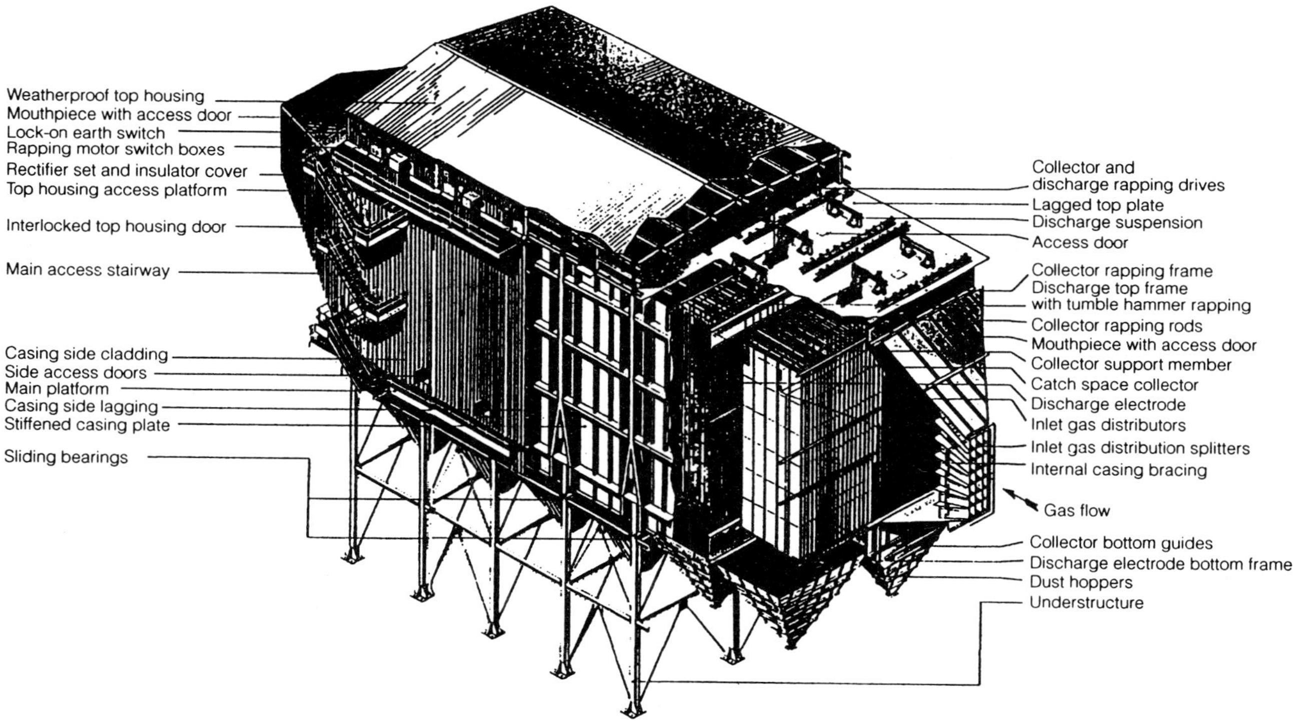

Figure 11.1 Typical sectionalised view of large steel cased precipitator.

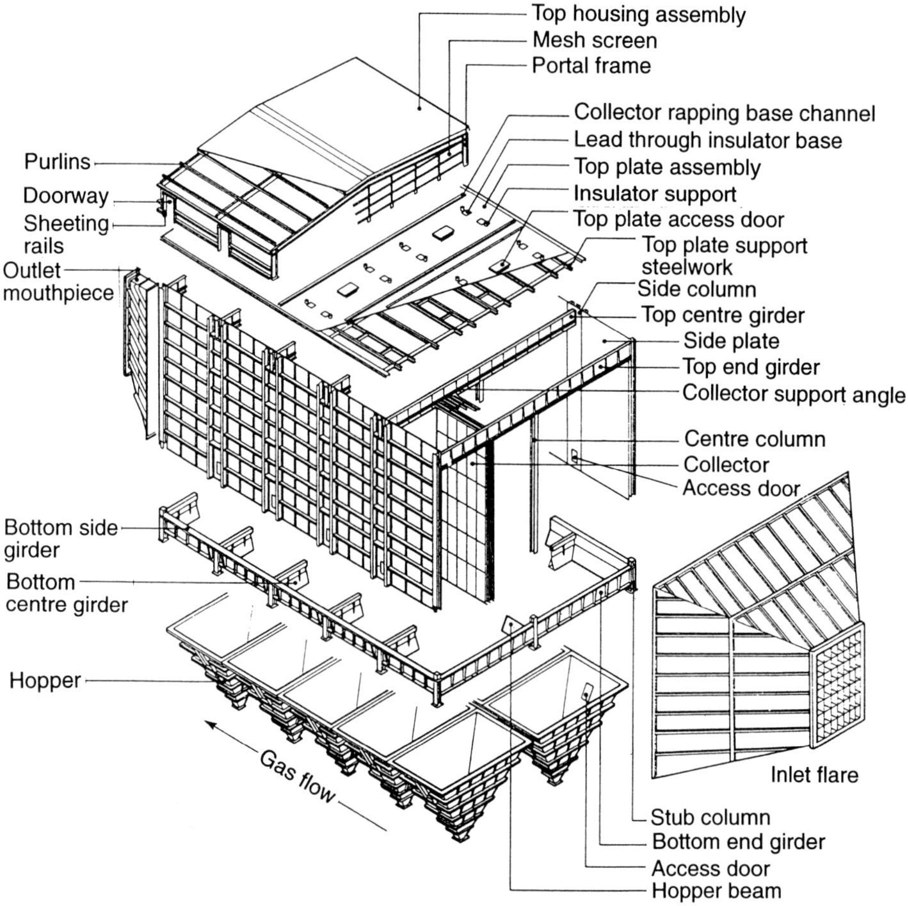

Figure 11.2 Exploded view of typical precipitator casing.

permit the casing proper to be built off horizontal datum points. Referring to Table 11.1 will indicate the requirements and how best to achieve them.

Placed on top of the understructure will be the sliding bearings which will accommodate any movement of the casing relative to it. One support column, which is located as close to the centre of the footprint of the filter as possible, is used as a fixed point. The casing is bolted to it at this point and any movement caused by thermal expansion will be away from it. By allowing the casing to slide over the understructure by equal amounts in all directions away from it, potential structural failure is avoided. The diagrams shown in Figure 11.3 highlight the above points.

Table 11.1 Typical tolerances necessary in civil engineering construction and mechanical stages of building ESP and instruments to obtain them

Item	Measuring device	Tolerance
Foundations	Tape measure, Theodolite, Spirit level	±5 mm
Columns–Verticality	Tape measure, Theodolite,	±2 mm
–Diagonals	Spirit level	±5 mm
Ring beam	Tape measue, Spirit level, Water level	±1 mm on overall length
Bottom girders	Tape measure, Spirit level, Water level	±1 mm on overall length
Casing walls	Theodolite, Plumb line	±1 mm
Top girders	Tape measure, Spirit level, Water level	±1 mm on overall length
Roof plate	Tape measure, Spirit level, Water level	±1 mm on overall length
Collector support channels	Tape measure, Spirit level, Water level	±5 mm
Collector–spacing	Tape measure, Plumb line	±2 mm
Collector squareness	Tape measure	±6 mm
Discharge electrodes	Tape measure, Plumb line	±2 mm
C/E to D/E spacing	Tape measure	±10 mm
D/E suspension supports	Tape measure, Plumb line, Spirit level	±3 mm
D/E suspension/carrying frame	Tape measure, Plumb line, Spirit level	±5 mm on all dimensions
Rapping drive chains	String line, Spirit level, Tape measure	±1 mm
Rapping drive bolted coupling	String line, Spirit level, Tape measure	±1 mm
Rapping drive flexible coupling	String line, Spirit level, Tape measure	±1 mm
Hoppers	Tape measure	±3 mm
Flares	Tape measure	±3 mm

There are, however, some units which can be built using a 'monolithic' design. This is a system whereby all expansion is taken in the substructure columns. The use of sliding bearings is, therefore, not needed for this application. It must be remembered, nevertheless, that the same setting-out procedure for the substructure must be undertaken.

The ring beam, shown in Figure 11.3, which consists of girders positioned longitudinally and laterally, creates a rigid framework onto which the dust hoppers and superstructure of the precipitator are attached. Once again it is essential to ensure that the diagonal measurements of these beams are verified (see Table 11.1.) If this operation is not carried out, then difficulties may be experienced during later stages of the commissioning in terms of bad alignment of the internal components. In extreme cases this situation may require major surgery to correct the problem at a time in the commissioning

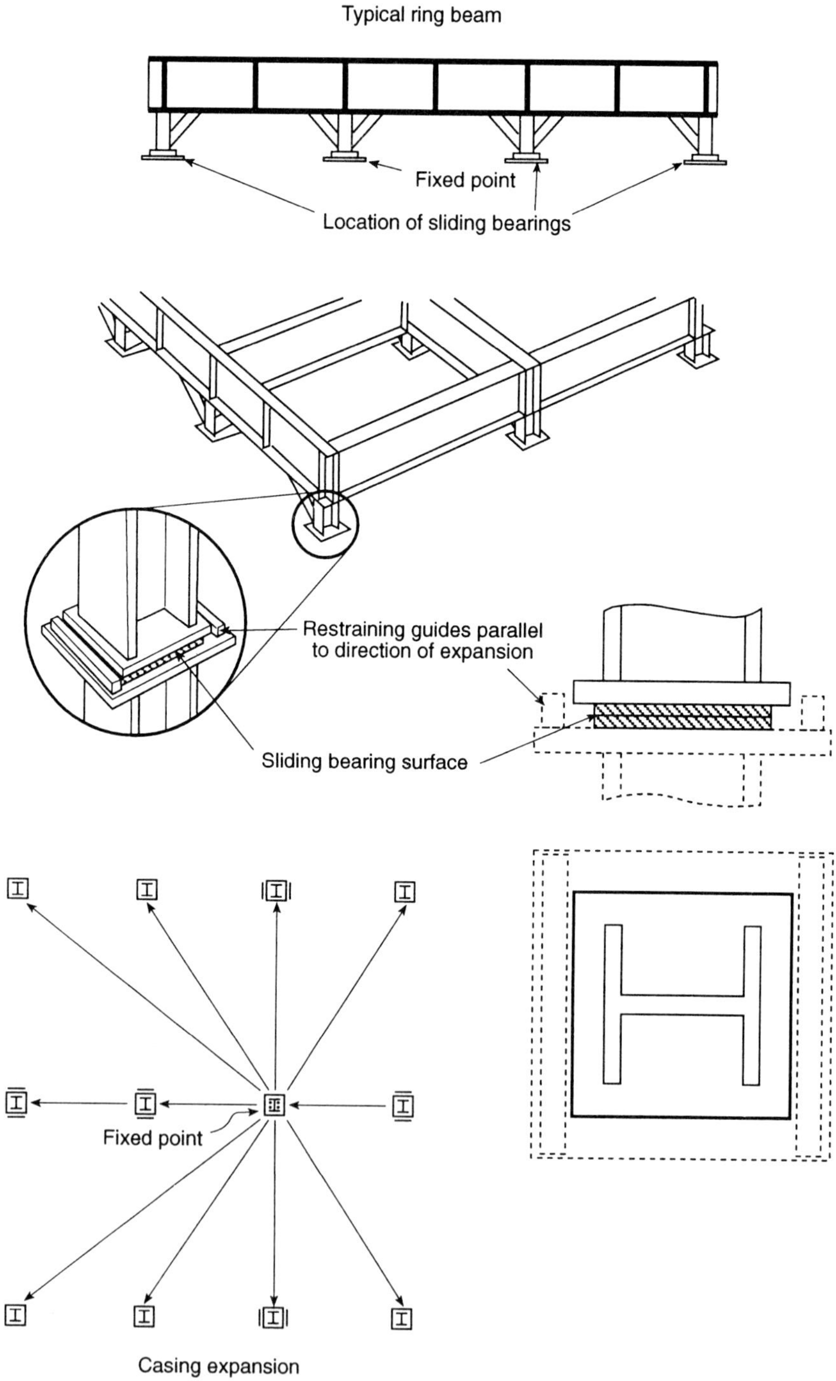

Figure 11.3 Bottom ring beam and sliding bearing details.

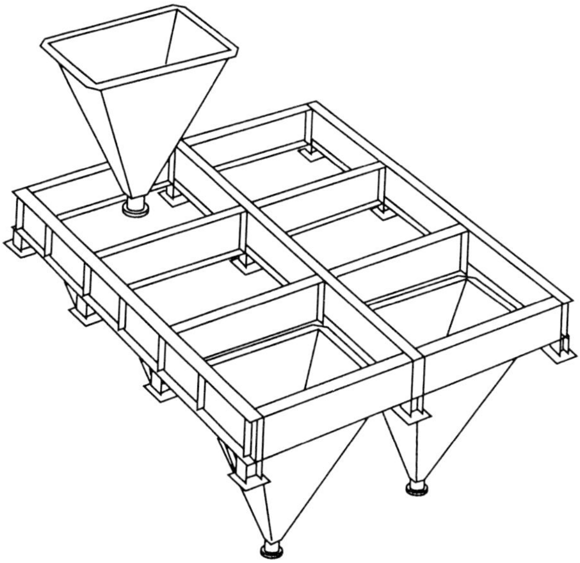

Figure 11.4 Relationship between ring beams and hoppers.

programme when access is difficult and time is at a premium. Figure 11.4 illustrates the relationship between the ring beam and the hoppers.

While access is available to all points of the casing and hoppers on their internal surfaces, the commissioning engineer must take the opportunity to inspect and confirm the integrity of all weld runs which are regarded as seal welds. These are the welds which stop the ingress of ambient air or the egress of process gas once the system is put into service.

The same degree of care should be taken when erecting the side walls and adding the top girders, for the same reasons as have been indicated in previous paragraphs. Failure to implement checks at this stage can have similar repercussions, particularly where designs of the internals involve the use of gravity to achieve and maintain good electrical clearances between the discharge and collecting electrodes. Table 11.1 indicates the degree of care required.

11.2.1.2 Internal components. Many designs of electrofilter use the top girders to suspend the collecting electrodes. Figure 11.5 depicts the way in which they can be lowered into the casing and Figure 11.6 their location on the top girder (see Table 11.1).

Small departures from the manufacturer's tolerances can be absorbed and are those which, under any other condition, could cause a reduction in

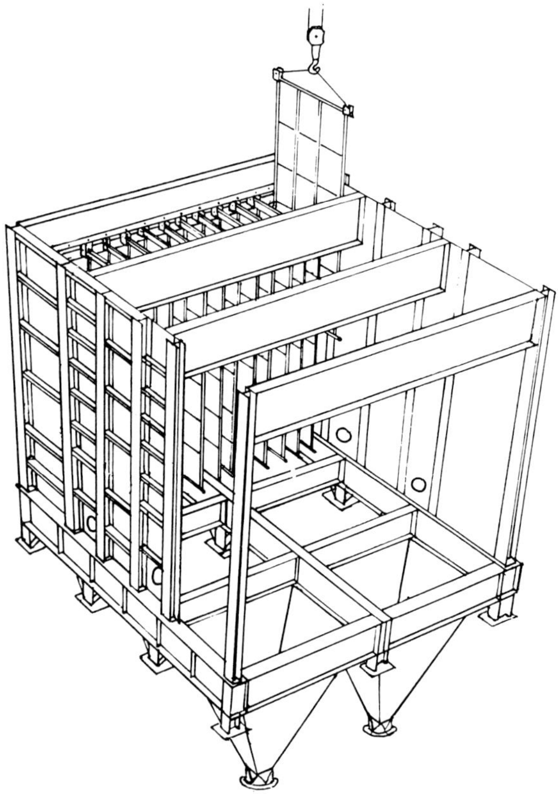

Figure 11.5 Lowering of finished collector plates into casing.

electrical clearances. This is because most types of 'plate' electrode are located at the top level and have adjustable guides at the lower level. However, the discharge electrode system often does not have this luxury. It may require gravity to maintain its alignment and verticality between the collector plates. The hackneyed industrial phrase 'One out. All out' may well apply to this situation, should the control criteria not be undertaken.

It is important when positioning the high tension discharge electrode support system, located at the roof plate level, to once again ensure that alignment will easily be achieved. It is imperative that the commissioning engineer has completed his inspection of the finished plant and granted provisional acceptance of its construction before there is any move to weld into position or make any permanent fixture of this emission electrode support equipment. Table 11.1 indicates the means to achieve this. An arrangement showing the post support insulators of a rigid mast type of discharge electrode system can be seen in Figure 11.7.

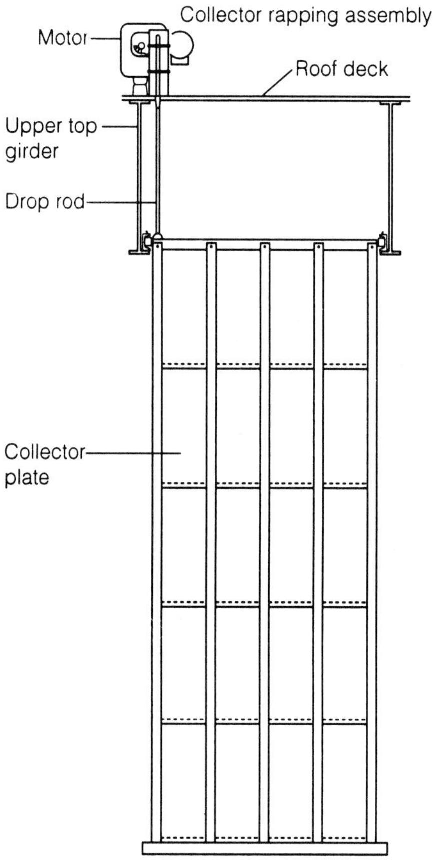

Figure 11.6 Location of collector plate on top girder.

Where the type of precipitator requires it, the rapping equipment for the gas distribution system and both discharge and collecting electrodes must be checked over. Again, depending on the manufacturer, these systems can be located entirely within the casing, with only the drives passing through it, or be a combination of these two, whereby certain parts can be found within and others attached to either the casing walls or roof plate. Figures 11.1, 11.6, 11.7, 11.8, 11.9 and 11.10 give varying views of typical rapping devices.

It is essential that all rapping drive shafts have not only been connected in their correct sequence, according to the construction drawings, but that they are aligned across the full width of the precipitator. This is usually done by string line (see Table 11.1). Where flexible couplings or some form of dog

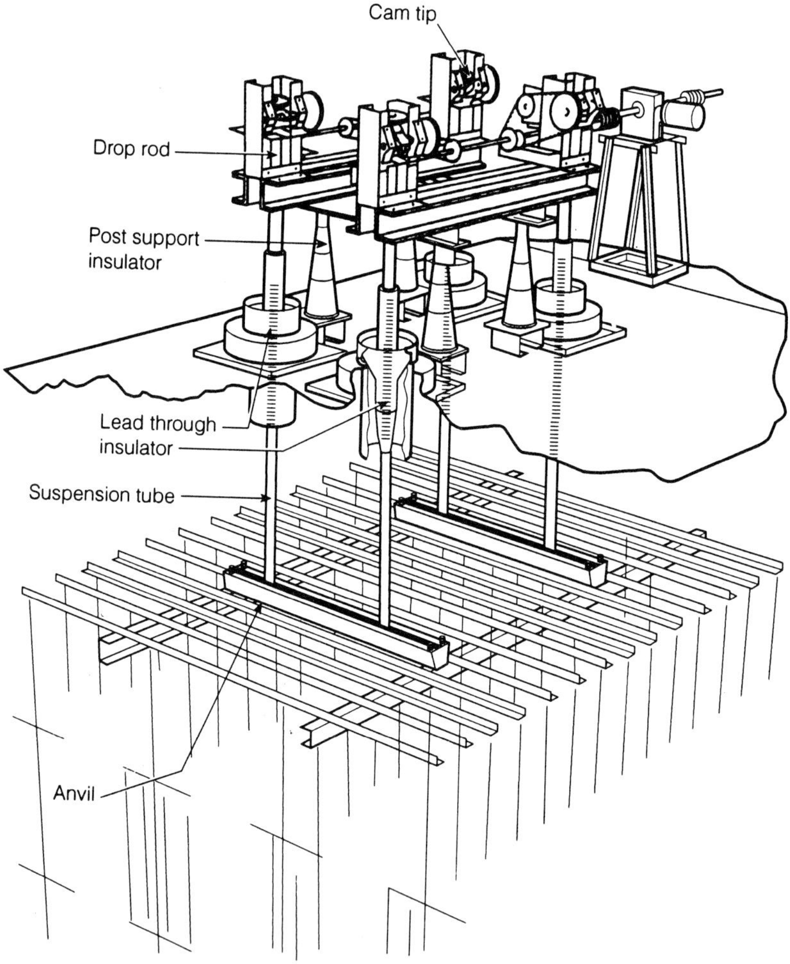

Figure 11.7 Discharge electrode assembly.

tooth drive are used to connect the individual sections of the shaft, alignment is still required but is not such a problem. If, however, the system relies on the relative position of the drive shaft to the rapping hammer, then alignment becomes far more important. Specifically, this refers to the type whereby a cam attached to a rotating shaft lifts and drops the rapper. The

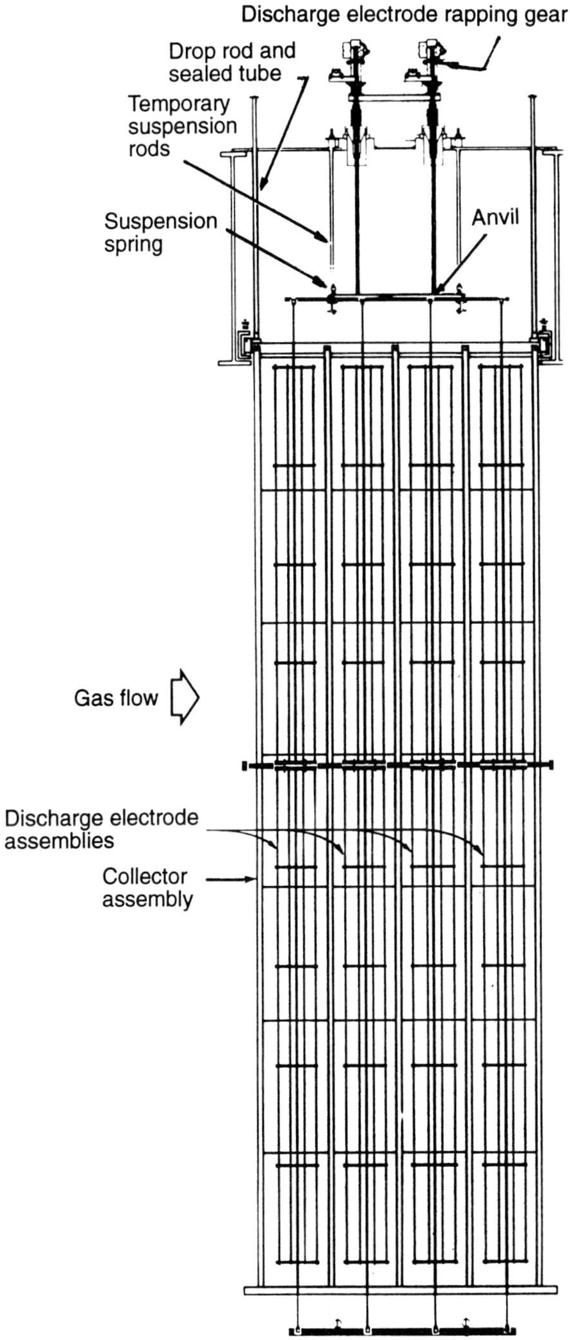

Figure 11.8 Discharge and collector electrode rapping gear.

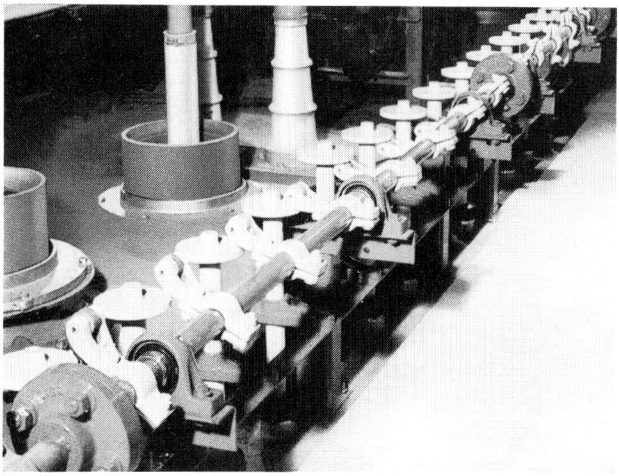

Figure 11.9 Collector rapping assembly.

relative position of cam to rapper is critical because the height of the lift and, therefore, the potential energy imparted into the rapper is directly linked to this geometric relationship.

11.2.1.3 Hoppers. By the very nature of their shape, hoppers and, most notably, the pyramid type, can be the cause of blockages to the flow of dust

Figure 11.10 Assembled precipitator internals showing clearance space for inspection and maintenance.

whilst they are being dedusted. The most common reason is damp dust in the bottom of these hoppers.

At their extremities, hoppers have a large surface area but a very low contained volume. Any hot dust retained in the hopper bottoms would not have the heat capacity to maintain its temperature, or indeed, the temperature of the hopper platework, at a level on or above the water dew-point of the process gas. This fact would be of particular concern when the process involved has a high moisture content, or it cycles through the dew-point frequently. Commissioning engineers would recognise the problem of cycling as a common feature on initial start-up of new process plant!

To overcome the potential for hopper blockages, heating devices are usually installed, covering their outer surfaces for several metres up from their dust discharge point. Further, to maintain the highest thermal efficiency possible, the heaters, whether they are of the electrical or steam coil type, and the remainder of the hopper surface area, will be lagged. It is, therefore, a prerequisite in the commissioning programme to be able to power up the heaters to verify that they are functioning prior to them being lagged over. It is a disheartening sight to see new cladding being ripped off to remedy a fault that could have been rectified before its installation!

11.2.1.4 Flares and gas distribution equipment. Both the inlet and outlet flares, sometimes referred to as nozzles, contain the means of achieving good gas distribution within the electrical fields and, as such, require initial, detailed checks (see Table 11.1). This equipment may also have rapping gear attached to it, as a precaution against dust build-up and the resulting maldistribution of gas. The gas distribution system can take on several designs. Probably the most common is the holed screen type, where several screens, in series, are installed in the inlet flare. Another type involves the use of splitters and gas turning vanes. Figure 11.11 illustrates several views of these types of gas distribution equipment. Whichever form has been used and with respect to the rapping gear, the same comments apply as for those made for the collector and discharge electrode rapping, described earlier in this chapter.

11.2.1.5 Lagging, cladding and heat sinks. Although not directly concerned with the initial start-up of the precipitators on processes where lagging and cladding of the casing and hoppers is required, it should be inspected for integrity and efficacy. Ideally this would be whilst access is still available, before scaffolding is removed.

Where the material of construction of the casing is steel, it would not be immediately apparent that a problem, which resulted from badly installed lagging, existed. The engineer who is inspecting the lagging and cladding should bear in mind that there will be a time lag of some months before any evidence will show itself. Particular attention should, therefore, be paid to areas where there are likely to be large heat sinks through the casing. The

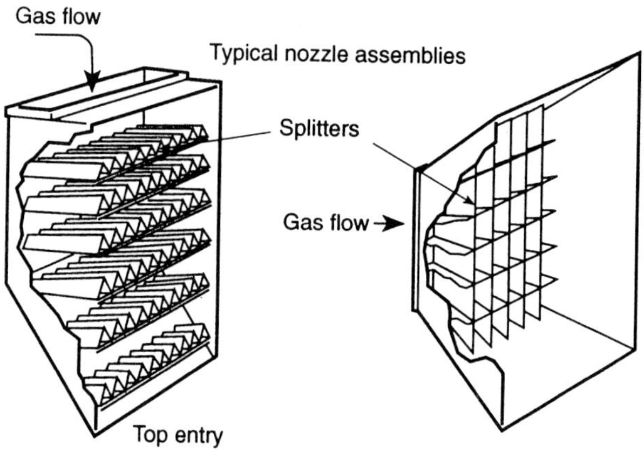

Figure 11.11 Precipitator inlet gas distribution devices.

predominant points are where items of equipment are positioned on the outside of, and in direct contact with, the casing, or the extreme corners of the electrofilter.

Whilst it is a common design feature to install some form of heat barrier between any parts that would contact the casing and the casing itself, it is not unheard of to find that this vital piece of the precipitator has been missed out.

11.2.1.6 Plant access. The final checks that have to be made in this phase of the commissioning are in respect of the access to the filter. These items would be all the staircases, cat ladders, landings, the treads, grating or plating used to walk on and the associated hand railing. Typical location of the access can be seen in Figure 11.1. With increasing legislation occurring on the topic of Health and Safety at Work it cannot be over-emphasized that all the above-mentioned components have the integrity of all of their fixings checked whilst site construction staff and alternative means of access, if required, are still available.

11.2.1.7 Casing pressure test. There have been occasions when end-users have requested a pressure test of the filter casing. In cases where the process involved generates toxic gas which passes through the precipitator, it is not an unreasonable one. However, where toxic gases are not involved, there is not really any need to perform this test. Furthermore, it is most unlikely that the manufacturer of the electrofilter would have either designed it to be totally gas tight or even budgeted to cater for an eventuality such as this. Indeed, for these types of application, it is common to expect a small percentage change in volume from inlet flare to outlet flare, as a result of insulator purging or slight egress or ingress of gas or ambient air! With this in view, a test of this nature is usually considered as a gross waste of time, money and effort for both the client and the manufacturer.

Should it be unavoidable, however, for whatever reason, it is best carried out whilst the casing is empty but otherwise finished. It will entail the complete sealing of all orifii, including flares, access doors, all points where any piece of equipment passes through the casing, the dust extraction ports at the hopper bottoms and any poke rod holes in them. As access may be required to parts of the casing to which, under normal conditions, it would not be possible to get to, the construction programme and possibly the method, would have to be rearranged to cater for this test. An example of this would be the collector plates. Many designs are capable of being constructed outside of the casing and then lifted and lowered into their final working position as a finished article. Where a pressure test has to be part of the construction programme, the casing roof plate would have had to have been put into place and totally sealed. The construction of the collector plates, therefore, would have to take place within the casing, so that the gas tight seal, created between the roof plate and the rest of the casing, is not broken.

Having reached the completion of the 'construction' phase of the commissioning, the period which links the cold and hot commissioning operations starts. This is best described as the 'final inspection' stage.

11.2.2 Post construction stage

11.2.2.1 Final inspections. This is the point in the programme at which all of the component parts of the precipitator are confirmed to be in their correct positions and that the mechanical fixings and welding required to maintain this state are completed to the specification laid down in the drawings and in the contract documentation.

In an ideal world, no further inspections should be required as the attention given to the filter should have been such that the whole plant had been confirmed as complete during the staged inspections whilst being constructed! Unfortunately, there is no such thing as an ideal world! The reason for this sad state of affairs is simple. It is nothing more than homo sapiens!

With the best will in the world, it is a practical impossibility for any individual to inspect a piece of equipment of such complexity and not miss something that may or may not have some direct bearing on the ultimate performance of the end product. However, on a more serious note, there is another, almost uncontrollable, element which could affect its condition. This is the fact that it is impossible to isolate the component parts from either human or mechanical contact after their completion within the construction schedule, if only because other sections still have to be attached to them.

In the final inspection programme, it will be necessary to recheck many of the actions already carried out in the construction phase. Although it is a repeat action, at least the commissioning engineer now has far more control over the plant as a whole. He will usually be able to specify where other members of the team can have access to the finished unit, such that there will not be any activities taking place that would be detrimental to the overall integrity of the precipitator.

The areas of specific concern for this portion of the programme are the discharge electrode and collecting electrode systems, their respective rapping gear, where fitted, sprays and washdown equipment found on wet plant, high voltage insulators and the gas distribution equipment, situated in both the inlet and outlet flares and their rapping equipment, if fitted.

11.2.2.2 Discharge and collector electrodes. In the first instance, the discharge electrodes must be checked to ensure that they have not been subjected to any physical damage. Broken or slack emitting wires or strips or distortion of the framework in which the emitters are installed should all be attended to, be they of rigid mast, bedstead frame or weighted wire type. Any broken wires or strips should be removed. Provided that there are not too many pieces in a zone, several metres in the many thousands that are installed in a precipitator will not cause a shortfall in the efficiency by not permitting the full potential current input to be realised.

Frames or rigid masts found to be creating a reduction in the electrical clearances between them and the collector plates, or any other part which is at earth potential, must be realigned. Where the verticality of the discharge electrode is achieved by the use of gravity, i.e. a wire and weight arrangement, any misalignment can be more difficult to remedy. The questions that would immediately come to mind in this situation would be:

(a) Are the collectors plumb, or worse?
(b) Are they and/or the casing out of plumb?

The first would be relatively easy to correct but the second highlights the necessity to ensure that the casing has been constructed, fully inspected and passed before any of the internals are added to it. This procedure has been described previously in this chapter.

It is an important fact to remember that in the process of making any corrections, it is not inconceivable that the alignment in areas already checked and approved as satisfactory could be moved out of tolerance.

Even though the collector electrode system is, by the very nature of its physical construction, more robust than the discharge electrode system, it, too, can be vulnerable. The more commonly found problems can be caused by such things as scaffolding equipment being moved or taken in or out of the casing. Often lengths of up to 3 m are being swung around and if they happen to connect with plates which are of a thinner section–in the order of 1.2 to 1.5 mm–it is not difficult to imagine that, at best, distortion could occur and, at worst, the plate could be punctured. The danger with this latter point is that it could produce jagged edges. Should they not be found, or are just left because it is thought that this is of no real consequence, reduced power input could result or, even worse, premature electrical breakdown, once the process is on-line, rendering the bus section inefficient.

Thin or thick plates can be susceptible to distortion at their mid-height, particularly on the very tall collectors. It is not difficult to understand that, at this point, a small induced load in a direction across the gas flow can create a large deflection and possibly permanent set because of the high leverage rates obtainable. Should any activity have been carried out at this level, following their installation, the problem of reduced electrical clearances becomes a distinct possibility!

11.2.2.3 Discharge and collector electrode rapping gear. With respect to the rapping equipment, the manual checks which are described below are essential actions.

Before any form of rotational checks are undertaken, tests should be made to ensure that all hammers are free to swing on their pivot pins or, where the rappers are in rod form, should be able to drop freely under the influence of gravity. If there is any evidence of frictional resistance to their action, its cause must be sought and remedied. Failure to do this will result in inefficient rapping of the electrodes, collector or discharge, and will allow unacceptably large amounts of dust to remain on them which, in turn, will impact on performance.

Upon completion of this stage, hand barring of the rapping gear can be done by mechanically turning the rapping drive shafts as close to the motor drive as possible. This usually entails disconnecting any drive chains or direct couplings to the motor and then, by use of a large wrench, turning the shaft slowly so that the action of all rappers can be observed through several revolutions.

Other types of rapping, which involve the lifting and dropping of either a complete discharge electrode frame or a complete collector sheet, or even both, may require the availability of power before this check can be made. Whilst it is possible to carry out this action manually, it is very arduous because of the masses of steel involved.

Yet another type is the magnetically induced lift and drop rapping. This operates in a manner similar to the mechanical drop rod but uses a solenoid device and employs one per rapping point. For obvious reasons, rotational checks will not be needed for this application. However, power will be required before the full operation can be assessed.

11.2.2.4 Irrigation and flushing sprays. Where sprays are used as the means to remove the dust from the internals, it is advisable to flush the pipework out prior to fitting the spray nozzles. All new sections of pipe are subject to the accumulation of weld spatter, bits of paper, matchsticks, *etcetera*, somewhere in their length. As some of these are rather bulky, compared with the spray nozzle orifice, it is vitally important to ensure that the pipes are clean and debris free. Several methods are available which will rid the system of these unwanted articles. One would be to remove the blanked-off end of every pipe and then turn on the spray pumps. Another would be to backflush the pipes fitted within the scope of the contract back to the point where supply commenced, by means of a separately piped source. Venting of the waste water using this method should always be at the lowest point available in the pipe network and through an orifice as large as possible. This because the water put in the pipes to clean them will be at a relatively low pressure, compared with that generated by the spray pumps and will not, therefore, have the same cleaning or moving power.

Once clear uncontaminated water is seen to be flowing from the pipework, the joints, end caps, *etcetera*, can be refitted and the spray nozzles installed. Before finally boxing up the plant, the sprays should be inspected with the pumps on, so that the spray operation can be verified against its design characteristics, such as volumetric throughput, spray shape and whether the pump will generate the pressure required at the spray level.

11.2.2.5 High voltage insulators. Some of the more commonly used materials for insulator manufacture, such as silica or porcelain, are very susceptible to breakage or cracking if given the lightest of taps in the wrong place. It must also be remembered that porcelain has an ability to absorb water. Therefore, if any of the insulators that are manufactured from porcelain are found to have any form of defect, action must be taken to either create an electrically stable seal or make a complete replacement where the exposure areas are too large. The main point of concern is where penetration of the salt glaze has occurred and the problem spots are open to the atmosphere or process gas.

As many insulators could be exposed to human contact after their installation, a close inspection of their integrity is essential. Many construction engineers will ensure that as soon as the insulators have been positioned, they will be given protection by creating some sort of box around them. These boxes would not then be removed until the commencement of the final inspection phase.

In some respects, the gas distribution equipment, which is located in the inlet and, sometimes, the outlet flares, can be subjected to the same problems as the collector plates during the construction programme. The examples given previously with respect to the collector plates are equally valid here. As good gas distribution has a vital part to play in ensuring that the efficiency of the electrofilter is achieved, any damage that is noted must be rectified.

11.2.3 Cold commissioning

At this junction, there is a need for both the mechanical and electrical commissioning programmes to come together. From a mechanical commissioning point of view, the immediate requirement is that the rotation of the moving parts, such as the rapping gear and insulator purge air fans, can be checked. From now on, the two programmes will be running in parallel and interfacing with each other on an ever increasing degree of necessity.

With the provision of both low and high tension power to the electrofilter, the basic function of the unit can be verified.

11.2.3.1 Rotation and function tests. Once rotational checks have been completed on all drive motors, the mechanical operation of the various moving components can be observed. In all cases, the driven parts must be watched through several cycles to ensure that their performance is as seen during manual tests made earlier in the commissioning programme. If there is a drive where the motor would be running for long periods when the filter is on-line, a proving period of 24 h is advisable. This is particularly important for all rapping gear when both the driving and driven items can be closely observed.

11.2.3.2 Gas distribution tests. Some manufacturers of precipitators will model test the installation prior to its construction in order to establish the positions of any gas distribution devices that may be required. Others will wait until this stage of the commissioning to do tests within the casing itself. Whichever method is adopted, it is important to know that the full cross-section of the filter is being subjected to an even distribution of process gas. A matrix of points across the casing area and at several positions within its working length can be assessed with the use of an anemometer, once the fans have been run-up. Personnel can be boxed into the unit (provided that the fans are only moving ambient air, of course) to ensure that this situation does exist. People involved in these tests are not under any risks by being enclosed in the casing. There may be an initial sensation in the ears when the fans are first switched on which is similar to that of being in an aircraft.

There will be a slight change in pressure within the ear which, provided the individuals concerned do not have any health difficulties with their ears, should not present any problems. If the tests are carried out at a time when the ambient temperature is low, warm clothing should be worn. The converse will obviously apply should the tests be in hot weather. A telephone or radio link must be available so that the people inside the casing can not only contact each other but also speak with the safetymen and even the control room.

Whilst in the casing for the distribution tests, it is also an opportune time to verify that there is no sneakage of gas into the hoppers or, where it is a design feature, into the dead space above the field. This can be done with the use of smoke pellets.

11.2.3.3 Casing integrity. Before the process is started, the integrity of the casing and access points can be assured by running-up the fans, induced or forced draught, so that a complete check of all locations can be made where either process gas can egress or ambient air ingress, from or to the precipitator, respectively. This is best done with the fans producing a gas pressure which, ideally, would be near to the operational level.

The importance of this action cannot be underestimated. Where an electrofilter is being used on a process which involves high gas moisture levels in conjunction with temperatures above water dew-point, corrosive gas or even toxic gas, it does not require a great deal of imagination to realise that process gas must be kept in and ambient air out of the casing. In the instances of high moisture and temperature, severe corrosion can take place in an amazingly short period of time if ambient air is allowed in. The ingress of dry and relatively cold air will create moisture as it meets the hotter, damper gases. The water which condenses out will usually deposit itself on some local steelwork, casing, door frames, collector sheets, *etcetera.* The resulting problems could be structural failure of the casing or high voltage electrical instability if the ingress occurs after the process has gone on-line.

Three of the simplest methods of performing this task are to either listen at each location, move one's hand around the joints or gently blow smoke at the seal. In the first method, it is common to hear a whistling noise as air passes from one side of the seal to the other if its integrity is not 100%. The second method relies on the fact that the hand would detect a cooling sensation where air is striking it, blowing onto the palm if it were egressing or between the fingers if it were ingressing through a bad seal. The third will be fairly obvious. If the casing is under suction, the smoke will disappear into the offending joint or get blown away very quickly from a pressurised casing.

11.2.3.4 Pre hot commissioning inspection. At this point, it is advisable for the commissioning engineer to make one final inspection of the whole plant. It must be made clear to all personnel involved in the construction and commissioning of the precipitator that the commissioning engineer and the person nominated as safety officer are to be the only people who are to enter any space which, when the process is on-line, is not accessible. These areas would typically include the casing proper and, where applicable, penthouses or high voltage insulator compartments. Should the process require a pressure test of the casing, the final inspection should be made prior to the test. This will be because once the seals of access doors have been made gas tight they should not be broken until there is a real need for access into the casing.

11.2.3.5 Final casing integrity test. No explanation is really needed to obviate the inherent dangers associated with the lethal combination of a leaking, pressurized casing and toxic process gas. However, as an example, if the casing of a filter cleaning blast furnace gas is not gas tight, highly toxic carbon monoxide could collect in a pocket where personnel have access. Very severe headaches at best, or death at the worst, could result! With this in mind, a pressure test of the casing is required. It will often revolve around the criterion that a pressure, approximately 1.5 times that of the process pressure, should be maintained for a period of several hours with only a relatively small percentage loss. It must be remembered that a pressure test, conducted in conditions where the ambient temperature is fluctuating or the sun is shining onto the casing, could easily give a very false idea as to the real state of its ability to contain the gases. The variations can be quite dramatic, particularly when there is a large casing surface area to gas volume ratio and the surfaces are unlagged. Indeed, the differences can be so great as to invalidate any test which is meant to verify the integrity of the casing. The periods which are the best for conducting such tests are either on a cloudy day or during the night, after the time when the ambient temperature has stabilised.

All potential leakage points should be identified and assessed by the commissioning engineer prior to the test, then whilst it is in progress these and other areas can be monitored. A common method of monitoring is by the use of soap and water spread liberally around the joint or suspect area. If any leaks are occurring, bubbles will very quickly form and be seen. Remedial action will then have to be taken to rectify the problem until a complete seal is achieved.

Once back outside the casing and having accounted for all personnel, the engineer can 'box' the precipitator up by completely securing all access points and doors through the interlock system.

11.2.4 Hot commissioning

Hot commissioning of the precipitator is the final stage of putting the plant on line. As the procedure is intimately bound up with Process Commissioning, its description will be found under the heading, later in this chapter.

11.3 Electrical commissioning

11.3.1 An overview

At first sight the process of electrical commissioning occurs at the end of a long line of operations resulting in the building of a precipitator. It could be said that the electrical commissioning engineer only switches on the finished product that so many people have had a hand in producing.

This concept of electrical commissioning is a long way from the reality. The act of electrical commissioning relies on a process of groundwork and long hours which lay down the foundations of an installation that is right first time. It requires a logical approach to the commissioning process and final checking and optimising of the finished product.

The commissioning process relies on the following functions:

(a) Familiarity – it is important that the engineer is fully familiar with all the component parts of the control systems that operate the precipitator, and understands how these systems function.
(b) Monitoring – the engineer must monitor the installation of the cabling, control systems, motors, rectifiers and all the electrical equipment required to operate the precipitator.
(c) Experience – the engineer must also have a working knowledge of the factors affecting the performance of the precipitator or 'know a man' that does.

11.3.2 Managers and commissioning

Generally managers do not commission physically, but they facilitate the process and as such can help or hinder the commissioning. At this stage, it is worth pointing out a number of management prerequisites that are needed in order to ensure a smooth electrical commissioning phase.

1. Once electrical commissioning engineers have been assigned to a contract they must feel that they have responsibility for all the commissioning activities. They should know their areas of responsibility and know whom to contact in the event of problems or when requiring advice.
2. They should be given the time and resources to prepare for the electrical commissioning process well in advance and be made available for the entire commissioning period. Swopping and changing engineers leads to inconsistency of approach, often with communication failures between

the various people and companies providing services, leading to a general reduction in motivation and ownership.
3. They must be well briefed and provided with all the relevant information, or at least all that is available at the time.

11.3.3 Familiarity revisited

The first stage of the electrical commissioning is undertaken back in the office, well before the job is erected. During this period the engineer must become absolutely familiar with the electrical schematic and layout drawings detailing the circuitry and its operating logic. Should the drawings not be completed at this phase of the contract, it is a prerequisite that the engineer makes contact and holds discussions with the electrical contract engineer. This will enable as full an understanding of the 'system' as is possible to be achieved. Only in this way can the equipment installation procedure be monitored effectively, because this is what the majority of the commissioning engineers' function comprises.

It is during this stage that quality rears its head. Nearly all customers, worldwide, require a quality plan of some sort to be developed. (In the UK this is essential if the company is BSI approved.) A quality plan defines the activities to be carried out, the type and extent of checks to be made, and the controlling documents or check sheets to be used to record the outcome of the checks. Quality plans allow each process relating to the contract to be monitored against measurable goals. The electrical commissioning quality plan should be written by the commissioning engineer responsible for the job and approved by the departmental head. In this way ownership of, and familiarity with, the commissioning process is instilled into the commissioning engineer at an early stage in the process.

In general the quality plan can be split into two areas broadly defined as follows:

1. Pre-energisation checks where each component of the system is checked off-line (electrically isolated) to ensure that what is supplied conforms with the material specification, that it has been installed correctly and that it has not been damaged during installation.
2. Live testing where equipment is powered up and checked to see that it functions in accordance with its operating specification and logic.

These areas will be covered in more detail later in this section.

11.3.4 The programme – who writes the programme and when?

The process of erecting a precipitator requires the programming of a large number of activities, in a logical order, to achieve a cost-effective and efficient build. A significant part of the electrical commissioning activities comes toward the end of the contract and as such can be subject to a number of factors which can lead to programme delays. This can be critical, in that most of the final commissioning checks involve live testing of equipment against specific dates for plant start-up, and any delays to programme can incur financial penalties.

In order for a relevant commissioning programme to be developed, the commissioning engineer must be involved in, or be familiar with, the following:

(a) Have a comprehensive understanding of the requirements of the contract, specifically relating to the electrical control and layout of the systems and components involved.
(b) A knowledge of the delivery and completion dates relating to all of the major electrical components and erection activities that will effect the commissioning processes.
(c) Be involved in the final inspection of major electrical components, such as control panels and rectifier sets. If the commissioning engineers have been thorough in their preparation this will have been written into the quality plan previously discussed. This process increases job ownership and responsibility. There is no one better to check a component than the person who will ultimately be responsible for making it work.

There are no excuses if the component does not work when it gets to site and this can really focus the commissioning engineer's mind.

The first stages of the programme development are provided by others and usually provide for an overall erection programme showing the start and finish dates for all the major activities and their relationship, but little detail. The commissioning engineer must be involved in the detailed development of the commissioning programme. This commissioning programme must fit around the precipitator build and mechanical commissioning functions and a great deal of discussion is necessary with all those involved in order to achieve a credible programme. At this stage communication between departments is critical.

The commissioning engineer must have a thorough understanding of the programme layout and requirements in order to identify the areas which are likely to cause delays. Late deliveries, construction delays and bad weather can all add up and mean that the programme falls behind schedule. All activities then tend to slip back and the final commissioning gets sandwiched between this date and the immovable plant start-up data. If this situation is left, commissioning activities tend to be concentrated into a mad rush of 24 h working and frayed tempers.

Commissioning engineers must study carefully the progress of all facets of the erection and evaluate the effects of these delays on their commissioning phases. They can then adjust their programme accordingly to make up for the delays.

11.3.5 Monitoring

11.3.5.1 Phase 1. Getting to know you. At this time all parties should be aware of the names of site engineers, through which all verbal and written communication and instructions should be channelled so that any misunderstandings can be avoided.

The first phase of the electrical commissioning engineer's monitoring process starts with a walk around the site with the electrical contractors site engineer responsible for the cable and equipment installation. The walk around usually familiarises everybody with the site layout and builds a working relationship between the commissioning engineer, the contractor's engineers and the client. It should be based on the programme, the specification and the quality assurance plan. From this there should start to develop the methods by which new cables and equipment should be installed on new installations, or alternatively, how existing cables and equipment should be removed and new cables and equipment installed on retrofit situations, within the timescales available.

The commissioning engineer should bear in mind that this should include the introduction of a safe working practice by the electrical contractor, with safety audits being carried out for each activity within the programme. Safety is everyone's responsibility.

As the installation proceeds the commissioning engineer's activities generally concern checking the following:

(a) Monitoring the progress of the electrical contractors in relation to the agreed programme. Identifying shortfalls and delays and developing actions, in association with the electrical contractors, in order to make up for lost time. Ensuring that the contractor has enough suitably qualified staff on the site to achieve the programme dates. Contractors sometimes take on too much work and can be tempted to remove personnel from site in order to fulfil other commitments. It is essential that clear agreed targets are developed and that the engineer is firm in the commitment to achieve then.

(b) Ensuring that the work is carried out in accordance with the contract conditions, particularly with respect to quality and safety. If the programme falls behind, the temptation to cut corners increases and if the electrical commissioning engineer is not careful, both quality and safety can suffer. During this period various problems will occur relating to the provisions within the contract. In many cases the original contract will not cover all aspects of a strip out and installation and so the dreaded 'claim of extras' will arise. This is where a clear head and a thorough knowledge of the specification is essential. All the groundwork carried out in the office should now pay off.

Commissioning engineers should keep in mind that it is not their responsibility to authorise extras to contract. The engineers should submit a report to the contracts engineer detailing the nature of the claim for extras and their comments on the validity or otherwise of the claim. Bear in mind that arguments on-site can affect the progress of the work and could cause delays. It is usual that these extras claims be acknowledged without prejudice and an agreement be obtained with the electrical contractor whereby he submits any claims for extras at the end of the job.

(c) Progress meetings with the client in order to brief him on current situation with respect to the electrical work. Again the electrical commissioning engineer will be part of a team fulfilling the contract and cooperation will be important in achieving the targets and satisfying the client.

During this period the commissioning engineer's presence on site should generally be 1 or 2 days per week. This will allow them to keep track of the progress. Status reporting back to head office is essential during this period. There are many other functions occurring at this time and a clear overview of the current build status by the contract engineer is essential for the smooth running of the overall contract. These reports should be of the exception type, i.e. only identifying deviations either from the programme, which may have knock-on effects, or problems which have occurred. The maxim should be 'be brief and concise'.

11.3.5.2 Inspections. Usually during this period manufactured equipment is becoming ready for shipment to site. This is when inspections occur and are usually at suppliers' premises. The most important equipment requiring inspection by the electrical commissioning engineer consists of the rectifier sets and control panels. These are critical to the commissioning programme and a thorough inspection is advisable and should be mandatory on the quality assurance plan. This inspection is carried out by the commissioning engineer and representatives from the suppliers and sometimes the ultimate client.

11.3.5.3 Inspecting the rectifier sets and control panels. It is usual for the supplier of the equipment to carry out the inspection checks which are witnessed by the commissioning engineer. These checks are usually based on 100% inspection, which means that every panel and rectifier is tested. The panels and rectifiers are usually inspected as matched pairs and clearly marked up as such for transportation to site. The procedure for these inspections should be agreed prior to them being carried out. Electrical commissioning engineers should familiarise themselves with the inspection procedures and ensure that remedial actions in the event of faults are clearly defined. At this time delays could jeopardise the build programme and there are too many contractors waiting to claim lost time because of late deliveries.

The inspections should generally include the pre-energisation checks set out as follows.

(a) Control panels

1. Carry out a dimensional check of the cubicle to ensure it is to drawing, remembering that this panel will eventually have to fit in a prepared opening on site. Check also that it complies with the specification in respect of dustproofing. A large number of control panels have to be sited in operational areas where the level of airborne dust is high and the necessity for sealed control panels therefore becomes important.
2. Check that the safety interlocks are fitted and operate as intended and that door alignment and seals are acceptable.
3. Check the painted surfaces for compliance with the specification and finish quality, checking for damage to the panels or scratches in the paintwork incurred during the build.
4. Carry out a cable and component check to ensure that the specified cable sizes and components have been used.
5. Check the layout of the components to ensure that they have been positioned in accordance with the backplate layout drawing. This should include any provision for segregation of different power supplies.
6. Check that the cables have been terminated correctly and are securely fixed. Loose connections can at best burn out components and at worst cause a fire.
7. Check that all cables are identified in accordance with the electrical layout drawings using specified ferrules and markings. The client will usually provide a cable number inventory and terminations and the equipment should be checked to ensure the numbering conforms with that specified.

(b) Rectifier sets

1. Check that the rectifier set dimensions and paint finish comply with drawings. If cooling fins are fitted, check that they conform to the required radiated area requirements.
2. Check that the LV terminals are secure and undamaged.
3. Check that the HV insulators are undamaged and fitted correctly.
4. Check for oil leaks at joints. It is usual to pressure test the rectifier tank overnight.
5. Ensure that a sample of rectifier oil has been checked for dielectric strength and moisture levels. Excess moisture can lead to premature arcing within the rectifier tank and can cause failure. A rectifier set cannot easily be repaired on site, because replacement of the internals requires a

clean environment; this can be an expensive option, requiring a sheeted area set aside for the work.

6. Check that the rating plates are as specified.

11.3.5.4 Live testing. This section is not intended to provide a blow-by-blow checklist for the live testing of the rectifier and control panels. It is intended, however, to identify the major components of the testing process.

These live tests must be carried out in accordance with a written test procedure agreed with the supplier beforehand. They must be performed in an area set aside for this work, which is safety interlocked to restrict access during the tests.

The tests should be carried out after the equipment has been thoroughly inspected in line with the visual and dimensional checks given above. A record of these test results should be made on the relevant quality assurance check sheets as the tests proceed.

The engineer responsible for the testing must ensure that all connections to the equipment are adequately protected by earthed screening or isolated in safety interlocked areas and all accessible equipment is fully earthed before the equipment is energised. It should be remembered that very high voltages are present (typically 100 kV peak voltages).

The control panel contains a combination of load carrying and electronic circuitry. In order that high voltage insulation tests can be carried out, all wires connected to electronic equipment such as the automatic voltage control unit and the thyristor control cards must be disconnected to prevent damage to the electronic circuitry. These wires should be securely isolated and 'made-safe'. Only after these precautions are complete should the cables be meggered and flash tests carried out.

Applied primary voltages should be checked and recorded together with output voltages from the transformers used within the control panel. The range and operation of the kV and Ma meters should be checked by injecting a 4–20 mA signal. The range and operation of the ammeter should be checked by injecting 0–1 amp signal.

The panel usually provides for a number of local and remote control functions. All these should be simulated in order to check that they function as required. All fuse and link connections should be checked by removing the appropriate fuse or link and ensuring that the correct components are de-energised. Operational checks on the panel and transformer/rectifier equipment can then proceed, during which the stop/start, trips and automatic voltage control functions are checked to ensure correct operation.

Once all these tests are complete and the relevant test sheets have been signed off, the panels and rectifier sets are ready to dispatch to site. It is usually a good idea to check the packaging to ensure that it is adequate for the journey. Damaged equipment can cause delays on site which could be expensive. The on-site storage of the control panels and rectifiers should be considered. They will require a sheltered area, under cover, where they will not be damaged by passing traffic.

11.3.5.5 Site checks as the installation proceeds. During the precipitator build stages, checks on the installation of the electrical equipment and the cabling are necessary. These checks ensure that the equipment has been correctly installed and covers the following items.

TR sets and control panels. After these items have been installed and wired, a visual inspection is required in order to check for any damage during transit or installation. These are intended as a final re-inspection, designed primarily to establish that all items are installed correctly and are to specification. They cover functional checks of the panel interlocks and door locks, together with an inspection of the cabinet and paintwork for scratches or dents to establish whether any rectification is required. Checking for rust at this stage is also recommended as this can be an indicator of inferior paint finish which can be identified to the supplier for rectification.

Site electrical checks should cover the following:

Ensure that all the switchgear, contactors, relays, thyristors and automatic voltage control equipment within the panel are installed to drawing and in line with the specification.

Check that the meters are correctly installed and have the correct ranges and units.

Check that the correct indication lamps are installed.

Ensure that the equipment within the panel is correctly terminated, checking particularly for any loose connections which could cause overheating during operation.

Check the type and version of control software installed to ensure it complies with the specification.

Visually inspect the data recorded on the transformer panels, to ensure it is as specified.

Check the level of the transformer oil. If this is low then a leak is most likely the cause. Visually inspect the rectifier for evidence of oil leaks and rectify if possible. Leaks from flanges and gaskets can usually be stopped by pinching up the retaining bolts. Further checks will be required to ensure that the leaks have stopped; if they have not, then a new seal gasket may be required. Remember that leaks usually get worse as the unit comes on-line and the rectifier temperature increases. Changing a gasket during the commissioning stage is preferable to having the unit off-line to do so later.

Check that all the rectifier transformer temperature trips are set and function as required. Finally check that the rating label is fitted as specified.

11.3.5.6 Rapping motors. Once the rapping motors are finally installed, the oil levels within the gearbox should be checked and topped-up as necessary. If the oil level is very low, check for leaks or damage and report if found. Check all stop and start buttons, door switches and shear pin switches for correct operation using a multimeter; replace any sticking or damaged switches as necessary.

As cables are pulled in, usually many metres, mistakes can be made and the cables can be terminated incorrectly. The commissioning engineer should ensure that all cables are numbered and ferrules fitted as they are 'belled' (for electronic equipment) or megger tested (for power cabling) by the electrical contractor. It is critical that at this stage all cores are identified and recorded as this will help in the event of problems later. The rapping motor resistance to earth should be checked and recorded together with the resistance between the motor windings. Finally check that the data on the motor rating plates is as specified and that the plates themselves are securely fitted.

The rapping motors are normally controlled from a motor control centre (MCC). If new, this panel should be checked to ensure it is placed as required by the contract and that the equipment within the panel complies with the specification.

During this stage it is possible that modifications to existing MCCs have had to be made to accommodate new equipment. Every effort is usually made to ensure that the site drawings are up to data, but undocumented changes can take place over the years and the commissioning engineer should check thoroughly that the existing MCC wiring is to drawing. If not, the drawings should be modified to reflect the 'as built' status. It is important that drawings are kept up to date as the job proceeds. It becomes a mammoth task to detail, or even remember, all the wiring changes carried out throughout the electrical installation and commissioning stage. A simple red line diagram will suffice, which can then be used to update the final 'as built' drawings supplied to the client at the end of the contract.

11.3.5.7 Insulator and hopper heating. With most plants, it is usual to provide electrical heating to ensure that the lead through insulators and dust hoppers are pre-heated before the plant goes on-line. This prevents (1) thermal shock and condensation forming on the insulators which lead to arcing and subsequent cracking, and (2) blockages in the hoppers due to damp dust forming as a result of condensation.

The heaters are controlled by monitor panels and the engineer should check that these are sited as specified. Again, the panel equipment should be checked to ensure that it conforms with that specified and all the cables are correctly terminated and numbered in line with the wiring layout drawing. Check to confirm that all the cables have been meggered or bell tested and that records are available for the quality assurance manual. The resistance of the heater elements and fuse ratings should be checked and recorded. Finally the heater thermostats should be set and recorded.

11.3.5.8 Lifting hoists. In most cases where the rectifiers are sited at the top of the precipitator, lifting hoists are provided to remove the transformers for maintenance. The cables to these hoists have to be checked to ensure they are correctly installed and have been meggered out. The electrical controls and limit switch operation must also be checked and recorded using a multimeter.

11.3.5.9 Earthing. The final checks relate to the main earthing system. This usually consists of heavy copper strip/tape clipped directly to the structure of the precipitator and linking all the current carrying components to earth. The joints in the copper tapes are brazed or soldered together and every joint should be ducter tested to ensure continuity to earth is maintained. The final connections are made to the plant earth system, therefore tying everything down to the same potential. After testing, each joint should be permanently marked, and a record kept of the tape routing on a general arrangement of the plant. Ducter test results should be recorded on a check sheet.

A second earthing system consists of the portable earths, which are used when the precipitator plant is off-line and access to the internals is required for maintenance. These are usually earthing balls connected directly to the electrode frames and sockets which are attached to poles and copper earthing tapes which are in turn connected to the main earth system described above. The engineer should check that the balls are fitted in such a way to ensure that the electrical clearance between the electrode and collector plate is maintained as specified. The portable earth assembly should be checked to ensure that it fits together correctly and that the earth pole location points for storage of the equipment during plant operation are sited correctly. These portable earths are a lifeline for personnel entering the precipitator and should be checked thoroughly to ensure that they function correctly.

11.3.5.10 Safety checks before energising. Having checked that the plant is erected correctly and to specification, a number of safety checks have to be carried out before the unit can be energised. A final inspection of the internal chambers, hoppers and rapping rooms should be carried out in order to check that no debris remains left over for the erection phase. Once the inspection is complete, the interlocking system should be checked. Every interlock should be checked to ensure the system gives safe access to the precipitator internals and that none of them can be defeated.

All earths should be in place and all switchgear should be isolated and locked off to prevent any attempt at switching on. The system should then be closed off in order to check that the interlock system effectively bars personnel from every access to the unit at this point (this includes all access doors, top houses, panels, switchgear cubicles) and that it cannot be defeated in any way, that all keys are correctly sited and that no duplicates are available. If there are problems at this stage, under no circumstances should one proceed to energise the unit.

This is the time that all the check sheets, filled in so conscientiously during the build, are checked, the quality assurance manual is inspected and any checks not completed are carried out.

11.3.5.11 Live testing. Live testing is perhaps the most satisfying and potentially the most dangerous part of the commissioning process, when all the equipment, that has taken so long to erect, is put into operation.

The rapping drive chains are disconnected and the rapping motors powered up. A visual check is made to ensure that the motors are rotating in the correct direction. Again this is important as the system will jam up and break the shear pin if the rotation is not correct; should the rotation be incorrect, it is a simple matter to switch off the power to the rapping motors and change two of the phases over. After the rotation checks are complete, the local and remote stop/start switches, shear pins and rapping room door switches are checked to ensure they are operational.

After this, the drive chains can be refitted and the rapping motors energised over a 24 h period to bed-in the rapping shaft bearings to prove the system. On completion of the 24 h trial, and before the rapping system is switched off, the motors are checked to ensure that they are drawing the correct power consumptions. High loads could indicate resistance in the drive system caused by misalignment of the rapping drive shaft; this must be investigated and corrected. If the problem is left unresolved, there is a likelihood of motor failure within a short period of time from initial operation. It is not easy to gain access to the internals or rapping room once the precipitator is on-line, because it is a high voltage area and hence is covered by the Plant Safe Access Permit system. Access times are usually after midnight in the small hours and are not pleasant, so fixing the problem at this stage is far more convenient.

Once the rapping system is fully functional, all the motor control circuits, lamps, push buttons, timers can be tested to ensure they are operational. It is becoming more common that the rapping control is effected by a PLC system that monitors and modifies the speed of the rapping shaft rotation and hence the frequency of the rapping blows. These systems should be checked to ensure that they function as specified and deliver the correct rapping speeds.

In order that the panels can be live tested, a lamp load must be applied across the transformer terminals in order to simulate the transformer to check the operation of the panel control circuits, the automatic voltage control and thyristor circuits. (The actual transformer primary must be disconnected for this test.)

The process allows the panels to be checked, without applying high voltages *via* the transformer rectifier, and permits testing of the basic control functions of the AVC units. At this time, all the overloads can be set to the specified levels and the values recorded together with calibration and operational checks of the panel kV and mA meters. Once the panel has tested out satisfactorily, the transformer load can be reconnected and load and short circuit tests carried out, checking that the panels and TR sets energise the correct fields. This is critical in that if the high voltage cables are fed to the wrong field, then the interlock systems are defeated and there is a serious danger of personnel electrocuting themselves by gaining access to energised fields unknowingly. The panels should then be electrically isolated until still air tests.

Prior to the precipitator being put on-line, i.e. process gases passed through it, the insulator and hopper heating must be energised. The heaters should be switched on and the monitor panel circuits checked to ensure they operate as expected. It is worthwhile checking that the insulators are getting hot and that the thermostats are operational. During this time the heater currents should be taken and checked against specifications to ensure that no excessive loads are being drawn. Again if there is a discrepancy it must be followed through and resolved.

The lift hoist should be checked to ensure that the davit moves freely and that the controls operate correctly. It is most important to check that the limit switches function and control correctly and that the rail stops are in place in the event of an overshoot. Finally, the hoist should be checked to ensure that it gives the correct length of lift and that it has a Load Test Certificate for the specified duty required.

11.3.5.12 Precipitator still air testing. The final checking to be done is the still air testing. This is coordinated with the mechanical commissioning functions. Only when the unit has been boxed up, in cooperation with the mechanical commissioning engineer, can this be carried out in order to prove that both the electrical functions and the mechanical alignment of the internals of the precipitator are satisfactory.

With the main power to the control panels on, switch the sets to local and the AVC to manual, then energise the panel to provide power to the transformer. Using the manual raise selector, increase the power to the AVC unit and observe the primary and secondary voltage and current readings on the panel meters. Readings at a number of steps, up to set maximum rating or flashover condition, should be taken and recorded. These secondary readings should be plotted on a linear graph to check the relationship between the voltage and the current and the results filed.

Generally as the voltage rises toward flashover the rate of ionisation, and hence current flow, should increase exponentially until full transformer load is reached or flashover occurs. During this period care should be taken to raise the power slowly and steadily in order not to cause premature flashover within the casing. If the panels cannot be energised or trip on start-up, it is probable that there is an internal fault causing an electrical short. Hard arcing can usually be located by listening to the sound produced by the flashover. In any event there is no alternative but to re-enter the precipitator and find the fault, sometimes a long frustrating job but essential for the unit to be brought on line.

Once all the fields have been energised and voltage current characteristic readings have been taken satisfactorily the unit is ready for service.

The still air for each field should be almost identical, provided the same type of discharge electrode, size of bus section and energisation method are constant. Any deviation in the curves, assuming that the panel meter readings have been checked as satisfactory, can only be associated with the mechanical alignment of the internals or a peripheral reduced clearance.

When the precipitator is in operation, the V/I characteristics will change as a result of temperature, gas composition and, more important, dust loading. The dust loading will give rise to either corona suppression or, more likely, space charge effects. The effects of these are discussed in detail in earlier chapters.

Typical V/I characteristics are illustrated, along with the still air results in Figure 11.12, for a three-field precipitator handling a fairly easy duty. The

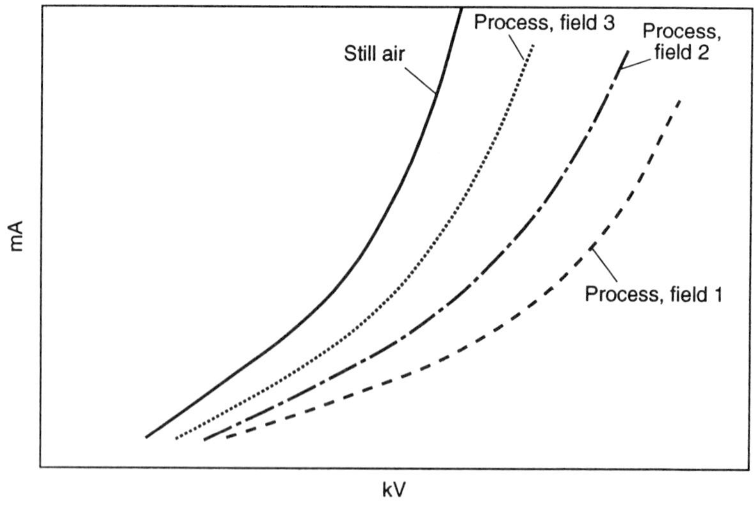

Figure 11.12 Typical voltage/current characteristics for still air and process conditions.

effect of space charge is to reduce the corona current flow mainly on the inlet field with a corresponding increase in voltage because the space charge effect reduces as the gas becomes cleaner, the outlet field corona current increases and the voltage reduces.

Taking and recording these characteristics, as soon as the process reaches some form of normal operation, provides a useful check on any future precipitator performance. Since any change could indicate unacceptable levels of dust deposition on the internals, a mechanical or electrical fault of some kind or a change in process conditions, for example a fuel change on a boiler plant. As such the curves provide the precipitation engineer with a simple comparison tool by which to judge operation.

11.4 Process commissioning

This part of the commissioning programme is probably the most interesting of all. It is the point at which, to paraphrase a well-known expression, 'The proof of the pudding will, if everything has gone alright, be in the heating.' The culmination of the combined efforts of designers, draughtsmen, project and contract engineers, construction staff and commissioning engineers should now be ready for the 'acid test'. Will it do the job it was built to do?

Whilst being interesting, this can also be a very frustrating part as well. Up until this period, the commissioning engineer has had to maintain a level of progress against which the only outside influence has been a construction programme. With the start of the process commissioning stage, the precipitator no longer remains centre stage, as it were. It becomes only one piece of equipment in the many which go to make up the complete process line. At this stage, its importance in that line, and thus the importance of place that the commissioning engineer has in the overall process commissioning picture, is measured by how much influence it has in the continuity of the operation of the process. The less importance that the filter has in the system, the higher the degree of frustration. This is because, should there be a need to make any alterations or rectifications, they may be found impossible to do because the process cannot be 'halted' for some period of time. The frustration is born from the loss of control previously held but now not being able to 'pull rank', as it were, and instigate the actions required immediately. Perhaps a gauge of its importance can be assessed by considering what happens if, for some reason, the filter fails. What is the knock-on effect? Does it completely stop everything? Can parts or all of the process carry on until such time as there is a window in the programme which will allow access to the filter?

For obvious reasons, if the unit goes on-line with no problems, it can be very satisfying. Only when it does not, do the varying degrees of frustration start to show. Is it the filter that has caused the problem and can it be rectified quickly? Has some other part of the process caused it and does it have to wait until such time as others say it can be rectified? Typical situations in the everyday story of commissioning folk!

To limit the potential for external sources being the cause of filter problems, it is essential that when bringing the plant on-line, no short cuts are taken to expedite or shorten the initial start-up period. If there is an agreed method for this procedure, it is of paramount importance that this is adhered to. Not all processes can be commissioned in exactly the same way. This section is aimed at helping to ensure that it is not the precipitator that is the cause of any procedural hitch in the process commissioning programme.

11.4.1 Hot commissioning

As with the previous sections, the operations will be taken in the order that they would normally occur when putting the plant on-line.

11.4.1.1 Initial activities. Depending on the process concerned, some preparation may be required in the electrofilter prior to process gas being admitted to the chamber.

Where rapping is used as the means of removing dust from the gas distribution equipment and both discharge and collector electrodes, it is advisable to switch them on as soon as power is available to the plant.

This applies to all heaters, when fitted, irrespective of their location on the electrofilter. It is necessary to allow a period of several hours to elapse from this action to that of allowing process gas into the chamber. This will ensure that the parts of the plant that are being heated are up to their operating temperature and as such will give the required protection for which they were installed. At the same time that heaters are energised on other components of the filter, any purge air fans and, where fitted, their respective heaters, should also be powered up.

11.4.1.2 Precipitator purging. In cases where the gas to be cleaned is potentially explosive, the casing must first be purged of ambient air. There are several ways of accomplishing this. One is to use dirty process gas whilst the other is to fill the chamber with an inert gas, such as nitrogen.

The purging process can be a time-consuming activity but unless it is carried out correctly, observing all the necessary safety precautions, the filter is a potentially explosive device. It is commonplace to ensure that all potential pockets within the gas pass should be fitted with vent pipes. This will allow air to be replaced by the purge gas. In designing the purge pipes, however, the differences in densities between air and the purge gas has to be known. It would be ridiculous to have vent pipes pointing upwards from the highest extremities of the pockets being purged if the purge gas is less dense than the air it is supposed to replace. It would merely vent straight to atmosphere leaving the air still in place.

An extra short length of pipework, which will include an isolation valve, is spurred off each vent line. The object of this is that it is possible to sample the gases being driven off and thus establish the level of oxygen still remaining in the chamber. It will be necessary, of course, to be able to close off the line venting to atmosphere by the use of a further valve so that the gas to be sampled is redirected to the sampling point. Once the predetermined safe level of oxygen has been achieved on several consecutive readings from all test points, then, and only then, can it be considered safe to continue with the commissioning programme.

It is worth repeating that there must be an oxygen level which is recorded as lower than the maximum allowed in the chamber and that it has been found at each and every sample point from at least two consecutive tests. It must also be borne in mind that one test will be a complete circuit of all test points, so at least two circuits must be made in order to verify the oxygen content. A typical purge pipe arrangement to be found on a coke oven gas detarrer, tube type precipitator can be seen in Figure 11.13. Points at which the tests referred to above can be carried out should be adjacent to the valves labelled A, B, D, E, G, H and J.

As a precursor to the admittance of process gas to the precipitator, it will be necessary to ensure that the dust extraction equipment is operational. Even though it is unlikely that there will be anywhere near full-load conditions prevailing, there will be dust being carried forward in the gas stream. If the filter is attached to a process which has a high moisture content in its gas, it is imperative that the dust is removed as quickly as possible. This will avoid the possibility of it sticking to the collector plates, hopper walls or even blocking the hopper discharge point.

11.4.1.3 Warm-up and condensation. Where there is any form of heat input into the process which is involved in producing the dust, there will inevitably be moisture as well. Typical examples are cement making processes which could produce gas with as much as 30% moisture in it, on a volume per volume basis. Municipal waste incinerators can have similar levels whilst power station electrofilters, controlling emissions from boilers fired with pulverised coal, can have between 8 and 15% moisture.

If it is considered that in some processes the volumes of gas involved, albeit at full production rates, can be as high as four million actual cubic metres per hour, it can be appreciated that this could mean a tremendous amount of water being bound up in the gas stream. It can and often does run into many tonnes per hour. Should the moist gas condense out on any of the internal surfaces of the filter, it will form a very good, damp surface on which to allow dust to adhere.

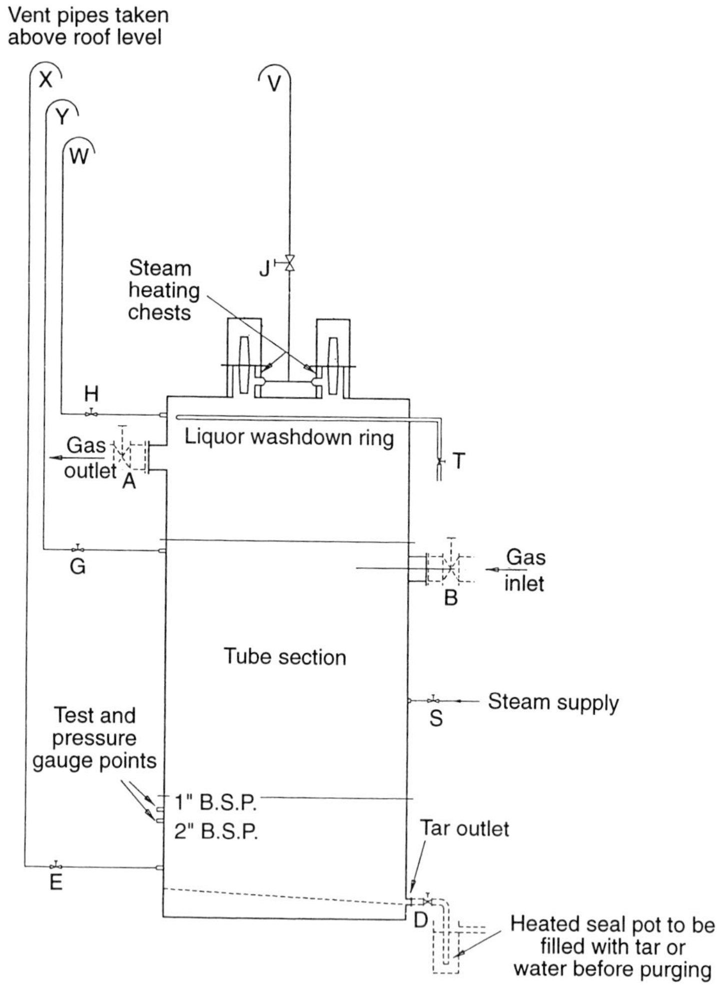

Figure 11.13 Detarrer purge pipe arrangement.

From this, it will very quickly be realised that as soon as gas is passed into the precipitator chamber, it will be impinging on surfaces that will almost certainly be at a temperature below the gas dew-point. With some processes, water, in abundance, can then be seen running in rivers from places like hopper outlets or expansion joints until such time as the dew-point temperature has been exceeded by all components in the casing. To alleviate the potential for water to build up in the hoppers or in dust conveyor casings, inspection plates could be left open at the lowest point available. When moisture has stopped coming from them, they can be boxed back up. This will also allow them to become warm and reach a temperature above the dew-point.

It is not uncommon for a time in excess of 4–6 h to be required to reach a point where the internals and casing have obtained a temperature which would not cause condensation when hot, moist gas comes into contact with them. If the process involves the necessarily slow build-up of temperature, perhaps because brickwork or refractory has to be dried out in a kiln of some sort, then this period of 4–6 hours could well stretch to several days.

11.4.1.4 Preheat. With current environmental legislation, many processes will not have the luxury of being able to bypass the filter for the initial start-up period. Indeed, they may well have to energise the precipitator at the same time that gas is passed through it, should dust be generated during this curing period. If this is the case, then it would be advisable, if at all possible, to have a system which can give some sort of preheat to the casing and internals, prior to the admission of process gas. This could take the form of a closed circuit heating system which is continually passing the gas content of the precipitator over some heating source. Obviously, this would only be sensible if the inlet and outlet flares are blanked off by good dampers. The contained gas would then be drawn from the outlet flare, ducted over the heating bank and returned to the casing through the inlet flare. By maintaining the internals at some temperature as close as possible to the dew-point of the process gas, the prospect of dust build-up is minimised at worst and eliminated at best. The dampers would remain closed until the process gas started to flow. Even though it would not have reached the dew-point at that particular point in time, this would be tempered by the fact that the internals were already hot enough to cope with the lower temperature.

11.4.1.5 Gas bypass. In the event that bypassing is allowable, it can only be possible on two counts. Firstly, and definitely most importantly, the local pollution control authorities will have to have sanctioned the action. This would mean that the maximum emissions would be at a concentration which did not infringe any preset legislative values. Secondly, some suitable ductwork has to have been built into the system. If this latter route is not possible, another alternative would be an emergency vent stack which can be sited just before the inlet to the filter. As will be discussed at a later point in this chapter, this type of stack has a more important role to play in the operation of the electrofilter.

To aid not only the commissioning engineer but also the plant operative, after hand-over has taken place, it is useful to position thermocouples in the ductwork immediately prior to the inlet flare and further ones in the outlet flare. The benefits to be had from these instruments are that they give a good indication as to the temperature of the internals of the precipitor. By noting the temperature differential across it, a close approximation to the real temperature can be made. From this information a better judgement can result in assessing when to start energising the rectifiers and thus minimise any potential problems which would occur from powering up when the internals were still wet.

11.4.1.6 Electrofilter energisation. Having reached a temperature whereby the plant internals will be dry, the high tension can be switched on. It is common practice to phase in the transformer rectifier sets such that the inlet field will be energised first, followed by the second and then subsequent sets.

With modern *PLC* controls and given the fact that most new precipitators are now computer managed, it is a relatively easy task to link the thermocouples into the control loops and use them to automatically bring in the sets at predetermined temperatures. Indeed, it is also possible to hold the power levels down until such time as the temperature would allow an increase and the dust burden being presented requires it.

11.4.1.7 Thermal expansion and thermal stresses. As the temperature of the process gas increases, there will be physical changes occurring to and within the casing and ductwork alike. The two most important of these changes are expansion and stress relieving.

With respect to the expansion aspect, there are a number of things that will occur. Considering ductwork first, there is the relationship of either inlet or outlet ducts to their respective precipitator flares. Whilst it is normal design practice to include expansion joints immediately before the inlet and after the outlet flares, provided any movement of the ducts relative to the flares does not impose any stresses onto them there will not be any problems. However, should due consideration of this relative movement not have been given, serious damage can occur. A typical example could be the approach angle of the inlet duct. In particular, if it comes in at some angle which is oblique to the axis of the filter, tremendous side loads could be exerted on the inlet flare. Because of their relatively low mass, when compared with that of the precipitator, ducts will achieve the temperature of the gas much more quickly. Therefore, relative movement will also occur as a faster rate. Strategically placed expansion joints should eliminate this problem. Nevertheless, as the heating-up process continues, regular checks should be made to ensure that the joints are fulfilling their duty. For instance, if a sliding joint is installed, is it sliding freely? If one that is flexible is used, has enough space been allowed for between the two solid parts, so that either it does not get pinched and holed or the two sections that it is separating do not touch?

There are some designs of rapping which involve the location of drives outside of the casing. It is feasible that there could be considerable differential expansion between the casing and any driven shaft. This would be of particular importance in situations in which there is a big difference in temperature between the process gas and that of the ambient existing in the locale of the drives. On applications where the gas temperature is between 300 °C and 400 °C, it is not inconceivable that the differential could be as high as 200 °C. This would mean that at stable temperatures, a solid one-piece rapping drive shaft could be around 20 mm 'shorter' than the casing where the casing is 10 m wide. If this design were installed, it would be extremely detrimental to the rapping efficiency of the lift and drop rod style. The cam lifter for the rod would be so far off-centre that the lift, and thus the potential energy available for the rap, would be greatly reduced.

The example highlights the necessity to split the shaft into smaller pieces and have them linked together *via* a coupling which can absorb the expansion without moving the shaft. Provided that the individual sections have a fixed point located half-way along their respective lengths, the effects of differential expansion can be minimised.

Collector plates can also be vulnerable if they have not been installed correctly. Because of their thin section, they can reach the temperature of the gas extremely quickly and thus their expansion can be quite large relative to their surrounding, more heavily sectioned, components. Clearances which will allow for these differentials should be verified during the earlier inspection stages.

One further case for consideration is the way in which the precipitator casing is mounted on its sliding bearings. As has been previously mentioned, allowance has to be made for the relative differences in expansion between the casing and its understructure. Although it is most unusual to experience any problems with this part of the filter, it should be included as part of the checking procedure required during the warming-up period. Unfortunately, these bearings are sited in the most inaccessible areas and, to exacerbate this even further, the individual bearings will normally have some sort of protective skirt around them, making them even more difficult to see. However, in the unlikely event of there being any problems, the results would become obvious. Skirts could become torn at their welds or the casing appearing to be sitting skewed on its understructure are several potential examples. If this does occur, causes must be found immediately.

The aspect of stress relieving can be addressed from several fronts. Firstly, maintaining the theme of high temperatures from the previous few paragraphs, there will be the potential for the thermal relief of welded joints or stresses built into plates as a result of fabrication. Unfortunately, the results will be quite difficult to find. However, the fortunate aspects are that damage very rarely occurs from this cause and, if it does, it is not usually of a magnitude that will be deleterious to the efficiency of the filter. Should anything happen that could affect the operation of the electrofilter, it would appear as plate buckling and/or weld tearing. The former will usually be contained and effectively limited by the casing stiffeners on a well-designed unit. The latter case will be the more serious result. If it does occur, the most effective way to trace it is to listen to the casing! Not as silly as it may at first sound but, as has previously been described, egress or ingress of gas from or into a casing can cause an audible sound which is readily traceable. It may not be easy to get at but at least its location can be traced! It is worth repeating the warnings, given earlier in the chapter, of the potential major problems which could arise as a result of a leakage casing.

The other type of stress relieving stems from the vibrations induced by the rapping gear. As with thermal stress relieving, the potential for it to cause any problems which are serious enough to affect the performance are quite remote. Settlement of mechanical fixings is the major concern here. Fraze which has not been removed before fitting bolts and their washers and nuts, will loosen as it gets compressed. Bolts that have not been tightened or torqued up correctly during the construction phase undo. Often the first that is known about these events is when the components of the fixings start appearing from the hoppers. If this sort of thing does occur, it should be investigated at the earliest possible time – before any, much larger, parts of the filter start 'relocating' to the hoppers!

11.4.1.8 Equipment optimisation. Once full load has been achieved on the process, optimisation of such equipment as rapping and automatic voltage controllers can commence. The former will be in conjunction with the electrical engineer whilst the latter will form part of the electrical commissioning sequence.

11.4.1.9 Rapping. The optimisation of rapping timings is usually a function which is carried out over a fairly prolonged period of several days, at best, but it could run into weeks. The object of this exercise is to ensure that stack emissions are kept to the best achievable level, whilst maintaining the cleanest possible internals without creating severe 'rapping spikes'. Rapping spikes are the result of a surge of re-entrained dust, rapped from collectors, going forward into the stack and being detected on stack dust monitors. They will show up on the trace as a severe peak on what would, it is hoped, be an otherwise consistently low recorded emission level. Should a stack monitor not be fitted, the results of the spikes can also be recognised by regular, short, sharp periods of discoloration in the plume emitted from the stack.

For processes that have equipment in their train where thermal stabilisation is not required, the procedure for bringing the filter on-line is a distinctly less protracted affair. Precipitators that are cleaning saturated gases, for instance, can be put straight onto maximum duty, provided that the essential precursors have been observed. Typical of these would be that all heaters had been on for sufficient time to allow the hardware that they are protecting to achieve a temperature that is at least equal to the water dew-point of the process gas.

11.4.1.10 Wash-down sprays. On plant that relies on water or a 'liquor' wash-down to clean the internals, a procedure which is analogous to rapping optimisation is required. The difference here will be that it is the frequency of flushing that has to be assessed. Often this will be an intermittent event which can be carried out at some convenient point during the operation of the process. Where this is not possible, however, the experienced engineer will soon start to recognise the telltale signs of efficiency reduction and from that be able to judge when to go into the cleaning cycle. As the depth of dust on the internal components starts to increase, so a gradual decline in the secondary current, or increased flashover, will be witnessed.

It is quite common for the vendor to suggest a frequency of wash-down which will be greater than is actually needed as a starting point. Over a period of time this frequency can be regularly assessed and adjusted to suit the prevailing conditions. As with rapping optimisation, a regime which is best suited to the process conditions can take a long time to establish.

11.4.1.11 Emergency dampers and stack. Earlier in the chapter, it was mentioned that an emergency stack has a very important role to play. To expand on this statement, imagine a scenario where there is a sudden emergency which requires that the electrofilter be physically isolated from the process and that the system is at, or close to, being at full production rates. The inertia of the process would normally be such that the only way to alleviate the situation would be to vent the gases to atmosphere. This would be done through a dampered stack at the inlet to the filter and sited either in the ductwork immediately prior to the flare or taken out of the top of the flare itself. A typical reason would be a case where the flue gas temperature exceeded the design maximum for the precipitator and therefore it would need diverting to eliminate the potential for mechanical damage. It these pieces of equipment are part of the system, the emergency damper will usually be activated from a thermocouple and controller situated upstream from the emergency stack. The commissioning of this portion of the plant would be a parallel action of both mechanical and electrical disciplines.

11.4.1.12 Instruction in operation and maintenance. There is an increasing awareness on the part of the end-user to ensure that all personnel that are likely to have any form of contact with the filter are fully conversant with the techniques of running and maintaining it. To this end, it is becoming more common for the commissioning engineer to be asked to give some sort of 'teach in' to the operatives. For obvious reasons it is best left until the end of the commissioning period. Only then can most of the quirks of the filter be known and thus passed on as pertinent information, along with all other relevant operating details. The opportunity should also be taken to highlight maintenance procedures and routines.

11.4.2 Back to the real world!

Once the process has reached and held the design load for several days, all its associated plant has reached thermal stability and there have not been any problems with the electrofilter, the commissioning engineer can consider reverting back into a normal human being and all that that entails: sleeping in a bed, not dozing in a chair in some control room in a remote part of the site; eating breakfast at breakfast time, not three o'clock in the afternoon; recalling how good a pint of beer tastes, when drunk out of a glass in a pub, not out of a tin at four in the morning in a hotel bedroom!

12 Dry type precipitator applications

K. PORLE AND K.R. PARKER

12.1 Introduction

Dry precipitator applications far outnumber the use of wet or mist precipitators. One of the advantages of dry type separation is that the particulates, removed from the process gases, are recovered in a more easily handled state. The precipitator itself, as the gas temperature is normally above dew-point, can be constructed from ordinary carbon steel, and the higher gas temperature also means that chimney discharges are adequately dispersed without resorting to expensive reheating strategies.

The largest single application is undoubtedly in the removal of fly ash arising from the combustion of carbonaceous fuels in the power industry. The usage is worldwide and precipitators fitted to a 2000 MW coal-fired station would collect in excess of 1 million tonnes of fly ash per annum from a 20% ash coal.

As indicated in chapter 1, the dry precipitator can be designed to satisfy the most stringent legislative regulations, in fact on numerous processes, emissions of less than 10 mg/Nm^3 have been measured, for example see Figure 1.3, chapter 1.

While the applications using dry precipitation are too numerous to address in detail, this chapter will review the main types and examine the various factors which determine the size, form and operation of the precipitator in order to meet the design performance levels.

12.2 Power generation industry

Without electricity, modern-day man would find life very different, particularly in the industrialised nations where one expects power to be readily available at the flick of a switch. In spite of the gravest fears of the environmentalists that the world is irrevocably polluting itself, power produced from the combustion of carbonaceous fuels far exceeds all other generation sources.

The continual demand for power and the need for countries to be self-sufficient has led to all forms of fuel being used for large-scale electricity generation, from low rank peat through to high rank anthracites, plus oil (hydrocarbon) based derivatives. This wide range of fuel types has meant that special designs of combustor and boiler have been developed to satisfactorily fire a specific fuel. Each fuel produces different waste gas conditions, in terms of temperature, chemical composition (analysis) and dust loadings, which will affect the size, design and possibly the operation of any precipitator.

Although some combustor designs can satisfactorily fire more than one basic fuel type, generally each design is specific to a single type or source. Because of the differing gas conditions, the size and design of precipitator will vary to optimise the performance for any specific fuel type.

Some countries which are rich in fuel reserves have very viable and successful export markets for both oil and coal. The internationally traded coals are mainly from Australia, South Africa, Colombia, Indonesia, Canada, the United States and Poland. There are other exporting nations; China, for example, with its vast coal reserves, will become an important player in a few years' time; however, its own expanding usage could mean China will become a nett importer in the early part of the next century.

The coals which are internationally traded tend to be typical steam-raising coals, having a high calorific value, lowish ash and usually a low sulphur content. These fuels can readily replace/supplement coals traditionally fired in many countries on existing boiler plant.

The greatest precipitator experience has been obtained from units firing steam-raising class coals. In the past two decades, however, different countries have used indigenous lower quality fuels to satisfy increasing power demands, for example brown coals in Australia and Eastern Europe and subbituminous coals in western USA and Canada. Other geographic areas having significant quantities of local coal have also developed plant to be self-sufficient. These different fuels have enabled the precipitator suppliers to extend their databank knowledge significantly and the ultimate user should have every confidence that a precipitator installation, provided by one of the major suppliers having similar relevant experience, should function entirely satisfactorily in terms of performance and availability.

In addition to the foregoing coals, the power industry also uses hydrocarbon-derived fuels, such as fuel oil, orimulsion and refinery residues; all produce waste gas conditions which will have an impact on the size and design of the precipitator. These changes will also be reviewed.

12.2.1 Bituminous coals

These coals represent the greatest fuel usage in pulverised coal (PC) power generation and are typified by a reasonable Hardgrove Grinding Index, 45–64, a high calorific value and contain a high volatile content 18–34%, which assists and stabilises combustion.

Dependent on the country of origin and form of mining, the fuel, as fired, can contain as much as 45% ash, e.g. India, but many mines use benification methods, such as coal washing, to improve the quality of the coal, particularly where large quantity shipments are involved.

A 2000 MW station would consume some 5 to 6 million tonnes of coal annually depending on the quality, usually denoted by its calorific value. Not only does coal washing reduce the amount of ash, but, dependent on the benification method adopted, can also remove iron pyrites and hence gives a lower sulphur content, an important feature when considering limiting the acid gas emissions on environmental grounds.

12.2.1.1 Medium sulphur bituminous coals. It will be appreciated from chapter 6, section 6, that the performance, or required size of any precipitator, is affected by the quantity of sulphur trioxide present in the flue gas, which acts as a self-conditioning agent for lowering the electrical resistivity of the fly ash particles.

For coals having a sulphur content of around 1.5%, which is typical for UK coals, the precipitator is of fairly standard design (see chapter 4) and its size or design migration velocity can be readily and confidently predicted from the typical curve presented in chapter 6, Figure 6.15, after making due allowance for required efficiency, temperature, dust loading and collector spacing effects.

12.2.1.2 High sulphur bituminous coals. For coals having sulphur contents in excess of say 2.5%, such as those from eastern US deep mine sources, the quantity of sulphur trioxide produced during combustion is more than sufficient to self-condition the fly ash particulates. In fact it is the excess sulphur trioxide which can give rise to difficulties and can influence the design of the precipitator.

A high sulphur content will, dependent on the excess air, combustion conditions and temperature within the system, produce potentially high acid dew-points, which can result in the acid condensation should the temperature fall below dew-point temperature in the precipitator. If this occurs then acid particles will be precipitated along with the fly ash and could produce very cohesive deposits. These will require improved and effective rapping to ensure that the build-up will not interfere with the precipitator electrical operating conditions.

Other areas requiring special consideration are those associated with potential acid deposition, where difficulties have to be resolved in order to improve precipitator availability; e.g. (a) hopper evacuation: as gas bypassing is controlled in a well-designed precipitator, the only heat arriving is by radiation plus some conduction; therefore, to avoid dedusting difficulties the hoppers should have steeper valley angles and the externally supplied heat should be increased to ease the situation; (b) to minimise insulator tracking problems, it is important that the insulator heating and purging system is fully effective; (c) casing corrosion: the thermal insulation of the casing needs special attention during installation; not only must all heat bridges be designed out, but areas around access doors and all corners, e.g. roof and hopper joints, must be adequately lagged and weatherproofed to eliminate cold spots, which would result in severe local corrosion.

Another problem, which can affect the chimney emission is that of acid condensation within the outlet ductwork or in the chimney itself. To minimise this it is important to maintain the gas temperature above acid dew-point by eliminating all areas of cold air ingress by ensuring that all seals are effectively maintained. Little can be done to avoid an acid plume as the gases leave the chimney, except to ensure adequate dispersion to reduce the visual impact.

12.2.1.3. Low sulphur coals. As the quantity of sulphur available in the coal reduces, the amount of sulphur trioxide for self-conditioning of the fly ash particles is lowered to a point where precipitation becomes difficult and the size of the unit to meet a specific duty will increase, as per the typical design curve of Figure 6.15, of chapter 6.

Generally, except for a more difficult fly ash to precipitate, the design features of the precipitator are fairly standard, except that the casing is much larger, as is the number of internal components. The main phenomena arising is the development of reverse or back-ionisation on the collectors, because of the higher resistivity of the fly ash, as described in chapter 6, section 6.5.

If, as a result of fuel changing to a lower sulphur coal on environmental grounds, an existing precipitator is found to be too small to perform satisfactorily, or there is a change in legislation demanding a higher efficiency, then it is possible to improve performance by the addition of small quantities of sulphur trioxide (see chapter 15) as a conditioning agent to reduce the fly ash resistivity.

For some Australian and other low sulphur fuels, which have an ash exceeding 90% silica plus alumina content, but low sodium etc., in order to improve fly ash coheson, while reducing its resistivity, dual conditioning using sulphur trioxide plus ammonia may be necessary, or in some instances only ammonia is injected, such as to optimise performance by eliminating/reducing re-entrainment of deposited material.

A theory arose amongst some suppliers and users, who appreciated that the problem of reverse ionisation was the result of high resistivity ash deposition on the collectors, that, if these could be kept free of dust the problem would disappear. This lead to the development of high '*g*' rapping specifications, which although assisting in some instances, did not fully resolve the problem, since with the general low cohesivity of the fly ash, the collectors were already relatively clean and all the higher rapping did was to increase the amount of potential re-entrainment and to result in higher maintenance demands on the rapping systems.

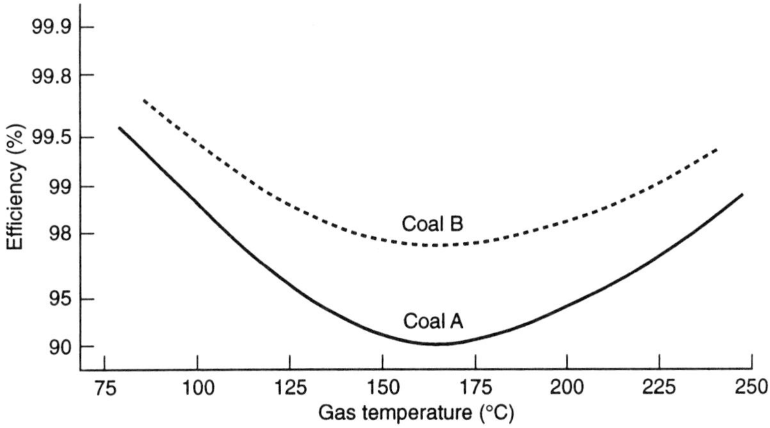

Figure 12.1 Effect of gas temperature on ESP efficiency.

Investigations in Australia, and elsewhere, into changing the temperature operating regime, have confirmed that when operating at a lower precipitator inlet temperature, the resistivity can be lowered by a factor (see Figure 6.16). The effect of operating at a temperature of 100 °C, because of a significant reduction in reverse ionisation, decreases the emission by almost a factor as indicated in Figure 12.1.

Other methods, for improving the performance of a precipitator firing low sulphur coal are related to changing the method by which the precipitator is energised, such as the use of intermittent energisation or some form of pulse charging [1]. These systems, which reduce the reverse ionisation by controlling the amount of 'charge' arriving at the collector in a given time interval, are fully described in chapter 8 dealing with the electrics.

12.2.2 Anthracite coals

This coal has a low volatile content, e.g. 6–15%, and is usually very hard and, as such, is difficult to successfully grind and fire. These properties mean that combustion is more difficult and high carbon carryover with the fly ash, of up to 8%, is the norm. This increase in carbon, where a normal boiler would have a carryover of less then 2%, can result in severe carbon re-entrainment or 'slip' which impacts on the overall emission from the precipitator. In extreme cases particulate emissions have been found to comprise up to 90% carbon.

To minimise the effect of this carbon re-entrainment or rejection, the designer/supplier of the precipitator should consider the following measures:

(a) configure the precipitator to have a lower gas velocity than 1.5 m/s, (b) ensure the HT rectifier equipment and discharge electrode system can provide sufficient power to 'clamp' the particles to the collectors once they have been deposited, (c) possibly skew the gas distribution to collect as much carbon material as possible in the inlet field (see chapter 5) and (d) if high carbon carryover is experienced in operation, apply ammonia conditioning to increase the cohesivity of the particles (see chapter 6, section 6.4).

While the above apply to the precipitator design and operation, improving the coal grinding efficiency, by the addition of dynamic classifiers to the mills, will improve combustion and hence reduce the carbon carryover and could possibly eliminate the necessity of the above precipitator modifications.

This latter approach is an important consideration, not only for anthracite coals, but with legislation demanding reduction in NOx emission, some low NOx burners, employing air staging and lower flame combustion temperatures, result in higher carbon carryover. So if difficulties are experienced on existing precipitators, after retrofitting with low NOx burners, then the fitting of dynamic classifiers to reduce the quantity of large coal particles has been found to significantly reduce the carbon carryover, which not only reduces fuel costs but, more important to the user, is the improved precipitator performance.

12.2.3 Subbituminous coals

These coals are typically strip-mined adjacent to the power plant and are normally of low sulphur content, but contain high moisture levels, 20–30%; those from the Western US can also have significant quantities of lime in the ash. The gas volume arising from the combustion of these fuels is larger than from bituminous coals, as a result of the higher moisture level, of some 12%, and increased excess air used for combustion.

In spite of the higher moisture content, which should have aided precipitation, at one stage the ash arising from the combustion of these coals was believed, by some US manufacturers, to be non-precipitable, or at least not with an economically sized precipitator, as a result of the severe reverse ionisation and re-entrainment difficulties exhibited by these fuels. Pilot precipitator investigations by European suppliers, however, produced design data from trial burns of these fuels which exploded this myth, and there are now many major power plants fitted with electrostatic precipitators very successfully operating in the western and central parts of the United States and other areas of the world firing these and similar subbituminous coals.

Possibly except for the larger size of precipitator demanded by the larger gas volume, fuel and ash analyses, the precipitators themselves do not contain any special design features. To accommodate the larger precipitator, its layout, particularly on retrofitted plant, is difficult, especially where limited boiler centres prevent a conventional 'in line' arrangement. With the larger units, the configuration, to avoid expensive double decking, take the form of a 'chevron', where the precipitators are angled to the boiler centre line, or 'back to back' arrangements, where the flow is at 90° to the boiler centre line. Difficulties with ensuring uniform dust distribution arose with the 'chevron' configuration, so many suppliers favour the latter 'back to back' approach.

As a cost alternative to the larger cold-side precipitators, there was a trend in the mid-1970s to install the precipitators upstream of the air heater and operate at a temperature of above 300 °C to overcome the effects of the reverse ionisation, etc. Engineering difficulties and performance problems of sodium depletion with the hot-side units [2], together with the reduced thermal efficiency of the units and higher power consumption, has resulted in a number of original hot-side units being recently converted to conventional cold-side installations.

12.2.4 Brown coals

These are much 'younger' coals being relatively soft and can contain, as mined, in excess of 60% water; they have a low ash content but can be high in soluble salts, typically sodium sulphate. The major reserves of brown coal are in Australia and Europe, where large areas are strip-mined to supply local power plants.

The fuel, because of its softness, cannot be ground in conventional coal mills and is only shredded, either in pin mills or more commonly in a fan having breaker bars fitted to the internal walls of the casing. Earlier boiler plant of the chain grate type used predried brown coal briquettes, but more recent plants are basically of the PC design, employing hot primary air to predry the fuel in the shredders before feeding the burners.

The waste gases from a brown coal or peat-fired station, with the large quantity of water intrinsically bound with the fuel, have moisture levels exceeding 20% and tend to have gas temperatures of around 200 °C; consequently, the gas volume approaches double that of a bituminous coal-fired unit.

Although the ash content is fairly low, the problems of comminution usually means that combustion, in spite of a high volatile content, is far from complete and results in a high carryover of carbon in the form of char. This char, being devolatilised coal particles, has a low mass but large surface area so is subject to potential re-entrainment.

To counter this, the design gas velocity should be limited to say less than 1.4 m/s and the minimum aspect ratio, i.e. field length to collector height, should be around unity. In addition (a) the TR equipment must be generously rated, not only because of the higher voltages resulting from the high gas moisture levels, but also to provide additional clamping forces for the char particles, (b) the rapping regime is also critical so that rapping spikes are minimised. Rapping spikes, because of the size and colour of the char, are very visible and are often referred to as 'Indian Smoke Signals' where re-entrainment is prevalent. With the char being readily re-entrainable, some installations employ complete flow isolation during the rapping period, using so-called 'off-flow' rapping (see chapter 4, section 4.7) to combat this.

Although the ash content is fairly low, the silica to alumina ratio is different to that from most other coals, and, where the alumina is high, e.g. 40%, as opposed to a more normal 20%, a large proportion of the ash particles are expanded to produce a very voided structure, often referred to as 'popcorn', which, again because of the low mass/large area, is readily re-entrainable, so the control of rapping and gas velocity is of paramount importance for the successful operation of these units; otherwise precipitation, *per se*, is very easy.

12.2.5 Lignites

This class of fuel falls midway between brown coal and bituminous forms; again it is a young coal and is typically strip-mined adjacent to the power plant. It can be found fairly universally, Germany, Greece, North America and Australia having possibly the largest of the known reserves.

Lignites being somewhat older than brown coals are 'harder' and can be reasonably ground in conventional milling systems and with their high volatile content are not difficult to fire. The fuel as mined, has a moisture up to 50% and sometimes a high ash, often containing a lime content of 30%, with occasionally a significant sodium chloride level.

Generally the gas volume from a lignite-fired unit, because of the higher back end temperature and moisture level, is larger than from a bituminous coal unit. Dependent on the lime and salt content of the ash, the precipitability can vary from moderate to difficult. The lime reacts with any free SO_3 produced and the salt tends to volatilise and pass through the precipitator as submicron condensed fume, or, more likely can be deposited on the discharge electrodes to give symptoms of corona suppression, i.e. reduced charging corona current flow.

Both these situations detract from the potential collecting efficiency based on coal and ash analysis predictions. By making due allowance for the high calcium and salt by fitting controlled corona or high emission electrodes, precipitator performance and reliability is completely satisfactory.

12.2.6 Oil-based fuels

12.2.6.1 Fuel oils. A number of fuel oils are fired in the power industry, ranging from light oil containing 0.5% sulphur, through to heavy (3600 second) or higher oils, containing a minimum of 3% sulphur. The ash content is very low, typically <0.1%, so particulate pollution from the ash rarely arises. The need to fit precipitators on heavy fuel oil installations results from the carryover of carbon and the production of SO_3 during combustion.

With light fuel oils the additional treatment at the refinery means that the final product is lower in sulphur and has a much lower viscosity. This results in improved atomisation and combustion, such that SO_3 and carbon carryover are significantly reduced and with only the true ash loading, further treatment is often not necessary, in many countries, to meet legislation.

In order to fire heavy fuel oil successfully, its viscosity must be reduced by heating the oil to a temperature of 170 °C, usually by steam, so that it can be pumped and sprayed through the burner system. While good atomisation can be obtained with most spraying systems, there are always some large droplets produced which fail to burn out completely, some giving rise to a coke-like structure, others forming 'cenospheres', which are really gas bubbles in an ash surround. The formation of these cenospheres is not completely understood, but since they are hollow, having a very small mass but a large surface area, like the coke particles, are subject to re-entrainment from the precipitator collecting plates.

The high sulphur content of the heavy fuel oils can, dependent on the excess air in the combustion zone, produce high acid dew-points, so the air heater exit temperatures around 160 °C tend to be the norm; this increases both the volume and viscosity of the gases to be treated by the precipitator.

With the low dust loadings normally arising, $\ll 500\,mg/Nm^3$, but containing up to 80% carbon, the design precipitator gas velocity needs to be moderate for high efficiency collection, with an aspect ratio of around unity. If the gas approaches dew-point temperature, the ash becomes very cohesive, so in spite of potential re-entrainment difficulties, the rapping needs to be effective, albeit at a fairly low impact frequency.

This is particularly true for installations which employ either magnesia or ammonia to control the free SO_3 levels in the discharge gases. The injection rate of ammonia is critical to prevent the formation of ammonium bisulphate/bisulphite, which will either fuse within the precipitator giving adhesive ash or pass through the unit as a vapour to condense later, giving rise to an unacceptable whitish/blue plume.

With correct control of the injection rate, e.g. a nominal stoichiometry of around 2 (the actual value being temperature dependent), this mode of operation is desirous, since not only does it effectively lower the acid dew-point temperature, but it also reduces the potential quantity of acid which can be adhering to the ash and carbon particles, making rapping and hopper dedusting easier. To cater for possible hopper dedusting difficulties, the hoppers must be adequately heated and thermally insulated and all areas of exposed steel work attached to the hoppers must be covered.

To avoid insulator tracking problems, with the potential high acid dew-point and carbon, it is imperative that the design of the insulator/

support system has an adequate and effective heating and controlled purging to ensure maximum plant availability.

Potential spontaneous combustion of the fine carbon particles on exposure to oxygen can be minimised by eliminating all points of air ingress, particularly into the hoppers where large quantities of hot carbon dust are collected. When the dust is removed from the hoppers, ideally it should be stored in closed containers or fed back to the combustor under nitrogen. In bringing the unit off-line as much dust as possible should be removed from the internals and hoppers to reduce any fire risk, which could result in serious mechanical damage to the precipitator internals.

12.2.6.2 Orimulsion. Orimulsion is a new fuel and is a 70:30 mixture of Orinoco Basin bitumen in water. The particle size of the bitumen is around 15 μm in the emulsion and as such creates few combustion difficulties in terms of 'burn out'. The ash content although higher than from fuel oils is still low at around 250 mg/Nm^3 but contains high vanadium, nickel and magnesium components.

With the high sulphur content, around 2.7% as fired, the quantity of SO_3 produced is very dependent on combustion zone oxygen levels, i.e. carbon monoxide in the waste gases. The amount of SO_3 in the gases produces two distinct ash forms, a light buff and a green colour, the green colour being associated with vanadyl sulphate, which has a high specific gravity. The high SO_3 levels give potentially high dew-points, so effective heating and purging of the insulator systems is essential to avoid tracking problems.

The fine droplet size produces a fine ash, some 70% being submicron [3] and, as such, high emission or controlled corona electrodes are required to avoid corona suppression effects detracting from performance. For the capture of the fine fume, the gas velocity, although not too critical, should be no higher than 1.5 m/s. Normally the precipitator would have a minimum of three separately energised series precipitation fields for an emission lower than 35 mg/Nm^3.

While the fine fume can create space/corona suppression problems, the operating gas velocity, temperature, etc., means that the TRs have to be adequately rated, particularly at the outlet end of the precipitator, where minimal space charge, as a result of the 'clean gas' conditions, can result in high discharge currents being consumed.

Although E.C. legislation is presently demanding an emission of 50 mg/Nm^3 dry at 6% O_2, the fine particle sizing requires an actual emission of around 25 mg/Nm^3 to reduce possible environmental complaints regarding high chimney emissions.

Hopper dedusting can, if air is allowed to ingress, cause difficulties, particularly when the oxygen levels at the combustor are such as to produce deliquescent vanadyl sulphate. This can rapidly absorb atmospheric moisture to produce a gelatinous material which is difficult to handle. The dust itself, being high in vanadium, is classed as a 'hazardous material' and requires 'safe disposal', either to special lined lagoons after conditioning or, better, to a chemical plant for the recovery of the vanadium.

12.2.6.3 Other-refinery-byproduct derivatives. This class of fuel is becoming of interest to power producers and refineries as power generating plants offer a safe means of disposal for these products, following legislation preventing their destruction by simple incinceration at the refinery.

The products fall roughly into two main forms, petroleum coke, which is solid at atmospheric temperature, and 'other residues', which like pitch, are really very high viscosity fluids, and although appearing to be solid, slowly flow depending on the temperature conditions.

The petroleum coke can be readily ground, although its hardness can affect mill maintenance requirements, and fed into the combustion zone as a pulverised fuel. The 'cokes' normally contain very little ash, have a low volatile content but a high sulphur content, up to 9% being found. Although as a pulverised fuel, combustion is acceptable, the high sulphur content can lead to excessively high acid dew-point temperatures and it is normal to find petroleum coke used to supplement other fuels having a lower sulphur. Some units have been developed to fire only petroleum coke and have operated successfully with a high back end temperature; however, with the high sulphur, desulphurisation of the flue gas is mandatory for this form of plant.

The use of 'other refinery waste residues' has been considered as an alternative fuel for PC units, but the general consistency makes it more suitable for fluid bed combustors, which will be described later.

12.3 The cement industry

Dry precipitators are used in the cement processing industry on a large variety of different applications. Their use dates back to 1912, when the first application was to reduce the emissions from the wet process kilns at the US Riverside Cement Works (see chapter 2). In the search for overall improved thermal efficiency, the wet process and the later semi-wet/semi-dry systems have been superseded by large throughput completely dry process plants.

12.3.1 Wet process manufacture

In the wet process, the feed material is mixed, ground and metered to the kiln as a slurry containing about 40% water. The slurry is dried as the material works its way down the kiln and the waste gas exiting the smoke box has a moisture content of around 30%, a temperature of 200 °C and a dust loading of some 10 g/Nm^3.

Precipitation of this ash is relatively easy, since at 200 °C and high moisture level the resistivity of the ash is around 10^{10} Ω-cm. The major process problem which can arise is the result of refeeding all precipitated hopper material back to the kiln, particularly if the raw material contains fairly high chlorides, which volatilise in the kiln and come forward as submicron particles and form 'snow flakes' as described in chapter 6, section 6.7. To overcome this difficulty a continual bleed-off of the outlet hopper dust catch is required [3].

12.3.2 Semi-wet processing

In the case of semi-wet/semi-dry processing, the slurry after mixing and grinding is partially dewatered by some form of pressure filter to give a cake having a moisture level of around 20%. The cake is shredded and fed on to a two-stage moving grate; the first section dries the cake and in the second, a precalcining stage, the temperature is raised from about 300 °C to 800 °C. The precalcined material then passes through a short rotating kiln for ultimate calcination to produce the clinker.

The gases leave the drying stage at a temperature between 100 and 160 °C and a moisture content of around 12%. The dust loading, because the bed acts as a barrier filter, is relatively low, so very high collection efficiency is not required to meet legislation.

At the lower gas temperatures, precipitation is relatively easy, but below 110 °C, severe corrosion of ordinary carbon steel components can rapidly develop, leading to reduced plant availability. To counter this situation, internals fabricated from aluminium/magnesium alloy have been found to give a satisfactory life.

At slightly higher temperatures, stainless steel is an alternative material, but pitting corrosion with high chloride raw materials may necessitate the use of high grade 317L stainless.

If the kiln feed is reduced or the cake has a low moisture level, then with temperatures approaching 160 °C, since precipitator performance is dependent on the relative humidity of the gas (resistivity effect), it may be necessary to use water conditioning to lower the temperature and increase the moisture level.

12.3.3 Dry process production plant

For modern high output plants, dry processing of the materials is now the preferred method of cement production (Figure 12.2). In this, the raw material feed is mixed, dried and ground in a mill using the process gas as both the carrier and drying media. The ground material is then passed

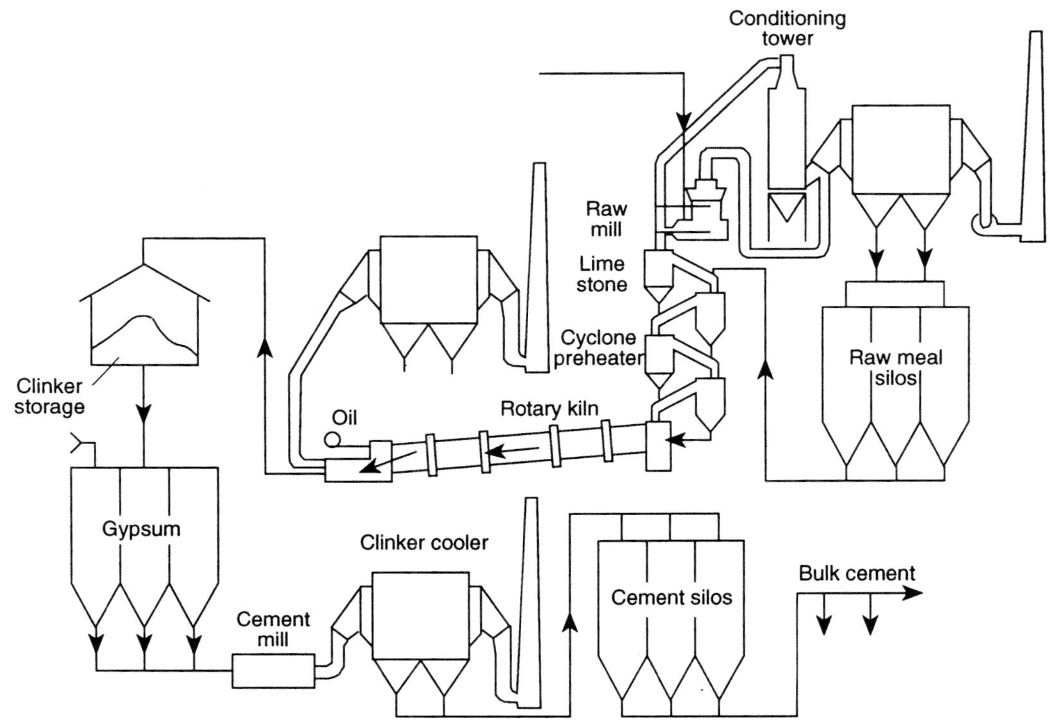

Figure 12.2 Rotary kiln with supension preheater dry process production.

through a series of cyclones, in the form of a preheater/precalcining tower, prior to entering a short rotary kiln for final calcination.

Two conditions arise necessitating treatment of the gases by electrostatic precipitators; the first is for the gases leaving the raw meal mill, carrying up to 1000 g/Nm^3 of material, i.e. the full raw material supply to the plant, and the second, the gases leaving the preheater cyclone tower installation, when the mill is out of service for maintenance, etc.

The gases leaving the raw mill, which originate from the cyclone tower, are at a temperature of around 90 °C with a relatively high moisture level. On many installations, where a spindle type of mill is fully gas swept, dust concentrations can approach 1000 g/Nm^3. To reduce the possible effects of severe space charge with this dust loading, some form of mechanical precollector can be fitted upstream of the precipitator, usually in the mouthpiece, to reduce the dust concentration to more manageable levels. Except for the mechanical precollector, the precipitator is of fairly standard construction, designed for an efficiency of some 99.95% plus.

For the kiln only gas condition, which occurs when the mill is out of service for routine maintenance, the gases are at a temperature of around 350 °C, a moisture level of 8% and a dust loading of some 80 g/Nm^3. The initial plants treated these gases in a 'hot' precipitator and although a large number performed satisfactorily, the transient period of bringing the mill off-line led to resistivity problems, with the dust remaining on the collectors giving transient high emissions. Once the dust on the internals had been conditioned by the changing process gas the precipitator usually worked well.

To eliminate this transient situation, when emissions could well exceed statutory levels, most modern plant now have cooling/conditioning towers to reduce the gas temperature to 150 °C, which at the same time reduces the dust resistivity to overcome the transient changes.

On some installations, the application of pulse charging techniques (see chapter 8, section 8.8) has been successfully applied to improve this transient situation [4]; however, the costs of these units are high and there is, unfortunately, no guarantee that pulse energisation will universally resolve the problem.

The design and operation of the conditioning tower is critical to plant operation and availability, since it is imperative that the water injected for cooling is fully evaporated to eliminate a wet tower bottom and the carryover of droplets into the field area. If this occurs, then the bottom of the tower could build up, ultimately effecting gas throughput and water carryover to the field would result in a cemented build up which the rapping would not be able to remove.

The types of spray most suited for this conditioning are either two fluid atomizers or high pressure spill back nozzles. Even though these will produce the small droplets required for rapid evaporation, the gas flow through the tower must be uniform, and to the same standard as for a precipitator (see chapter 5) to avoid wall wetting and ineffective conditioning.

The precipitator is often fitted with controlled high emission electrodes, in at least the first and possibly the second fields, to minimise the risk of possible corona suppression and space charge effects, arising from the volatilisation of alkali-type materials present in the feed or the high dust loadings. The TR equipment is typically fitted with intermittent energisation facilities in order to optimise both performance and power consumption.

With the high dust capture rates the design and sizing of both the hopper and dedusting systems is critical and must be carefully evaluated to avoid the hoppers overfilling to such an extent that the internals are damaged. Hopper sealing against air inleakage is important, since with the high suctions experienced under normal mill operation of up to 1000 mm w.g., any ingress would re-entrain significant quantities of hopper dust, and in the outlet hopper region would result in excessively high chimney emissions.

12.3.4 Alkali bypass plant

To comply with cement product specifications, it is necessary on some installations to remove/reduce the level of alkali materials in the final product. This is achieved by drawing off some of the gases at the kiln smoke box; at this point the gas is around 800 °C and the alkali components are largely in a gaseous phase. After leaving the smoke box the gases are cooled by peripheral air injection, to around 300 °C, thereby condensing the alkali compounds.

To optimise the performance of the alkali bypass precipitator it is necessary to further cool and condition the gases, by means of a cooling tower, to a temperature of 150 °C. This conditioning/cooling tower, although usually smaller than the main kiln tower, because of the lower gas volume being handled by the bleed-off system, must nevertheless be carefully designed along the same lines as for the main process plant.

The alkali bypass precipitator is of a standard design, except that to avoid corona suppression from the condensed alkali material, the discharge electrodes are preferably of the high emission type. Generally the same electrodes are fitted to all fields. Dust extraction from the hoppers can be more difficult and extra care must be exercised in the hopper design; also the dust extracted from the hoppers must be handled separately for safe disposal, possibly to a fertiliser manufacturer where the alkali components are potassium based.

12.3.5 Clinker cooler precipitators

In the measures taken to improve the thermal efficiency of cement production, heat from the calcined clinker, which was originally wasted as the clinker was cooled in planetary air coolers situated around the kiln, is now recovered. Cooling is achieved by feeding the hot clinker onto a moving grate, where the clinker, at a temperature of some 1000 °C, is cooled by undergrate air supplied by an FD fan system, as indicated in Figure 12.2.

Part of the heated air is used as primary combustion air, while any excess is vented through precipitators to atmosphere. Although this air is hot, the amount of heat is variable and many operations cannot justify further thermal recovery plant.

The hot clinker fed on to the grate is very variable in terms of quantity and depends on the kiln discharge rate; if a kiln 'slip' should occur, then the mass of clinker can change by more than a factor very rapidly.

During this situation, the amount of heat extracted by the air is very high and tempertures can change from a normal 120 °C to over 400 °C in a short time interval, during which, because of the high gas velocities arising through the high temperatures, large increases in dust carryover can occur, the loading changing from some 5 to over 25 g/Nm^3.

Originally, it was thought that the clinker dust entrained in the cooler gases would be very difficult to precipitate; not only has the dust been exposed to temperatures in excess of 1000 °C during calcination which would have destroyed all self-conditioning aspects, but the only moisture available for reconditioning is that contained in the atmospheric cooling air.

This would mean that the maximum moisture level would be unlikely to exceed 2% and generally a very much lower figure would prevail. Therefore, from Figure 6.10, chapter 6, one would expect a dust resistivity of some 10^{13} Ω-cm and above, which should give a difficult dust, i.e. reverse ionisation conditions, within the precipitator. Pilot precipitator investigations showed, however, that in spite of the high resistivity, the pilot operated satisfactorily without serious back-ionisation manifesting itself. This was attributed to the dust being fairly coarse, having a high specific gravity and a low cohesivity, and as such, was readily removed from the collectors preventing the full impact of back-corona.

Although the pilot work indicated that the dust was precipitable, the initial full-scale installations were fitted with water conditioning sprays in the cooler outlet ducting, but the short contact times and dust build-up problems quickly led to the water injection being discontinued. Unfortunately, these units did show some increase in emission during kiln slipages, because of the rapid increase in gas volume and temperature impacting on the electrical operation, a factor which the spraying system should have resolved. Later plants designed to cater for this high volume/temperature condition, however, have worked well and are probably the most cost-effective option to handle the rapidly changing conditions.

12.3.6 Cement mill precipitators

In order to have a saleable cement product, the clinker after calcining and cooling has to be ground to form cement as most people know it. Grinding is either achieved in a traditional ball mill or, more recently, in vertical spindle mills of some type. The gas volume from both is relatively small and consequently the precipitator is usually of a standard modular size rather than a custom-designed larger unit.

In the case of the ball mill, although most are air swept to entrain the ground material, the gases first pass through cyclones which reduce the effective dust loading by up to 90% before reaching the downstream arrestment plant. Most spindle mills are not fitted with cyclones and the whole of the product arrives at the downstream collection plant.

Like the raw meal installations, to reduce the dust loading into the first precipitation stage, many cement mill precipitators are fitted with simple mechanical impact collectors in the precipitator mouthpiece. Again because of the potential corona suppression, high emission discharge electrodes are often fitted throughout the unit.

To minimise the amount of heat developed within the mill, water sprays are normally fitted to limit the temperature. This water effectively conditions the mill dust and many precipitators operate with a temperature of 90 °C, but with a dew-point of around 60 °C. These conditions are ideal for precipitation, so few problems exist, unless the water cooling system becomes ineffective for some reason. The main difficulty is associated with failure to remove dust from the hopper, which because of the small size of precipitator does not have much storage capacity; therefore, overfilling can lead to mechanical and electrical problems which impact on performance.

12.4 General steam-raising plant

Although some of the following applications produce steam for either process heating or electricity generation, the size of the combustion plant is much smaller than those used on power stations, many units being less than 100 MW thermal output.

These combustors are fired by a wide range of fuels having very different characteristics and heating values. To ensure these burn satisfactorily, special designs of burner and associated firing zones have been developed to optimise combustion and, hence, thermal efficiency.

The types of fuel include colliery and oil refinery wastes, bark, straw and other biomass products and conventional municipal wastes. To successfully fire these fuels, although the chain grate stoker can be used, reciprocating/cyclic grates and fluidised systems, such as bubbling, deep and recirculating bed designs, offer improved and more efficient combustion for some fuels.

Some of these combustor designs have offered various challenges to the precipitator supplier and designer, but the performance of these precipitators, when correctly sized and designed, can satisfy all emission regulations.

12.4.1 Moving grate combustors (chain grate, reciprocating and cyclic beds)

The combustion zone, on a moving bed grate system, is established about two-thirds of the way along its length. Raw fuel, usually in large pieces, is fed on to the moving bed and is slowly moved toward the combustion zone. Combustion is achieved by control of both undergrate and overgrate air, the overgrate air acting as secondary air to burn off carbon monoxide and any carbon particles. The majority of the ash from the fuel remains on the grate to fall off into an ash pit at the remote end of the bed.

The waste gases from the grate tend to have a fairly high oxygen and a temperature of around 180 °C with a moisture level mainly depending on the moisture and hydrogen content of the fuel. Ash carryover, because of the low velocity in the combustion zone, is typically $2\,g/Nm^3$ with a fairly narrow size range. For an efficient combustion control system, the ratio of undergrate to overgrate air is critical; too little overgrate air can result in poor combustion and the carryover of carbon particles, which can impact on the performance of the precipitator because of potential re-entrainment problems.

Generally with optimised and efficient combustion, few precipitator difficulties arise handling fly ash from a coal-fired moving grate type installation. Not only is the dust loading low, but the particle size tends to be larger than from a PC type combustor. Also, with the different combustion conditions, the quantity of free SO_3 arriving at the precipitator is proportionately higher, so self-conditioning of the particles to reduce their resistivity is rarely a problem.

12.4.2 Fluidised bed units

These fall into two main designs having different operating regimes, either at atmospheric or high pressure. The latter is a special case and will be covered in chapter 16, dealing with new and future designs and applications.

In the case of atmospheric designs, some retain the material, typically sand, fuel and limestone, close to the bed itself as in the bubbling bed, or, as in the case of the high turbulence deep bed approach, in the freeboard area. While others, e.g. the recirculating bed, actually ensure that the material is carried out of the freeboard area completely, passing into a cyclone, the underfeed from which recirculates the material, after dedusting, back to the bed.

An advantage of the fluidised bed over other combustors is that it can accept a wide range of feed stock types and sizings and with the highly turbulent bed situation, lime or limestone added to the bed material acts as an effective desulphuriser. While this can influence the size of the precipitator, which is usually decided by the calcium salt carryover, this, in its own right, limits the resistivity effects of only partly conditioned fly ash.

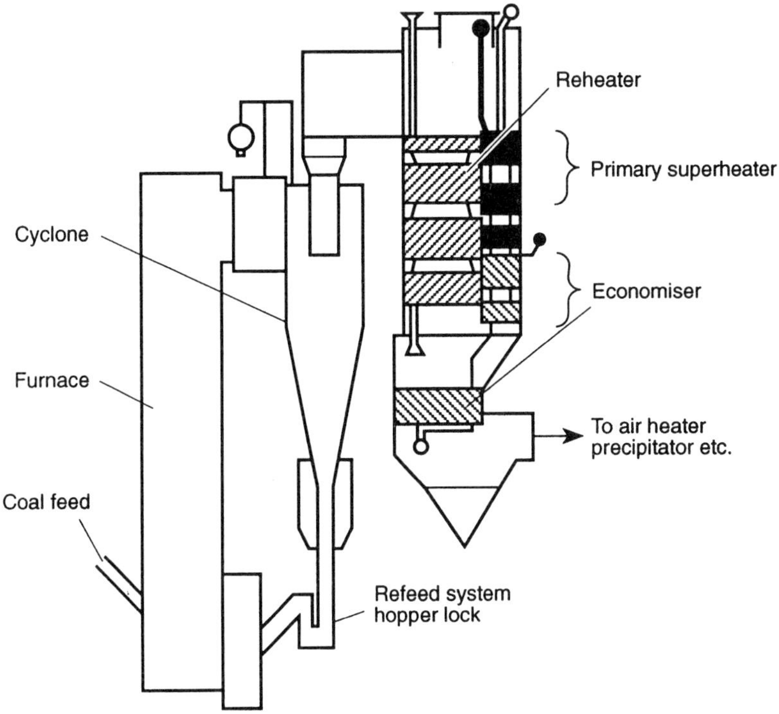

Figure 12.3 Principle of a circulating fluid bed combustor.

In the case of the bubbling bed, the main combustion occurs within or just above the bed, so little dust is carried out of the freeboard area whereas, with the deep bed approach, the major combustion takes place within the freeboard area and, because the velocities are higher, results in an increased loading but still below a typical PC operation. As combustion, particularly in the bubbling bed, is largely substoichiometric the volatiles tend to react and combine with the fly ash and bed material, so are retained and not carried forward with the waste gases.

With the circulating bed (Figure 12.3) only the cyclone 'slip' passes forward to the downstream arrestment plant and while the particles are cyclone slip, they are invariable coarser than the d_{50} of a PC installation, so present little problem to effective precipitation, even when limestone is added for desulphurisation.

12.5 Biomass-fired steam-raising plants

In attempting to reduce dependency on fossil fuels, there are many installations now using biomass or 'renewable' fuels for combined heat and power

(CHP) steam-raising duties. The type of combustors are similar to those described in the preceding section but, depending on the properties of the biomass fuel, some prior to entering the combustion zone require preparation, e.g. drying or comminution to give a greater surface area.

12.5.1 Wood chip combustion

As one would expect, wood chips are an ideal source of fuel but can present the precipitator with several problems; the most onerous is the volatilisation of various organic materials which can coat the normal particles with an insulating layer promoting back-ionisation. This condition, although very obvious from the precipitator electrics, is difficult to predict since it depends on combustion conditions within the plant; if it should arise, the precipitation engineer must cooperate with the supplier of the combustion equipment to modify the combustion conditions.

Generally the particle size of the particles from a wood chip fired installation are very fine and have a low bulk density. In addition to possible corona suppression, hopper sizing and dedusting must be carefully assessed during the design stage to eliminate potential difficulties.

The carryover of still smouldering or burning particles can also create a fire risk, but again this is dependent on the overall combustion conditions. The high moisture content of the gases associated with wood chip combustion, if one can avoid the problem of volatiles, is ideal for precipitation and many successful precipitators are in operation worldwide.

12.5.2 Chicken litter etc.

With the trend to raise all types of poultry in battery conditions, considerable quantities of waste material have to be disposed of. The large amount of straw and litter arising from the batteries has been recently used as a biomass fuel and there are a number of CHP plants designed to burn this type of material. The resultant fine ash, like that from wood chip installations, need the hopper and dedusting systems to be carefully designed; otherwise the precipitator is of a fairly standard configuration.

Again because of the high moisture content of the furnace gases, in spite of the fine particle sizing, requiring high emission electrodes, the precipitability of the ash presents few actual precipitation difficulties. The only real area of concern, particularly with chicken litter, is the high concentration of alkali salts, such as potassium chlorides and phosphates, which can result in ‘snow flakes’, which tend to settle out in the roof area and may short-circuit the electrics.

12.5.3 Municipal wastes

Most countries favour the idea of incineration to dispose of wastes arising from normal living, rather than landfill which can create its own problems.

The idea of combustion is not new and during the 1960s and early 1970s a large number of continuous grates were built to incinerate municipal waste. Many of these did not have energy recovery but the present economic trend is to use waste heat boilers for steam-raising as CHP.

The early units cooled the hot gases, often with temperatures of 1000 °C, by cooling/conditioning towers to achieve an inlet temperature to the precipitator of around 300 °C. At this temperature no serious precipitator difficulties occurred when combustion was reasonable. The only problem which arose was in the carryover of unburnt flakes of char from incompletely combusted paper. This was overcome by the addition of simple char collectors, either in front, or in the outlet flare of the precipitator.

An unexpected difficulty did arise from operating at 300 °C, in that if the chloride content of the waste was high, then the deposition of ferrous chloride, at this temperature, is fused and reacted with the carbon steel to form the higher ferric chloride by attacking the steel; temperature control and operation became critical to avoid this situation.

An operating condition which manifested itself was that, on some installations which only fired waste for 5 days, there was a tendency for the last operating shift to just switch off the fans and electrics when combustion was still taking place on the grate. This led to a situation where the downstream equipment was left over the weekend full of gases below dew-point (nominally 70 °C), which resulted in general corrosion of the plant. By introducing a proper purging system and leaving the electrics on, this situation with proper attention to thermal lagging and weatherproofing was overcome.

All recent and modified plants now include thermal recovery by passing the gases through a waste heat boiler. This serves to reduce the temperature to below 200 °C, and in spite of the lower moisture content of the gases, precipitation is still excellent and few difficulties are experienced [5].

Recent legislation governing the emission of acid gases and heavy metals has meant that the municipal incinerator, because of the potential high chloride and heavy metal content of the waste, is subject to severe emission regulations. Most plants are now fitted with some form of scrubber system, either ahead of, or behind, the precipitator. For those units employing either upstream spray driers or dry scrubbers, precipitation is still at an acceptable level, although because of the increased absorbent loading, the size of unit has been increased. Again the inlet conditions to the precipitator are still acceptable and no performance difficulties have been encountered.

12.6 Iron and steel works

12.6.1 Sinter plants

To improve the overall production efficiency of the blast furnace, the charge, which originally comprised loose basic raw material, coke and flux, is now partially reduced by a sintering process before feeding the furnace. In this process, the raw materials are premixed, wetted and pelletised before being fed on to a moving grate in a controlled manner. The bed moves forward to pass beneath a combustion zone, which is usually fired by a mixture of coke oven and blast furnace gas. This combustion zone serves to dry and ignite the bed which, with the high temperatures, results in some reduction of the iron ore by the coke, etc., to produce a basic sponge iron sinter product. The sinter then proceeds through a controlled cooling section of the grate to strengthen the product, prior to its discharge from the grate into a breaker and screening system. The hot screened and sized sinter then passes to air-type coolers before going to stock. The undersized particles are returned for reprocessing on the sinter plant.

Assuming the raw product emissions have been controlled, the pollution arising from the sinter plant itself can be considered as arising from two main areas. The first is that from the hot grate bed, where combustion gases are drawn through the grate under high suction, e.g. 1000 mm w.g. or more. The second is that arising from the sinter breaker, screens, coolers and general handling operations. In addition to the high suction conditions, the main strand gases are hot and comprise the combustion gases plus air, while that from the second source is predominantly atmospheric entraining air.

12.6.1.1 Main strand gases. Dust arrestment, in the case of the main strand gases, it now almost exclusively achieved with electrostatic precipitators to reach the emission levels now demanded by legislation. Originally only cyclones were installed, but even with the low dust loadings of 2 g/Nm^3, cyclone 'slip' not only resulted in severe fan wear, producing out of balance conditions on the high duty fans, but also the high density of the particulates up to 4000 kg/m^3 caused serious local fall-out difficulties.

The electrostatic precipitator used for this duty is of a fairly standard three-field configuration, except that the casing must be designed to cater for the high fan suction, e.g. 1000 mm w.g. plus possible temperatures up to 300 °C. The high ID fan suctions also impose restrictions on the hopper sealing device, since if this leaks, inleakage air can re-entrain large quantities of hopper dust which will raise the average emission levels.

The dust loading at the precipitator is typically low and 'fine', since with most sinter plant designs there is always a large 'wind box' area beneath the grate where any heavier and coarser particles inertially fall out. The quantity of dust, because the bed acts as a barrier filter, is small comprising mainly raw materials together with volatiles driven off during combustion. Some of these volatiles will condense into submicron-sized alkali fume before the precipitator, possibly resulting in corona suppression conditions which will impact on precipitator performance.

Although the gas temperatures can vary widely during 'upset' conditions, the general operation of the plant is fairly stable. Moisture levels, with the water from the mix and from combustion, gives a concentration of around

6% v/v in the waste gases, which in addition, comprise some 6% carbon dioxide, around 9% oxygen and a nitrogen balance. Provided the gas moisture is maintained at 6%, few precipitator performance problems are met under normal operation.

During short-term strand stoppages, the main ID fans should be damper isolated to maintain the relative humidity within the precipitator. If air is allowed to be drawn through the system it serves to 'dry' the material on the collector plates and electrical re-entrainment can result. This is unacceptable, since not only does this result in a short-term emission situation, as the 'dust sheds', but when the unit returns to service any dust remaining on the plates has to be reconditioned with the strand gases for the precipitator to reach optimum performance.

The size of precipitator, or design constant (EMV), is governed by the amount of lime added to the mix; this not only depends on the ore composition but whether the furnace is to operate with a basic or acid slag in the bosch (hearth). The terminology used is the sinter basicity, as the number increases above unity, i.e. to give a basic flux, the amount of lime added increases and, as some comes off the bed as CaO, the precipitator design constant must be reduced. Another aspect of adding lime is that, during processing, the lime tends to drive off alkali materials from the mix as sodium and potassium based salts, which increases the quantity of submicron fume passing through to the precipitator.

A further operating problem on installations which have attempted to use petroleum or finely ground coke, is that of carbon being pulled off the bed and passing through to the precipitator This can result in selective carbon re-entrainment, which affects the emission, or in the worst scenario, a carbon fire in the hoppers which will mechanically damage the intervals.

12.6.1.2 Sinter-plant dedusting. Although the entraining gas in this application is predominantly air, it can contain a dust loading of some 20 g/Nm^3. This dust can contain very sharp pieces of sinter, which could cause destruction of any bag filter media during normal operation or in cleaning; this possibility makes the electrostatic precipitator suitable for this application.

Originally, since the process gas is air with only ambient humidity, but heated by the various processing stages, it was considered that sinter dedusting operating conditions would make precipitation uneconomic because of the inherent high resistivity of the particulates. Pilot test investigations, however, showed that the dust could be readily precipitated. This was attributed to the high specific gravity and poor cohesivity of the dust, which when precipitated easily shed from the collectors when rapped thereby preventing reverse ionisation or back-corona conditions being established. Following these findings, many successful dedusting precipitator installations have been supplied and are the preferred method of air pollution control.

The electrostatic precipitator is of conventional design with usually three fields. Unlike the main strand operating conditions, suction at the precipitator is lower, which reduces the amount of casing stiffening required and simplifies hopper sealing requirements in terms of inleakage, although effective sealing is still mandatory to eliminate hopper dust re-entrainment. Since the gas is basically warm ambient air, the precipitator casing is generally left without thermal insulation, which significantly reduces installation costs.

If the installed precipitator should suffer from any performance shortfall, usually because of extending the extraction system to include basic raw material dedusting operations, e.g. lime into the system, then precipitability can be improved by increasing the relative humidity slightly. The minimum moisture level for good precipitation should be around 1.5% v/v, i.e. a dew point of 13 °C, so the amount of water to be evaporated to achieve this is only some 2 l/s even for large installations.

12.6.2 Pelletising plants

A few iron and steelworks employ a mixture of sinter product plus pelletised raw material as the blast furnace feed to further improve production output. The quantity of pellets used depends on the permeability of the furnace shaft; too low a permeability will limit gas throughput and, hence, iron production.

Without describing the actual pelletising plant itself, the basic principle is that the blast furnace raw feed material is mixed and formed into a 'green' pellet. This green pellet is fed into the pelletiser for drying, hardening and finally cooling. Similar to the sinter plant, the drying and hardening stage is carried out under combustion conditions and the final cooling by atmospheric air, so separate electrostatic precipitators are used and similar operational comments apply as for the sinter plant precipitators.

12.6.3 Steel making

The older form of decarburising the pig iron was to use open hearth furnaces having air as the oxygen source; later, however, oxygen enrichment of the air offered improved efficiency. The next development was to replace the open hearth furnace by a rotating vessel which was oxygen blown, e.g. the Kaldo process, where again electrostatic precipitators could effectively handle the emissions.

Most modern steel making employs direct injection of oxygen into the hot pig iron contained in a pivoting ladle, i.e. BOS converter in major steelworks, or into an electric arc furnace in the case of mini steel plants, where scrap is melted by electrical means to provide the hot metal.

12.6.3.1 BOS converter steelmaking. Carburisation is achieved by injecting oxygen into hot metal to burn off the carbon contained in the pig iron as carbon monoxide. In the earlier plants, the carbon monoxide was burnt off in the mouth of the converter giving rise to large volumes of very hot gas which had to be treated/cooled before being cleaned.

Although heat was recovered by both a water-cooled hood, extracting the waste gases over the converter mouth, and waste heat boilers, the gas was finally cooled to a temperature of between 150 and 250 °C, by means of a cooling/conditioning tower ahead of a dry precipitator, which with the well-conditioned dust operated highly successfully.

Later process developments reduced the total gas volume to be treated by using controlled combustion approaches over the mouth of the converter, where only part of the combustible gas is burnt off to form a 'dead burnt' gas zone around the main central gas core. This gas can contain up to 70% carbon monoxide and is a valuable fuel source when cleaned. The length of time during which this CO-rich gas exists is about 10 min, or some 50% of the total processing time for a typical converter operation.

Since the gas composition rapidly changes during the processing, from air through to very rich CO gas, reducing CO content and finally back to air, control of the gas flow control valves, either delivering gas to the gas holder, if BOS gas recovery is practised, or to the flare stack is critical.

Cleaning of the main converter off gases can be achieved by electrostatic precipitators, but to cater for the potential explosive conditions that can arise, the precipitator casing is designed as a cylindrical pressure vessel having a special dust evacuation system. To further minimise risks, in addition to strict control of any bypass valves, etc., the HT supply to the precipitator is closely related to the process oxygen injection timings to reduce the risk of sparking which might initiate an explosion.

12.6.3.2 Secondary pollution control. In addition to the above major pollution source, significant secondary pollution can arise during the following converter operations:

(a) initial charging of scrap into the converter
(b) charging of hot metal into the converter
(c) sampling and alloy additions to the converter
(d) pouring of the finished steel from the converter.

These operations require the convertor to be tilted from beneath the main extraction hood and the need not to impede access, for the cranes feeding the scrap or hot metal, severely limits the positioning and location of extraction hoods. Although local hoods are fitted, in spite of these operating under high suctions, much of the fume escapes into the building to create not only an unacceptable working environment, but unless captured gives rise to external pollution.

The capture of these secondary fumes is normally achieved by converting the root area immediately above the converter into a hood; the gases, usually after being mixed with the converter local hood gases, can then be

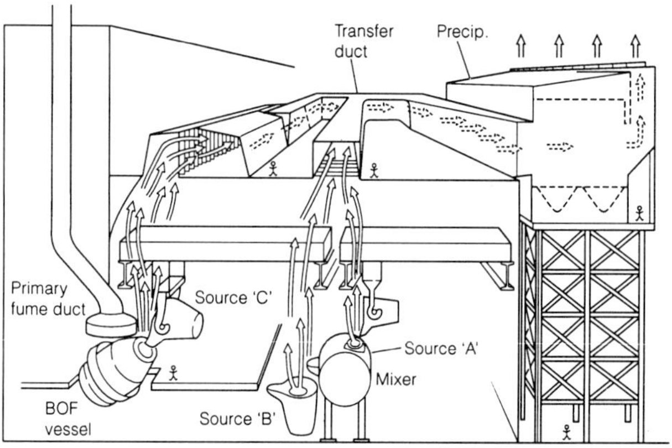

Figure 12.4 Secondary fume collection and cleaning – BOS plant.

passed through an electrostatic precipitator to remove the fine iron oxide fume. As the dust concentration is usually less that 1 g/Nm^3, the electrostatic precipitator is more economic than a bag filter, since it only has to achieve a mid-90% efficiency, and with the large extraction volumes, the fan power cost of a bag filter is extremely high. In the case of precipitators the fan power costs are minimised by mounting the precipitator at high level to use the thermal lift of the gases themselves, since the pressure loss through the precipitator is only a few mm w.g. as indicated in Figure 12.4 [6]. The local extraction hoods are handled by their own smaller fans, which assist the flow through the total plant.

12.6.3.3 Electric arc steelmaking. For mini steel plants, the hot metal is usually derived by melting scrap in an electric arc furnace; the furnace has carbon-type electrodes supplying a three-phase current or high frequency DC to the scrap causing it to melt. The main furnace gas extraction is accomplished by means of a 'fourth' or 'second' hole in the furnace roof; the gases from this point traditionally pass through a combustion area, where any volatiles or unburnt carbon monoxide are completely combusted.

This combustion process gives rise to very high temperatures, and the original gas cleaning plant, which treated only these furnace off gases, employed precipitators following cooling/conditioning towers to reduce the gas temperatures down to a suitable condition for effective precipitation.

Although this satisfied the main gas condition, it did nothing to prevent the secondary emissions which arise when the furnace roof is off for charging, alloying and pouring. To collect these fumes, a large hood is built above the crane rails, used for charging the furnace, which is operated under suction to evacuate the hood as it fills with secondary fume. As this containment volume is very much larger than the 'fourth or 'second' hole gas flow, depending on whether the furnace is AC or HFDC operation, the latest approach is to mix and treat these two gas streams together. By thorough mixing of the gas streams, some plants dispense with the combustion stage and rely on the dilution effect of the hood gases to minimise any explosion risk from volatiles or carbon monoxide, i.e. the final gas concentrations are too lean to ignite. As an alternative to the ESP, many plants now use bag filters for this combined duty.

12.6.4 Operations involving the casting of hot metal

The hot metal route following any necessary alloying, which is carried out at a small ladle station, is either to produce a cast slab or billet, or a continuous casting machine.

For mini steel plants, although most of the decarburisation is accomplished in the electric furnace, if special treatment is necessary, this is usually done in small ladles using molten steel from the arc furnace. In this situation, if a combined furnace hood and main gas extraction system is employed, it would be possible to duct the converter ladle station off gases into the electrostatic precipitator. As the ladle off gases are unlikely to have controlled combustion hooding, this approach would give a 'dead burnt' gas containing little or no carbon monoxide. A similar situation arises on major steelworks, but the volumes emanating from the ladle station normally justify their own gas cleaning plant. In this case either electrostatic precipitators or bag filters can be used, but with the higher gas temperatures and possible difficult fume, the former is the preferred system.

Where the steel is cast into slabs, prior to further operations, e.g. rolling, the slab undergoes a scarfing or scalping operation, where the surface is removed by passing the slab beneath an oxy-hydrogen burner. The temperature of the burner is sufficient to locally melt the surface and volatilise the metal; the resultant fume comprises a large quantity of iron oxide fume together with iron particles plus other tramp elements from the surface.

The machine in which the scarfing occurs is fully enclosed and placed under suction to contain the fume. This fume was originally treated, after quench scrubbing to eliminate the hot iron particles, before final cleaning in a wet precipitator. Later plants employ a dry precipitator which, although somewhat larger in size, offers the advantage that the discharge gases are hot and buoyant and so eliminates the cold steam plume from the wet precipitator.

12.7 Non-ferrous industries

There are many dry precipitator applications in the non-ferrous field; these differ from those in the iron and steel industry as a result of a generally much lower metal concentration in the ores. The ores usually require some form of pretreatment before the metals can be extracted; this can involve the use of water benification where the heavier metal-rich components settle as deposited material, and many undergo some form of chemical treatment. The following outlines some of the processes involved in the 'winning' of copper and nickel metal, which can be considered as being typical of other non-ferrous processing.

12.7.1 Copper and nickel recovery

There are numerous applications from the drying of the concentrates, through primary extraction and finally refining. While the drying and refining equipment is fairly standard, the methods of smelting differ from plant to plant.

One of the major differences is that the copper and nickel ores are mainly used as an oxide or frequently a sulphur bearing concentrate. While this is of assistance in the roasting operation, in that sulphur is exothermic when converted to sulphur dioxide, the waste gas will contain high concentrations of sulphur dioxide, e.g. 11–15%, plus high levels of free SO_3 resulting in high acid dew-point temperatures. These can well exceed 200 °C, depending on the SO_2, gas moisture and excess air available. These high acid dew-points mean that, to avoid acid deposition within the precipitator leading to adhesive dusts and corrosion, the operating temperatures are usually held in excess of 300 °C.

Except for the additional design duties for the casing, the possible fitting of controlled corona-type electrodes to eliminate corona suppression, and good rapping, the precipitator itself comprises fairly standard components when used for roasting applications.

A typical process route is (a) to produce the concentrate, (b) dry the product, (c) roasting or calcining to develop a matte and (d) converting the matte to a raw product.

12.7.1.1 Concentrate driers. Because of the high sulphur content of the concentrate, drier operation is critical, in terms not only of how the heat is applied for drying, but also the temperature to which the concentrate is exposed and the oxygen concentration of the waste gases if fires and other problems are to be avoided.

The waste gases would typically be at a temperature of around 120 °C, a moisture of 20% and a dust loading up to several hundred g/Nm^3. The main difficulty with this application can lie with the hopper dedusting operation; not only is the loading high, but the dust also has a high angle of repose. The avoidance of air ingress, which would result in a fire, must also be of concern to both the designer and user.

12.7.1.2 Roasters. In the roasting operation, the dried concentrate is allowed to fire under controlled conditions, such that the heat of reaction produces a calcined product ready for smelting. Most roasters, because of the high exit temperatures, have waste heat boilers to reduce the temperature to 350 °C. The gases then pass through conventional 'high temperature' precipitators to remove the dust, which can be up to 40 g/Nm^3, depending on the process, before it passes on to a sulphuric acid production unit.

12.7.1.3 Roasting/smelting. After drying the concentrate, the next step is to produce a calcine, either by roasting in a multihearth furnace or fluid bed system. The calcined product is then fed into a fossil fed reverberatory furnace, similar to the open hearth furnace used in steel making, where a matte containing some 50% copper is produced. To improve thermal efficiency and the dependence on fossil fuels some plants use an electric furnace, similar to the electric arc furnace used in steel making.

Whether the calcine smelting, to produce the matte, is carried out in a reverberatory or electric furnace, the waste gas moisture and SO_2 levels at 10% and 1%, respectively, are much lower than from the roaster. In this case, the waste heat boiler can be more efficient, as the exit temperature can be below 300 °C, rather than 350 °C before the precipitator as in the case of roasting.

For processing using the Outokumpo or Inco 'Flash Smelter', which basically differs only in shape, the dried concentrate and flux are injected into the furnace with the combustion air/oxygen, which then reacts with the concentrate. The heavier solids drop into the slag/matte in the base of the furnace while the waste gases pass over the surface to leave the furnace at high level. The gases leaving the flash smelter have a much lower volume, because of the more controlled combustion conditions, and contain between 50 and 150 g/Nm^3 of dust. This is carried in the gas, which has a relatively low moisture and some 10–15% SO_2. The gases are again cooled by means of waste heat boilers to 350 °C before passing through a precipitator ahead of an acid plant.

12.7.1.4 Converters. The matte is converted to the raw copper, 95% or higher, usually in a Pierce Smith Converter. This is basically a tilting furnace through which air or oxygen-enriched air plus flux is blown into the molten matte.

In converting to the basic refined metal, the molten matte undergoes various stages of refining by the oxygen. The flux, usually added with the air, is used to protect the copper and to collect the oxidised iron and other materials released from the matte. After a certain time the slag is run off by tilting the converter, and further refining occurs with possible intermediate slag run-off opertions. The 'blowing' is continued until the required purity metal has been achieved. The converter is then normally tilted backwards to run off the 'blister' copper.

Because of the various processing stages and vessel rotational requirements, the thermal properties of the waste gas are variable and while waste heat boilers are sometimes used, other plants employ cooling/conditioning towers for this duty. Again the temperature of precipitator operation is around 350 °C with the gas containing some 10 g/Nm^3 of dust. For the waste heat boiler scenario, the moisture is around ambient level, as compared with 15% when a cooling tower is employed, but in either case effective precipitation is achieved. With a sulphur dioxide concentration of up to 12%, the cleaned gas normally passes to a sulphuric acid plant.

12.7.1.5 Arsenic recovery. On plants which have ores containing high levels of arsenic, after cleaning at 350 °C, to remove the heavy metal components, the gases are further cooled to condense the vapour-phase arsenic trioxide by a cooling/conditioning tower to around 105 °C before the gas enters another dry precipitator, or a bag filter, to recover a fairly pure arsenic trioxide.

The design and operation of this section of the plant is critical because of the toxicity of the arsenic. Special procedures are necessary if the unit is to be internally examined or maintained, so absolute reliability is essential.

12.7.1.6 Further refining. If further refining is necessary, this is normally carried out in smaller units, which can be directly or indirectly fired, electric arc or induction-type furnaces. These can be individually hooded and the extracted gases passed through a dry precipitator or bag filter of conventional design, operating at a temperature of around 100 °C.

12.8 Aluminium smelting

Aluminium is normally extracted from a melt of fused sodium aluminium fluoride by electrolysis using a high current low voltage DC supply. The anodes are composed of carbon/pitch and the cathode, carbon blocks lining the furnace or cell. Pollution arises during the electrolysis, which takes place at a reasonably high temperature, and the waste gases collected from each cell contain fairly high loadings of alumina together with hydrogen fluoride. These waste gases are initially treated by injecting an absorbent into the gas stream, ahead of a dry precipitator, which not only collects the alumina, but also the absorbed HF as a sodium aluminium fluoride which is returned to the cell for reprocessing.

While this method of treatment is only on the edge of the true non-ferrous applications, it nevertheless can be widely used for other acid gas removal systems, where an absorbent is injected upstream of the precipitator, which collects the offending gas as a dry product.

12.9 Paper and pulp industry

Processing of raw timber into paper is a complex operation, involving a large number of applications to which the dry precipitator is eminently suitable. Again, without presenting all the facets of the complete process, most of the applications are related to pollution arising from combustion processes which provide steam employed in the manufacturing of the final product.

12.9.1 Bark firing

Bark from the timber used for further processing, is fired, after shredding as a 'hog' fuel, on either moving grate or fluidised bed combustors. Similar to wood chip combustors, there is a risk of burning or smouldering particles being carried forward and it is not unusual to find a cyclone upstream of the precipitator to collect this coarse material, which from a moving grate can comprise up to 70% carbon. With the fluidised bed system the improved combustion produces much less carbon so cyclones are not absolutely necessary.

The gas temperature is typically 200 °C, has a moisture level of some 17% and a dust loading of 3 g/Nm^3 from the moving grate, and 10–15 g/Nm^3 from the fluid bed units. Except for the fire risk and possible re-entrainment of carbon particles, few precipitation difficulties are encountered.

12.9.2 Cellulose pulp production

During the manufacture of pulp, the chemicals used for digesting the wood are expensive and need to be recovered on economic grounds. The chemicals, caustic soda and sodium sulphite, are used in a strong solution, referred to as cooking liquor, which, after digestion, turns black from the organics, mainly lignin, contained in the timber.

To recover these chemicals from the used liquor, it is first concentrated by evaporation to produce a water content of 20–40%. This is then sprayed into an oil-fired recovery boiler, where the resultant smelted product, mainly sodium carbonate and sulphite, collects in the bottom of the furnace and is tapped off to be ultimately re-used after treatment as the cooking liquor. Some of the sodium from the smelt combines with the fuel gas phase sulphur products to form sodium sulphate, or carbonate if there is insufficient sulphur present. Due to impurities, the waste gas will contain carbon particles, sodium chloride and potassium salts, which are collected and recycled back to the dissolving tanks.

There are two main precipitator locations, either upstream of the economiser where the temperature is some 320 °C, or downstream at a temperature of 160 °C. In both cases, the dust loading of largely volatilised salts is around 15 g/Nm^3, which means that the precipitator must be fitted with high corona emission electrodes. Although the precipitator internals are rapped with conventional impact systems, two hopper arrangements can be found: in the US the trend is to use a 'wet' bottom, where the dust falls into a bath of molten black liquor, or in Europe where wide dry scraper chain conveyors are favoured.

A further process modification can also be found, where the hot dirty flue cases are brought into contact with the black liquor in a direct contact evaporator (DCE) ahead of the precipitator, where the evaporator takes the form of a rotating tube bundle passing through the black liquor to provide an adhesive impact surface to the dust carried in the hot gases; another approach is where the black liquor is sprayed into a 'wet' cyclone, which acts as both an impact collector and spray drier. Either method produces a gas temperature of around 180 °C and a final dust loading of 10 g/Nm^3 ahead of the precipitator. Other differences to the normal approach are a higher moisture content because of evaporation and a much lower plant odour.

Good thermal insulation and sometimes preheating of the precipitator casing, if of steel construction, is necessary to minimise corrosion and sticky dust difficulties. To avoid these difficulties, particularly corrosion, some casings are constructed from lined reinforced concrete.

12.9.3 Lime sludge burning

In the black liquor recovery system, the sodium-based products from the smelt are dissolved to form a 'green' liquor, to which quick lime is added to produce the 'white' liquor used in the cooking process. The resultant calcium carbonate deposited from the 'green' liquor stage is extracted as a lime sludge or lime mud. This mud is then used as a feed material for a conventional rotating kiln for conversion back to CaO. The impurities driven off during calcination can contain a high proportion of sodium salts which are initially volatilised and then condense in the cooler feed end of the kiln. The waste gases to be treated by the precipitator are typical of a wet process cement kiln, i.e. a temperture of some 180 to 200 °C, a moisture content of 33% and a dust loading of 13 g/Nm^3. Except for the problems which may arise because of the condensed submicron sodium salts, precipittion is not difficult.

12.9.4 Magnesium sulphate burning

As an alternative to the black liquor route, some processes employ a magnesium or 'red' liquor approach to digestion. The preparation and firing of this used liquor is similar to that used for the black liquor, except that during combustion the furnace is operated to maintain a dry bottom, i.e. all products are carried forward with the waste gases. The fly ash comprises mainly magnesium oxide and some sulphate, carried in a gas stream having 30% moisture and, to ensure satisfactory precipitation, a controlled gas temperature of 160 °C. The dust collected in the hoppers is reacted with warm water to form magnesium hydroxide, which is then used to scrub the SO_2 from the gases, the resultant bisulphate being returned to the cooking liquor stage.

12.10 Conclusions

An attempt has been made in this chapter to indicate some of the applications to which dry electrostatic precipitators have been applied and to give some of the operational characteristics which can impact on the efficiency of the installed plant.

The listing above by no means covers all the applications to be found, but it is hoped that by examining those most widely used, a general impression will be gained of the sort of factors which need to be considered in the sizing, design and operation of any dry precipitator application.

References

1. Porle, K. and Karlsson, R. (1995) Long-term experience with pulsed energisation of ESPs at a Danish power station, EPRI/DOE International Conference on Managing Hazardous and Particulate Air Pollutants, May 15–17th, Toronto, Canada, in print.
2. Bicklehaupt, R.E. (1980) An Interpretation of the deteriorative performance of hot-side precipators, JACPA, **30**, 812.
3. Parker, K.R. and Cottingham, C.R. (1993) An electrostatic precipitator designed specially to collect orimulsion fly ash. Proc. 10th Particulate Control Symposium and 5th International Conference on Electrostatic Precipitation, April, Washington, USA, pp. 7.1–16, EPRI TR 103048 **2**, Palo Alto, CA, 1993.
4. Darby, K. and Parker, K.R. (1990) The use of electrostatic precipitators in the cement manufacturing industry for the control of dust emissions. Proc. 4th International Conference on Electrostatic Precipitation, September, Beijing, China, **3**, pp. 173–86, International Academic Publishers, Beijing, 1991.
5. Parker, K.R. and Russell-Jones, A. (1985) Control of Refuse Incinerator Emissions to the Atmosphere. Conference on Energy Recovery from Incineraton, 6th Feb., pp. A1–18. Inst. Mech. Engs, London, U.K.
6. Darby, K. and Parker, K.R. (1977) Problems of secondary ventilation in the iron and steel industry. Proc. 4th International Clean Air Conference, May, Tokyo, Japan, Clean Air Soc. of Japan, Tokyo, Japan.

13 The wet electrostatic precipitator: design and applications

K.R. PARKER

13.1 Introduction

Although the wet electrostatic precipitator or WESP, operates on the same basic principles as the dry type of unit, the major difference is that the charged particulates, on arriving at the collector, are removed by a flushing liquid rather than by mechanical rapping. Otherwise they both function in the same manner. The wet collector interface makes the wet precipitator ideal for the collection of particulates that are either sticky or are carried in a gas stream which is close to, or at, saturation temperature. Under these circumstances the dry form of precipitator would rapidly build-up with deposited material, which could not be removed by mechanical means, and hence its performance would deteriorate.

The history of the wet precipitator, as will be appreciated from chapter 2, dates back to the mid-1910s, when difficulties were experienced in the dry collection of various materials, either because of electrical resistivity problems, or the material to be collected was in a gaseous phase which condensed on cooling in the atmosphere to produce an undesirable plume.

Possibly the first form of wet precipitator was with a wetted collector, where water was allowed to run down the collector surface as a film, upon which the dust was deposited and removed with the water. This approach was developed to reduce the drop in precipitator performance when handling dusts having high electrical resistivity.

On some of the US metallurgical smelters, it was found that after removing the metallic solid-phase materials at relatively high temperatures, there was still an air pollution problem resulting from condensed volatile compounds, such as arsenic and cadmium. Whilst it was appreciated that the dry precipitator could collect these materials in a condensed form, the practical problems of maintaining a specific gas stream temperature, by means of water sprays, or air cooling (which further impacts on the size of the precipitator), led to the use of wet precipitators, usually following a quench tower or upstream scrubber used to cool and saturate the incoming gases.

13.2 Design considerations

Whilst the principle of operation is basically identical to the dry unit, there are significant differences in the practice of wet electrostatic precipitation which impacts on the engineering design of the plant. The following discusses the dominant aspects of these differences.

13.2.1 Dust deposition and removal

There are basically three approaches used for removing dust deposits in wet precipitation. The first, as indicated, is a film of water flowing over the surface of the collector. The second, which may be used in conjunction with the film flow approach, is simply to periodically sluice the field with water from large capacity sprays situated above the electrode system, and finally a fully irrigated plant, where a multiplicity of small medium pressure sprays, operating continuously, results in the total electrode system being wetted.

13.2.1.1 Collector film flow. The modification of the early vertical tube type units to film flow, rather then mechanical rapping, was fairly simple. As the tubes were suspended from a header plate, forming weirs at the top was not too difficult. These weirs are essential to ensure that each of the tubes is fully wetted. A typical flow rate for a 150 mm diameter tube would be in the order of 9 l/min.

To avoid the electric field pulling liquor draining from the bottom of the tubes to the discharge electrodes, the bottom of the tubes were belled outwards and cut at an angle to lead the liquor away from the discharge electrode. Had this not been done, then flashover or short-circuiting would have resulted. To further minimise this effect, the electrode emission characteristics were locally reduced by having the electrode wires ferruled at both ends to increase their diameter at the point where they pass into the collector tube. These points are illustrated in Figure 13.1a.

With the plate form of precipitator, the simple weir approach proves impractical, since not only is one attempting to uniformly cover the full collector length but also equally cover both sides. This is particularly so for large collector plates, which are used to minimise costs for large gas flow treatment. In practice, almost regardless of the design used, as one is relying on surface tension to give reasonably uniform flow, collectors having dimensions greater than 3 m in length and 6 m in height can give difficulties in maintaining complete wetting. By significantly increasing the amount of liquor flowing over the plates it is, however, possible to use plates having heights greater than 6 m; some plants in fact have plates up to 12 m.

To minimise the difficulty in maintaining complete film flow along a typical thin collector, many designs use a thicker, or double-sided collector, lined up with the water distribution arrangement along the top. A typical

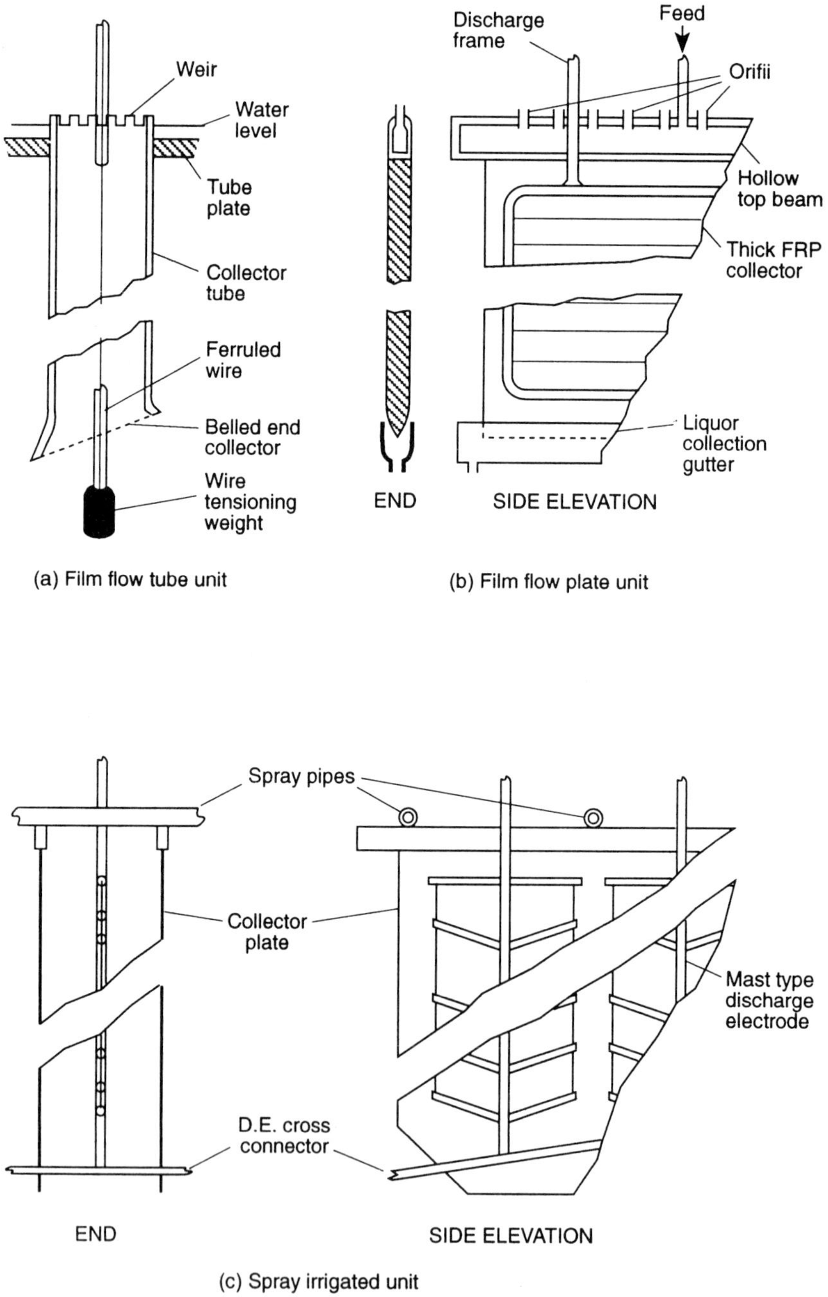

Figure 13.1 Different water washing approaches. (a) Film flow tube unit. (b) Film flow plate unit. (c) Spray-irrigated unit.

wet collector would have some form of hollow top beam which would have numerous orifice inserts in the beam, as indicated in Figure 13.1b. These inserts are adjustable in height to cater for differential water pressure along the collector length together with fabrication and erection tolerances, the height and hence flow rate being set once the plant is ready for operation.

The nozzle orifii are normally 2 to 3 mm diameter to accept recycled dirty water usage and operate with a head of around 5 mm w.g.; consequently, flow control along the collector length is difficult, not only to set up, but to maintain in practice.

A typical flow rate, for a 2.5 m long collector, would be some 40 l/min; again, with this largish water flow, drainage from the collector bottom presents difficulties if it is not to be pulled from the collector surface by the electric field. Many designs incorporate a sloping collector bottom edge which, with surface tension effects, leads the water to a safe position, i.e. away from the influence of the electric field, before discharging it into the hopper or sump.

13.2.1.2 Wash-down plant. In an attempt to reduce the problems in maintaining a continuous film of water over both faces of an entire collector, the full wash-down approach was developed. A typical application would be for a gas above saturation temperature creating dry spots which would result in operational problems.

In this design, large-capacity low-pressure sprays are mounted above the collector and discharge system, each spray having a covering capability and throughput of some 200 l/min/m^2 at a distance of 1500 mm below the spray. Having such throughout and operating at low pressure provides sufficient large coarse droplets to completely cover the system to flush off deposits from both collector and discharge electrodes. With this large quantity of liquid passing into the field it is sometimes necessary to electrically isolate the section being washed down and also, in some instances, to reduce power on the immediate downstream field. This is because liquor splashing and carryover can lead to electrical instability. Using the latest control technology, however, means that the power to the TRs can be automatically reduced so as to eliminate the need to electrically isolate the field being washed down.

Although the need to lead the liquor to a position outside the influence of the field is not as critical with this approach, as the field is not effectively energised during the wash-down period, there is a finite time to be considered before full optimum energisation can be re-established. There are many plants which have been installed using this form of wash-down, but as there is always a drop in efficiency during and immediately following a wash-down, if the unit has to meet a specific maximum emission, then this loss of performance must be allowed for in the plant sizing.

While this loss of efficiency during the wash-down period has a cost disadvantage in terms of sizing, the approach does mean that the discharge electrodes and collectors are simultaneously washed, and therefore the performance following a wash-down is optimised.

With film flow designs, unless off-line wash-down is also incorporated in the approach, there are no means of removing deposits from the discharge electrodes except during a plant outage when manual washing is possible. Deposits on the discharge electrode system can be readily identified by a gradual deterioration in performance, reducing corona current and, ultimately, increased flashover. In practice it is normal to find both forms of wash-down system being provided to cater for this event.

13.2.1.3 Full-spray irrigation. To eliminate the off-line and reduced performance scenario associated with the wash-down approach, it was found that by injecting atomised water droplets into an electric field it was possible to maintain energisation without running into a flashover situation. Investigations showed that only moderate atomisation was necessary and involute type sprays, having orifii of 2 mm diameter operating at 0.5 bar g pressure, were more than adequate. These sprays are mounted above the collectors, typically with the sprays on the duct centre line, such that both sides are wetted. Some spray droplets also reach the discharge electrodes thus helping to keep them clean.

Since the droplets are quickly precipitated onto the collectors, it is necessary to have a multiplicity of sprays along each collector to ensure full liquid coverage (see Figure 13.1c). A typical coverage rate is in the order of 4 l/min/m of collector.

Again, because of water draining continuously from both collectors and discharge electrodes it is necessary to control this flow, such that breakdown, resulting from continuous filaments of liquid being pulled by the electric field off either electrode system, is minimised. Typically, some form of 'shedder' is included in the design for this purpose (see Figure 13.1c).

13.3 Discharge electrodes

The form of discharge element found on wet precipitators, because of the relative ease by which corona normally arises, tends to be sturdier than that used in the dry precipitator design. The actual profile of discharge electrodes can be round, square or twisted square. In the case of potential corona suppression applications and those likely to suffer from high space charge effects, star or other high emission type electrodes are used (see chapter 3).

The collector height and width limitation generally means that the design of the discharge system is smaller and simpler, whether it be a weighted wire, mast or a bedstead frame type of mounting. However, like the collector, the design should allow for safe water shedding to avoid electrical breakdown problems.

In many designs, with the limitation on collector size, the top frame supporting the discharge system is of fairly light construction. With full wash-down, this offers less masking, so there is a more effective coverage. This avoids 'dry spot' formation and, hence, efficiency and reliability are improved.

13.4 HT insulators

The wet saturated gas conditions found on many wet precipitator applications exclude the use of the dirty gas insulator design used on some dry installations. With 'wet conditions', the HT insulators supporting the discharge electrode frame are normally mounted outside the gas stream in heated chests. This ensures the insulator temperature is maintained well above the water dew-point temperature of the gases. Many designs also incorporate a gas purging system to prevent process gas contacting the insulator itself.

A typical design would employ either steam or electric heating coils surrounding the lower end of the insulator in a sealed, thermally insulated compartment, mounted either above the discharge suspension system or outboard on the side of the casing, as indicated in Figure 13.2.

The insulator material is typically of high density porcelain or alumina. This would be used as either lead through bushings or 'flower pots' for inboard mounting, or as vertical support insulators for outboard designs. With the outboard design, these usually have hot gas purging to keep the process gases from entering the chest and depositing material in base or on the insulator surfaces.

As many wet precipitators are used in potentially explosive gas conditions, such as cleaning of blast furnace gases, the insulators for these applications are of the lead through type and are normally mounted inboard and above the discharge system and do not incorporate purging. In addition to being mounted vertically and above the main gas flow, to avoid deposition problems on the insulator most use an extended bus ring below the chest to extend the hot area. The use of a lead through type of design simplifies the insulator gas sealing requirements which are essential for explosive/toxic gas applications.

13.5 Casing/hopper design

With the exception of the lower bottom section, the casing design is similar to that for a dry unit. The major difference lies in the hopper area, which

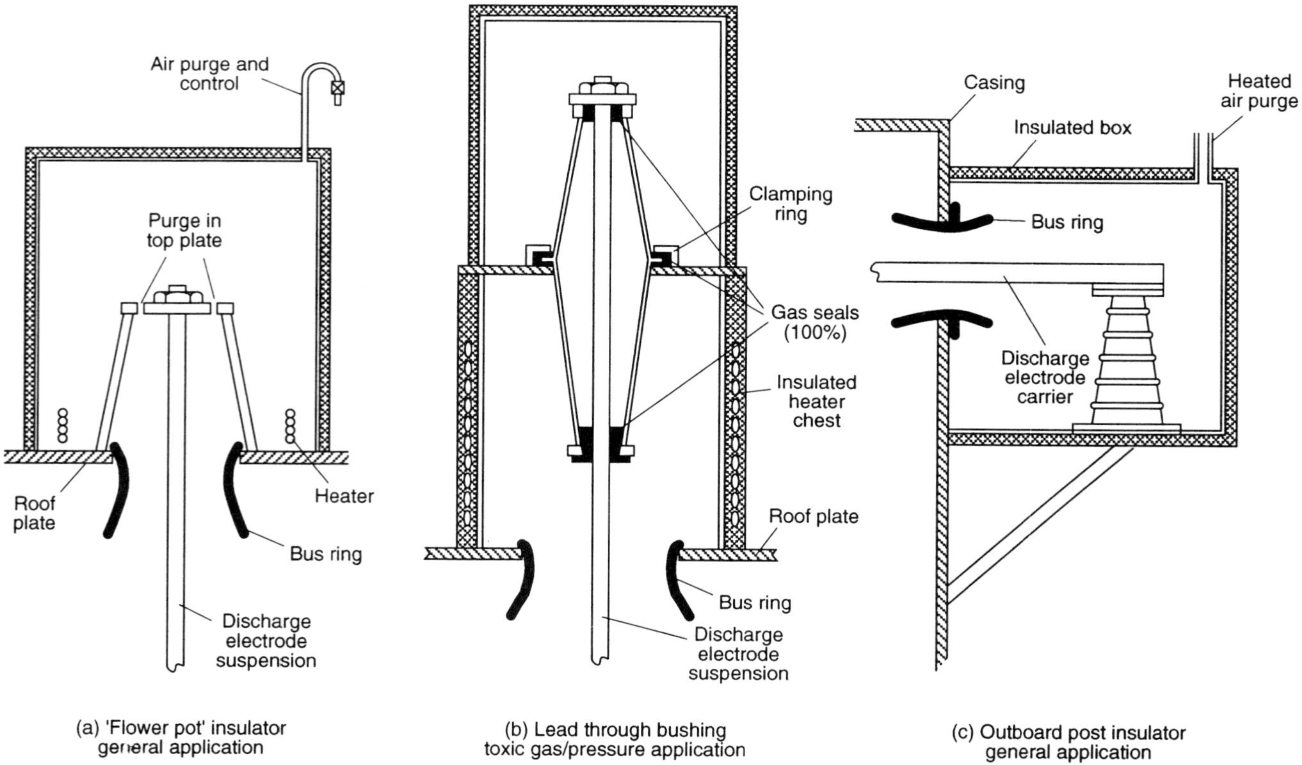

Figure 13.2 Typical discharge electrode system support insulator arrangements. (a) 'Flower pot' insulator – general application. (b) Lead through bushing – toxic gas/pressure application. (c) Outboard post insulator – general application.

usually takes the form of a shallow trough incorporating a single end discharge point. A shallow trough, as opposed to a pyramid hopper, means that the amount and weight of liquid that can be retained is limited, so dead loads are minimised.

Evacuation of the liquor can be effected by either overflow from a hydraulic leg (water seal arrangement) or by pumping between high and low level detectors in a sealed sump, which forms part of the trough. Whilst it is desirable to operate with low particulate to liquid ratios, if the particulate is of high density and likely to settle out in the trough, additional liquor is injected into the base to minimise/prevent settlement. This usually takes the form of irrigation of the trough surfaces which are most likely to suffer from sludge build-up.

With the lower temperature of operation most wet precipitators tend to have no thermal insulation on the casing, except for the most arduous ambient environments. In this case, the insulation would be to prevent freezing of the flushing liquor system or to minimise external casing corrosion.

In the case of pressure type applications, the casing, in addition to the dead loads of the internals, can be subjected to high process gas pressures.

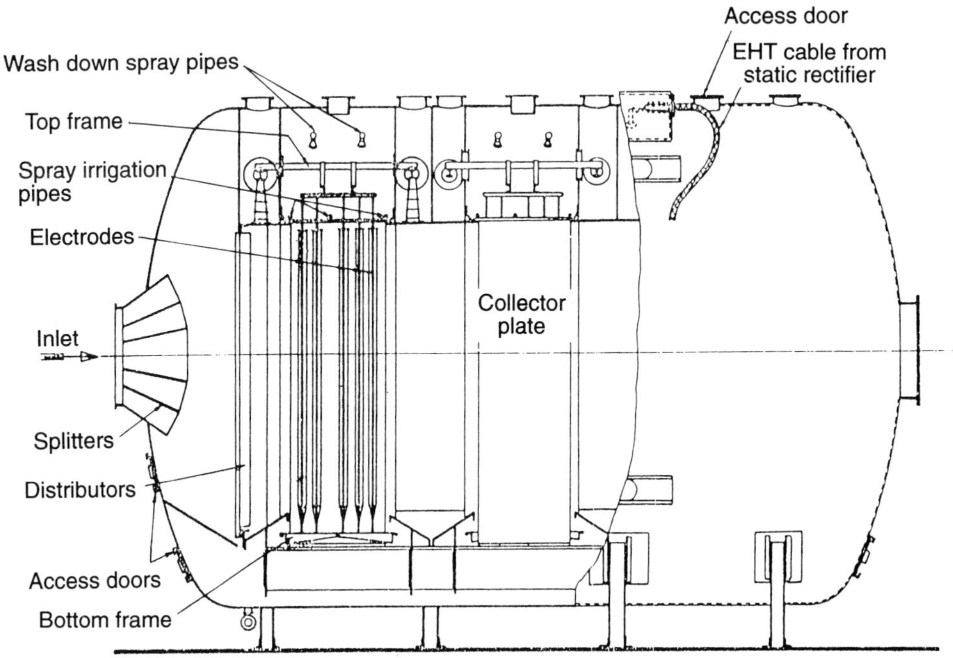

Figure 13.3 Typical electrofilter for cleaning high pressure gases from blast furnace (up to 2 atm.).

Two typical exmples of this pressure induced stress would be high pressure blast furnace gas cleaning and oxygen blown steel making processes; both can produce potentially explosive gas mixtures. For this type of application the casing takes the form of a horizontal cylindrical pressure vessel which, whilst designed to withstand the maximum required pressures at the lowest casing cost, the packing of the internals is not as economic as an ovoid (Figure 13.3) or flat-sided vessel. There is, however, a cost advantage in that the space between the outer walls of the precipitator and cylinder shell can accommodate the quench or scrubber system.

13.6 Water treatment

As considerable quantities of liquor, usually water, are involved on any wet precipitator installation, regardless of design, it is essential to recycle the liquor at rates of several thousand literes per min, to minimise operating costs. By using sprays having large orifii, the liquor only needs to be relatively clean; concentrations of up to 500 ppm of total suspended solids (TSS) can be accepted. Lower concentrations are obviously advantageous in reducing possible blockages within the system, but the use of clean water or further cleaning has a significant cost disadvantage.

Typically the liquor extracted from the sump is initially cleaned in a 'hot well'. Here the denser, coarser particles are allowed to settle and are then removed by scraper chain from the bottom of the well. From there, the water on large installations, such as blast furnace gas cleaning duties, will pass to clarifiers where the overflow concentration is reduced to below $100\,mg/m^3$ (100 ppm). Clarifiers are circular ponds of some 20 m in diameter. The underflow from the clarifier is usually treated in hydroclones, the sludge passing to a belt or vacuum filter for further treatment, while the hydroclone overflow, depdending on its cleanliness, is re-introduced back into the water circuit. A typical water treatment circuit is illustrated in Figure 13.4.

Blockages within pipework are generally the result of suspended solids. If, however, the dissolved solids/pH of the liquor results in the depositon of calcium salts, a more difficult situation will have arisen as the deposition can occur anywhere in the system. The total dissolved solids in suspension (TDS) can be controlled by reacting the circulating liquor with lime or limestone in a separate vessel to precipitate calcium salts. Once these salts have been brought out of solution they can then be removed.

The type, size and, hence, cost of the water treatment system is very site specific. For the complex system indicated, however, the cost is a very significant proportion of the complete installation cost, at times being more than 50% of the total.

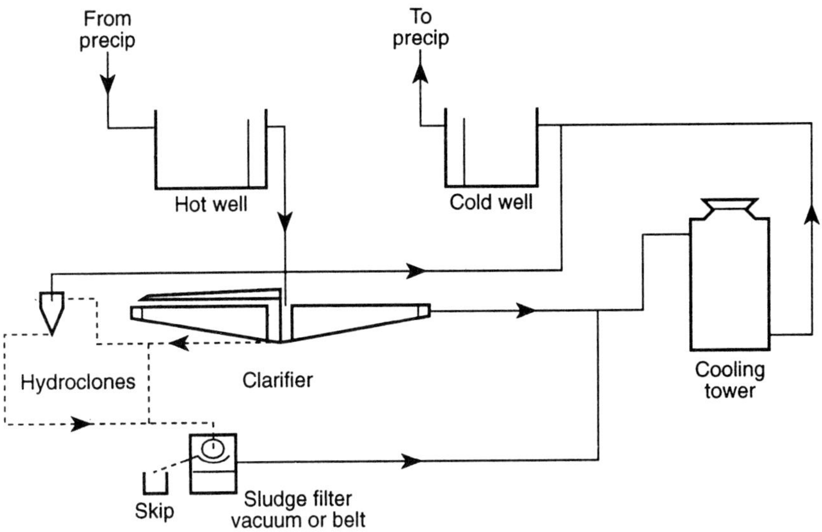

Figure 13.4 Typical water treatment plant – schematic.

13.7 Materials of construction

The majority of wet electrostatic precipitators operate at, or close to, the gas saturation temperature, often with the gas containing relatively high oxygen levels. This means that the choice of construction material is of paramount importance if the plant is to have a reasonable life without suffering from accelerated corrosion.

In applications where gases are deficient in oxygen, such as blast furnace gas, A36 carbon steel can give many years satisfactory life. However the liquor circuit may need to be treated with a biocide to eliminate any sulphur reducing bacteria. These have been known to attack carbon steel, resulting in expensive component replacement in a very short time frame.

For casings, since corrosion is normally from the inside only, it is common practice for larger plants to be of a lined/protected carbon steel construction. This protection has over the years taken many forms; examples are FRP, glass flake, chlorobutyl rubber, silcone rubber and 'wall papered' stainless steel or Hastelloy coatings. For wall papering, the base metal casing has battens of the higher grade material welded to it. On top of these battens there is a complete overlay of the higher grade material, typically 1.5 mm thick, which is welded to itself using the alloy as filler to reduce possible weld corrosion.

With FRP and rubber type coatings it is essential to ensure that the preparation of the base material is good and the adhesion sound. Any material imperfections, porosity, cracks or penetration through fabrication or site damage, will let liquor through to the base material and allow corrosion to proceed unabated. The 'wall papering' approach was developed to overcome these difficulties, particularly with material porosity and poor adhesion. If, however, the weld metal is attacked and liquor penetration occurs then the casing will be damaged, so correct weld preparation and welding techniques are essential to minimise these risks.

For precipitators where very high concentrations of, say, hydrogen chloride or fluoride are to be found in the carrier gases, reinforced concrete casings lined with acid resistant tiles and adhesives have been used to protect against corrosion. The advent of acid gas legislation has, however, eliminated the need for these special approaches, since it is now usual to remove these gases upstream of the precipitator in a separate system.

Because the weight of the internals in a large precipitator can be significant, the choice of material is critical to the procurement and fabrication cost. If one takes ordinary carbon steel as unit price, then the 316 series of stainless steel would increase the cost by a factor of 6. The higher molybdenum 317 series would have a multiplier of around 10, whilst that of the high nickel alloys, e.g. Hastelloys, would be 15 or more, depending on grade. Even higher values would result if pure metals were considered, e.g. titanium. Consequently, the overall cost of a wet precipitator can vary by an order of magnitude, dependent on the duty, the materials used, and the corrosion resistance required.

The selection of materials for the internals of a wet precipitator is a complex process. In order that pitting and crevice corrosion are avoided, consideration has to be given to the oxygen and chloride concentrations in the process gas and wash-down liquors. With respect to the liquor, it must be borne in mind that the pH is controlled by the process technology of the on-site water treatment plant.

In the context of corrosion resistance the following table shows a selection of materials and the range of chloride concentrations under which they can successfully operate. The final choice, however, may well depend on the supplier's previous experience for similar applications or data from corrosion test spools. This is always provided there has been sufficient 'up-front' time to establish the data.

Chloride concentration	Material selection
Below 14 000 ppm	316 Series stainless, e.g. 316L
Up to 14 000 ppm	317 Series stainless, e.g. 317LMN
Up to 24 000 ppm	Duplex stainless, e.g. Alloy 255
30 000 ppm and above	High Nickel Alloys, e.g. C22/C276

The procurement and fabrication costs increase, as previously indicated, the more alloyed the material. Unlike the mist type precipitator, which is dealt with in the following chapter, the wet precipitator does not handle particularly corrosive fluids, such as strong acids, and therefore alloys can be used to reduce the overall costs.

One area to be aware of in the use of alloys is in selecting the best method of fabrication. Ideally, any welding should be carried out using filler of identical composition. If composite fixings are used, they should be electrically insulated to avoid the formation of electrolytic cells and, of course, the fittings must be corrosion resistant. Although sacrificial, very heavy carbon steel structural beams are sometimes used in lieu of lighter but more expensive alloys. The down-time for replacement of the sacrificial components must be carefully assessed in any life-time costing analysis prior to making a decision on materials.

The collectors, on some irrigated plate designs, take the form of a sandwich layer of FRP approximately 75 mm thick, so that a uniform flow of flushing liquor can be obtained from the water distribution system, without the need of a double skin metallic plate format.

On intermittent wash-down plant, carbon fibre coated conductive FRP has been successful and, on fully irrigated plant, glass veiled FRP plates have been used. PVC, polypropylene and similar materials have also been applied; these, however, although showing excellent corrosion resistance, can suffer from thermal distortion and flashover-induced fires. Consequently they do not have a wide usage.

13.8 Electrical energisation

The system of energisation is similar to that employed on the dry type of plant. However, as the fields tend to be smaller, the TR set capacity tends to be lower because the total plate area energised is reduced. The 'wet operating conditions', however, produce a slghtly higher operating voltage and corona current than would be found on dry plants. For precipitators handling potentially explosive gases, the supply to the field is often *via* a HT cable, but for other wet precipitators, the feed can be either HT cable or bus duct connection.

With the intermittent full wash-down form of plant the HT and wash-down valves are typically electrically interlocked, such that during wash-down the TR is switched off or operates at a very reduced power. Power is only slowly returned to optimum after a programmed time delay, any instability arising in the adjacent fields being catered for by the TR automatic voltage control systems.

In the case of potentially explosive gases, the gas stream is monitored by an oxygen analyser which, if a certain value is exceeded, automatically switches off the TR power supplies and earths down the fields. This action prevents any residual charge on the electrode system, or electrical spark, initiating an explosion.

13.9 Typical applications of wet precipitators

In the precipitation industry, the distinction between the wet and mist precipitator application is sometimes blurred. Often the process gases are at the saturation temperature and water droplet deposition will occur in both. For the purpose of this chapter, it is assumed that wet precipitators are installed to handle gases containing solid phase or adhesive liquid particulates which require a liquid film to remove them. The mist precipitator, described in the following chapter, basically handles condensed liquid phase particulates which are self-draining from the collectors. Nothing, however, is so simple! Many mist precipitator application gases contain some solid phase material, so it is usual for the mist unit to be fitted with some form of intermittent wash-down system to periodically remove any agglomerated material.

13.9.1 Applications in the iron and steel making fields

13.9.1.1 Blast furnace gas cleaning. One of the earliest large-scale applications of wet precipitators was in the cleaning of blast furnace gases. However, with the advent of high pressure blast furnace practice, the gases can be satisfactorily cleaned by downstream high energy scrubbers; therefore the need for wet precipitators has disappeared. Although the precipitation industry responded with designs capable of handling pressures up to several bar, the economics of the scrubber for high pressure installations wins out against the precipitator, as the latter needs electrical power and obviously has a higher capital cost and space requirement.

Both ovoid and cylindrical casings can be found in the field and both have pumped liquor evacuation systems because of the pressure situation. To avoid frequently handling the large quantity of liquid associated with intermittent wash-down, the high pressure design either incorporates full-spray irrigation or film flow collector systems, backed by intermittent wash-down for very infrequent cleaning.

In the past, with low pressure applications, i.e. <1000 mm w.g., the casings were flat sided and intermittent wash-down was invariably used for cleaning. The liquor from the wash-down was received in an open sump, which acted as a hydraulic leg seal.

With gas flow rates often in excess of 100 Nm^3/s, dust loadings of 1 g/Nm^3 and emission requirements of <10 mg/Nm^3, it is normal to find an arrangement of several parallel multifield units serving a single furnace. Each field is normally supplied by its own transformer rectifier set for operational and performance implications.

The commissioning of any potentially explosive gas plant is complex and must be carried out by competent engineers. A preset programme of purging must be implicitly followed to ensure that all areas of the precipitator and associated ductwork are free of oxygen, i.e. $<0.5\%$ v/v, before electrically energising the unit. A similar situation occurs in bringing the unit off-line, when the reverse is practised. In this case, as man access usually follows, the unit must be completely purged of all process gas. With this in mind, the designer must ensure that there is no potential for trapping dangerous gas in pockets within the casing.

The need to arrange for insulators and access doors to be adequately sealed is of prime importance and the provision of explosion relief doors on the lower pressure applications must be considered. The risk on modern high pressure applications is significantly reduced and explosion relief doors are not normally supplied.

13.9.1.2 Cleaning of gases from oxygen steel making

(*A*) *Ladle processes.* Mention has already been made of the use of wet precipitators for oxygen blown ladle steel making, employing controlled or suppressed combustion. In this case the CO- and H_2-rich gases are drawn off the central section of the ladle hood and are treated in a cylindrical shelled wet precipitator usually having high energy wet scrubbers in the space between the collector side walls and the vessel shell. In addition to the 'special design', the control of the total extraction system is critical to the safe operation of such an installation. A typical BOF blowing cycle would be 3 min for scrap and hot metal charging of the ladle, 2 min reheat, 10 min oxygen lancing, 2 min for sampling/alloying, possibly a further short period of oxygen lancing and then 2 min pouring of the finished steel. During this cycle only some 8/9 min produces gas which is cleaned by the wet precipitator before passing to the gas holders. At other times the gas, after scrubbing, passes through flare stacks to the atmosphere; hence, the timing and sequence of valve operation is critical to the safe operation of the system.

(*B*) *Secondary BOS gas cleaning.* The gases arising during suppressed combustion are only partially cleaned by the high energy scrubbers during normal operation and contain up to 100 mg/Nm^3 of particulates. This gas is stored in gas holders and before further use is normally subjected to further treatment to achieve a cleanliness of better than 10 mg/Nm^3.

This cleaning is carried out by wet precipitators located downstream of the gas holder. As the pressure is less than 1000 mm w.g. the precipitator casing is normally flat sided. Even though the lower pressure simplifies sealing of all casing penetrations, the same basic design principles apply because of the potential risk of explosive gas concentrations due to possible air ingress. To minimise this, the units are fitted with oxygen monitoring systems which, in the event of detecting oxygen levels in excess of say, 1% v/v, will switch off the HT and earth down the fields to reduce any possibility of flashover triggering off an explosion.

Again because of the explosive nature of the gases, a strict purging and operating procedure has to be followed to ensure the 'safe operation' of such plant.

These units can either be of the intermittent wash-down or irrigated type. The usual inlet dust concentration is <100 mg/Nm^3, so the frequency of wash-down is very much reduced compared with the blast furnace gas cleaning plant, where inlet loadings of 1000 mg/Nm^3 are commonly encountered.

(C) Electric arc furnace gases. A number of wet precipitator installations can be found on electric arc furnace plants. These tend to treat both the primary gases leaving the furnace together with the secondary emission gases contained by hoods above the furnace. As the secondary gases form the largest gas load, any CO or H_2 generated from the smelting and not burnt off is heavily diluted; therefore, there is little risk of explosive gas mixtures being generated. This simplifies the design and construction of the precipitator in spite of the oxygen lancing situation.

13.9.1.3 Scarfing and scalping operations. To remove scale and other impurities from the surface of steel billets prior to rolling and further production operations, they are normally subjected to scarfing or scalping. In this process an oxy-hydrogen burner is used to remove the top surface of the billet. With the high water content of the combustion gases and the low temperatures created by high levels of excess air, the traditional system was to use wet precipitators. The modern approach, however, to avoid the water pollution aspect, is to use dry precipitators for this application.

With the low suctions used to capture the fume, the wet precipitator is of flat-sided construction and it is normal for these units to have intermittent wash-down. The process is such that the scarfing usually occupies a time of no more than 20 min and then several minutes elapse while changing or turning the billet. The wash-down usually takes place during this down period, so that the precipitator performance is always optimised for the commencement of the fresh billet face.

Depending on the quality and chloride content of the wash-down liquor, corrosion risks in the case of these units in spite of the high oxygen levels can be handled either with low grade stainless, or by using heavy gauge sacrificial replaceable components.

13.9.2 Applications in the chemical industries

There are numerous applications of wet precipitators to be found in the chemical industries. These units generally follow scrubbers for the polishing and removal of toxics before discharge to atmosphere. Some are used, however, for super cleaning of process gases, e.g. chlorine from electrolytic cells or ethylene from oil refinery operations.

The following indicates some of the typical applications. Whilst it is not intended to be a full listing, it serves to illustrate the more unusual applications.

13.9.2.1 Lead refinery operations. Because of the high toxicity and fine particle sizing of condensed lead fume, strict limits are laid down with regard to discharge levels and, in many operations, the wet precipitator is used to remove and polish the exit gases from scrubbers.

The size of these units tends to be fairly small and they are positioned on the downstream side of the scrubber fan. Consequently they have flat sides and, as polishers, have either one or two fields. In order to maintain optimum efficiency these units are normally spray irrigated to avoid down-time periods for cleaning.

Generally the gases and liquors can have high acidic components, such as chlorides, so the internals are normally constructed from high grade stainless, whilst the casing, if carbon steel, is coated with a rubber lining.

As the gases being treated are largely combustion gases plus air, there are no special requirements, except the normal safety precautions necessary for the operation of all precipitator installations.

13.9.2.2 Cleaning of chlorine gases from electrolytic cells. Although this was a special application, the construction details are of interest. The unit was of a vertical tube type, where the collector tubes were fabricated from hard rubber, ebonite, and the whole of the internal casing was coated with rubber. The discharge electrodes were of titanium weighted wire format. These measures were necessary to protect against chlorine gas corrosion. The collectors were film flow irrigated to remove any deposits carried over with the chlorine. With the chlorine gas, strict safety procedures were required for the safe operation of the plant.

13.9.2.3 Ethylene gas cleaning. The type of wet precipitator used for this duty is virtually identical to that used for high pressure blast furnace gas cleaning, so the plant construction details are similar to those described in 13.9.1.1.

Because the precipitator is usually located within an oil refinery, special features, particularly those associated with the electrics, have to be incorporated to protect against fire and possible explosion hazards. In all other aspects, the operation is the same as for other forms of wet precipitator handling explosive gases. The electrical safety features are fairly standard, but increase the capital costs of the installation.

13.9.3 Applications following acid gas scrubbers

With the recent interest in air toxics, wet precipitators are being applied as polishers and demisters following acid gas scrubber installations on many processes. These include the removal of SO_3 mist from power station gypsum FGD plants, ammonia slip and salts from gaseous ammonia or ammonia liquor injection systems used for DeNOx and DeSOx installations.

13.9.3.1 Power station applications. Because the gases from a gypsum FGD plant leave the scrubber saturated, the application of wet precipitation is ideal to remove trace toxic condensed vapours and SO_3 mist which pass through the scrubber system. The gas flow rates on this type of application can be up to 1000 Nm^3/s and, as such, the design and construction details are of paramount importance, not only to give performance reliability, but also to produce a long life at the lowest capital and maintenance costs. Most power plant have half lives in excess of 20 years, with only short maintenance outages every 4 years or so.

The materials of construction for this type of application tend to follow the guidelines listed in the section on materials (13.7) and are dependent on the quality of the liquor, mainly its chloride content. Ideally the wash-down should be of the fully irrigated or film flow type to limit the liquor load on the common precipitator/wet scrubber system.

On installations which fire high sulphur petroleum coke and other residues, high acid mist concentrations will be encountered giving rise to a more hazardous corrosion scenario. The construction materials on one particular installation have composite FRP film flow collectors with a Hastelloy discharge electrode system contained within a FRP solid casing. This approach, having three such series fields to achieve SO_3 emissions of less than 5 mg/Nm^3, makes such installations expensive.

13.9.3.2 Gaseous ammonia and ammonia liquor scrubber installations. In this type of installation the gas flow rates are high and, consequently, to contain capital costs, the materials used for the internals are selected depending on the quality of the liquors to be handled. These are normally high in ammonium salts, sulphites/chlorides, depending on the process being considered. A general material used is either a carbon fibre coated FRP for the collectors or, if film flow is being used, straight veiled FRP.

For the former, the electrical resistivity of the deposits of ammonium salts/water are sufficient to allow intermittent wash-down to be used for cleaning purposes. However, in both it is necessary to adopt intermittent wash-down to ensure cleanliness of the discharge system and to remove any agglomerates from the collector which could lead to flashover damage to the FRP. If the chloride levels in the wash-down liquor can be controlled, it is possible to use stainless 317L type steel for both collectors and discharge electrodes. Normally the precipitator is arranged in the circuit such that the casing can be of flat-sided construction in order to minimise cost.

With the large gas volumes being handled from, say, an iron ore sinter plant or a power station, it is usual to find the plant comprising a number of multifield parallel precipitators treating the gases from a process unit.

13.9.4 Incineration-type processes

With the growing concern regarding the emission of heavy metals and generally Pm_{10}-type particles, the processes giving rise to the highest concentrations are those associated with municipal, clinical, sewage sludge and hazardous waste incinerators. The size of these units is generally much smaller than other industrial processes, but since the feed material can contain significant quantities of heavy metal compounds, there is a growing need to 'polish' the outlet gases to comply with legislation. This is particularly so for the gases following wet scrubbers which have been traditionally used to reduce the acid gas constituents to an acceptable level. The scrubber in cooling the gases condenses the volatile heavy metal compounds which are then readily removable by a wet precipitator.

To ease the problem of cleaning the water/liquor used by the scrubbers, it is normal for the gases to be precleaned before they arrive at the scrubber. This precleaning usually implies that the gases contain very few particulates and any subsequent condensation material is too small to be removed by the installed low pressure wet scrubbers. With ever-reducing emission legislation, it is now normal practice to find wet precipitator retrofits behind scrubbers on incineration plant to reduce the heavy metal and Pm_{10} particulates to the required emission level.

These wet precipitators are fairly standard in terms of design and construction. Some, however, employ corrosion resistant type materials if the precipitator is to use the scrubber blow-down liquor for wash-down purposes, the materials of construction being dependent on the chemical analysis of the liquor. The precipitators themselves are fairly small and have either one or two fields or zones, depending on the required performance levels, which can vary from 90 to 99+% in terms of efficiency. The final liquid effluent from these precipitators, being rich in terms of heavy metals, needs careful processing to avoid increased water pollution, which in many instances is a more difficult problem than the air discharges.

13.9.5 HAC discharges

There are a number of processes, associated with oil refinery type applications, which fire waste oil products giving rise to a characteristic blue haze discharge from the stack. This discharge, although small in terms of mass concentration, is very visible and results from the condensation of hydrocarbon radicals on cooling with atmospheric air. The blue haze is synonymous with particles in the submicron range, typically 0.15 μm diameter, which can be readily collected by a standard wet precipitator.

The total gas cleaning plant normally comprises an upstream scrubber/quench system to control the acid gases, followed by a wet precipitator to control these HAC emissions.

13.9.6 Glass manufacturing

In the production of glass, the feed materials are smelted in a furnace at a temperature well in excess of 1200 °C and many impurities and alloying additions are volatilised from the furnace to be carried forward with the waste gases. During the production of leaded crystal glass, in particular, a high proportion of the added lead compounds are volatilised, and with the legislation regarding lead emissions, wet precipitators are used to polish the final discharge gases. In addition to the lead, other volatile materials, e.g. boron, are also captured by this unit.

13.9.7 Other applications

The foregoing sections only indicate some of the more usual applications and are by no means complete. It is safe to comment that ignoring cost, size for size, the wet precipitator installation is generally superior performance-wise to the dry unit. The wet precipitator has no re-entrainment problems and resistivity effects are negated by the water film so that achieving low outlet concentrations is easier. With the reduced operating temperature the electrical operation is optimised and the gas flow is lower, hence, size is minimised. Commercially, however, one must consider the higher cost of the wet plant, particularly if corrosion-resistant materials are employed, the cost of water and its clean-up, together with the more difficult disposal of a sludge rather than a dry dust.

The final choice of wet *versus* dry precipitation must be based on a careful life-time costing of both systems taking all the above into account. For some applications it is not possible to consider dry precipitation and in this case it is important to optimise the costs of the wet unit.

Even so, some other typical applications for which wet precipitators have been used, as both main cleaning and/or polishers, can be summarised as follows:

- non-ferrous smelters
- paper pulp industries
- carbon black processing
- carbon anode baking furnaces
- fibreglass curing oven discharges
- vehicular automatic spray paint booths.

14 The mist precipitator: design and applications

K.R. PARKER

14.1 Introduction

The electrostatic precipitator design used for the collection of condensed liquid particulates is similar to the wet precipitator described in chapter 13. The main difference lies in the frequency of water washing since the condensed liquid droplets are self-draining from the electrode system, and hence, the possibility of troublesome build-up is significantly reduced.

In most industrial processes which give rise to condensed liquid/mist droplets, there is invariably some solid matter which is carried over to the precipitator, and for this reason the mist precipitator is usually fitted with some form of wash-down system. Generally, the quantity of solid material is only a few percent of the total, so the wash-down is infrequent and only applied if there is evidence of unstable electrical operation, for example, a reduction in the operating voltage or corona current, arising because of deposition of solid material reducing electrode clearances.

For some applications, the carrier gases may be above or only close to the water saturation temperature, so the precipitator will be collecting undiluted product in liquid form. In others, the precipitator is preceded by a scrubber or quench system to ensure the maximum product condensation; in these circumstances, the mist unit will collect both water and condensation products together; hence, there could be a dilution effect on the collected products where the liquids are miscible.

Although the operating temperature and dew-point condition does not effect the precipitation duty, it can impact on the materials of construction; for example, in the case of acid mist precipitators installed on a contact process sulphuric acid plant, the acid mist can be oleum, SO_3 in sulphuric acid, which is a highly concentrated acid, while on other plants operating at water-dew point, the acid can be diluted because of the simultaneous collection of water droplets.

The true mist precipitator collecting condensed liquid particulates, unlike a wet precipitator used for water demisting following, for example, a gypsum scrubber on a power station, tends to be relatively small and handles gas volumes well below 100 Nm^3/s, usually 10–20 Nm^3/s. This generally means that the cost of using sophisticated alloys or pure metals for corrosion-resistant precipitator internals is more acceptable as the total weight of component material is low.

Although the mist precipitator only requires infrequent wash-down to ensure the plant is clean, there are designs, using FRP as a corrosion-resistant material, which have film flow collector plates to ensure material conductivity and also to remove any collected particulates to minimise the possibility of electrical flashover which may damage the FRP.

In practice, the actual number of true mist precipitator applications is relatively small, although, since the wet precipitator is used for demisting (water droplet carryover) on a large number of installations, the distinction between true mist precipitators and those used for water droplet collection, is often blurred, as previously mentioned.

On a number of applications the precipitator is of the vertical upward flow type to facilitate rapid draining of the collected droplets, which are free flowing at the temperature of operation. Horizontal flow plate type units are installed on other plants and these work equally well and, having horizontal flow, enable series precipitation fields to be used for increased collection efficiency, without the complicated ducting runs associated with the vertical tube approach. The final selection of upward flow, tubular or plate type, or horizontal flow plate precipitators, is a matter for the supplier and ultimate user and all types can be found in practice.

A disadvantage of the upward vertical flow is that, if more than one field is required to achieve the very low emissions now being demanded for atmospheric discharge, as against the precleaning of the process gas prior to further usage, double decking of the units or the ducting for series field connection become expensive. There is a plate form of design, having liquor flowing over a fabricated FRP thick collector, where the liquor is collected from the bottom of each collector and led away to safety. This design, as illustrated in Figure 14.1, means that it is possible to have multifield series field operation while still maintaing upward vertical flow.

With the traditional form of tubular unit there is a disadvantage in that only one side of the tube is used for collection. Although the collecting efficiency, plate area for plate area, is higher, as a result of the better electric field distribution, the overall cost tends to be greater than other forms of plant.

To retain the high strength of the cylindrical casing, synonymous with the tube-type unit, a concentric ring collector approach was developed and is widely used. In fact, for small gas flows there is no reason why this form of plant should not be used more extensively for general wet precipitator duties. In this design, the collectors take the form of accurately fabricated concentric cylinders having the discharge electrodes mounted from a concentric ring top and bottom locating frame arrangement.

An alternative to the circular tube is the hexagonal and the plate type of collector. These and the concentric ring form of collector do not have as good an electric field distribution as the tube type and hence require slightly longer contact times to achieve the same efficiency. The cross-section of

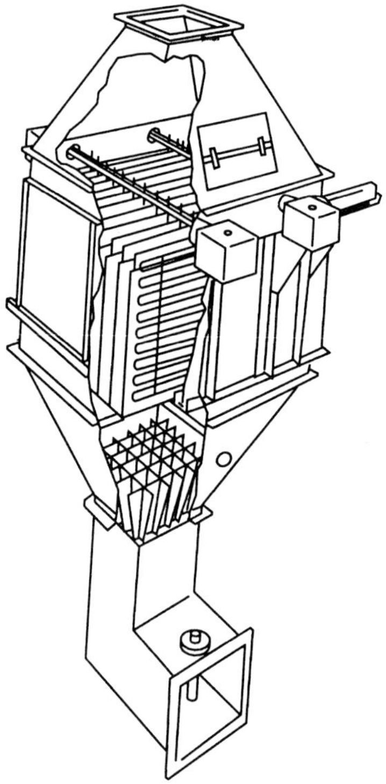

Figure 14.1 Arrangement of vertical flow precipitator (single field). US Patent No. 4,362,538 1982.

these collectors and the discharge electrode location are illustrated in Figure 14.2.

Gas distribution within the field area is important to maximise the efficiency, as shown in chapter 5. With the tube form of collector, the distribution is critical, since once the gas enters the tube it cannot redistribute itself, as can occur to some extent, in plate-type units. As most mist units have mainly horizontal single inlet and outlet duct connections, their location and orientation on the unit is important and splitters may be necessary in the outlet area to prevent preferential one-sided pulling of the bulk flow.

With the traditional tube unit, illustrated in Figure 14.3, the collectors are held between an upper header plate and a spacer at the bottom. The gas inlet connection is usually between the plates and the initial distribution is achieved by the tubes themselves, although in most instances for ultimate flow control, there is a perforated plate arrangement in the bottom tube spacer area. Not having a tube nest in either the concentric ring or

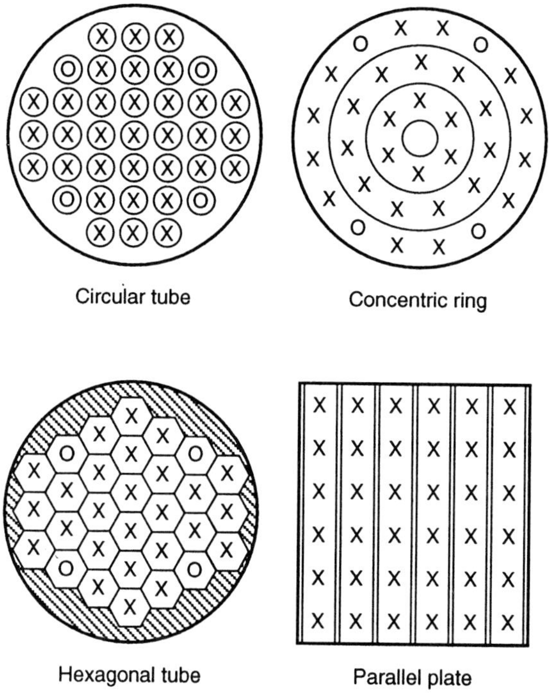

Figure 14.2 Cross-section through different collector arrangements (× electrode; ○ stabilizer).

hexagonal tube form of unit in which to initially turn and distribute the gases, the gas distribution system is more complex and is achieved by splitters, etc., in the area beneath the collectors.

One type of gas distributor is indicated in Figure 14.4; this comprises a horizontal tube running across the bottom of the unit. This tube contains slots of differing widths by which to control the gas flow. Although the mist unit is relatively small and not much time would be involved in site gas distribution testing, if the flow requires to be corrected, the materials of construction mean that modifications are complicated. For this reason a simple large-scale model can pay dividends, both in terms of cost and time, particularly, for example, if the precipitator has been in service collecting acid, where, prior to any access into the unit, it must be thoroughly washed and all acid removed and neutralised.

As previously indicated, the actual design of the mist precipitator is similar in many details to that previously outlined for the wet plant. The mist precipitator design and form will be best described by reference to the following typical applications.

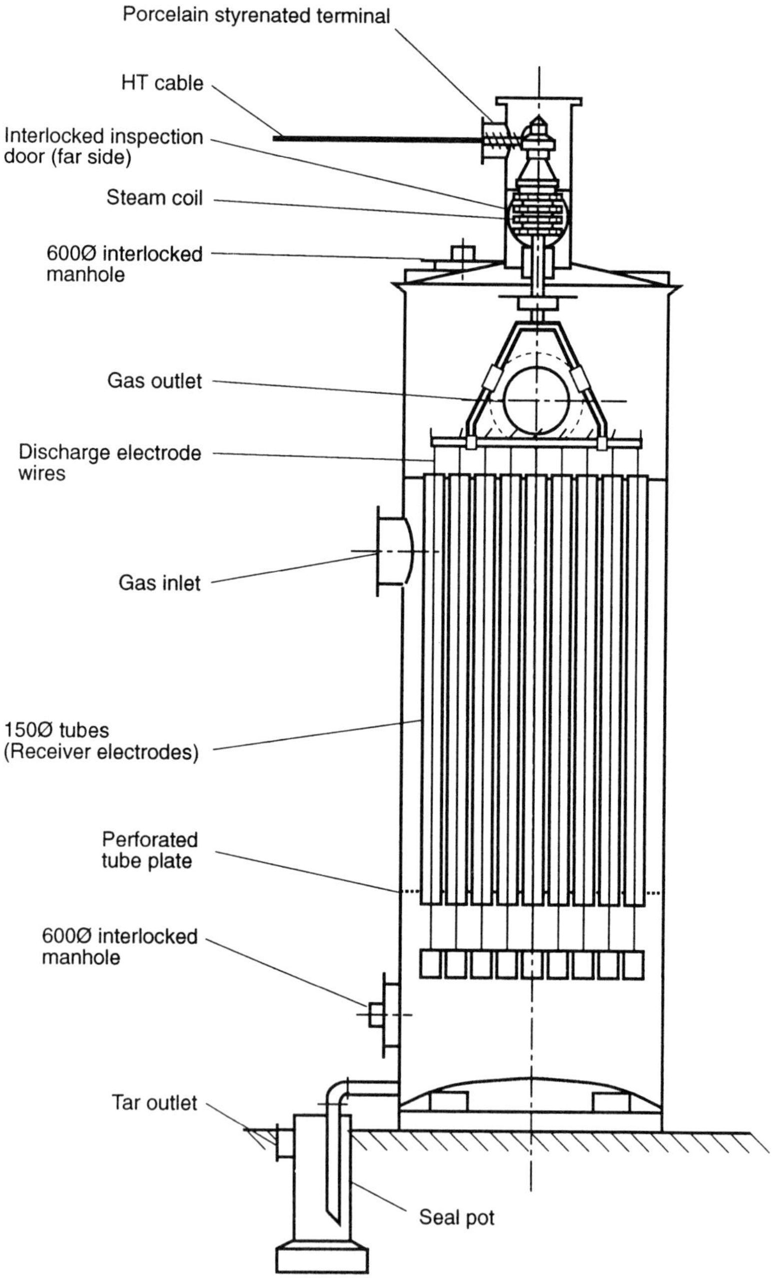

Figure 14.3 Cross-section of tube-type detarrer.

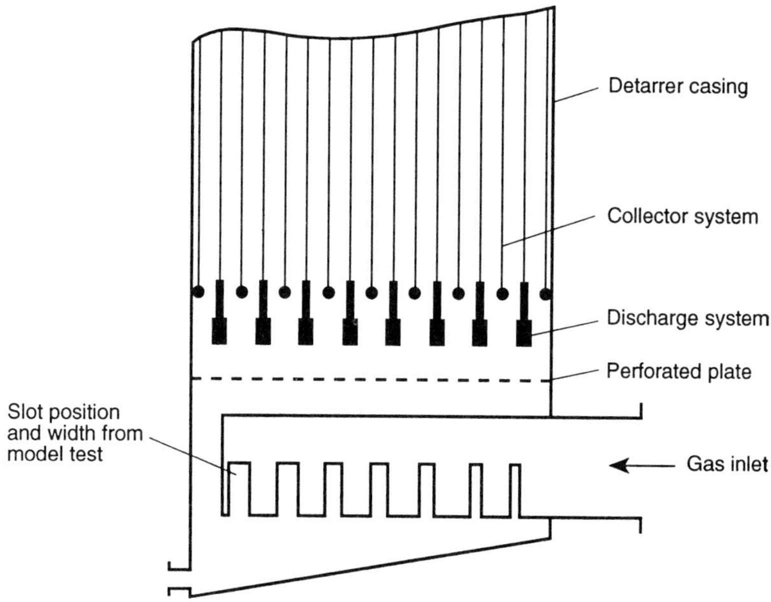

Figure 14.4 Slotted tube distribution system.

14.2 Applications of mist precipitators

14.2.1 Collection of sulphuric acid mist

The control of acid mist from metallurgical smelters was probably the first real attempt to use electrostatic precipitation industrially. The initial commercial installation, following pilot investigations by Dr F. G. Cottrell (see chapter 2) used a mist precipitator to combat atmospheric pollution from smelters around the San Francisco basin.

The mean, or d_{50}, particle sizing of sulphuric acid mist is around 0.15 μm, and although relatively soluble in dilute acid, there is the difficulty of actually contacting the droplets, even in venturi-type scrubbers. The electrostatic precipitator, as indicated earlier, is ideal for this particle size range and there are numerous mist precipitators in operation around the world successfully collecting sulphuric acid mists, with efficiencies well in excess of 99%.

In addition to metallurgical smelters, particularly those roasting sulphide ores, the mist precipitator finds widespread use in the manufacture of sulphuric acid. Depending on the process design and cleaning need, there are two situations which arise; the first is where the gases are spray cooled toward their adiabatic saturation temperature and the second where the gas temperature is further reduced by contact cooling (super cooled) to ensure maximum condensation. The boiling point of SO_3 is some 44 °C, as compared with sulphuric acid at 360 °C, but with the high saturation conditions there is always acid mist present in the cooled gas stream.

As with all mist precipitators, the material collected is usually free flowing from the collectors and both vertical and horizontal flow units can be found. The precipitated acid mist, which can vary as a collected material from dilute to very strong, means that the choice of material for design and fabrication is very important as regards the final installation costs.

To cater for the wide range of acid concentrations, antimonial lead was a traditional constructional material and is still used for the internal components. Nowadays, the problem of exposure to lead fumes arising during plant fabrication and the dearth of experienced lead burners has meant that other materials are being more extensively considered for this application. High nickel and chrome alloys, for example Hastelloys, have been used, but suffer a cost disadvantage against lead fabrication.

Casings have been constructed from lead-lined A36-type steel, acid brick construction and more recently straight FRP. Where the precipitator is installed behind a venturi scrubber, but upstream of a fan, it is imperative that if lined mild steel is used for the casing, then the lining is pressure relief vented, to prevent suction damaging or pulling off the lining. The advantage therefore rests with FRP construction but, even so, it has to be thick and robust to withstand the process suction, which can exceed 1000 mm w.g.

The electrical energisation of a mist precipitator is typically on a unit basis, where one rectifier set feeds a single field. Modern installations have full solid state controls fitted with AVC units, which can be connected to the plant control room computer, as explained in chapter 8.

The gas composition on most acid mist applications contains a relatively high moisture content plus sulphur dioxide, together with a large quantity of submicron fume. This means that it is necessary not only to use high emission type electrodes, e.g. star lead, but also a higher rated voltage rectifier unit to overcome space charge effects and provide sufficient field to enable the precipitator to perform satisfactorily. For a 250 mm diameter tube unit, having star lead discharge electrodes, a typical operating voltage would be in the order of 75 kV, giving a field of 6 kV/cm, and a current density approaching 1.0 mA/m^2 of collector surface.

The discharge suspension insulators for an acid mist type precipitator are similar to those described in chapter 13 for the wet precipitator and illustrated in Figure 14.5. The main difference lies in the need to maintain a higher temperature in the surrounding area or an increased purge air temperature because of the higher dew-point situation.

If acid should become deposited, its conductivity, even at low concentrations, and the high boiling point, usually means that the insulator will

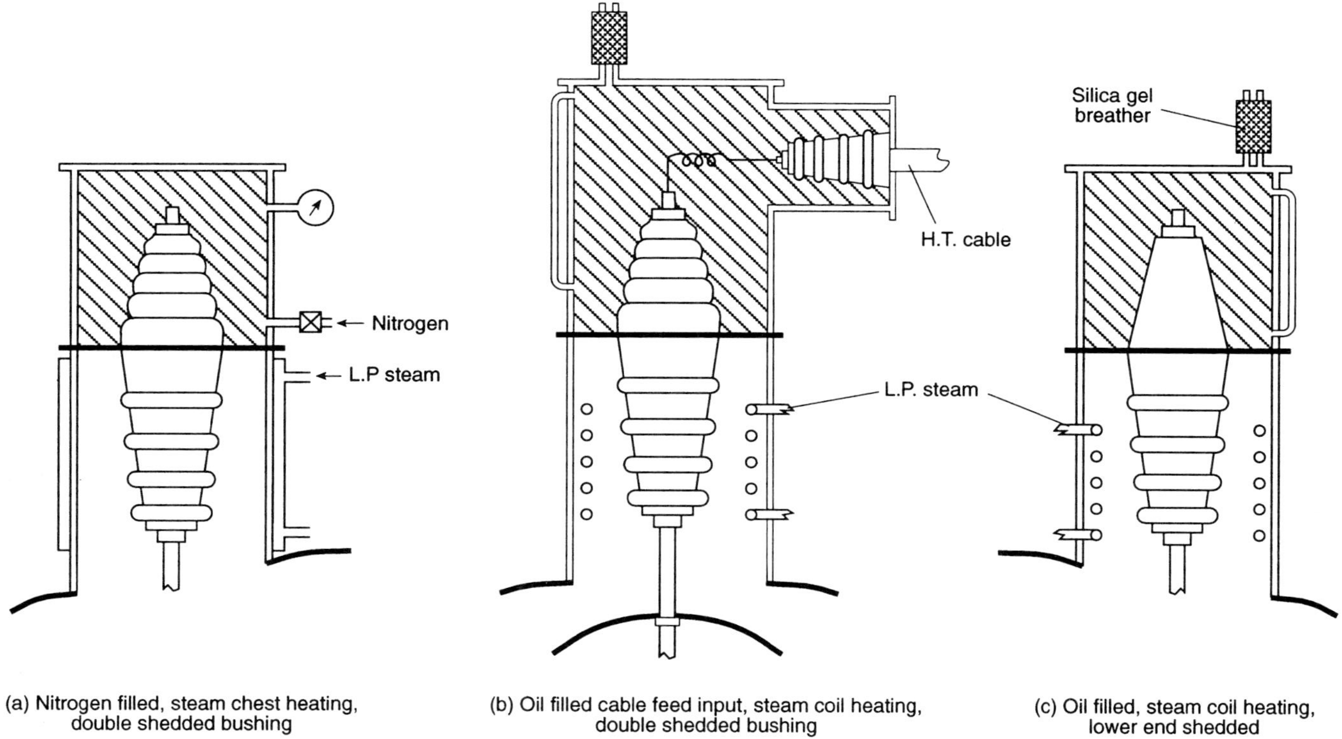

Figure 14.5 Typical suspension insulator arrangements. Note – components will vary with supplier. (a) Nitrogen filled, steam chest heating, double shedded bushing. (b) Oil filled, cable feed input, steam coil heating, double shedded bushing. (c) Oil filled, steam coil heating, lower end shedded.

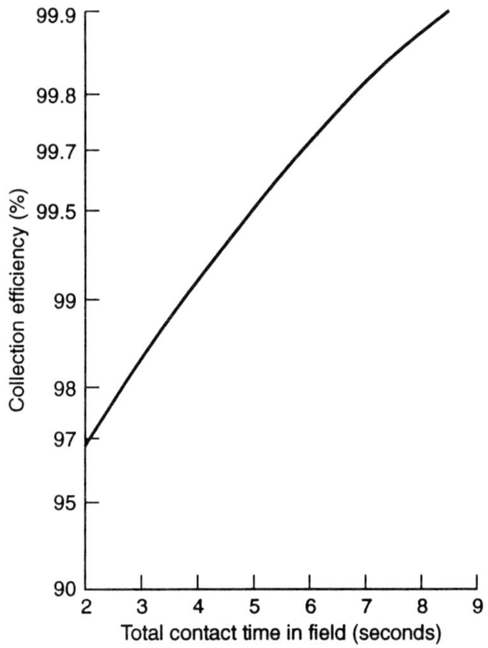

Figure 14.6 Acid mist typical design curve for tubular-type mist precipitators.

quickly fail electrically, or perhaps mechanically, as a result of power arcing if the AVC does not step back to prevent it. In either case, the performance and efficiency will fall dramatically, particularly with a single field configuration.

As indicated earlier, the collection of mists is readily precipitated and is not subject to rapping re-entrainment. This enables higher bulk velocities to be used advantageously and units with mean velocities of 2 m/s and above can still obtain efficiencies well in excess of 99%.

For tube-type units, although each manufacturer will have a preferred modular configuration, the precipitator sizing is based for convenience on a gas retention or contact time, rather than a plate area or migration velocity. A typical design curve is shown in Figure 14.6. This must be judiciously used, since each supplier will have some variation associated with his particular design.

The acid which is deposited on the collectors flows down and either drips or is led into the base of the unit, which acts as both storage and hydraulic seal. On some installations suctions in excess of 1000 mm w.g. can be experienced. The acid in the base can be used as wash-down liquor if required, or it can be pumped to another part of the installation for reprocessing.

14.2.2 Gas detarring

Coal tar is one of the byproducts resulting from the distillation of carbonaceous fuels, particularly during the production of coke used in iron making, and from synthetic gas production plants. Tar is a generic term covering a wide range of organic materials having molecular weights from 100 upwards.

These tars are initially volatilised from the coal at the high temperature of processing, and leave the coke ovens or retorts as vapour which condenses following cooling and initial scrubbing of the hot gases. Dependent on the molecular weight, the tar condenses to give particles in the size range 0.1 to 100 μm.

The tar, from a typical coke oven installation (Figure 14.7) can, depending on the coal's volatile content, reach concentrations of 15 g/Nm^3 or more and unless this can be satisfactorily removed, it causes build-up and fouling problems in downstream equipment, such as ammonia scrubbers, naphthalene recovery, gas purifying plant and the final burners themselves.

The removal of the tar, because of the wide particle size range, is difficult; high efficiency scrubbers can only achieve some 60% removal and, for this reason, electrostatic precipitators, called detarrers, are used and can readily achieve efficiencies in excess of 99%.

The position of the detarrer in a coke oven installation is normally on the downstream side of the compressor used to drive the gas through the byproduct plant itself, as indicated in Figure 14.7. Although ideally an upstream location would prevent fouling of the compressor, the risk of air ingress forming an explosive mixture within the precipitator means that the downstream location is preferred. The gas passing the compressor is normally at a high enough temperature to minimise condensation of the tar, so little difficulty is encountered by the downstream positioning.

Detarrers have been used for this particular application since the early 1930s following some years of successful operation cleaning 'Towns Gas'. In spite of its long usage, the basic design approach has not significantly changed, except for the HT equipment, which is now solid state and is fitted with an automatic voltage control unit of some form to fully optimise performance.

One of the most common forms of detarrer to be found is still the vertical tube and coaxial wire type of unit. As the size of the coke oven has increased, so has the number of tubes required for a specific contact time. As an economic alternative to a multiplicity of small units, the very much larger later designs are fitted with three suspension insulators to support the discharge electrode frame, as illustrated in Figure 14.8. Having three insulators, instead of one, not only reduces their specific loadings, but also helps stabilise the frame. This minimises any tendency for the frame to swing, which would be detrimental to optimum performance, since any electrode movement would reduce the electrical operating conditions.

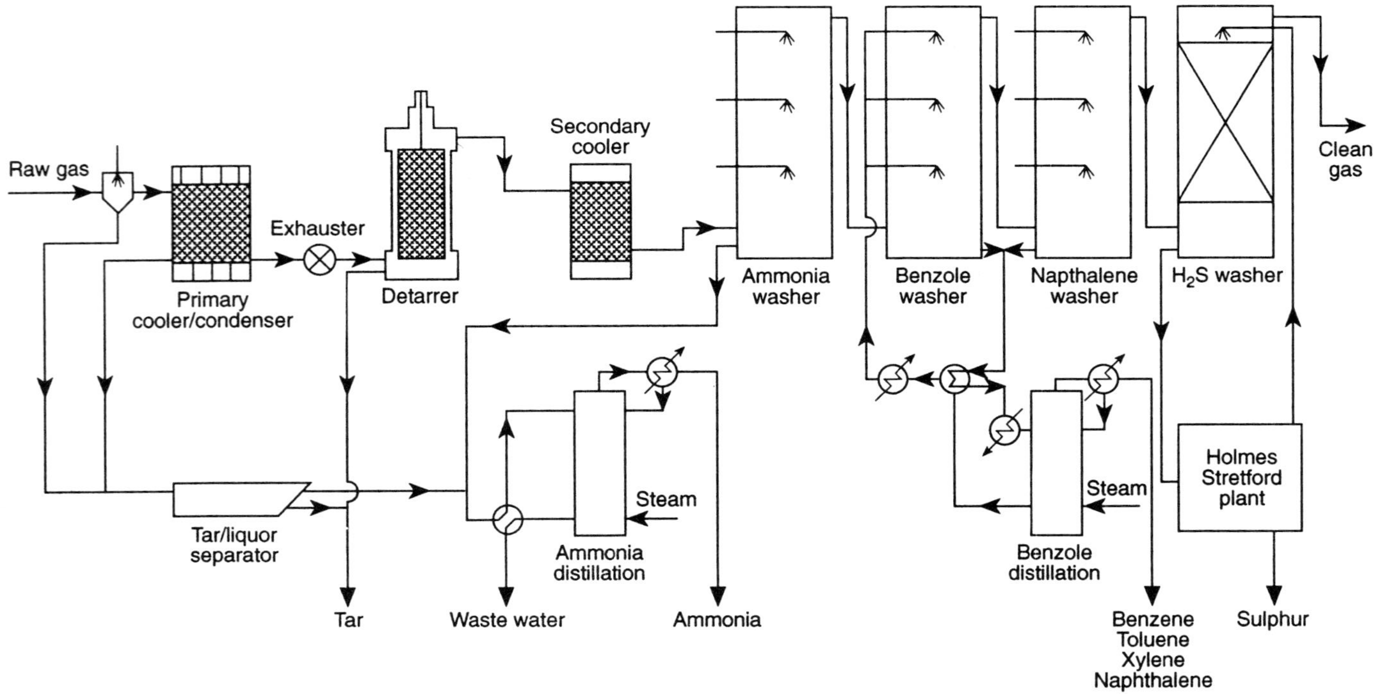

Figure 14.7 Schematic flow diagram of a coal gas processing plant (courtesy Lodge Sturtevant Ltd).

Generally, the low oxygen concentration of coke oven gas minimises corrosion risks in spite of the warm humid conditions within the detarrer itself. This means that ordinary A36, mild steel fabrication can provide an acceptable life for the unit, particularly as it is normal practice during plant start-up and purging to pass the gas through the unit for some time, so that sufficient tar is deposited on the internal faces of the unit to form a self-protective coating.

Although the risk of normal corrosion is reduced, there have been incidents of accelerated corrosion being experienced on detarrer precipitators as the result of 'sulphate reducing bacteria' being present in the wash and scrubbing liquors. It is essential therefore that the liquors be treated with biocides to eliminate this risk.

The casing is normally fabricated from 10 or 12 mm plate which provides a corrosion allowance for a long maintenance-free life. To facilitate transportation of the larger site erected installations, the casing is normally split into several sections. The operating pressure for a detarrer is fairly low at 0.25 bar g, so does not present too much of a design problem and both cylindrical and flat-sided vessels can be found.

Other mechanically loaded components, such as the top and bottom discharge electrode frames and gas distribution splitters, etc., are fabricated from 6 or 8 mm plate, and although subject to gas wetting on both faces, are thick enough so that corrosion is not a significant concern. The suspension rods carrying the discharge frame are usually 38- or 45-mm diameter rods depending on the design and supplier. To avoid tar reaching and impacting on the insulators, 'Chinamans' Hats' are usually installed on the suspension rods beneath the bus ring (Figure 14.8).

The collectors in a vertical tube detarrer, are fabricated from 3 or 4 mm plate with a top header plate 10 or 12 mm thick; this thickness plate is necessary since most of its mechanical strength has been removed because of the holes. Concentric ring or conventional horizontal-type collectors are usually fabricated from 6 or 8 mm plate, which allows sufficient flatness and mechanical rigidity to be self-supporting and the collectors themselves only need to be located in position.

The discharge electrodes in a detarrer are normally small in cross-section to minimise corona suppression effects, and are usually made from ferritic or austenitic stainless material for long life. Although ordinary mild steel can be used, these electrodes should then be considered as replaceable items. For tubular-type units the electrodes are of the weighted wire type and on plate plants the electrode can be located and tensioned in a frame.

The base of the detarrer, whether it be circular or rectangular, is normally sloping and the bottom is heated to reduce the viscosity of the tar so that it is free flowing. The tar is generally evacuated from the vessel through a hydraulic leg. Any vessel stiffeners required to withstand the design pressure would be located on the outside faces to ensure unimpeded flow of the collected products.

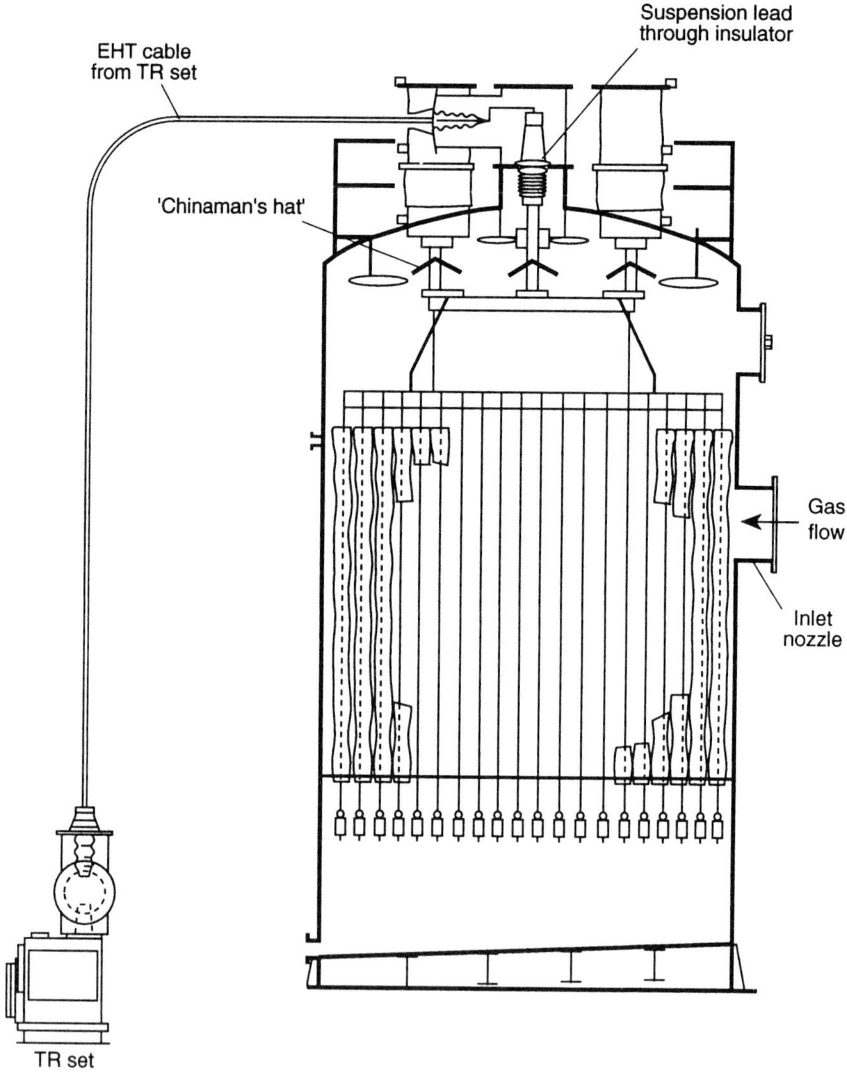

Figure 14.8 Large tube-type detarrer – 3 point suspension.

With the gas being both toxic and inflammable it is imperative that all access doors, duct connections and casing penetrations are capable of being sealed 'gas tight'. The insulators used for supporting the top discharge electrode frame are usually of the bushing or lead through type, shown in Figure 14.5. To enable the connection of the HT, one of the insulator support boxes is fitted with an oil or gas filled extension, carrying a bushing-type insulator, to which the HT cable is connected. Inert nitrogen containment of the 'none gas' side of the insulator is also possible, where a change in the nitrogen pressure would indicate that an insulator has failed.

The HT equipment for detarring is similar to that used on other forms of precipitator and generally operates at around a field strength of 4.5 kV/cm and a current density of up to 1.0 mA/m^2. Again because of the none re-entrainment scenario, gas velocities can be higher than for dry precipitators and sizing to meet a specific efficiency can be conveniently based on contact time, similar to the acid mist curve shown in Figure 14.6.

Because of the potential explosive nature of the coke oven gas when mixed with air, the start-up and shut-down of the unit is critical. A strict purging sequence, as described in chapter 11, must be carried out before energising the plant to ensure that all air has been expelled. With shut-down, a similar but opposite purging routine is necessary, to ensure there is no gas remaining inside the unit, particularly if man access is required. Any remaining tar deposits can be steamed off, or a naphtha solvent can be used to wash down the unit before air purging commences.

14.2.3 Collection of radioactive particles

In common with all operations in the nuclear industry, all processing is carried out in a fully contained negative pressure environment to ensure retention and control of all products. The gases extracted from the containment area have then to be filtered through very high efficiency HEPA filters before they can be discharged. Although the HEPA filter, being a positive membrane device, is extremely effective in retaining all particle sizes, its pressure drop increases with the material collected and, hence, has a definite useful life before it needs to be replaced to ensure that gas throughout is maintained.

The main problem facing the nuclear industry, is not so much in terms of the cost of replacing the filter, as much as what to do with the contaminated used unit. The only recognised or acceptable method is to store the exposed filter in a safe containment area which could be better utilised. To increase the useful life of the HEPA filter and to reduce handling and storage difficulties when contaminated, mist precipitators have been fitted upstream to reduce its loading by at least 99%.

In order that the precipitator functions satisfactorily, the inlet gases must be below their saturation temperature to produce liquid droplets, which then act as condensation nuclei for the radioactive particles. In the field area of the precipitator the operating corona current must be maintained high enough to neutralise and recharge any positive ions or particles so they can be precipitated in the normal manner.

The precipitator used for this application is of a fairly standard approach, but there are certain differences as a result of the nature of the application. To avoid the unwanted collection and hence build-up of radioactive material, there are no horizontal surfaces in the design, the gas typically entering from beneath the field area. The inlet duct is fitted with a water seal, not only for isolation purposes, but also to collect the precipitated products in a concentrated form, i.e. a low volume for 'safe' containment.

To ensure a long maintenance-free life, special attention is paid to the component fabrication; not only is the material corrosion-resistant, but all welding is checked for penetration and any areas where crevice corrosion can be initiated are eliminated in the design.

The HT equipment, although standard, is considered to have a definite life and to enable this to be changed or maintained, the rectifier is located outside the active area and HT cabling is used to connect the unit electrically. The discharge electrodes, because of the need to possibly neutralise positive particles, are specially designed to have a higher than normal corona emission. With essentially an air load, the electrical operating conditions are a field voltage requirement of 6 kV/cm and a corona current of 1.0 mA/m^2 of collector surface.

A number of such plants have been in service for some 25 years and have performed entirely satisfactorily, justifying their installation costs, in terms of reduced usage of HEPA filters and their ultimate storage demands.

14.2.4 Other mist precipitator applications

Although the foregoing describes some of the more usual and interesting applications, the mist precipitator can be found in the manufacture of nitric acid from the oxidation of ammonia at high pressure, and in producing phosphoric acid from phosphate rock by means of an electric furnace followed by oxidation of the phosphorous and hydration to form the acid. Both applications follow condensation/stripping vessels so the precipitator collects product plus water droplets as dilute acid phase.

Another application is in the collection of oil mists produced during operations, such as rolling or spinning, or in the production of glass fibre blocks used for thermal insulation, where oil and phenolic binder fumes are evaporated from the furnace during the final processing. The alternative form of two-stage precipitator, as basically used for 'air cleaning', is often found on this type of application; however, as the two-stage precipitator is outside the scope of this publication, these processes will not be reviewed.

14.3 Conclusions

Compared with other forms of device that might be used for the collection of submicron particles the mist precipitator offers the following advantages:

(1) the ability to handle a wide range of inlet conditions, in terms of temperature, particulate loading and particle size
(2) capable of meeting very high efficiencies
(3) low operating costs for energisation and pumping
(4) low pressure loss < 25 mm w.g.
(5) fully automatic in operation
(6) minimum maintenance requirement.

Finally the use of modern engineering materials in the design and construction ensures a long and trouble-free life.

15 Upgrading of existing precipitator efficiencies

K.R. PARKER and H. KRIGMONT

15A Modifications/changes to existing plant

With the ever-increasing legislative pressures to reduce discharges from all sources, the owner/operator of a specific process plant fitted with an electrostatic precipitator, or any form of particulate collection system, is faced with the dilemma as to how to improve the performance of his particular plant to comply with these changes of emission, without necessarily completely replacing the collector and with the minimum down-time or interruption in process production, while changes are made to the existing installation.

In addition to improvements required for legislative measures, there are other instances needing precipitator efficiency enhancement, e.g. a change in production rate leading to higher particulate carryover, higher temperatures, gas flow rates, plus uncontrolled air inleakage producing severe volume overloading, an alteration in the fuel supplied to a pulverised fuel installation, or the existing plant is not in a serviceable condition as a result of corrosion and mechanical damage.

The approach one must adopt, regardless of the need for improvement, is similar in all instances. Firstly, one must evaluate the current performance of the existing unit, to establish the precipitability of the particulates, or migration velocity, and the current operating volumes, temperatures, dust loadings, etc. These would be determined from a test or series of tests, which should follow a planned outage, when not only should the precipitator be brought back into a good mechanical and electrical condition (see chapter 11) but also the gas distribution should be corrected to the standards outlined in chapter 5. It is assumed that the precipitator electrics will have been modified earlier, or during the present outage, to include one of the latest forms of automatic voltage control, so that the performance is fully optimised electrically and also rapping re-entrainment has been minimised prior to the tests.

Assuming the existing precipitator has been optimised as above, the next step is to determine the actual collection efficiency, using one of the methods given in chapter 10, under what, hopefully, will be representative of the future operating conditions. These tests, in addition to determining inlet and outlet particulate concentrations, must also include full gas velocity, temperature and analytical surveys. The sampling of the gases for dust concentration determinations, should also extract sufficient sample to enable complete chemical and particle size determinations to be made. In the case of boiler plant, one should also arrange for the fuel to be analysed and a record compiled of the actual boiler conditions under which the tests were performed. On other process plants a log of the plant operating conditions should be made for future reference.

15A.1 Assessment of required performance improvement

The information from these tests will not only provide the hopefully updated plant operating conditions, but will enable the precipitability of the particulates to be determined in the form of a migration velocity/modified migration velocity. This will allow a curve to be drawn of efficiency versus specific collecting area (SCA), similar to that illustrated in Figure 15A.1, from which the SCA required for the new efficiency can be established, as will be illustrated in the following procedure.

After establishing the new/required SCA, the total plate area or contact time can be calculated for the upgraded plant to meet the required efficiency. From this, one can decide the economics of providing a complete precipitator of the correct size or adding the additional required surface in series or parallel configuration. Whether the extension is connected in a series, or in parallel, to the existing installation depends on various factors relating to the site and plant operating requirements.

In order to obtain a full understanding of the procedure, the following example illustrates how the performance data are treated for a boiler plant precipitator which already meets the UK Large Combustion Plant Directive, i.e. an emission of 50 mg/Nm^3 dry at 6% O_2, but for economic reasons, the client wishes to increase the generated output by some 8%. Although in

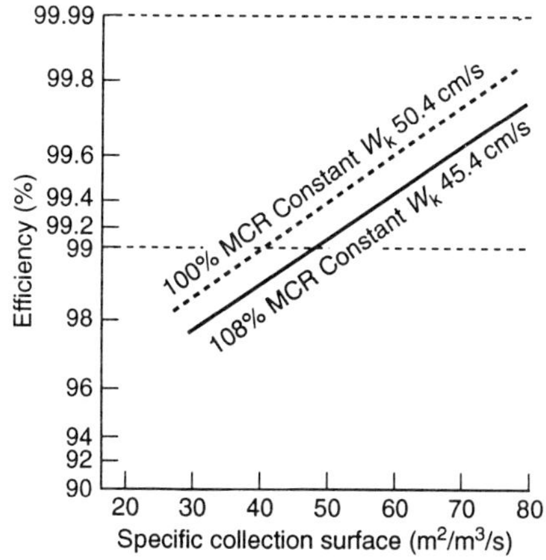

Figure 15A.1 Precipitator performance line. Constant w_k *versus* specific collecting area.

the example the existing precipitator meets the Large Combustion Plant Directive, the same procedure and methodology would be adopted for any precipitator requiring upgrading for any reason.

For this worked example, it is assumed that the existing fuel supplies will be maintained over the next 5 years or so, having the following typical analysis.

Moisture	9.0%
Ash	20.0%
Sulphur in coal	1.42%
Na_2O in ash	0.42%
Vol Matter	24.0%
Cal Value	24.0 mJ/kg

The existing precipitator configuration is as follows: two parallel flows, with conventional straight inlet and outlet duct connections. Each flow has three series fields, containing collector plates of 5000 × 11 000 mm, and each field has 40 ducts at 300-mm centres. The total plate area as installed is 26 400 m^2.

For this application, two series of efficiency tests have been carried out, the first to re-establish the design performance and the second to determine the precipitator operating conditions under the higher plant output.

Table 15A.1 Measured test data

	100% MCR rating	MCR + 8% rating
Gas flow (am^3/s)	400	440
Gas temp. (°C)	140	150
Duct pressure. mm w.g. (suction)	300	350
Oxygen at inlet (% *d* v/v)	5	5
Moisture at inlet (% *d* v/v)	8	8
Inlet dust conc.[a]	16.0	18.0
Emission[a]	50	97
Efficiency (%)	99.67	99.46
Gas velocity (m/s)	1.51	1.67
Contact time (s)	9.9	9.0
Deutsch EMV (cm/s)	8.74	8.70
Modified Deutsch EMV (cm/s)	50.41	45.43
SCA (m^2/m^3/s)	66.0	60.0

[a]Inlet dust loading and emission corrected to mg/Nm^3 *d* 6% O_2.

From the derived modified Deutsch EMVs, one can evaluate the anticipated efficiency for only two of the three fields in service. This gives the following figures: MCR operation with two fields only, efficiency 99.1%, and for the increased duty, 98.6%. These values have been used to construct the curves shown in Figure 15A.1, from which the SCA required for the new efficiency of 99.72% can be read off using the derived modified Deutsch number of 45.43 cm/s. A reason for the slightly lower figure under the proposed future operating conditions is attributed to the higher operating temperature at the increased boiler loading.

15A.2 Performance improvement modification options

Applying the modified Deutsch number, 45.43 cm/s, and new efficiency of 99.72%, directly into the Matts–Ohnfeldt equation [1] produces an SCA of 76.1 $m^2/m^3/s$. This gives a total plate area of 33 487 m^2, i.e. an increase over the existing area of 7087 m^2, based on a collector separation of 300 mm.

As a series field extension to the existing precipitator, the additional field length required would be 4000 mm. The precipitator would then be 19 000 mm overall and have a contact time of 11.4 s. Although not essential if the extension is directly coupled and maintains the same collector height and spacing, one can assume that the gas distribution will remain satisfactory and the general outline of the precipitator will be unaltered.

Although changing the internals to fit higher collectors is probably not justified in this case, because of the excellent test results, the required total plate area could be achieved by removing the roof and fitting collectors having a height of 13 950 mm. This would reduce the gas velocity to 1.31 m/s and raise the contact time to 11.4 s. If such an approach was to be considered, it might be more economic to install only 30 ducts per flow at 400-mm centres, while maintaining the 11.4 s contact time, but in either case, one must reconsider the gas distribution to fully utilise the taller plant internals.

The additional plate area could take the form of a parallel unit, which keeping the same size of collector and number of fields would have 22 ducts at 300-mm centres, or 17 ducts at 400-mm centres. To meet the overall efficiency/emission, the flow through the existing unit would have to be reduced to 347 am^3/s, which would reduce the gas velocity through the existing unit to 1.31 m/s, with the remainder passing through the parallel unit again with a contact time of 11.4 s.

A different approach might be to reduce the flow through the existing precipitator to some 347 am^3/s and treat the remainder through a parallel bag filter. Assuming a filtering rate of say 1.4 m/min on a gas flow of 93 am^3/s would imply a filter media area of some 4000 m^2. This could be arranged in four sections and pulse jet cleaned. Because of the increased pressure drop across the bag filter of say 200 mm w.g. total, it would require an addition fan of 250 kVA rating to pass the gas through the bag filter and return the cleaned gas to the upstream side of the main plant ID fans.

In practice, as the bag filter should give a lower emission than 50 mg/Nm^3 dry at 6% O_2, the gas volume passing through the bag filter may be reduced with cost savings in terms of fan power, and the existing precipitator allowed to accept a higher gas volume than 347 am^3/s, provided the total emission remains within design 50 mg/Nm^3 limit.

While one might consider replacing the internals by a bag filter, which is unlikely because of the satisfactory performance of the existing precipitator, on this unit the available area of the top plates would be insufficient to support/accommodate around 19 000 m^2 of filter material (assuming a filtering rate of 1.4 m/min), and therefore in this instance such an approach would be totally uneconomic.

After completing the foregoing analysis and deciding the various options, one then needs to assess the most economic approach to modify the plant to comply with legislation. The economics of such an appraisal is, however, very site specific, so will not be further explored. There are, however, various scenarios which impact on the economics and need to be considered.

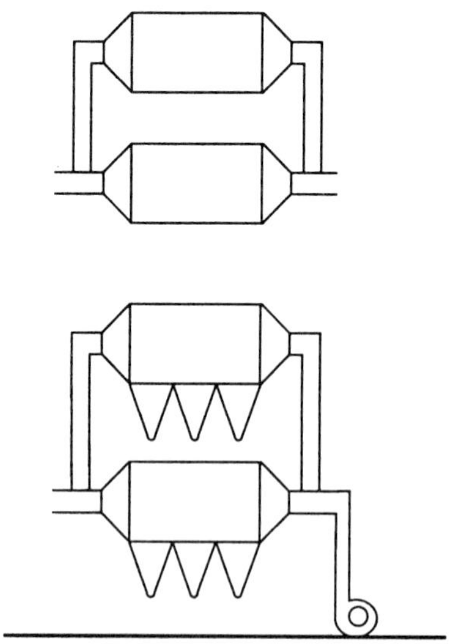

Figure 15A.2 Parallel units.

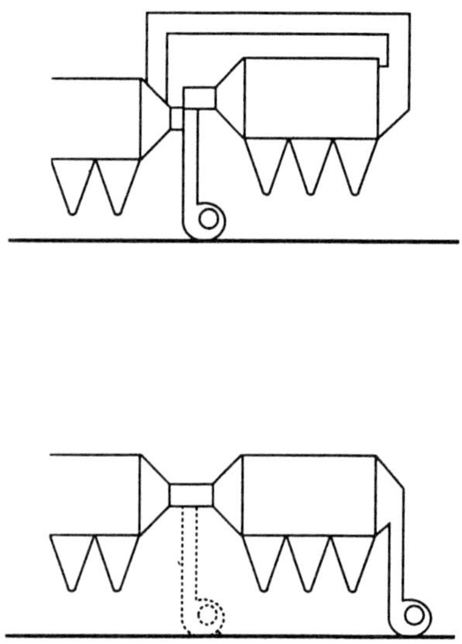

Figure 15A.3 Series units.

There are advantages with the parallel arrangement (see Figure 15A.2) in that the final operating gas velocity through the existing unit is reduced and the new precipitator, or bag filter, can normally be fully installed without the main plant being off-line, except for the final break-in, so process production is maintained.

The series connection (Figure 15A.3), on the other hand, offers an in-line arrangement, so there is less problem in balancing gas flows and possibly simpler connection to the existing dedusting system, but its main problem lies in the down-time for building the extension in most instances, unless it is erected as a separate downstream unit, when the only outage necessary is for the final duct connections to be made.

15A.3 Alternative solutions

15A.3.1 Electrical

Reference has previously been made in chapter 8, dealing with electrical operation of precipitators, as to the effect of applying intermittent energisation or pulse charging to precipitators handling 'difficult' fly ashes. In these situations the alternative energisation serves to modify the particle charging to minimise the impact of reverse or back-corona.

For intermittent energisation, as most modern TR controller equipment incorporates this facility, or can be easily modified, the application of this approach can be most beneficial in improving the performance under difficult conditions; again whether or not the final emission will satisfy present regulations depends on the precipitator design and operating conditions.

Pulse energisation can have remarkable effects on precipitator improvements [2]; however, not only is the equipment expensive to install, but the design of the precipitator and operating conditions may limit its general application. For an operating plant, it would be pertinent to install the equipment as a single field trial, rather than commit the system on a permanent basis, to establish if the expenditure on pulse charging equipment will bring about the required degree of improvement for this particular application and duty.

15A.3.2 Mechanical changes

For some designs of precipitator installed prior to 1970, it has been feasible to increase the SCA by up to 15%, but generally 10%, by fitting extended collector plates and discharge systems into the inter-field area. On others, a larger SCA increase has been possible by lifting the roof and fitting taller collector plates within the existing casing. These approaches were normally

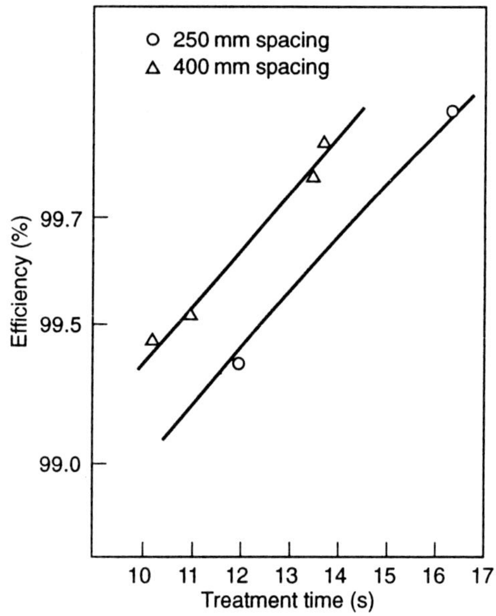

Figure 15A.4 Improvements in performance as a result of change to 400 mm collector spacing.

associated with a planned internal replacement programme, during which the latest precipitator improvements, such as gas distribution, rapping and the latest TR controls, should also be included. Although successful in improving the overall performance, whether or not the final plant will meet the required emission level is dependent on whether the total plate area is sufficient to meet the uprated efficiency.

As both the collector plate height increase or lengthening implies a complete internal replacement, it may be more economic to consider fitting out the plant with a collector spacing of 400 mm and save on component costs. The change to 400 mm spacing needs careful consideration of clearances outside as well as within the field area. Some plants have experienced voltage limitations in adopting the wide collector spacing approach, because of insufficient electrical clearance in bus ducts, bus rings, support insulators and discharge electrode rapping shaft areas, all creating lower than optimum voltage operation. Where the application has been successfully carried out significant improvements in performance have been achieved as illustrated in Figure 15A.4, from Darby [3], when comparing before and after effects for the same contact time as a result of the general improvements in electric field and current distribution.

15B Precipitator improvements achieved by changing the electrical resistivity of the particulates

15B.1 Change of temperature/relative humidity of the gases

The use of water injection ahead of dry precipitators in the cement, metallurgical and other process fields has been mentioned in chapter 12. The injection is not only to reduce the temperature and volume of the gases, but more importantly to increase the relative humidity of the gases, such that the electrical resistivity of the particles is reduced, as illustrated in Figure 6.10. Theory, coupled with operational experience, has shown that generally the higher the relative humidity, the better the performance of the precipitator. Generally the quantity of water that can be evaporated is limited by the incoming temperature, liquid droplet size and available contact time.

In the collection of particles and fume from a comprehensive sinter strand dedusting system, precipitator improvements were obtained by increasing the gas moisture level by only 1.2% v/v. The emissions decreased from 70 mg/Nm^3 down to 40 mg/Nm^3. The increase in moisture, on what is basically a low temperature/entrained air application, was achieved by spraying water into the hot discharge duct leading to the sinter breaker at the discharge end of the strand. The effect of the added moisture was to increase both the operating voltage and current into the precipitator, which resulted in the efficiency increasing from 99.63% to 99.79%. In this case, the performance change was achieved without significantly reducing the gas volume or temperature of the incoming gases.

As an alternative to water injection, to increase the relative humidity, the same effect can be achieved by cooling the gases by means of an upstream heat exchanger. The effect of temperature on electrostatic precipitator performance is, as previously illustrated in Figure 12.1. One advantage of such an approach is that the gas volume is automatically reduced serving to increase the specific collection area (SCA).

Investigations have been carried out in Japan, where the gases immediately upstream of a pilot precipitator have been further cooled by means of a water/gas heat exchanger, rather than a conventional gas/gas exchanger, to limit the amount of air inleakage (usually around 10% for a conventional gas/gas exchanger) [4]. The application can be limited in its general application, however, by the potential re-entrainment of low resistivity particles and hopper dust handling problems, plus possible acid dew-point difficulties unless associated with an acid gas absorption system.

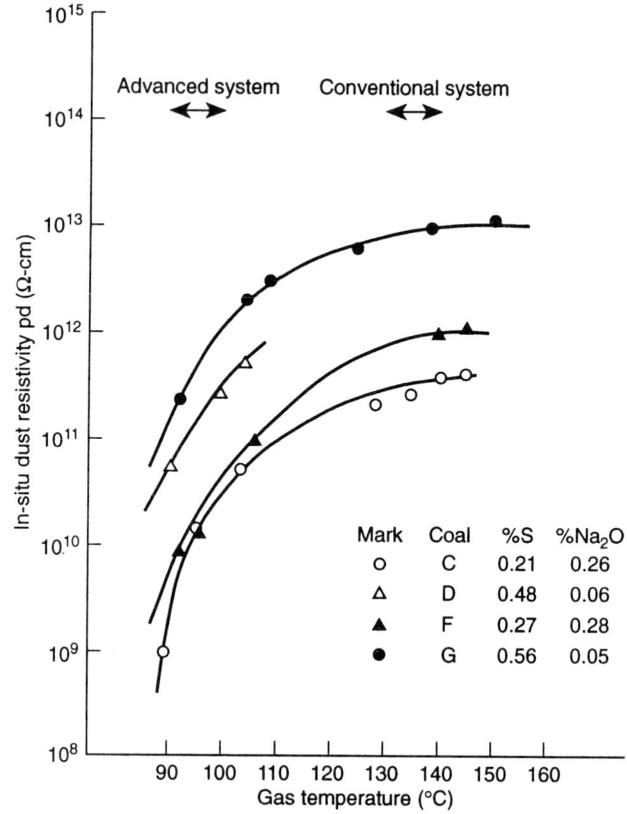

Figure 15B.1 *In-situ* dust resistivity of ESP inlet fly ash.

The results of this investigation, carried out on very difficult imported fuels, has shown some significant performance improvements when operating at 90 °C, compared with the conventional temperature of 130 °C. Figure 15B.1, from Tanaka *et al.* [4], indicates that the electrical resistivity of the particles is lowered by a factor of two by operating at the reduced temperature for the particular coals investigated.

15B.2 Flue gas additives to improve performance

For boiler plants, which fire low sulphur fuels, not only is it possible to increase the particulate collection efficiency by the use of conditioning agents, such as sulphur trioxide gas injection, but by the use of intermittent pulse mode electrical energisation plus SO_3, it may be possible to further

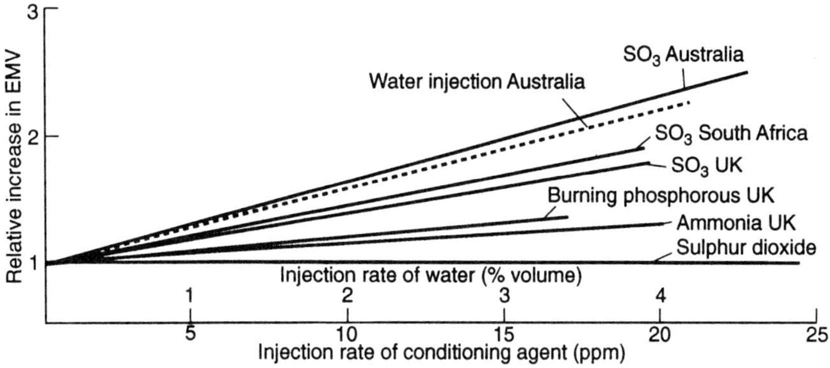

Figure 15B.2 Relative increase in migration velocity as a result of conditioning.

improve the performance of an existing precipitator. In this situation a planned testing programme could identify this possibility. The effect of SO_3 gas conditioning with high sulphur coals would not be expected to result in a significant performance improvement, regardless of the method of energisation; in fact additional sulphur trioxide injection may bring about an increase in opacity due to excess SO_3 passing through the precipitators, to condense as H_2SO_4 in the cooler back end regions of the plant.

The application of flue gas conditioning systems is not new and was recognised and used in the early days of precipitation. The modern approach dates back to the early 1960s to improve the collection efficiency of precipitators handling fly ash arising from low sulphur fuels [5]. The initial installations were pseudo pilot plant evaporating stabilised liquid sulphur trioxide and injecting it with completely dry air, having a dew-point of $-40\,°C$ to prevent the formation of sulphuric acid in the injection pipework.

These conditioning plants, although performing reasonably well, were not acceptable commercial installations, since they required unacceptable levels of maintenance. The results of the conditioning were remarkable, however, as illustrated in Figure 15B.2 from Busby and Darby [5]. These were obtained from full-scale testing of power plant precipitators in several countries, all firing low sulphur coals producing various degrees of back-ionisation which limited precipitator performance to an unacceptable level. At this stage there was less pressure on environmental control and the conditioning units were allowed to fall into disrepair through lack of necessary maintenance.

It was not until the early 1970s when various countries enacted legislation controlling emissions, that flue gas conditioning was 'rediscovered' and led to the development of conditioning plants, based on those used for sulphuric acid manufacture but on a much smaller scale, and these have now become part of the overall plant air pollution control system in meeting legislation. These conditioning plants are now used worldwide for improving precipitator performance when handling low sulphur coals giving rise to high resistivity fly ash.

In addition to sulphur trioxide flue gas conditioning, some plants need to employ dual conditioning of sulphur trioxide plus gaseous ammonia to not only reduce resistivity but also to increase particle cohesivity when handling fly ash comprising mainly 90% plus silica and alumina compounds. Such fly ash was found to be subject to severe re-entrainment losses, which could be minimised by the addition of gaseous ammonia to produce ammonium salts, which increased the general cohesivity of the collected material.

With the more recent demand for NOx reduction, units fitted with low NOx burners have experienced high carbon carryover from the furnace; these carbon particles, particularly if in the form of 'coke' having a low resistivity but large surface area, tend to be easily re-entrained and impact on the overall emission from the precipitator. In these cases, the use of ammonia gas conditioning has been used to increase the particle cohesivity and thereby reduce re-entrainment such that the emissions now comply with legislation.

Because flue gas conditioning is now very much associated with electrostatic precipitation engineering, whether it be used to improve the performance of existing units, or initially installed to reduce the size and capital cost of the plant by having a smaller precipitator; the methodology is important and the following section of this chapter is devoted to flue gas conditioning in its entirety. It has been prepared by Dr Henry Krigmont, who spent many years with one of the main conditioning plant suppliers and has tackled the subject, considering both the theory and the practical side of the equipment, plus how the injection rates can be optimised to minimise operational costs.

References (15A and 15B)

1. Matts, S. and Ohnfeldt, P.O. (1963–1964) Efficient gas cleaning with SF electrostatic precipitators. *Flakt Rev.*, **6,7**, 105–22.
2. Parker, K.R. (1987) The impact of pulsing on electrostatic precipitator performance. *Inst. of Physics Symposium on Electrostatics*, Oxford, Institute of Physics, London, UK.
3. Darby, K. and Novogaratz, D. (1990) Increased plate spacing in electrostatic precipitators. *8th EPRI/EP Symposium on the Transfer and Utilisation of Particulate Control Technology*, San Diego, USA, Session 6A EPRI, Palo Alto, CA, USA.
4. Tanaka, T., Fujishima, H. and Tsuchiya, Y. (1993) Development of advanced dust collecting system for coal-fired power plant. *5th International Conference of the Electrostatic Precipitation Society*, Washington, USA, April, pp. 35.1–17, EPRI TR 103048 **2**, Palo Alto, CA, USA.
5. Busby, H.G.T. and Darby, K. (1963) Efficiency of electrostatic precipitators as affected by the properties and combustion of coal. *J. Inst. Fuel*, May, **36**, 184–97.

15C Theory, principles of operation, equipment and applications of flue gas conditioning

15C.1 Introduction

Acceptance of the process called flue gas conditioning (FGC) as an engineering solution to a common environmental problem has come slowly. To start with, the process is mis-named. It is not the gas which is conditioned, but the particles of dust in the gas or, in the most common application, the fly ash. Adjusting the resistivity of fly ash particles by application of a dilute acid solution to their surfaces is an easily understood and straightforward task. The very idea that the behavior of a gas stream might be adjusted in any useful way by the addition of a few parts per million of anything boggles the mind.

Some of the earliest applications of the electrostatic precipitation process were in the collection of dust and fume from non-ferrous smelters in the western part of the United States and Canada [1]. In 1912 it was found that dust from the copper-converter furnaces at Garfield, Utah, was easily collected when sulfitic ores were processed, but that an insulating layer of dust formed on the collecting electrodes and the collection was effectively shut-off when the sulfur content of the ores was low. It was reasoned that the absence of a minimum quantity of sulfur trioxide in the flue gases made the dust non-conductive, causing the 'shut-off' effect. This theory was subsequently verified by evaporating sulfuric acid in the high-temperature gases and observing that normal precipitator operation was restored. This appears to be the earliest documented recognition of the importance of dust resistivity in ESP operation, as well as the first use of flue gas conditioning. It is interesting that control of resistivity by sulfur trioxide was also recognized at this time.

Electrical resistivity is one of the critical parameters influencing fly ash collection by electrostatic precipitators. The electrical resistivity of fly ash depends on the chemical composition of the ash, the constituents of flue gases, and the temperature. Fly ash composition is largely determined by the type and composition of the coal being burned, and the furnace operating conditions. At lower temperatures, fly ash resistivity is determined by electrical conduction over the surface of the particles. The latter is produced by the movement of ions in molecular thickness coatings on the particles, and is termed surface conductivity. High temperature resistivity, when plotted against inverse absolute temperature, is a straight line, illustrating conduction through the bodies of the particles, called 'bulk' or volume conductivity.

Surface conductivity is dependent on interaction between the flue gas and the ash. Environmental factors include temperature and the concentration of gaseous and condensed phases in contact with ash. Flue gas temperature influences the concentration of water vapor, the existence of a condensed phase, and the reactivity between the ash and the environment. The surface conductivity is also sensitive to the chemical composition of the fly ash, because alkali metals serve as charge carriers. Alkali metal ions have been identified as the positive charge carriers. Lithium and sodium are particularly effective. However, sodium is the principal migrating ion due to its larger concentration. Although the mechanism of charge transfer is nearly the same for surface and volume conductivity, the former dominates at low temperatures because of acid and moisture adsorption, while the latter dominates at high temperatures. Conduction may be considered as taking place through two parallel paths, one dependent on the volume conductivity and the other on surface conductivity.

15C.2 Electrical resistivity

The resistivity of the collected particulate layer has a significant importance on the operation of electrostatic precipitators. The amount of useful electrical power that can be supplied to a precipitator is limited for the most part by the electrical resistivity of the fly ash layer of the collecting plates.

Corona current from the discharge electrodes must pass through the collected dust layer on the plates to reach grounded collecting plate surfaces. Passage of the current builds up a voltage across a dust layer in accordance with Ohm's law. Theory and experience indicate that when the dust resistivity exceeds a critical value of about mid-10^{10} Ω-cm, corona current is limited by electrical breakdown in the collected dust layer. This, in turn, limits useful operating voltage and reduces precipitator efficiency. The loss in performance increases quite rapidly for resistivity greater than 10^{10} Ω-cm, and resistivity is therefore a major factor in precipitator design [2].

The following is the generally accepted dust resistivity classification in Ω-cm:

10^4–10^8	Conductive
10^8–10^{10}	Normal
10^{10}–10^{11}	Moderate
10^{11}–10^{13}	High

There are two conduction mechanisms which determine the resistivity of the collected dust layer. These mechanisms are termed volume and surface conduction. The former relates to the bulk composition of the material,

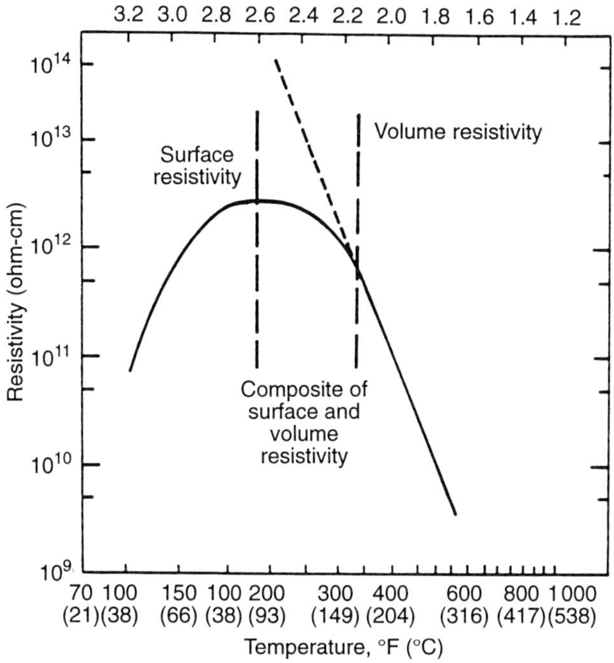

Figure 15C.1 Typical temperature–resistivity relationship for fly ash.

while the latter depends primarily on a film adsorbed on the particle surface and is related to the composition of the gas and dust surface (Figure 15C.1).

Fly ash resistivity depends primarily on the chemical composition of the ash, the ambient flue gas temperature, the water vapor and sulfur trioxide content of the flue gas. At air preheater outlet gas temperatures common for cold-side utility applications 120–175 °C (248–347 °F), surface conduction over the fly ash particles is the prevailing mode and the conductivity depends mainly on the amounts of SO_3 and water vapor adsorbed on the particles. At higher gas temperatures ahead of the air preheater, so-called hot-side installations 316–427 °C (600–800 °F), volume conduction through the particles prevails.

In volume conduction, electric charge is transferred through the bulk material that comprises the collected dust layer. In most materials volume conduction occurs by means of electron carriers within the materials and is dependent upon the thermal excitation of the electrons in the molecular structure of the materials.

The voltage drop across the precipitated dust layer is dependent upon the corona current density, the electrical resistivity of the dust, and the thickness of the dust deposit. For high resistivity dusts, the voltage drop across the collected dust layer can be significant. Obviously, the electrical energization equipment must be capable of providing sufficient voltage to accommodate this voltage drop, while maintaining adequate voltage across the inter-electrode spaces.

Another important consideration of the effects of dust on corona generation concerns the deposits formed on both collection and discharge electrodes. On the collection electrode, dust deposits alter the electric field and sparking conditions as a result of the voltage drop within the dust layer. This effect limits the voltage and current at which the precipitator can operate.

The effect of the resistivity of the collected dust layer or performance, however, is more severe than its influence on power supply voltage. The electric field in the dust layer can be quite high for high resistivity dust leading to sparkover of the precipitator at lower applied voltage and current density. If dust resistivity is further increased, a diffuse corona glow may appear over large areas of the dust surface. Under these conditions the positive ion production by the reverse corona may be sufficient to completely disrupt the charging process, and effective precipitation is not possible under these conditions.

Dust deposits also form on the discharge electrodes of operating precipitators. These deposits can be quite heavy in the case of some types of dust. The effect on corona can be considerable depending on the nature of the deposit and the electrical properties of the dust [4]. If the dust resistivity is reasonably low, the effect of the deposits will be to effectively increase the diameter of the discharge electrode. This results in higher voltage required for corona initiation or reduced corona current for a given voltage. If the dust deposits are uneven, an uneven distribution of the corona along the length of the wire may result. If the dust resistivity is high, the effect generally would be to reduce corona current for a given voltage. However, if the deposits are somewhat porous, breakdown of the gas in the interstitial region can occur and the effect of the deposits may be reduced.

15C.2.1 Prediction of fly ash resistivity

It is frequently necessary to estimate the resistivity to be expected from fly ash from coals which may not be available for such measurements until the ESP and/or FGC system have been sized, designed and installed. The US EPA sponsored a development by Dr Roy E. Bickelhaupt at the Southern Research Institute (SoRI) to produce a method for fly ash resistivity prediction using detailed analysis of the coal and ash [5]. Resistivities calculated by this method are in good agreement with actual field measurements, and form the basis for estimation of ESP performance. Figure 15C.2 illustrates a typical result of this procedure, showing predicted resistivity as a function of temperature.

The right-hand portion of the trace in Figure 15C.2 is essentially a straight line descending from left to right, and represents the effect of volume

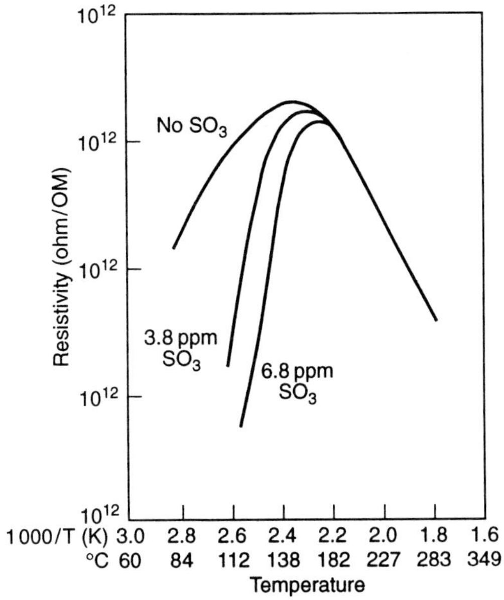

Figure 15C.2 Typical fly ash resistivity curves.

conduction through the body of the fly ash material. Three lines are shown which descend from right to left. The left-most line shows the effect of water vapor present in the flue gas, depicted as though no SO_3 were present. The inner lines illustrate the added effect of SO_3 for two different concentrations. It is seen that the chemical composition of the ash and the water vapor content of the flue gas act to determine an upper boundary for the resistivity values, and the SO_3 quantities present reduce the resistivity below the boundary values. Thus, if the resistivity, resulting from the combination of flue gas and ash properties with the amount of SO_3 naturally produced by combustion, is higher than desired or best ESP operation at a particular temperature, a suitable lower value can be obtained by the introduction of additional SO_3.

15C.2.2 Resistivity effects in ESPs

Back-corona is one of the major problems that leads to non-ideal behavior in electrostatic precipitators. Back-corona describes the condition where a local electrical breakdown is taking place in the high resistivity dust layer. This condition initiates at the top surface of the dust layer and leads to the formation of negative ions that are driven to the positive collection elec-

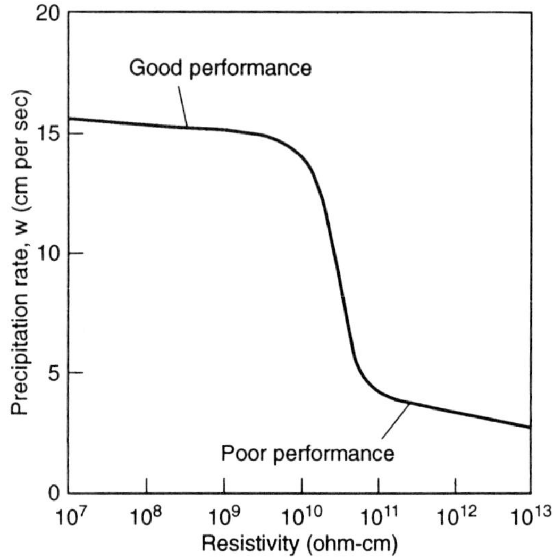

Figure 15C.3 Precipitation rate (migration velocity) for laboratory pilot ESP.

trode. A corona here generates positive ions that are projected into the interelectrode space and driven toward the corona electrode.

As positive ions flow into the interelectrode space, they encounter negatively charged particles and negative ions. The electric field from the charged particle exceeds that from an ion at reasonable distances; therefore, the majority of the positive ions will flow to the negatively charged dust particles, neutralizing their charge. This causes a proportionate reduction in the electrical force acting to collect these particles. Additionally, back-corona disrupts the space charge that contributes to the electric field adjacent to the collection electrode.

Figure 15C.3 represents the effect of resistivity on dust collection for a small laboratory precipitator [6]. The vertical scale of this figure displays the migration velocity, a measure of the rate of dust collection, and the horizontal scale shows the apparent bulk resistivity of the collected dust. Because of the gas, dust, and electrical conditions are relatively uniform throughout such a unit; the transition from high collection rates on the low resistivity side of the chart to significantly lower collection rates at high resistivity giving the abrupt change between the two operating states becomes readily evident.

A full-scale precipitator may be considered to be made up of a number of laboratory-scale units arranged in series and parallel. The exact operating

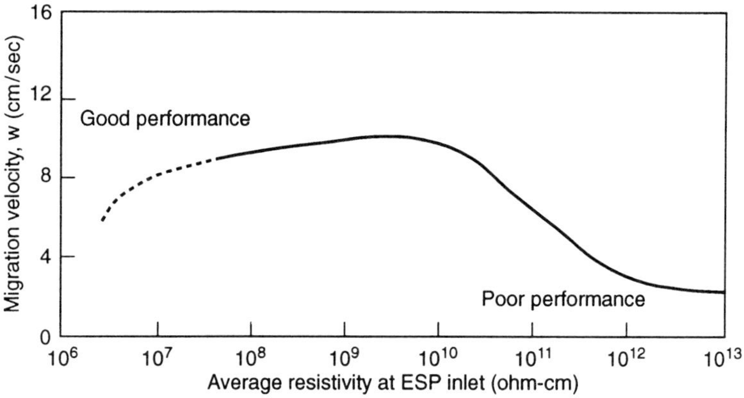

Figure 15C.4 Typical resistivity effect. Full-scale ESP.

conditions in each of these hypothetical small portions differs from the others so that the drop of performance, as the average resistivity of the dust at the precipitator inlet increases, occurs at different times, giving a less abrupt transition from good to poor performance as shown by Figure 15C.4 [7]. It is clear that obtaining the maximum dust collection rate from the ESP requires that the dust resistivity fall in an optimum range, and it is also apparent that improvement in performance is available from proper adjustment of the resistivity if the unit is operating at resistivity levels above the optimum range.

Figure 15C.4 also indicates that reduced dust collection rates are attained at low values of resistivity. The electrical forces holding the dust to the collecting plates decreases as resistivity is reduced and this causes increased losses due to rapping and re-entrainment by the gas stream.

The presence of carbon particles (very low resistivity) in the fly ash, in amounts greater than 10%, usually creates excessive re-entrainment and particle losses (Figure 15C.5) [8]. Those losses are predominantly due to a so-called 'dancing particles' effect caused by low electrical resistivity (less than $10^4\ \Omega$-cm), where, as the particles approach the collector they are repelled back into the gas stream to be recharged and the process repeats. As they reach the outlet they can escape as uncharged particles with the outlet gases.

On the other hand, when the carbon particles are retained in the dust layer (with the aid of dual flue gas conditioning, which will be discussed later), the resultant effect is similar to lowering the dust cake resistivity. The carbon acts by a physical admixture effect in which the current is carried by the carbonaceous particles in the matrix of the fly ash and unburnt carbon as a whole.

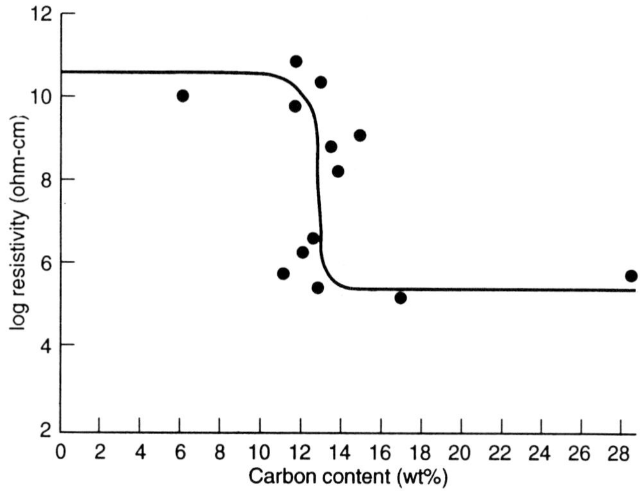

Figure 15C.5 Fly ash resistivity *versus* carbon content of ash.

15C.3 Flue gas conditioning

The use of additives to ESP inlet gas streams to improve particulate collection rates originated shortly after the first commercial use of ESPs was begun early in this century. Evaporation of acid in smelter gases to reduce dust resistivity occurred in 1912 [1]; water injection into gases from cement kilns and steel refining vessels has been used for many years, and ammonia treatment of catalyst dust in petroleum refineries was begun in the 1940s.

Introducing additives into flue gas to enhance the performance of electrostatic precipitators is well known and has been applied in commercial installations since the early part of this century [1,7]. Commercial conditioning of fly ash usually involves the injection of a chemical reagent into the flue gas stream to control the electrical resistivity of a dust and thus improve its collection. Currently, however, this term is assuming a much broader meaning, for conditioning involves processes other than resistivity modification (i.e. coagulation, agglomeration, cohesivity improvement, etc.).

Flue gas conditioning of fly ash relies on one or more of the following effects:

- modifying the surface electrical conductivity of the dust;
- increasing the inter-electrode space charge, and/or
- increasing dust cohesivity to reduce losses during rapping.

15C.3.1 Conditioning by sulfur trioxide

The most common conditioning agent used for modifying high fly ash resistivity is sulfur trioxide [7]. Two major factors control ash resistivity: the sulfur content of the coal, and the overall elemental composition of the ash.

Sulfur occurs in coal as organic and inorganic compounds–notably pyrites and sulfate salts. When coal is burned, more than 95% of the pyrite sulfur appears in the flue gas in the form of sulfur dioxide. The kinetics of sulfur oxidation in boiler flue gas do not allow more than a small fraction of the oxides to appear as sulfur trioxide.

When the temperature of flue gas drops to around 300 °C (about 572 °F), a significant fraction of the sulfur trioxide gas reacts with water vapor to produce sulfuric acid vapor. This process is essentially complete at temperatures around 150 °C (about 300 °F), where electrostatic precipitators normally operate [9]. The sulfuric acid vapor is adsorbed or condensed on an otherwise poorly conducting fly ash surface and directly participates in the electrical conduction process. At very low concentrations of sulfuric acid vapor conduction in fly ash is principally influenced by the charge-carrying ability of the alkali metal ions as affected by the interaction of water vapor and the ash surface. At high concentrations of sulfuric acid vapor, the conduction process is principally controlled by the adsorbed acid. At intermediate concentrations of acid vapor, both mechanisms contribute to the conduction process. Thus, in a strict sense, in cold-side precipitators resistivity modification results from sulfuric acid, rather than sulfur trioxide, effects.

The acid may, however, react with basic constituents of the ash and thus undergo conversion to a poorly conducting layer of sulfate salts. Calcium oxide, a fairly common component of ash from many coals, is a probable cause of acid neutralization and its nullification as a conductor. Thus, two coals similar in sulfur content may produce similar concentrations of sulfur trioxide, but the one containing a more alkaline ash is likely to have a substantially higher resistivity.

Thus, sulfur trioxide conditioning is effective at flue gas temperatures below about 204 °C (400 °F), where surface conduction prevails. At higher temperatures, the effectiveness of sulfur trioxide conditioning is decreased because the ash resistivity is partly or wholly determined by volume conduction, and also because the fraction of the available acid which condenses on the particles is reduced due to vapor pressure and equilibrium characteristics.

The most commonly used method for generating sulfur trioxide is catalytic conversion from sulfur dioxide [7]. This method can use liquid sulfur dioxide as a feedstock; however, the majority of installations burn elemental sulfur to produce gaseous sulfur dioxide which is then catalytically converted to sulfur trioxide. Heated ambient air is normally used as a carrier gas in the injection system, with the sulfur trioxide being a small percentage of the carrier volume. The sulfur trioxide flow is modulated to match boiler load changes, fuel variations or other variables. Injection into the flue gas stream may be either upstream or downstream of the air heater, depending on convenience and accessibility.

15C.3.1.1 SO_3 injection rate prediction

(a) Discussion. Since the electrical resistivity has a pronounced effect on the electrostatic collectability of fly ash, it is desirable to have advanced knowledge regarding the magnitude of fly ash resistivity that one might expect from a given coal. Obviously the best source of this information would be *in-situ* resistivity measurements made during the burning of the subject coal in a commercial boiler. If the coal has not been used commercially, one has the option of burning the coal in small-scale pilot furnace and measuring the resistivity *in-situ* or in the laboratory, or one can utilize one of the methods for predicting fly ash resistivity.

The methods for predicting resistivity are based on correlations that have been established between resistivity and fly ash compositions for specific laboratory test conditions.

The key to better understanding and more accurate estimation of flue gas conditioning by SO_3 was developed by Dr R. E. Bickelhaupt at the Southern Research Institute under the sponsorship of the US Environmental Protection Agency [5]. Application of the findings of these studies results in resistivity *versus* temperature plots typified by Figure 15C.2. As discussed, the uppermost line on this figure represents the expected resistivity with no SO_3 present in the gas. A maximum resistivity value occurs at approximately 140 °C in this example, dropping to lower values at higher temperature due to increased electrical conduction through the body of the ash particles, and similarly dropping at lower temperatures from the effect of conduction through water attached to the ash surface. Two lines emanating from the upper curve show the effects to be expected from the presence of different amounts of SO_3 in the gas.

Since it is possible to make empirical estimates of the quantity of SO_3 naturally generated from fuel sulfur, the expected resistivity of the ash without the addition of a conditioning agent can be estimated, and a decision can be made regarding the necessity for reducing resistivity to improve collection.

The calculation methods developed by Dr Bickelhaupt were published in a form which estimates the resistivity of fly ash in the presence of a given concentration of sulfuric acid. Simple mathematical manipulation of the Bickelhaupt formulae makes it possible to calculate the acid concentration required to bring about a desired change in resistivity. This amount of acid has proven to be an important factor in the injection rate equation [10].

The acid values in the Bickelhaupt equations are equilibrium concentrations determined by laboratory experiments which maintain a fixed volume concentration of acid in the gas in contact with the fly ash sample over a relatively long period of time at a fixed temperature. This obviously differs from the situation in an operating boiler plant where the fly ash is exposed to injected SO_3 at concentrations which vary with respect to time, temperature and physical location, and where the time of exposure is quite short. Thus, the acid figures derived from the Bickelhaupt formulae are indicative of, but not directly convertible to, the required injection quantities.

Since the conditioning process is concerned with the application of an acid film to the surface of the fly ash particles, it is clear that the coal ash content will affect the amount of acid to be injected. A degree of compensation for ash in the fuel is contained in the Bickelhaupt calculation methods, but it is considered to be unlikely that this fully accounts for injection requirements arising from this factor. Thus, it appeared initially that the acid requirement calculated by the Bickelhaupt methods and the ash content of the coal would be two of the variables to be examined for inclusion in any injection rate equation.

Acquisition of an acid film by the fly ash particles is presumed to take place by a condensation/partial pressure deposition process and would therefore be expected to be temperature sensitive. Dr Frisch has shown that the surface composition of the ash particles can affect acid attachment [9]. Dr Bickelhaupt also showed that a 'threshold' effect was exhibited by certain ashes, evidenced by a lack of change in resistivity at ambient acid concentrations below the threshold level [5]. These phenomena are clearly related to ash chemistry, and call for inclusion of all possible combinations of ash chemical properties as variables to be tested for significance.

Intuitively, one would expect that ash particle size distribution, by determining ash surface area as a function of weight, would affect acid requirements. It has been indicated that the presence of elements which volatilize at flame temperature and the viscosity of the fluid ash in the furnace affect the quantity of fine fly ash particles generated [8]. To take the latter effect into account the hemi-spherical ash fusion temperatures for oxidizing and reducing conditions were calculated from ash chemistry and included in the list of trial variables. Other items were speculatively included, on the chance, however remote, that they might prove to have significant injection influences.

Over 20 legitimate data points (sites with known coal supplies, injection rates and precipitator performance data) were used in the development of the injection rate equation. They include coals ranging in rank from lignite to low-volatile bituminous, taken from mines in eastern and western USA locations, as well as South Africa and Australia. Ash contents range from 4 to 26% and flue gas temperatures from 140 to 216 °C. Nearly every type of pulverized coal firing arrangement except cyclone furnaces was represented [10].

(b) SO_3 injection rate equation. The Bickelhaupt method cannot be directly applied to determine SO_3 injection rates, because the SO_3 quantities used in the method are amounts in equilibrium with the gas-and-ash environment in a laboratory situation, and these differ from injection requirements for flue gases containing suspended fly ash. Having an estimate of the ash resistivity without conditioning and knowing from experience the desired final resistivity to obtain best ESP performance, both at the intended operating temperature, simple subtraction gives the value by which the resistivity has to be changed, and common sense says that the injection rate will be a function of this change. Similarly, the required injection will be a direct, but not necessarily linear, function of the quantity of ash to be treated. Other factors than these two will enter into the determination of the injection rates. As it was noted earlier, Dr N. W. Frisch pointed out that the surface condition of the ash will have a bearing on the ability of SO_3 to attach to the particles [9]. Computerized regression analysis was used to sort through a variety of possible factors and to choose those having significance in determining required SO_3 injection rates for a wide variety of coals.[10]. Ultimate analysis of the coal and chemical analysis of the ash are required as input data to the resulting equation for the calculation of injection rates.

The significance of the trial variables was tested by computerized stepwise regression analysis, first taking a single independent variable in the form

$$y = K_1 + K_2(X_1)^a \tag{15C.1}$$

where y is the dependent variable (injected SO_3 + natural SO_3 from coal sulfur), X_1 is the independent variable, and K_1, K_2, a are constants.

Regression analysis was applied to determine the value of the constants in the equation, while substituting for the independent variable in turn, each of the trial variables listed in Table 15C.1, plus their logarithmic and exponential values, their squares and square roots, and the inverse of all of these. The variable and its form having the highest correlation coefficient (R^2) was then chosen to replace the independent variable in the term tested, and another term was added to the equation in the same form as those preceding. All the trial variables were then tested in the new term, and the process continued. At each term addition, the new term and all those preceding were tested for statistical significance.

The detailed description and the final form of the derived injection rate equation is described in [10] and a typical SO_3 injection rate *vs.* temperature curve is presented in the Figure 15C.6.

It is seen that the required rate of SO_3 injection at temperatures below about 165 °C is essentially independent of temperature. Resistivity curves

Table 15C.1 Trial variables

Category	Symbol	Description
Ash	LI	Lithium oxide, percent weight
	NA	Sodium oxide, percent weight
	K	Potassium oxide, percent weight
	MG	Magnesium oxide, percent weight
	CA	Calcium oxide, percent weight
	FE	Iron oxide, percent weight
	AL	Aluminium oxide, percent weight
	SI	Silicon oxide, percent weight
	TI	Titanium oxide, percent weight
	P	Phosphorus pentoxide, percent weight
	S	Sulfur trioxide, percent weight
		(Plus all of the above taken as molecular percentages, plus all of the above taken as cation percentages)
Ash-derived	SL	Sodium plus lithium, cation and molecular percentages
	CM	Calcium plus magnesium, cation percent
	ALSIM	Alumina plus silica, molecular percentage
	HTO	Hemispherical temperature, oxidizing
	HTR	Hemispherical temperature, reducing
	BAR	Base/acid ratio, both weight and molecular percentage
	DR	Dolomite ratio
	OXF	Oxide factor
	ALK	Alkali ratio
Coal	SUL	Sulfur, percent weight
	ASH	Ash, percent weight
	HV	Heating value
Combustion	TEMP	Flue gas temperature
	ASHG	Total ash generated, wt./unit gas volume
	ASHC	Ash carried over, weight/unit gas volume
	ACIDDP	Flue acid dewpoint temp., coal SO_3
Other	ACIDB	Ppmv SO_3 to change ash resistivity at flue gas temperature from the no-injection value to the desired value as computed by reference [1] methods
	DPACID	Total acid for dewpoint at flue gas temperature, ppmv
	TVS	Temperature of interaction of volume and surface resistivity curves
	TVSA	Temperature of intersection of volume-plus-surface and acid resistivity curves

calculated by the Bickelhaupt method, such as those shown in Figure 15C.7 for a Western USA Powder River Basin (Wyoming) coal, have relatively steep slopes of resistivity with respect to temperature for a given equilibrium acid concentration.

(c) Application of the theoretical approach. Figure 15C.6 shows predicted injection rates for two coals of widely differing characteristics and origins. Each of the two plots shows a low temperature region which is essentially a reflection of the ash resistivity curve. As temperature is increased, both curves come to an inflection point at which transition is made to a rapidly

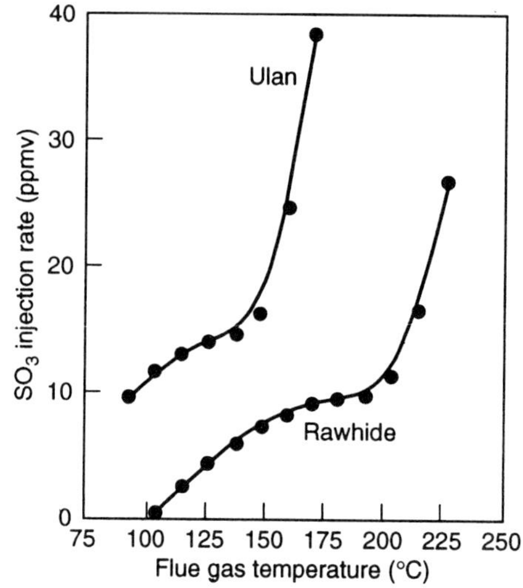

Figure 15C.6 Predicted SO_3 injection rates.

rising characteristic which approximates a straight line. At temperatures below the inflection point, virtually all of the injected SO_3 attaches to the ash. At higher temperatures, the amount of SO_3 injected needs to be in excess of the SO_3 level at the inflection point to provide the partial pressure for attachment of sufficient SO_3 to the ash to obtain the desired resisivity. The excess SO_3 above the inflection point level remains in the gas stream and, if the excess is about ten parts per million or more, may be visible as a blue–white condensation plume from the stack.

The trace on Figure 15C.6 having an inflection point at lower temperatures is representative of coal ashes having very low base-to-acid ratios, such as those from the Sydney Basin of Australia. The characteristic having the higher inflection point is for coals such as those from the Powder River Basin in the USA which have very high base-to-acid ratios. These are two extremes–nearly all the world's coals have inflection point temperatures intermediate to these two. Although it is convenient to refer to these effects in terms of base-to-acid ratio, it should be remembered that this is only a rough general index to the inflection point temperature for discussion purposes; other factors appearing in the injection rate equation must be included for accurate determinations.

At temperatures from about 165 to 190 °C, depending on ash analysis, an inflection point occurs in the injection rate curves, with the rate increasing with temperature above this point. It seems logical that the temperature at which inflection occurs is a function of ash surface conditions–a measure of susceptibility to attachment of acid, perhaps, in as much as the lower

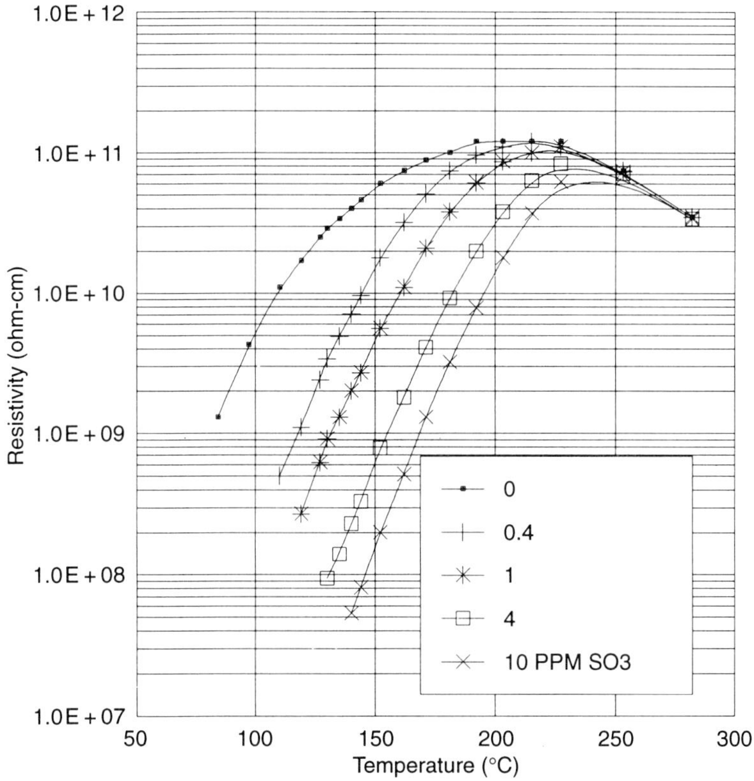

Figure 15C.7 Fly ash resistivity *versus* temperature for PRB coal.

inflection temperatures occur for ashes having low base-to-acid ratios and the higher points correspond to higher ratios.

The inflection point effect explains some of the difficulties various experimenters have encountered in attempting adjustment of ash resistivity with sulfur trioxide alone. In some low sulfur coals the percentage of acidic compounds in the ash (aluminum oxide, iron oxide, and silicon oxide) is high (above about 90%). Here, the injection of sulfur trioxide alone to reduce resistivity loses some of its effectiveness because the acidic sulfur trioxide cannot readily attach to the highly acidic ash at higher temperatures. Substantial amounts of excess sulfur trioxide must be injected while treating such ashes to produce a conductive film on the surface. Furthermore, if the operating temperature is much higher than the inflection point, no reasonable amount of injected SO_3 will be effective in reducing the resistivity.

The increased requirement for sulfur trioxide at temperatures above the inflection point indicates that an increase in the required partial pressure of

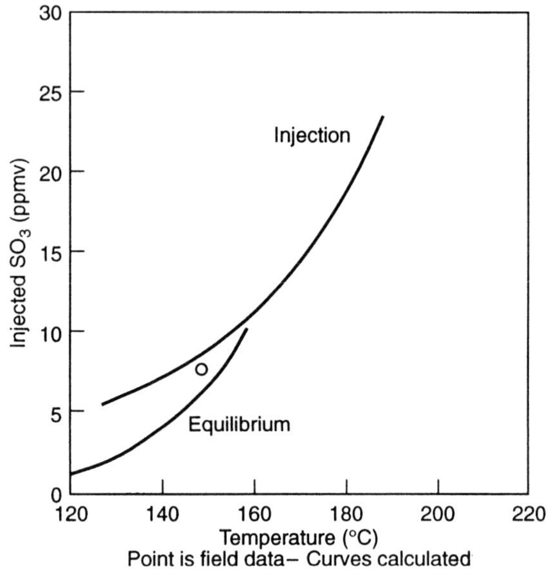

Figure 15C.8 Effect of temperature on injection rates and equilibrium curves.

sulfur trioxide is needed to attach sufficient sulfuric acid to the particles to obtain the desired ash resistivity. The difference between the injection level at low temperatures and that at a high temperature is a measure of excess sulfur trioxide which will escape from the stack at the higher temperature.

The temperatures for the constant target resistivity (for example, of $5 \times 10^9\ \Omega$-cm) may be plotted as an equilibrium (iso-resistivity) curve with the predicted injection rate curve for the same ash (Figure 15C.8). For comparable conditions of acid deposition on the ash it would be expected that the equilibrium curve, representing the acid remaining in the gas stream after deposition, would always fall below the injection curve. However, it is seen that the two curves intersect, and at higher temperatures the equilibrium value exceeds the injected amount. The difference cannot be explained as representing the SO_3 originating from combustion of coal sulfur because this would imply no deposition of acid on the ash, and it is known from the operation of actual boiler units in the higher portion of the temperature range that the ash resistivity is modified by injection, thereby verifying the attachment of acid to the ash. It must be concluded that the differences between constant-temperature laboratory deposition and the injection process tend to promote more effective transfer of the acid to the ash particles in the injection case.

It is thought that the higher transfer efficiency with injection results from the relatively high acid concentration of the injected stream. In the process of mixing the injected stream with the flue gas, at least some of the cooler ash particles will penetrate the injected stream and, thus, be subjected to a high driving force for acid attachment. This will result in the deposition of more acid than would be calulated from equilibrium (completely mixed) conditions.

Accurate prediction of injection rates for SO_3 flue gas conditioning systems is now possible for the majority of steam coals in use. The described procedure may be the first step in any FGC system optimization. Coal and ash analyses are required for input to the calculation procedure. Equipment size and feedstock requirements can be closely determined, given knowledge of the range of fuel properties to be accommodated.

As with any mathematical function derived by regression analysis, care must be taken in extending the results to areas beyond the limits of the original database. No cyclone boilers were included in the database, and it is therefore expected that application of rate prediction to that type of unit would require some compensation for the reduced carryover of ash to the precipitator, and possibly for changes in the ash analysis in the fraction carried over as compared with the bulk analysis. Also, some ashes are composed almost entirely of silicon and aluminum oxides, with very minor amounts of other elements. No ashes representing the extremes of this situation were included in the study so application of this theoretical approach to those types may be misleading.

15C.3.2 Ammonia conditioning

Numerous investigators have reported the use of ammonia conditioning with various degrees of success. Ammonia conditioning usually works best with the acidic fly ash. If injected ahead of the air heater (unless air heater temperatures were unusually high– >400 °F) the air heater was reported to have a tendency to plug as an additional, but unwanted, result. Ammonia injection downstream of the air heater was reported to be unpredictable [13].

The use of ammonia (NH_3) for modification of dust resistivity has not been developed to the same degree as SO_3. Some experiments have shown that ammonia injection improved the collection of fly ash in ESPs, but it is not clear whether the improvement resulted from resistivity modification or other effects [11]. Other trials have shown little or no change in fly ash resistivity or ESP performance [12].

No effect of the ammonia on fly ash resistivity has been reported with any degree of certainty. There has been much debate by many authors regarding the use of ammonia alone for modifying the electrical resistivity of fly ash. One argument suggests that for high resistivity ashes a reduction in resistivity might be produced, permitting higher voltages and currents to be maintained without sparking or back-corona. On the other hand, for lower resistivity ash an increase in resistivity, due to neutralization of excess acid, might permit a sufficient electric field to exist across the ash on the precipitator collection electrodes to overcome losses from rapping re-re-entrainment. The effect of ammonia appeared to be an enhancement of the space-charge component of the electric field. In addition, the increase of the cohesiveness of collected ash and a reduction in re-entrainment appears to be another 'major player' in the successful application of ammonia conditioning.

While successes have been reported using ammonia as the sole conditioning agent for some low sulfur Australian coals, the US experience has been improved precipitator performance for some low sulfur coals and not for others. It has been theorized that ammonia conditioning is more effective for acidic ashes, in terms of the relative concentrations of acidic and basic oxides present (acid/base ratio), but this has not been true in every case [10]. Ashes having high acid/base ratios with small to moderate amounts of sulfur trioxide present in the flue gas can have high resistivity and it may be that resistivity is reduced as a result of ammonium bisulfate acting to facilitate sulfur trioxide attachment. If little or no sulfur trioxide were present, ammonia alone might have little effect.

15C.3.2.1 Background. The value of ammonia as a conditioning agent was reportedly discovered in 1942 in efforts to deal with the high resistivity of catalyst dust in the petroleum industry. The use of the ammonia for treating fly ash in coal-fired power plants, on the other hand, is 'comparatively' new. First reports of the *intermittently* successful use of the ammonia as a conditioning agent came from Australia, by Watson and Blecher in 1966 [14]. The results of this study showed good results while using the ammonia as a conditioning agent. There the SO_3 flue gas conditioning was proved to be ineffective due to highly acidic ash (pH of 3.5–4) and, thus, speculations were that conditioning with a basic rather a thin acidic agent may prove more effective. The initial pilot plant trials with ammonia conditioning conducted at the Tallawarra plant in 1959 were found to be very successful. Similar promising results were obtained at the No. 5 Tallawarra precipitator. However, subsequent ammonia conditioning at the No. 6 precipitator at the Tallawarra did not produce the same encouraging results that were obtained on No. 5.

The electrical behavior of the precipitators with ammonia conditioning changed considerable within minutes after the introduction of ammonia into the flue gas. Increases of 50 to 100% in electrode potential and corresponding decreases in electrode current were maintained consistently and tended to persist for several hours after ammonia injection was stopped.

Resistivity measurements were also carried out; however, these presented rather perplexing results when ammonia-conditioned dust was evaluated. It is generally accepted that the increase in precipitation efficiency brought about by flue gas conditioning is caused by a reduction in the particle surface resistivity and the elimination of back-corona, a phenomenon reported to be quite common with New South Wales fly ash. The resistivity of the ammonia-conditioned ash, however, showed no appreciable difference from that of non-conditioned fly ash. Both were in the range of 10^{13} to $10^{14}\,\Omega$-cm which is certainly well above the critical level of $10^{10}\,\Omega$-cm suggested by Dr White [2].

Watson and Blecher [14] proposed a hypothesis that perhaps ammonia gas conditioning changed flue gas characteristics sufficiently to eliminate back-corona and/or the forces of particle adhesion and cohesion sufficiently to bring about cleaner electrodes and, hence, better performance.

Shortly thereafter, Baxter [15] reported success with ammonia conditioning of several plants in the United States, where the apparent effect was to lower ash resistivity in some instances and to increase it in others. Initially, he injected sulfur trioxide with unsatisfactory results. Performance improvement was attained only by the use of excessive conditioning which allowed an objectionable acid vapor plume to escape from the stack. The presence of the acid plume indicates that the necessary adsorption on the dust particles was not occurring. Where sulfur trioxide conditioning has been effective, stack measurements show no increase in effluent SO_3 concentration; rather the injected gaseous agent attaches to and is discharged with the dust collected in the precipitator, as reported by Darby and Heinrich [16]. Analyses of the collected ash indicated that SO_3 flue gas conditioning reduced laboratory measured resistivity from 10^{12} to $10^{11}\,\Omega$-cm and increased acidity of the dust from a pH of 5.5 to 3.5.

Subsequent trials with ammonia injection reportedly produced marked precipitator performance improvement. Dust analyses showed a decrease in resistivity, similar to that obtained with sulfur trioxide conditioning; however, the pH of the ash increased to a neutral condition. Although Baxter [14] reported varying success with precipitator performance, improvement has only been accompanied by air heater plugguage at each test location.

Finally, tests reported from a power plant with two identical cyclone fired boilers produced still unexplained results, i.e. the units reacted in a very explicitly different manner. One of the units responded well to the ammonia injection, while the other reacted rather poorly. Moreover, chemical and physical analyses of the dust produced no reasons for such a difference. The resistivity was measured and found to have actually increased. Dust pH also increased but remained quite acidic. Unfortunately the authors did not publish detailed analyses. Furthermore, it is worth noting that, according to the authors, 'No correlation can be established between effectiveness and initial ash resistivity, nor between ammonia injection and treated ash resistivity'.

In 1968, Reese and Greco [17] were successful in the use of ammonia to treat the fly ash from high sulfur coal and concluded that the effect of ammonia was to overcome an unfavorably low resistivity stemming from an excess of naturally occurring sulfur trioxide.

Some of the trials with ammonia flue gas conditioning have been successful, but others have not produced the desired results. One source of difficulty in the use of the ammonia conditioning has been the absence of basic information on mechanisms of conditioning with this compound and the circumstances in the power plant that favor its use.

15C.3.2.2 Chemical reactions involved in ammonia conditioning. The reaction of sulfuric acid mist with ammonia in the flue gas is believed to follow a course of absorption of ammonia into the sulfuric acid mist, evaporation of water, formation of solids, and solidification. There is no way to confirm the reaction of ammonia gas with sulfuric acid gas experimentally. However, if the theoretical tendency is considered qualitatively, the following changes can be visualized.

1. When ammonia is absorbed into sulfuric acid mist, ammonia ionizes in the water solution; this is because the sulfuric acid mist in the flue gas exists as an aqueous solution. When the reaction proceeds, the equilibrium partial pressure of the water vapor is lowered, and the water in the solution evaporates.
2. As the reaction and evaporation of the water progress, the viscosity of the solution increases, and its state proceeds from the state of ionized water solution to that of a mixture of molten salt.
3. When the reaction proceeds further (dissolving of ammonia), crystals in solid form are deposited in the solution. In this case, the solids deposited differ according to the gas temperature.
4. After the solids start to precipitate, crystals are deposited only in the amount corresponding to the quantity of the absorbed ammonia; finally, the solution is exhausted and solidifies.

Elementary explanations of ammonia conditioning effects have assumed that the chemical compounds formed when ammonia is injected, either concurrently with injection of sulfur trioxide, or when the latter is formed by 'natural' (combustion) processes in sufficient quantities, are only ammonium bisulfate NH_4HSO_4 or, if the stoichiometric ratio is appropriate, the normal sulfate $(NH_4)_2SO_4$. Besides these, the existence of intermediate chemical compounds is possible. The melting points change continuously according to the H_2SO_4/NH_3 stoichiometric ratio, and there is a possibility that compounds with very low melting points may be formed in the process. In addition, the ones which are stable in solid form at temperatures above 150 °C (300 °F) are those with stoichiometric ratios 1.2 and above.

The chemical reaction postulated by Rendle and Wildson [18] indicates that ammonia will react with any natural sulfur trioxide present plus moisture to produce ammonium bi-sulfate:

$$SO_{3_{gas}} + NH_{3_{gas}} + H_2O_{gas} \Rightarrow NH_4HSO_{4_{\text{liquid, acidic sulfuric ammonia(ammoniumbisulfate)}}} \qquad (15C.2)$$

NH_4HSO_4 has a melting point of 146.9 °C (296.4 °F), and deliquescent characteristics, i.e. it absorbs water readily. When the flue gas temperature is about 140 °C (284 °F), NH_4HSO_4 begins to solidify.

In the presence of excess ammonia, this reaction will slowly continue to produce ammonium sulfate:

$$NH_4HSO_4 + NH_3 \rightleftarrows (NH_4)_2SO_4 \tag{15C.3}$$

or

$$NH_{3_{gas}} + NH_{4_{liquid|vapor}} + HSO_{4_{liquid|vapor}} \Leftrightarrow (NH_4)_2SO_{4_{solid\ sulfur\ ammonia(ammonium\ sulfate)}} \tag{15C.4}$$

or

$$SO_3 + 2NH_3 + H_2O \Rightarrow (NH_4)_2SO_4 \tag{15C.5}$$

The melting point of $(NH_4)_2SO_4$ is 513 °C (955 °F). In the above equation (15C.2), $(NH_4)_2SO_4$ is dissociated at temperatures higher than 200 °C and decomposes into NH_4HSO_4.

Lacking excess ammonia, or as cooling occurs, the liquid bi-sulfate will freeze at temperatures below 147 °C (297 °F) into a sticky solid. It is this adherent nature of the resultant that has been a 'suspect' to play a major role in the precipitator performance improvement.

A variety of interactions of ammonia with sulfuric acid are possible as follows:

$$NH_{3_{gas}} + H_2SO_{4_{gas}} \Rightarrow NH_4HSO_{4_{solid}} \tag{15C.6}$$

$$NH_{3_{gas}} + NH_4HSO_{4_{gas}} \Rightarrow (NH_4)_2SO_{4_{gas}} \Rightarrow (NH_4)_2SO_{4_{solid}} \tag{15C.7}$$

$$NH_4HSO_{4_{solid}} + \text{nucleus} \Rightarrow NH_4HSO_{4_{particle}} \tag{15C.8}$$

$$(NH_4)_2SO_{4_{solid}} + \text{nucleus} \Rightarrow (NH_4)_2SO_{4_{particle}} \tag{15C.9}$$

$$NH_{3_{gas}} + NH_4HSO_{4_{particle}} \Rightarrow (NH_4)_2SO_{4_{particle}} \tag{15C.10}$$

$$H_2SO_{4_{gas}} + (NH_4)_2SO_{4_{particle}} \Rightarrow NH_4HSO_{4_{particle}} \tag{15C.11}$$

The reactions in (15C.6) and (15C.7) are collision reactions. Due to the low vapor pressures of NH_4HSO_4 and $(NH_4)_2SO_4$, these compounds usually solidify easily. The quantity of $(NH_4)_2SO_4$ formed increases as the quantity of ammonia increases. The reactions in (15C.10) and (15C.11) are subject to the effects of the diffusion of the gas currents of $NH_{3(gas)}$ and $H_2SO_{4(gas)}$. The respective diffusion constants are 0.37 cm^2/s and 0.15 cm^2/s in 140 °C (284 °F) air; therefore, the reaction in (15C.10) should be somewhat faster than in (15C.11).

15C.3.2.3 Mechanisms of the ammonia conditioning. Ammonia in the presence of sulfur trioxide may work in a variety of ways. Although the exact mechanisms responsible for improved precipitator performance are not completely defined, they appear to include such phenomena as an improved acid attachment, space-charge effects, agglomeration and increased cohesiveness of the fly ash.

(a) Effects of ammonia on the electrical resistivity. A study of the effects of the ammonia on the electrical resistivity of the silica-alumina catalysts may shed some light on the possible mechanism of the conductivity improvement on the one hand and, perhaps, answer numerous questions why ammonia is very unpredictable in its conditioning nature.

It has been demonstrated that sodium and potassium increase the conductivity of silica-alumina materials (note: fly ash is predominantly silica-alumina compound) at about 204 °C (400 °F) or higher. It was further speculated that the increased conductivity results from ionic conduction of sodium and potassium ions. Ammonia and amines are reported to markedly increase the resistivity of silica-alumina catalysts below 93 °C (200 °F). The increased conductivity is closely associated with large amounts of physical adsorption, which probably increases the mobilities of the current carriers. The irreversible adsorption of ammonia below 93 °C (200 °F) is independent of temperature and is a measure of the 'acid sites' on the particulate. The reversible adsorption increases with decreasing temperatures. The increase in conductivity depends on the properties of both adsorbate and the particulate.

The chemisorption of ammonia on silica-alumina compounds, like fly ash particles, results in the formation of ammonium ions from the reaction of ammonia with protons on the ash particle surface. This, perhaps, could be responsible for the lower resistivity over the temperature range of 150–260 °C (300–500 °F). With ammonia, it seems reasonable to conclude that most of the conductivity occurs on the surface. The physically adsorbed ammonia offers an excellent medium for the passage of ions. In addition, the relative size of the ammonia ion, the internal structure of silica-alumina and the temperature regions where the precipitators are operating are not favorable for ionic conduction (by ammonium ions) in the bulk. The exact mechanism of the increased surface conductivity resulting from ammonia adsorption is difficult to deduce, but one could speculate that the presence of traces of alkali metals would contribute appreciably to the conductivity of the fly ash particles.

Some evidence suggests that injection of ammonia promotes improved attachment of available sulfuric acid to fly ash particles, making it more effective for resistivity control. It creates a thin conductive film on the ash surface, believed to consist of low melting point ammonium sulfate products which contribute to the resistivity modification. It has been suggested that the addition of ammonia improves upon the efficiency of the sulfur trioxide conditioning by providing an additional charge carrier, the NH_4^+ ion [19].

Since the conditioning process is concerned with the application of an acid film to the surface of the fly ash particles, it is clear that the coal ash content will affect the amount of acid to be injected. Acquisition of an acid film by the fly ash particles is presumed to take place by a condensation/partial pressure deposition process and would, therefore, be expected to be temperature sensitive.

Those who have studied factors affecting the resistivity of fly ash in a flue gas atmosphere are very familiar with the hump-backed characteristic of resistivity as a function of temperature. Generally, resistivity has a maximum value at a temperature around 150 °C (300 °F) with decreasing values above and below the maximum point. In determining the amount of SO_3 to be required to reduce the ash resistivity to a desired lower value, one would think that the maximum rate would be required at the maximum unconditioned resistivity point, with decreasing amounts at higher and lower temperatures; that is, a hump-backed curve of injection rate reflecting the unconditioned resistivity characteristic would be expected. An interesting discovery made in the development of SO_3 flue gas conditioning is that the amount of SO_3 required to attain a desired level of resistivity follows the expected dome-shaped curve with respect to temperature only up to a point, after which it breaks off to a rapidly rising characteristic [20].

It has been found that the simultaneous injection of ammonia (NH_3) with the SO_3 will condition the ash surface to accept resistivity adjustment without excess SO_3 being required. Ammonia FGC tends to overcome this problem by allowing greater adhesion of the residual sulfur trioxide (again, one has to presume that there is a sufficient supply of *natural* SO_3 present) to the acidic ash particle surfaces to optimize its resistivity. This allows for proper fly ash resistivity adjustment without using excess sulfur trioxide. It seems logical that the temperature at which inflection occurs is a function of ash surface conditions – a measure of susceptibility to attachment of acid, perhaps, in as much as the lower inflection temperatures occur for ashes having low base-to-acid ratios and the higher points correspond to higher ratios.

(b) Space-charge effect. When ammonia is injected into a gas stream in the presence of the sulfur trioxide, a fine fume consisting of a variety of ammonium sulfate particles, all much less than 1 μm in size, is produced. These particles alter the electrical characteristics of the flue gas between the discharge and collecting electrodes and produce a space-charge enhancement of the electric field. This well-documented effect arises when fine fume is charged in the precipitator and the electric field is thereby increased. Moderate, strictly controlled quantities of fine particles would, therefore, increase the charge level of fly ash particles and the field near the collecting plates. The higher field increases collection efficiency. The space-charge effect is usually more pronounced in the front fields than in those following. This

happens because a large proportion of the particles are collected in the front fields and the remaining small quantity is not sufficient to produce a significant space charge.

(c) Fly ash cohesivity control. Ammonia can generate low melting point substances depending on the H_2SO_4/NH_3 stoichiometric ratio; at low stoichiometric ratios low melting point substances having high moisture absorbency are generated which promotes cohesion. This cohesion, applied properly and strictly controlled, improves overall precipitator efficiency.

The cohesive properties of fly ash influence stack opacity and precipitator efficiency by agglomerating particles and by reducing re-re-entrainment. To simplify, the ammonia combines with sulfur trioxide to form ammonium bisulfate. The melting point for this material, as discussed earlier, is about 147 °C (297 °F), so it is semi-liquid at typical flue gas temperatures and acts as a binding agent when mixed with fly ash. Because the agglomerated ash groupings are larger and heavier, rapping losses are minimized.

For high resistivity ashes it has been found that variations of cohesion due to additives are of minor importance to precipitation efficiency. For low resistivity ashes, where the use of an additive has little or no effect on resistivity, it has been found that changes in efficiency of precipitation are associated with variations of cohesion. Good cohesion has resulted in high efficiency due to reduced rapping losses and re-re-entrainment, or more effective initial capture of particles. The electrical forces of attraction between particles in an ash layer on the collecting electrode in an energized precipitator are the predominant component of the cohesive strength of a highly resistive ash layer and the variation of the mechanical component due to the additive is then relatively unimportant. With low resistivity ash the electrical forces are reduced and may even become repulsive, and under these circumstances the mechanical forces causing re-re-entrainment become important.

15C.3.3 Dual flue gas conditioning

Simultaneous injection of SO_3 and NH_3, arbitrarily called dual conditioning, was mentioned earlier as assisting in obtaining attachment of SO_3 to fly ash at temperatures above the inflection point of the injection rate curve. Therefore, one may define a *dual* FGC as a system which simultaneously and independently injects two conditioning agents, sulfur trioxide and anhydrous ammonia, into the flue gas stream. These agents work together in a variety of ways. It has been demonstrated that it may be used to control rapping and re-entrainment losses from ESPs. It appears that the process can be crudely visualized as reacting the two conditioning gases to form ammonium bisulfate which has a melting point close to typical ESP operating temperatures. This is thought to act as a binding agent (liquid glue?), increasing the cohesivity of the ash particles so that they adhere to each other and fall into the hopper with reduced dispersion into the gas stream.

A striking demonstration of this effect occurred at the Monroe Station of Detroit Edison where SO_3 alone permitted operation only up to 450 MW before the compliance limit on particulate emission was reached, but dual conditioning allowed the full 750 MW unit rating to be attained. In tests conducted by Ontario Hydro it was found that dual conditioning was sufficiently effective to obtain significant reduction of stack losses, caused by high unburned carbon levels in the ash.

From the preceding description of the inflection point, which appears in the injection-rate-*versus*-temperature curves, it is clear that one criterion for needing dual conditioning is flue gas temperature exceeding the inflection point temperature. Criteria for advance determination of a need for dual conditioning to suppress excess rapping and re-entrainment losses by means of fuel and ESP data have not yet been developed, partly because the factors affecting ash cohesivity in the ESP have not been explored or defined, and partly because the mechanical and fluid dynamic characteristics of the ESP, which affect the magnitude of rapping and re-entrainment losses, may differ radically from one installation to the next and have not been subjected to any investigative procedures capable of developing applicable loss factors. It seems odd that such a significant ESP loss mechanism should be so poorly characterized, but it is expected that continued research will help to clarify this situation in the future. In the meantime, the only recourse is the observation of trials with fuel, SO_3 and ESP conditions as nearly as possible the same as expected in service. Then, observation of the difference between recordings of flue gas opacity, with and without rapping, will immediately indicate, if the difference is more than a few percent and large rapping spikes appear on the instantaneous trace, that dual conditioning should be considered.

15C.3.4 Balance of plant impact

The addition of SO_3 alone to boiler flue gas streams is only necessary if the coal is low in sulfur, as otherwise the ash resistivity will be low enough not to require conditioning. Nearly all of the injected SO_3 is attached to the ash particles and is captured with them in the ESP. Without conditioning, 1–1.5% sulfur content in the coal would typically give the same ash resistivity as that obtained by injection of a low sulfur coal, meaning that the same amount of SO_3 is on the ash surface; however, an additional amount would still be present in the gas stream and would be emitted from the stack. This latter amount, typically about one-tenth percent of the SO_2 content of the gas stream, is not present in the emissions from the conditioned unit, unless operation at temperatures above the inflection point of the injection rate curve is attempted without the simultaneous injection of ammonia.

Fly ash from a conditioned system carrying acid on its surface, in about the same concentration as that from unconditioned combustion of one-plus percent sulfur coal, is not significantly different in handling requirements and environmental effects from the ashes dealt with in the past.

Ash handling systems which sluice SO_3-conditioned ash to a pond and recirculate the water may find that the pH of the pond will be depressed below acceptable levels for discharge to the environment, making neutralization treatment necessary.

Ammonia treatment of fly ash may cause release of objectionable odors from concrete made with cement in which the fly ash has been used as an additive. Determination of the maximum acceptable ammonia content for fly ash for this use is a trial-and-error process. The level has been found to be greater than zero in all cases to date, but the maximum appears to be influenced by cement and ash chemistry and ash handling conditions.

Ammonia and sulfur, as used by conditioning systems, are essentially pure chemicals which neither introduce nor generate any hazardous or toxic pollutants. Except for improved control of particulate emissions, the overall impact of the boiler system on the environment is essentially unchanged by the use of flue gas conditioning.

Some experimenters in the past found that the injection of NH_3 into flue gas streams containing appreciable quantities of SO_3 caused the ash to become 'sticky' enough to cause problems in ash handling systems. Experience with dual conditioning to date indicates that this problem arises when reatively large quantities of NH_3 and SO_3 are present as compared with the amount of ash. The quantities required to provide effective prevention of excessive rapping and re-entrainment losses are controlled to much lower values. None of the dual conditioning installations thus far (14 854 MW) have had problems with ash handling from increased ash cohesivity.

15C.4 Flue gas conditioning equipment

15C.4.1 SO_3 FGC systems

15C.4.1.1 Conventional sulfur trioxide generators. With very few exceptions, all commercial SO_3 FGC systems installed to date are based on catalytic conversion of sulfur dioxide (SO_2) which is either supplied in liquid form or obtained by burning elemental sulfur. Elemental sulfur is the preferred feedstock for long-term permanent operations because its costs are lower, but liquid SO_2 is used for trials and small or short-term situations, where reduced capital costs can offset increased feedstock expense. The catalytic conversion design was chosen over other methods by which SO_3 may be made available, partly because it is flexible and easily controllable, but mainly because it minimizes the quantity and the difficulty of handling hazardous materials resident in the system.

Figure 15C.9 is a diagrammatic representation of a typical sulfur-burning system for SO_3 conditioning. Molten sulfur, a common trade item in the chemical industry available throughout the US, is delivered by thermally insulated tank trucks fitted with steam coils for melt-out. In locations where sulfur cannot be delivered in molten form, bagged or bulk solid sulfur may be supplied for melting on site. The sulfur grade is designated as 'Bright Yellow' which contains very low levels of contaminating materials. The exact analysis varies slightly among suppliers, but completely lacks chemicals which could act as catalyst poisons and contains only very minute quantities of hydrocarbons. Storage in insulated steel tanks with steam-blanketing provisions for fire suppression is standard, but concrete-lined pits are sometimes used. Tanks and molten sulfur piping are heated by steam controlled to a saturation temperature of approximately 145 °C (290 °F) at which the sulfur has ideal flow characteristics. Steam tracing is strongly preferred because of the ease with which controllable highly uniform temperatures can be maintained throughout the system. Sulfur metering pumps are supplied in duplicate so that one may be serviced while the other is in operation. Even though extensive precautions are taken to maintain cleanliness of the sulfur and piping, it is impossible to eliminate all possible chances of contamination and, therefore, impossible to guarantee continuous unimpeded pump operation over long periods of time. In as much as emission compliance often depends upon continuous operation of the flue gas conditioning system, it is usually considered that the cost of an installed spare pump is justified.

Combustion of the sulfur to generate SO_2 is obtained by introducing sulfur into an air stream which has been electrically preheated for start-up purposes to the temperature at which the catalyst becomes active. Since this temperature exceeds the autoignition point of molten sulfur, burning is initiated immediately upon the introduction of sulfur and SO_2 is delivered to the catalyst for conversion to SO_3. Roughly 4000 BTUs of heat are generated per pound (2222 kcal/kg) of sulfur burned. This replaces a portion of the start-up electrical heat input, and at full system rating all the required heat to maintain the catalyst at operating temperature is supplied by sulfur combustion. Clearly, operation of the system in this manner allows the generation of any quantity of SO_3 from zero to full system rating as a function of any selected control signal used to determine the rate at which sulfur is delivered.

15C.4.1.2 Control systems. In the past, controls for SO_3 FGC installations consisted of safety and sequencing devices, plus load-following

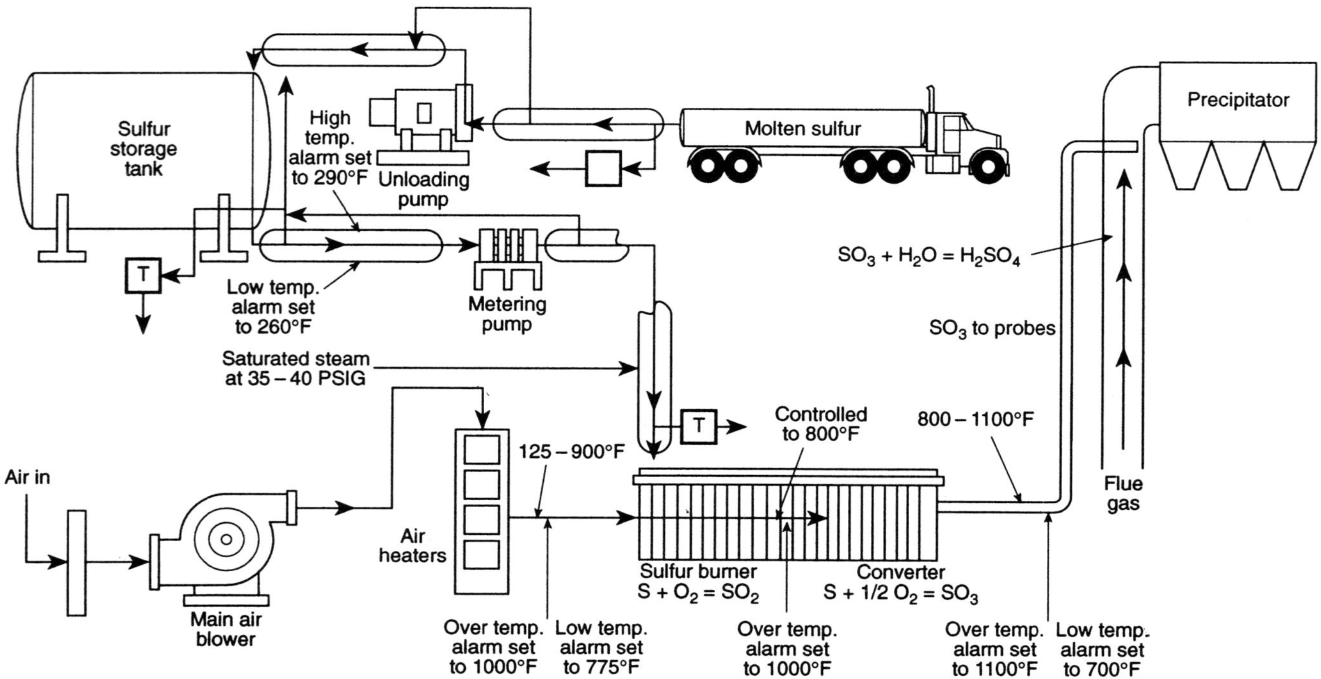

Figure 15C.9 Sulfur burner typical conditioning system.

apparatus to modulate the rate of delivery of SO_3 in proportion to boiler load. The level of SO_3 to suit a given coal was determined experimentally by the operator by manual adjustment of a proportioning factor between the boiler load signal and the SO_3 generation rate while observing the effect on ESP operation and emissions. It was generally found that the setting for a given coal, once determined, was seldom subject to much variation, so it was possible for most stations to make initial determinations for the various coals they fired and use those values with little need for subsequent adjustments. These control functions were implemented with relay logic and analog controls, and permitted operation with a minimum of attention from station personnel. They were reasonably satisfactory for most applications.

There were, however, instances where the coal being burned at any given time might be unidentified or mis-identified, so that previous history was of no value in determining the injection rate setting and there were some stations burning 'spot' coals received in small quantities where constant determination of new settings was required. There were also locations where difficulty was encountered in determining optimum settings through unfamiliarity with ESP operation. For these reasons, as well as the desirability of freeing the operators from the task of identifying which coal was being fed to the burners, and especially the times at which a particular coal started and ended, effort was directed to automatic determination of the optimum SO_3 rate. An initial attempt was based on the idea that the rate would be a function of the sulfur content of the coal, and that quantity could be inferred from measurement of flue gas SO_2. This proved to be partially workable, but did not adequately account for ash chemistry or ash quantity differences between coals with the same sulfur contents. Another aborted attempt relied on the increase in ESP power input accompanying SO_3 injection, but seemed to have an unbreakable habit of running to maximum system rating, whether that injection rate was optimum or not. Excess injection of SO_3 will tend to make the ash resistivity in the ESP too low, decreasing its collection efficiency, causing severe rapping puffs and an increase in stack opacity because of SO_3 carryover. If SO_3 injection is ahead of the air preheater, acid condensation and corrosion may occur in the cold end baskets.

Controls are now available, implemented with solid state apparatus, which digitally store the operating characteristics of the ESP as a function of the injection rate setting and, for any coal being fired, determine the characteristic in use and proceed to the optimum point of that curve. Periodic perturbations of the setting are made to determine that the set point remains optimum. If not, a new optimum is determined, and the new characteristic is stored for future reference. Besides freeing the operator from the necessity of closely monitoring the fuel and adjusting the set point, these controls retain a history of operation which can be retrieved for information on control and system performance.

15C.4.1.3 Catalyst for SO_3 flue gas conditioning applications. The catalyst for the conversion of SO_2 to SO_3 may be chosen from any of the types developed for the manufacture of sulfuric acid and similar applications. Vanadium pentoxide is the active ingredient in most of these, and is classified as a hazardous material. It appears that catalyst life in this service is on the order of 10 years, so that frequent handling is not required, but prevention of dust dispersion ingestion of, or contact with, the material, retention of the material in sealed containers, and disposal by qualified handlers, are necessary. In spite of its hazardous classification, servicing of the catalyst bed is neither particularly difficult nor expensive, but it is nevertheless desirable to do it as infrequently as possible. To this end care should be taken to exclude dust and water from the air intake to the maximum extent possible, and to service the air intake filter on a regular basis. Dust entering the system will tend to plug the inflow end of the catalyst bed and decrease the available air flow, eventually causing over temperature tripouts. Water has the effect of breaking down the physical structure of the catalyst pellets, so that interstices in the bed become plugged and air flow is restricted. Air intake locations which might be subject to overspray from cleaning hoses or windblown rain, for instance, should be suitably shielded or relocated to an absolutely dry position.

15C.4.1.4 SO_3 delivery to flue gas. The combustion air stream, bearing the generated SO_3, will exit from the catalyst at temperatures from 400 °C (750 °F) to about 540 °C (1000 °F), depending on the rate of SO_3 production. It is essential that this stream be held above its acid dew-point temperature all the way through the delivery manifold and injection probes until it actually issues into the flue gas. If the temperature falls below the dew-point, the acid will condense out in the manifold or the injection probes. This is undesirable for two reasons: firstly, none of the acid will reach the flue gas to do the intended conditioning job, and, secondly, the condensed acid will corrode the piping and nozzles. Since the SO_3 concentration is a few percent in this stream, the acid dew-point will be on the order of 240 °C (460 °F), but maintaining the delivery end of the system in a condensation-free state requires that the gas temperature, as it issues from the injection probe nozzles, be not less than 260 °C (500 °F). For this reason the distribution manifolds are heavily insulated, and in addition the injection probes, if installed on the cold side of the air preheater, are thermally insulated from the flue gas. Also, calculations of the expected temperatures at the most distant (or coldest) injection nozzle are made for every cold side installation.

It is probably obvious that avoidance of acid condensation in the manifold and probe areas requires preheating of the system with hot air when being started up from cold. It is not quite so obvious that a hot purge of the system during shut-down is required to remove sulfur products from the catalyst, so that acid will not be present in the warm-up gases at the next start-up. When proper start-up, operating and shut-down procedures are observed as described here, long manifold and probe life is obtained. Some early systems are approaching 20 years on-line with the original probe and manifold parts still in service. Steel sheathing over the insulation of cold-side probes is subject to the same erosive fly ash attack as other internal parts of the ductwork, and requires similar periodic maintenance and repair.

15C.4.1.5 Hot-side injection vs. cold-side. Because installation of injection probes is ordinarily more easily accomplished in the ESP-type ductwork on the cold side of the air preheater than in the boiler-type construction on the hot side, most of the present FGC installations inject on the cold side. It should be noted, however, that hot-side installation has the advantage of lacking any close approach to acid condensation temperatures in the probes, and provides excellent mixing and contact between the SO_3 and the fly ash as it passes through the air preheater. Hence, cases where long manifold runs may make it difficult to maintain sufficiently high probe temperatures, or where there is little cold-side gas travel length for mixing of the injected SO_3 with the flue gas, favor location of the injection probes on the hot side. In determining the desirability of putting the probes on the hot side, possible SO_3 condensation in the cold side of the air preheater must be considered. This is not usually a problem because the injected SO_3 quantity establishes air preheater conditions roughly equivalent to those expected from an unconditioned boiler firing coal containing 1.5% sulfur. If the average cold end temperature and basket material of the air preheater are compatible with that assumption, hot-side injection would ordinarily be acceptable.

Probes for hot-side installation are less expensive than cold side because no thermal insulation is required and the erosion-protective outer sheathing can sometimes be eliminated. The probe insertion point into the boiler on the hot side may expand/distort more than a cold-side point when the boiler is started up and shut down, and this can add to the difficulty of design and the expense of the manifold system, possibly offsetting savings in probe costs. As far as operation is concerned, every existing hot-side installation works as well as, or better than, equivalent cold-side units.

Note that the foregoing applies to injection of SO_3. Injection of ammonia, to be discussed later, is ordinarily not allowed on the hot side of the air preheater because of the tendency to plug the preheater with ammonium-sulfate compounds, unless there is little or no SO_3 in the boiler gas.

15C.4.1.6 One FGC system for multiple boilers. On the face of it, it seems like a good idea, considering the relatively high cost of the SO_3 generators, to use a single generator to feed conditioning gas to several boilers. The primary objection to such an arrangement is that, assuming operation of the FGC system is necessary for compliance with emission limits, all the boilers served by whatever portion of the system is common must be taken off-line simultaneously before the common portion can be shut down for service. The high availability figures compiled by existing installations are based on periodic (at least, annual) boiler outages during which accumulated maintenance requirements of the conditioner system can be met.

If, however, measures can be taken to maintain emission compliance with the FGC system off-line, or if simultaneous removal from service of all boilers on the common system is not objectionable, then a proportion or all of the SO_3 generator may be made common to more than one boiler. The most straightforward way to do this is by making the blower and air heating apparatus common and providing individual burners (if used), catalytic converters, manifolds and injection systems for each boiler. Control of such an arrangement is uncomplicated, and manifolding and injection requirements are identical to those for single boilers. The possible alternative arrangement where the burner and catalyst are also made common appears to offer cost savings, but is actually difficult to implement unless simplifying assumptions, such as operating all common boilers at the same output level, etc., can be made. To retain full flexibility in boiler operation it is necessary to provide valving in the manifold system. Careful heat tracing of the valves and the portions of the piping which may become dead-ended while still exposed to SO_3-containing gases is required to prevent acid condensation. Cold spots in valve bonnets are not easily avoided. Acid-resistant valves are expensive. Another less severe problem with this system concept is that fully flexible operation of the individual boilers requires a relatively complicated control system. In general, this approach is not recommended.

15C.4.1.7 Patterned or 'tailored' injection. Ideally, the amount of conditioning gas delivered to any single injection nozzle in the flue gas stream should be proportioned to the amount of ash passing through that nozzle's treatment area and adjusted for the flue gas temperature at that point. It is entirely possible to make comprehensive sets of measurements which determine these values, at least for one set of operating conditions, and it is also possible to divide the delivery of conditioning gas among the injection nozzles in accordance with such measurements. In practice, a full scientific treatment of this type is never done, and is rarely even approximated. The reason, aside from the variability of the target conditions, is that ESPs have a considerable degree of built-in tolerance for random variations. Witness to this fact is provided by many existing installations where flue gas conditioning was neither needed nor used. It is common in these units to find that a wide spread of temperatures across the air preheater outlets is passed through to the ESP inlet along with non-uniform dust concentrations, in spite of which the ESP exhibits excellent operation. In general, inlet conditions which meet industry criteria for satisfactory ESP operation allow treatment of a flue gas conditioning system on the basis of uniform temperature and flow distribution. Another way to state this is that situations which could require special precautions for distribution of conditining gases usually also require modifications to permit satisfactory ESP operation, and incorporation of the latter will usually eliminate the former.

15C.4.1.8 Load cycling. The burner and catalyst chambers of FGC systems are lined with refractory materials which can be damaged by thermal shocks. Recommended procedures for starting a system from cold conditions require a gradual warm-up procedure over several hours. Some modern boilers are capable of cycling from stand-by to full load in as little as 45 min. To permit the FGC system to follow this type of load change, it is held in stand-by condition with internal temperatures maintained at operating levels by the start-up heat source. For economy, air flow rates in stand-by mode may be reduced from normal operation, thereby reducing heat input. The FGC unit is capable of going from stand-by to full output in 30 min, and can therefore accommodate the requirements of boiler cycling. Because of the length of time required for a cold start it is normal practice to hold the FGC system in stand-by when the boiler is in a stand-by or banked-fire condition, and not go to a cold shut-down unless the boiler will be off-line for at least a few days.

15C.4.2 Ammonia FGC systems – design features

15C.4.2.1 Gaseous or anhydrous ammonia systems. The anhydrous ammonia generating systems generally consist of a liquid ammonia unloading station, combined storage/vaporizer tank, ammonia system skid and a manifolded injection system.

Liquid anhydrous ammonia is received in a tank truck. The ammonia is fed into the storage/vaporizer tank by means of a flexible transfer line. The anhydrous ammonia storage/vaporizer tank produces a supply of ammonia vapor at a constant pressure by maintaining boiling liquid ammonia at a preselected pressure in the tank. External electric immersion heaters with solid-state controls maintain a constant pressure in the tank as ammonia is vaporized. Upon withdrawing gaseous ammonia, the pressure and temperature will rapidly shift to the control point conditions of 4 Barg (60 psig) and 4.4 °C (40 °F). High and low pressure or liquid level will cause alarms. The vaporizer tank is fully automatic in operation and has a control range from 0 to 100% capacity.

The system skid includes a blower to provide the necessary filtered process air which mixes with ammonia gas from the vaporizer. Air discharging from the blower passes through an air flow indicator before mixing with the gaseous ammonia. Ammonia flow rate is adjusted automatically depending upon boiler load.

The ammonia/air mixture is fed to a separate manifold system with probes for distribution into the gas stream. Each probe will contain properly sized and spaced nozzles, which, in turn, inject the mixture at high velocity to ensure good mixing of ammonia and flue gas.

15C.4.2.2 Aqueous ammonia systems. Basically, the aqueous ammonia generating systems include an air blower, an air heater, controls and mixing chamber.

Liquid aqueous ammonia is supplied from the storage tank to the system skid by a metering pump. The flow of ammonia will then be controlled and delivered to the vaporizer vessel. A centrifugal blower, located on the system skid, provides the necessary filtered process air which is heated through an air heater box assembly, and injected into the mixing chamber with ammonia spray from the nozzles. Air discharging from the blower passes through an air flow indicator before entering the heaters. Ammonia flow rate is adjusted automatically depending upon a boiler load signal.

The ammonia/air mixture is fed to a manifold system consisting of probes for distribution into the gas stream. Each probe contains nozzles, which inject the mixture at high velocity to ensure good mixing of ammonia and flue gas.

15C.4.3 ESP power consumption

During most of the history of ESP collection of fly ash, it was considered that use of the full available output of the high voltage supplies indicated good operation and high efficiency collection, and this was highly desirable. In later years attempts to collect high resistivity dust, without resort to flue gas conditioning, lead to the construction of ESPs of monumental size, complete with row on row of hoppers, high voltage power supplies and ash handling apparatus. Whether or not these passed their guarantee test requirements, quite a few were eventually fitted with FGC which, among other things, promptly caused all the power supplies to go to full rated output. What once had been considered good ESP operation was now rightly considered outrageous consumption of station auxiliary power. Fortunately there is a cure for this situation.

Control units for the high voltage power supplies are now available, from most ESP manufacturers, which provide a feature called 'intermittent energization' or 'skip-cycle control'. Although the claim that this type of control materially improves collection of high resistivity dust has proved to be somewhat optimistic, there is a definite economic justification for its use in conjunction with FGC. Once the fly ash resistivity has been adjusted by FGC to an optimum value, high collection rates may be maintained while ESP power consumption is greatly reduced by use of the 'skip-cycle' feature. Power reductions to levels on the order of 20–30% of the unmodified power input have been obtained without noticeable increases in emission.

15C.4.4 Mixing requirements

The length of travel of the flue gas in the ductwork after the injection point, required to provide essentially complete mixing with the SO_3, is often glibly cited as 'one second mixing time' or 'ten times the nozzle spacing'. These rules of thumb are derived from experiments showing that in turbulent flue gas flow, complete mixing with another gas, injected through a bank of nozzles arrayed as a uniformly spaced grid, occurred at a distance downstream of the grid equal to about eight to ten times the nozzle spacing. In FGC systems a nominal grid spacing of 3 ft (94 mm) has been found to be a reasonable compromise. Ten times this spacing gives a mixing distance of 30 ft (9.14 m), which at typical duct velocities of 60 ft/s (18.3 m) allows only 0.5 s mixing time, and this is adequate provided you have a straight run of uniform size that long. Ordinarily you don't, hence the conservative quest for 1 s. Expansions, contractions and bends in the ductwork interfere with net mixing patterns. If the ESP ductwork is being modeled for flow visualization, it is desirable to include an examination of the mixing behavior which may be expected. Modification of the spacing, location and gas delivery rate of the nozzles are steps which may be taken to cope with less-than-optimum mixing situations.

15C.5 Application of flue gas conditioning in converting hot-side fly ash precipitators to cold-side operation

In some existing precipitator installations the difficulties associated with hot-side operation have become so great that complete revision of the emission control system is necessary to permit economical operation in compliance with emission limits. Studies have shown that conversion of the ductwork arrangement to reposition the precipitator to the cold side of the air preaheater, coupled with the use of sulfur trioxide injection to solve the high resistivity problem, is often the most economical alternative.

Increases in precipitator ash collection and boiler thermal efficiencies over the hot-side design basis becomes available. Net economic gains from such a conversion can provide payback of the conversion over reasonable time periods, especially, if the hot-side installation incurs generation losses from outage time for precipitator mechanical problems or from derating of output to meet emission limitations.

15C.5.1 Introduction

The ills which can afflict hot-side precipitators are well known. With all the recited problems, why in the world do we have so many units in our boiler plants? The underlying reason can be succinctly stated in two words – ash resistivity. In the 1970s the hot-side concept became temporarily popular because burning low-sulfur coals became common, either as a way to meet restrictions on sulfur dioxide emissions or because they were the only economical fuels available to the plant. There were at that time only two ways proven in utility service to successfully collect the resulting high resistivity fly ash: big cold-side precipitators, or smaller ones on the hot side.

Figure 15C.4 depicts the typical form of the response of precipitator ash collection rates to variations in ash resistivity. The dashed portion on the left indicates reduced collection due to rapping and re-entrainment losses when the resistivity is quite low. The position at which this effect appears depends on gas velocity and rapping practice. As resistivity increases a region of maximum collection rate is found, usually around 5×10^{10} Ω-cm. As resistivity increases further the collection rate drops off, until finally at high resistivities a level of low collection rate is found.

Figure 15C.2 is a typical plot of resistivity *versus* temperature for fly ash in the form developed by Dr R. E. Bickelhaupt [4]. It is clear that operation at cold-side temperatures will produce resistivities with this ash that are considerably higher than the 5×10^{10} Ω-cm value at which (for this set of curves) best collection rates would be obtained. One possible choice for collecting this ash, then, would be to build a precipitator large enough to achieve the desired efficiency at the reduced collection rate. Unfortunately, in the 1970s (and maybe now) selection of the correct precipitator size in this operating mode was somewhat uncertain and vendors' sizing margins for guaranteeability made for excessively large and expensive units. However, examination of Figure 15C.2 shows that the resistivity comes down very nicely to the optimum level if the temperature is raised above about 350 °C (650 °F). With care taken to postpone the onset of rapping and re-entrainment losses at low resistivities as much as possible, this was expected to be an area of high collection rates, handicapped only by the necessity of sizing the precipitator to accommodate the gas volume generated at the chosen temperature.

Not all hot-side precipitators exhibit severe problems. Some worked well as-built, and others as-modified. But some didn't. The assorted structural problems arise from improper execution of high temperature designs, not from the concept of hot-side operation. However, two major problems with the concept itself have surfaced. First, the 'sodium depletion' phenomenon was unknown when these units were designed. It was only after ash resistivities which were supposed to be low turned out to be drifting high as operating time was accumulated, that Dr Bickelhaupt and the Southern Research Institute proposed and documented this nasty bit of ash behavior in precipitators [21]. Secondly, the procurement specifications for these precipitators almost invariably spelled out performance requirements at full load, but were silent on the subject of reduced loads. When load drops, temperature decreases, resistivity increases, and the units tend to run out of the window of good operation and produce excess emissions at lower loads.

As the units age, cycling often becomes common, adding to the owners' woes [22].

In more than a few cases these problems with the hot-side concept, added to those of unsatisfactory execution, have led to consideration of the possibility of switching problem units to cold-side operation. If this is done, there are two basic approaches to attaining adequate collection of the fly ash. One is to accept the ash resistivity as produced by the boiler – generally high, the reason hot-side was chosen in the first place. Sufficient fly ash collection with this approach can be obtained in three ways: by building a new baghouse which is insensitive to resistivity; or by extending the existing precipitator, again hoping to make it big enough. In the 400 to 600 MW size range, current installed costs for baghouses or new precipitators are $25 million and up, mostly up.

The second approach for cold-side operation is to adjust the ash resistivity to an optimum value by sulfur trioxide flue gas conditioning (FGC). Current costs for complete conversion including FGC, again in the 400–500 MW size range, appear to be from $9 to $12 million per boiler unit. Factors involved in this choice are the subject of this chapter.

In some cases the existing precipitator, or that precipitator with an addition, may be large enough for cold-side collection of a particular ash without FGC. This occurs when the expected cold-side ash resistivity is low enough to be in the good operating range. Such situations should be thoroughly examined to be sure that the expected resistivity will be achieved over the full range of coal and ash analyses, and that reduction of resistivity to the desired point is not heavily dependent on the sodium content of the ash. The reason for concern with reliance on sodium content to produce the desired resistivity is that sodium in coal is believed to result from the salt content of water to which the coal was exposed during formation. Considerable variation often occurs as mining progresses. Also, the sodium depletion phenomenon is at work at low temperatures as well as high, and the familiar drift of resistivity with operating time can occur, resulting in high apparent ash resistivity. However, if these factors have been taken into account and, if flexibility with respect to coal sources is not a major consideration, the existing or enlarged precipitator may be adequate without FGC. Costs for this option are, of course, highly site specific.

15C.5.2 Flue gas conditioning

Table 15C.2 shows design figures typical of hot-side precipitators, compared with the expected performance on a very conservative basis for the same unit converted to cold-side with FGC. Because the decreased gas volume increases the operating SCA, and the increased gas density allows higher operating voltages, a significant reduction in emission occurs with the same ash loading at the precipitator inlet. Unit 1 at the Pleasant Prairie station

Table 15C.2 Typical factors – hot-to-cold conversion

	Hot	Cold w/FGC
Flue gas temperature (°F) (°C)	800 (427)	300 (149)
Flue gas volume (%)	100	60
SCA Ft^2/1000 ACFM ($m^2/m^3/s$)	300–350 (59–69)	500–580 (99–115)
ESP migration velocity		
w1.0 (cm/s)	8.5	6.5
w0.5 (cm/s)	44	50
ESP efficiency (%)	99.5	99.9

of Wisconsin Electric Power Co. has operated for nearly 7 years in this mode (converted to cold-side before start-up) and has reliably and continuously performed at higher migration velocities and lower emissions than shown by Table 15C.2 [23]. Unit 2 at the same plant has operated similarly for more than 1 year. Both units burn coal from the Powder River Basin of Wyoming, and are deep-cycled on a daily basis. FGC has made hot-to-cold conversion possible with a major gain in compliance margin for a fixed precipitator size.

15C.5.3 Installation design

15C.5.3.1 Probe location. A first consideration for FGC installations is location of the injection probes. In a hot-to-cold conversion there is some freedom of design of the new ductwork connections from the economizer to the air preheater and from the APH to the precipitator inlet. Advantage should be taken of this flexibility to avoid handicapping FGC operation through lack of reasonably satisfactory probe positions. As discussed, it is desirable that there be at least 1 s residence time of the injected SO_3 in the flue gas ahead of the precipitator face. At least half of this should be in non-expanding ducts ahead of any divisions. Vanes should be at least one equivalent duct diameter distant from the probes, both up and downstream. A sizeable temperature spread usually exists across the gas outlet of most APHs, so, where possible, mixing of streams from multiple APHs should be done so that the individual temperature distributions tend to cancel out and produce final streams of relatively uniform temperature; this is important since ash resistivity and SO_3 injection rates are temperature-related and more uniform temperature promotes more uniform precipitator operation. In plants where the primary APH(s) is not banked with the secondaries, similar precautions should be taken in positioning the ductwork connections which join its gas stream into the main gas flow.

In some cases it is impossible to establish a reasonably satisfactory probe location downstream of the APH. In these cases locating the probes on the hot side of the APH is an alternative. There are numerous existing installations where this arrangement has operated for a number of years without problems. Average cold end temperature considerations and soot blowing requirements for the APH then become roughly equivalent to those experienced by boilers burning coals having sulfur contents in the 1.5–2% range.

The preferred orientation for probe insertion is vertically downward. From the standpoint of system operation, there is no difference between vertical, horizontal or slanted orientations, and all have been used successfully. However, probe inspection and maintenance is easier with a top-mounted vertical arrangement.

Few, if any, of the FGC installations now in operation fully comply with all the probe location requirements set forth here. There is, therefore, some latitude available to accommodate the physical restrictions inherent in existing plants. Following these guidelines to the greatest reasonable degree will help in obtaining best operation and simplest maintenance.

15C.5.3.2 Burner-converter. Having fixed the probe location, the next step is locating the burner converter unit. This is typically about 8 ft wide by 8 ft (2.44 m) high by 20 ft (6.1 m) long, weighs about 13 tonnes, and is for indoor service. Three feet of space on all four sides should be provided for access. The first consideration in locating this unit is that it be as close as possible to the probe location, preferably within 17 to 24 m (50 to 80 ft) to minimize the installation expense and heat loss from the SO_3 hot gas piping, and, secondly, that the station operators should have easy access to the chosen position.

15C.5.3.3 Tank and pumps. The sulfur storage tank and the pumping unit are relatively free of restrictions on their location. They will be at grade, and the pumping unit requires weather protection. Truck access for sulfur delivery is, of course, required.

15C.5.3.4 Piping. The hot gas piping which conveys the SO_3 gas from the converter outlet to the injection probes is stainless steel, and is heavily insulated to prevent the SO_3 gas temperature from dropping below its aid dew-point before injection into the flue gas. This pipe run must be designed and supported to accommodate thermal expansion effects arising from temperature changes from ambient to 540 °C (1000 °F). Also, pipe routing and the positioning of anchor points and supports must be chosen to prevent overstressing due to ductwork movement, wind and seismic loadings.

Pipework for molten sulfur is steam-jacketed throughout its length. Its design is less demanding than hot gas piping, since its operating temperature is approximately 140 °C (285 °F), but care must be taken to provide for proper condensate drainage from the jacket. Concentric sulfur-inside-steam arrangements and straight-through valve jackets are used to minimize pressure drop in the steam system and to avoid the cold spots and steam flow restrictions which appear at flange-and-jumper joints. Selection of proper steam flow patterns and points of steam supply to the system is important so that freezing and thawing of the sulfur does not develop excessive internal pressures at pumps or instruments.

15C.5.3.5 Steam supply. A single 600 MW boiler installation will require about 70 kg/h (150 lb/h) of saturated steam at 2.4 bar (35 psig) for cold start conditions (frozen tank and lines) and about 23 kg/h (50 lb/h) during normal operation. Naturally these figures vary with tank size, pipe length, quality and type of insulation, etc. This is usually taken from a low-pressure station header. Desuperheating may be required, depending on source conditions. For these relatively small quantities saturator-type desuperheaters appear to be inexpensive and reliable. Small auxiliary boilers are sometimes supplied for start-up of FGC systems in isolated plants.

15C.5.4 FGC Operation

An objective of the design of SO_3 FGC systems has, from the start, been minimization of the amount of operator attention required. Most of the operator input for normal running takes place at the burner-converter control panel, with occasional operations at the pump unit control panel.

However, a cold start which occurs only after an extended outage requires that the steam supply system valves be opened to initiate thawing of sulfur about 4 days before the system is to be put into service. Then, about 2 days in advance of boiler operation, warm-up of the burner-converter unit is commenced by operating the 'Standby' push-button at the control panel. This starts an automatic program which gradually heats the refractory materials and catalyst chamber to operating temperature. When this is complete, SO_3 delivery may be begun by pushing the 'Run' button, and the system will go on-line. Adjustment of the SO_3 injection rate is made by trial-and-error variations, starting from an estimated value or from a setting determined by previous experience, and using opacity measurements or precipitator power input as a criterion for evaluating the effect of ongoing rate changes. To stop injection for a short outage the 'Standby' button is pushed and the unit reverts to a hot ready-to-run condition. For a complete shutdown the 'Purge' button is pushed and the FGC operation is terminated after automatically stopping sulfur flow and blowing all sulfur products out of the hot gas pipe and probes.

When fuel and ash analysis changes, a repeat of the trial-and-error procedure for determining the best injection rate is again necessary. In most installations there is a fairly broad range of injection rates at which ash resistivity and precipitator performance are near optimum, and determination of a satisfactory injection rate is easily and quickly accomplished. For stations burning a wide variety of coals, additional provisions for automatic adjustment of injection rate may be desirable.

15C.5.5 Economics

Experience indicates that the cost of a single-boiler FGC system is about 10–15% of the cost of converting the precipitator from hot to cold. The percentage decreases for multiple boiler systems because of shared pump skid, tank, and engineering costs. Where the existing air preheater can be used in the new system without relocation, the outage time for conversion of the existing precipitator with FGC is almost invariably less than that of any other option.

When the air preheater must be moved and/or replaced, longer outage times may be required, perhaps making other options competitive insofar as outage time to convert is concerned. In any case, minimum outage time requires careful and knowledgeable scheduling of prefabrication and installation operations.

Table 15C.3 Factors for preliminary FGC cost estimation sulfur-burning systems[a]

	300 MW	600 MW
Equipment cost[b]	1,200,000	1,550,000
Cost to install[c]	See note	See note
Sulfur usage lb/h (kg/h)		
Maximum[d]	101 (46)	205 (93)
Typical[e]	51 (23)	103 (47)
Electrical power (kW)		
Connected	170	350
Operating	97	195
Steam usage[f] lb/h (kg/h)		
Max., cold start	117 (53)	176 (80)
Operating	33 (25)	48 (22)
Operator (man-h/year)	300	300
Maintenance (man-h/year)	300	300
Spare parts cost/year ($)	5000	5000

[a]Single boiler, 0.4% sulfur coal, 10–20% ash, 20 ppm SO_3.
[b]1987 $ FAS East Coast port, includes supply of storage tank and preparation of preliminary installation design plus services to inspect construction and to supply start-up assistance, operator training, and instruction manuals. For additional boilers on same system add 65% of single boiler cost per added boiler.
[c]Installation cost for material and labor to install system and make piping and wiring connections between factory units and to plant utilities is approximately 35% of equipment cost.
[d]At full output capacity (20 ppm).
[e]At 10 ppm output.
[f]Steam at 35 psi g, saturated (2.4 bar).

A preliminary estimate of FGC system costs can be made from the figures in Table 15C.3. Note that the figures given include services to perform preliminary site surveys, prepare recommendations for equipment locations, and furnish piping support and routing information, intermittent construction advice on site, start-up assistance and operator training, instruction books and spares recommendations, along with on-site vendor-owned spares to support warranty requirements. Other services available include estimation of cold-side precipitator performance and emission opacity, and extra costs after start-up services, ranging from periodic inspection and/or maintenance, with or without parts supply, to continuous on-site operation.

SO_3 FGC does only one thing – it eliminates problems due to high ash resistivity. The conversion program should include funds to correct all the accumulated precipitator faults which have been recognized in previous operation, or have been found in internal inspections, so that trouble-free performance will be obtained on a long-term basis. Also, provision should be made for insulator venting and heating, and for hopper heating and level control – items which were often minimally present or omitted from hot-side designs. Some ash-handling systems have been based on the use of precipitator hoppers as ash storage bins. This has been found to be poor practice, and modifications to allow constant ash removal should be installed [24].

Precipitator power controllers should be state-of-the-art in every respect, especially in their ability to control sparking and to follow variations in operating conditions. Obsolete or marginal controllers should be replaced, and power savings possible through the use of precipitator energy management systems or 'intermittent energization' (IE) control units should be evaluated. With FGC all the power supplies downstream of the first two fields tend to go to maximum rated power input. Use of IE controllers, which interrupt conduction of power to a precipitator field during a fraction of the line-frequency cycles, permits a power saving of one-half or more of the total rated kilowatts of all the transformer rectifier units on the precipitator without sacrificing collection efficiency [25]. This is sufficient, in many cases, to pay off the investment in new or modified power controllers in an attractively short period of time.

Factors which contribute to amortization of the hot-to-cold conversion expense are the increase in thermal efficiency from decreased heat losses ahead of the air preheater (to the extent that heat can be recovered in the APH), possible decreased ID fan horsepower from decreased air inleakage (often high on hot-side units), decreased lost generation from derating for emission compliance or from wash-down outages, and, for some, elimination of penalties imposed for excess emissions. Other positive factors more difficult to quantify include decreased precipitator maintenance, better stack appearance, greater choice of fuel sources, and freedom to make load reductions or to cycle load without emission excursions.

15C.5.6 Conclusions

It is concluded that SO_3 FGC provides a solution to the ash resistivity problem which gave rise to the use of hot-side precipitators in the first place. Use of the existing precipitators converted to cold-side operation is, in many cases, more economical than possible alternative cold-side conversion options, as well as being more economical than continuing with the existing hot-side operation. Experience with more than 200 retrofitted boiler units shows that cold-side FGC can be relied upon. In summary:

1. FGC allows maximizing precipitator performance, independent of natural ash resistivity.
2. FGC allows a precipitator switched to the cold side to outperform its hot-side design basis.
3. FGC plus hot-to-cold conversion is often far more economical than other options.
4. Theory and operation of FGC has been repeatedly proven in both new and retrofit installations, and is both predictable and reliable.

15C.6 FGC systems – optimization

The main objective of the optimization methods is how to extract the maximum amount of unbiased information regrading the factors affecting an 'injection rates – opacity' relationship from as few tests as possible. A term 'optimization' for the purpose of this discussion is defined as a search for conditioning agent(s) injection rate(s) to provide the best ESP performance, i.e. lowest stack emissions and opacity.

Typically, the FGC systems allow for an independent adjustment of the SO_3 and/or NH_3 injection rates to evaluate their impact on the precipitator performance (stack opacity). Therefore, one could adjust the injection rates in a systematic manner to learn which factors have greatest impact on the resultant opacity. Using this information, the injection rates can be constantly improved until optimum performance is obtained.

There are three fundamentally different approaches to the FGC system optimization:

1. theoretical approach, based on the computer models for estimating a SO_3 *vs.* flue gas temperature relationship;
2. procedural approach to adjustment of one variable at a time; and
3. statistical methods of experimental design while working with more than one independent variable:
   (a) simplex method
   (b) central composite design

15C.6.1 General

One assigned person should be responsible for all the injection rates at any given site. The reason for this is that experience will make it possible to recognize the effects of change and the proper steps to be taken with the least amount of delay and experimentation.

It is important to make sure that the meters on the precipitator power supply controls are in good operating order. All power supply controls should be in good operating condition with current limits set to the primary current rating of the transformer-rectifier sets, and they should run under 'automatic' control. Also, opacity monitors should be in calibration and operating correctly. All this is necessary because these are the devices to be used for references in making injection rates adjustments.

15C.6.1.1 Definitions. Figure 15C.10 represents a specific case where two opacity meters register instantaneous opacity side-by-side.

Rapping spike	A single peak registered on the opacity chart. The rapping spikes are usually the best indication of rapping mechanism impacts.
Baseline	The line approximating the lowest portion of the opacity trace.
Baseline trend	The direction of the baseline (either downwards when opacity is decreasing or upwards when opacity is increasing).

15C.6.2 Theoretical approach

15C.6.2.1 Discussion. The calculation methods developed by Dr Bickelhaupt were published in a form which estimates the resistivity of fly ash in the presence of a given concentration of sulfuric acid. Simple mathematical manipulation of the Bickelhaupt formulae makes it possible to calculate the acid concentration required to bring about a desired change in resistivity. Using the latter as a starting point, Coe and Krigmont developed a method for SO_3 injection rate estimation [10]. Accurate prediction of injection rates for SO_3 flue gas conditioning systems is now possible for the majority of stem coals in use. The described procedure may be the first step in any FGC system optimization. Coal and ash analyses are required for input to the calculation procedure. Equipment size and feedstock requirements can be closely determined, given knowledge of the range of fuel properties to be accommodated.

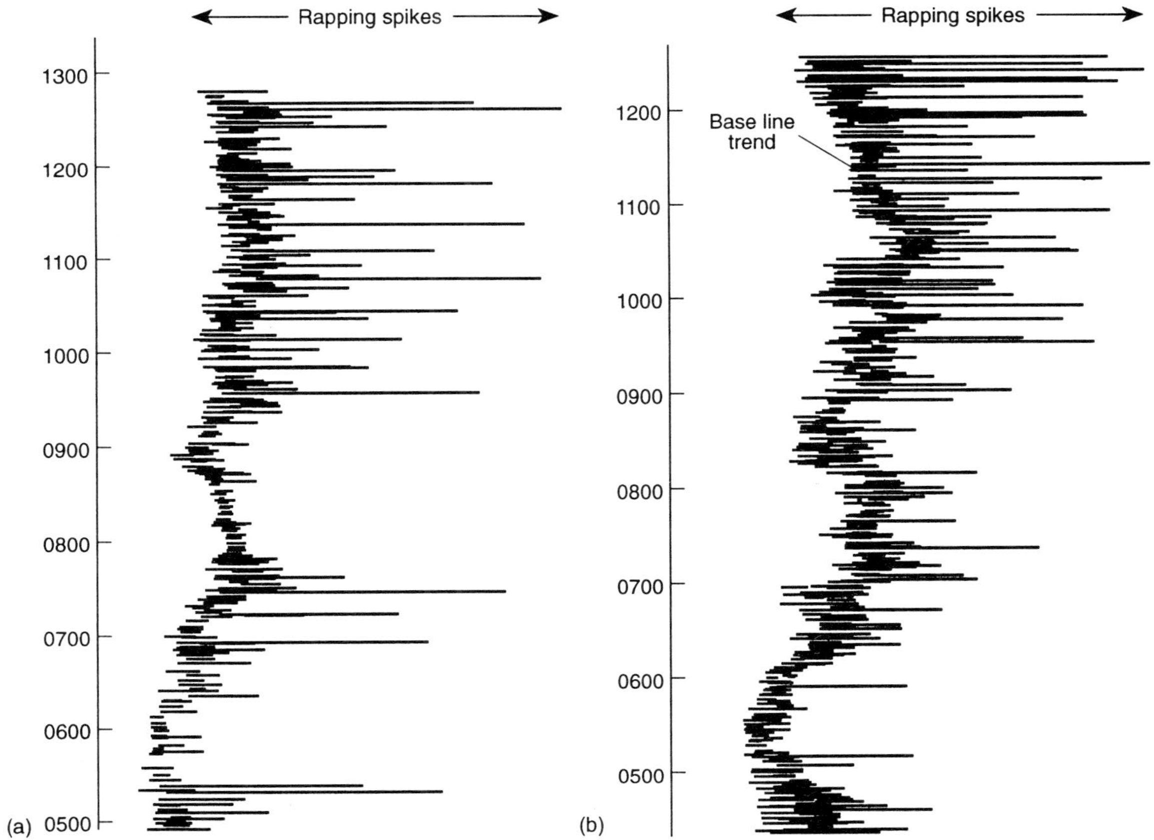

Figure 15C.10 Two opacity meters registering instantaneous opacity.

15C.6.3 Procedural approach

15C.6.3.1 SO_3 injection optimization. A need for SO_3 injection is usually shown by increasing opacity and decreasing power input to the precipitator, as compared with clean start-up conditions. The initial amount of SO_3 required may be obtained either from the computer models or assumed based on previous experience of the 'optimizer'. The former will be estimated based on the most recent coal and ash data and is not an exact value but should be approximately correct. The correct value is the one which gives the least emissions, to be determined during optimization trials. The ppm amount of SO_3 being injected is typically set by the ratio station.

Generally, fly ash resistivity increases with increasing temperature and decreases with a drop in temperature. Resistivity too high for best operation causes reduced precipitator power input and increased opacity. Resistivity which is too low can lead to excessive rapping losses. A proper SO_3 injection rate will eliminate the high resistivity effects without bringing about excessive rapping losses. Rapping losses show up as 'spikes' on the opacity chart. Some spiking may be present at the best injection setting, but should not be so great as to raise the average opacity more than a few points above the opacity base line.

To find this balance point it is suggested to start with the recommended SO_3 injection rate. This should begin to bring down the opacity and increase the current delivered by the power supplies. Where a layer of high resistivity ash has been laid down on the plates, there may be a considerable time delay before the SO_3 penetrates into the layer and reduces its resistivity. There have been delays as long as 5 days, but a change is usually seen within 8 h. If there is no change after 24 h and the opacity trace is not showing high rapping spikes, the injection rate should be increased in steps of about 20% of the previous rate, waiting 8 h after each change, until operation starts to improve or the rapping losses become excessive.

From here on, trial-and-error adjustments of the ppm rate are made, first upwards, and then, when total emissions (by stack opacity monitor) start to increase, start to back down the ppm rate until the total emissions start to rise. At this stage, each step of adjustment should be about 10% of the rate, or 1 ppm, whichever is smaller. On the first few steps watch the opacity monitors and see how long it takes after a step change for the precipitator operation to settle out to a constant emission level. Allow this length of time plus 1 h between changes of the rate. There is nothing wrong with a longer waiting time between steps, just be sure to wait long enough to see the full effects of the previous change. Usually 8 h per step is enough, but go by the monitors. The final result is to determine the lowest injection rate at which minimum emission is obtained.

15C.6.3.2 NH_3 injection optimization. A need for NH_3 injection is typically shown by excessive rapping losses (high rapping spikes on an opacity chart) remaining after SO_3 injection rate has been optimized (in the case of the dual FGC system), or existing due to the 'naturally' high amounts of the SO_3 present.

To find the best combination it is suggested to begin to increase the NH_3 injection rate in 1 ppm steps beginning with 1 ppm.

When NH_3 is introduced with SO_3 present, a fine fume consisting of different ammonia sulfate products is formed. The effect of these products can be observed in the increased space charge and improved ash coagulation and agglomeration characteristics. The former may be detected by increased voltage (in the first field predominately), while the latter should command reduction in the rapping losses (reduced rapping spikes on opacity charts).

The SO_3–NH_3 chemical reaction obviously will reduce the amount of SO_3 available for the resistivity modification, which may lead to reduction of precipitator power consumption and increase of the stack opacity. That will be a signal to increase the SO_3 injection rate by the same amount as the latest change of NH_3.

Further, continue to make trial-and-error adjustments, first to the NH_3 injection rate followed by the same ppm of SO_3 should the precipitator power consumption begin to decrease and/or stack opacity begin to increase. Follow the guidelines as described in 15C.6.3.1. The final result is to determine the best combination of SO_3 and NH_3 to obtain lowest possible average stack opacity by eliminating or significantly reducing rapping losses.

As long as the coal supply and boiler operating conditions remain relatively constant, the ppm rate should stay the same over a long period of time. However, occasional trials by raising and lowering the rate one step should be made to make sure operational changes have not occurred. For all such tests, both initially and later on, record boiler operating data and precipitator power supply readings at each ppm rate for future reference. Opacity monitors are the most convenient and informative references for this kind of adjustment, but precipitator power data can serve the same purpose.

15C.6.4 Statistical approach

The first step in finding the optimal combination of SO_3 and NH_3 injection rates is to specify the goal function. It must take into account the effects of all the important real life variables, with the exception of the independently controlled injection rates. As the first approximation it may be the stack average opacity itself. As for the second approximation, one may additionally incorporate boiler load change and specify the goal function as:

$$\text{Opacity factor} = \frac{\text{Constant}}{\text{Boiler load} \times \ln\left(\dfrac{\text{Opacity, \%}}{100}\right)} \qquad (15C.12)$$

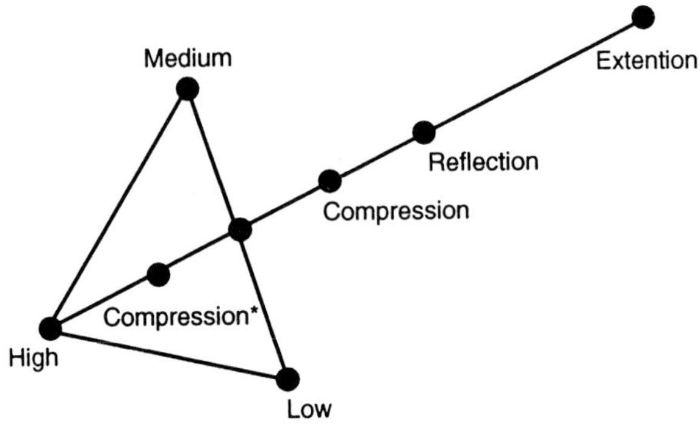

Figure 15C.11 Simplex method layout.

The boiler load and the stack opacity could be further normalized to account for the boiler maximum continuous rating (MCR) and established opacity limit. After the goal function is established, the optimization task is to minimize this function in the two-dimensional matrix ($n = 2$) where the coordinates are the SO_3 and NH_3 injection rates in ppm.

15C.6.4.1 Simplex method. In the most simplified form, the Simplex method for two independent variables (SO_3 and NH_3 rates, ppm) might be realized on the two-dimensional matrix (Figure 15C.11). At the starting position, one has first to find the value of the goal function (a function that has to be minimized – opacity or opacity factor) in two randomly selected points on the matrix. The difference between the appropriate injection rates for these points is usually chosen in the range 2–4 ppm. Next, one has to find the value of the opacity in the third point, which is located on the matrix so that it forms a nearly normal triangle with the two previous points. That is, it is equally distanced from the above-mentioned points. The three values of goal function (opacity) are then compared and the highest one is chosen. The point where this value was observed has to be excluded from the optimization process. To find the next, or 'third' point for the triangle (simplex) on the matrix, one moves in the direction opposite to the excluded point. The position of this new point may be easily found by reflecting the rejected point symmetrically with respect to the line connecting the two remaining points. When the value of the opacity in this point is found, the values of opacity for the three recent points are compared. Again the 'worst' point is located and rejected. This procedure is repeated continuously until the triangle will start to rotate on the same zone of the matrix. Then one may decrease the distance between the points (i.e. the dimensions of the triangle) to locate the optimum point with satisfactory accuracy.

One of the most efficient and reliable methods for the search of the optimal combination of these rates is the modified Simplex method (Nelder–Mead method). The idea of this method is to compare the values of the goal function in $n + 1$ (three in our case) points and to move the simplex (triangle in this case) in the direction of the optimal point. During this process the simplex (triangle) changes its form to reflect the specifics of the response surface shape – it extends in the direction of surface slope, rotates near the surface cavities and compresses itself near the optimum point.

In each iteration step (while using this method), one has to find a new triangle on the plane (SO_3, ppm–NH_3, ppm). Following the above procedure, at the start of the process the goal function values at three points (low, medium and high at Figure 15C.11) are determined. Next the coordinates of the reflection point are calculated and the value of the goal function in it is found. If this value (opacity reflection) is less than the lowest value for the three points used (opacity low), the coordinates of the extension point are calculated and in the next experiment the goal function value (opacity extension) is found.

In either case the opacity reflection is compared with the opacity medium and based on the results of this comparison we use this reference point, or go to a new point, called compression. The position of this compression point is calculated based on the sign of inequality between opacity factors in the reflection point and the 'highest' initial point. The next step gives us the next three points that are the basis for the following step. This iterative process is going on continuously with the duration of these steps equal to the time required for unit opacity stabilization on any new point.

Theoretically, the optimization process would stop itself when for a recently found triangle highest and lowest values of opacity factor became close enough (for a predetermined accuracy). But in real life, due to the possible changes in fuel, or in unit operating parameters, this process will continue to look for the current position of the optimum point.

15C.6.4.2 Central composite design method

(a) Overview. Basically, the yield of the dual FGC system optimization is a function of precipitator performance and, respectively, the stack opacity. Unfortunately, these two variables often do not affect the resultant opacity in linear fashion. In other words, it is not 'the more, the better'. Rather, both of these conditioning agents are related in a curvilinear fashion to the resultant stack opacity. Thus, our goal would be to *optimize* the yield *surface* that is created by the two independent variables: SO_3 and NH_3 conditioning agents, and the opacity as a dependent variable.

Figure 15C.12 represents an example of such a surface. Specifically, it is recommended to use a 'second-order central composite design' [26] to design an experiment that would allow one to estimate this surface. The two

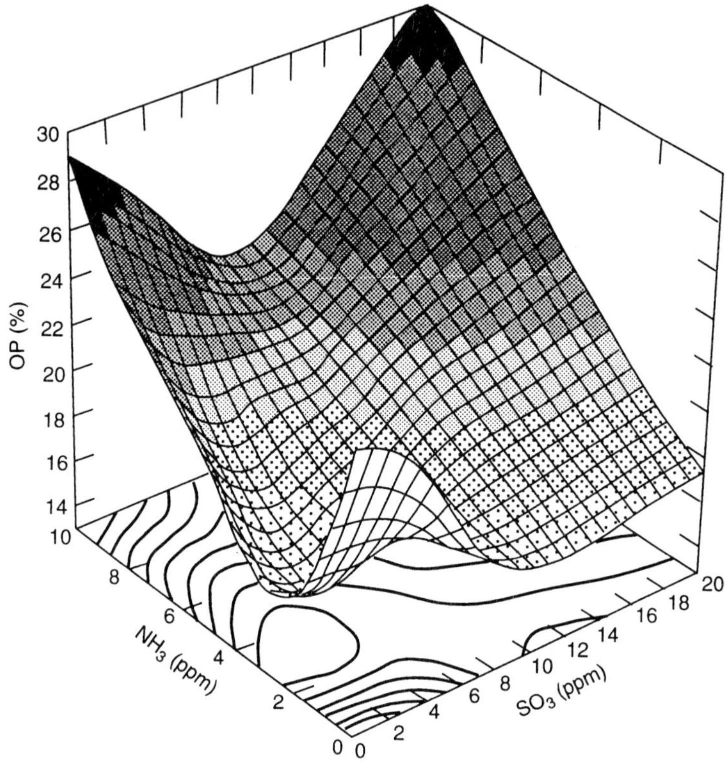

Figure 15C.12 Second-order central composite surface.

horizontal axes in the plot are the two process independent variables (SO_3 and NH_3 injection rates). The vertical axis shows the expected opacity. Clearly, resultant opacity is not a simple linear function of the two variables. However, given the information on the plot we would have a fairly good idea in which direction to proceed, i.e. how to change the injection rates in order to optimize the stack opacity.

(b) Basic description. Intuitively, it should be clear that in order to detect curvature in the relationship between the input factor (such as SO_3 injection rate) and the outcome-dependent variable of interest (stack opacity), one would need at least three levels for the respective factor (or, what is called in statistical science, three design points). Three points are: zero (0) or the middle point of the injection rate range, for example 10 ppm for a 20 ppm system; negative one (−1) or one-half of the range under evaluation, for example it could be 5 ppm for a 20 ppm system; and a positive one (+1),

for example keeping the same 20 ppm system, it would be 15 ppm value. Only then can it be decided whether, for example, the change from point −1 to 0 and from 0 to +1 is linear, resulting in proportional changes in the outcome variable, or whether the relationship is curvilinear, resulting in disproportional changes in the outcome variable. In the simplest experiment this is exactly what one would do, i.e. choose three design points (levels of the independent variables), and explicitly test the significance of the linear relationship and the curvilinear relationship.

This logic easily extends to multiple factors; however, in that case we would not have a simple (additive) relationship between two variables (SO_3 and NH_3 injection rates), but the outcome variable of interest (stack opacity) rather forms a surface over the ranges of the injection rates.

The surface in question is constructed (estimated) entirely from the points in the design (combinations of setting for different factors). There is the necessity to include at least three levels (settings for each factor in the study), otherwise the curvature of the surface could not be detected.

Therefore, to the design points forming square (only two dimensions) or 'cube' points, we have to add so-called 'center points' (coordinates 0,0) to allow for an explicit test for curvature.

Furthermore, in a second-order central composite design, the 'cube' is enhanced by a 'star', the design points required to estimate the nature of the curvature; specifically, to allow estimation of the second-degree polynomial or quadratic components of the relationship between the factors and the dependent variable.

The levels for the factors that make up the design (experimental) points may appear as arbitrary. In fact, these numbers are not at all arbitrary but rather are carefully chosen so that the 'star' points are arranged to form a circle, together with the 'cube' points around the center point. Therefore, the information functions, that is, the amount of information we can obtain from the design about the surface, is symmetrical about the center point. Consequently, the design can be rotated (around the center point) in any direction without gaining or losing any information.

Computationally, this type of the experiment can be analyzed *via* multiple regression procedures. The data is fitted in the following model:

$$\gamma = \beta_0 + \beta_1\chi_1 + \beta_2\chi_2 + \beta_{11}\chi_1^2 + \beta_{22}\chi_2^2 + \beta_{12}\chi_1\chi_2 + \varepsilon$$

Thus we estimate the following parameters:

β_0, a constant (intercept);
β_1, the parameter describing the linear relationship of y (the dependent variable opacity) with variable x_1 (SO_3 injection rate);
β_2, the parameter describing the linear relationship of y (opacity) variable with x_2 (NH_3 injection rate);

β_{11}, the parameter describing the curvilinear (quadratic) relationship of y with x_1 (this parameter is only estimated when analyzing a second-order central composite design);

β_{22}, the parameter describing the curvilinear (quadratic) relationship of y with x_2 (this parameter is only esimated when analyzing a second-order central composite design);

β_{12}, the parameter describing the (linear) interaction between x_1 and x_2 in their relation to y.

15C.6.5 Rapper adjustments

Once optimum injection rates are found, further reduction of emissions may sometimes be achieved by optimization of the timing and impact setting of the rapping system. Here again the opacity monitors will serve as a guide.

There is no clear rule of thumb with respect to the rapping frequencies. A factor of four (4) may be a good starting point. Using the 15 min first fields frequency extend each subsequent field four times longer than the

Table 15C.4 FGC system optimization data acquisition

		FGC system information	
Boiler data	Precipitator	SO_3 skid	NH_3 skid
Boiler load (MW)	Primary power consumption (A, V)	SO_3 ratio station setting	NH_3 ratio station setting
Coal flow (24 h period)	Secondary power consumption (mA, kV)	Boiler load signal (%)	Boiler load signal (%)
Coal proximate analyses	Spark rate	Air heater control setpoint (°F) °C	Ammonia tank pressure (psi) bar
Economizer outlet temperature (°F) °C	Rapping cycle schedule	Converter inlet temperature (°F) °C	Air flow (inch w.g.) mb
Air heater outlet temperature (°F) °C	Number power supplies in/or out of service	Converter outlet temperature (°F) °C	
Precip. outlet temp. (°F) °C		SO_3 skid air Fl. (scfm) sm^3/s	
Steam flow (lb/hr)		Sulfur flow (%, lb/h) kg/h	
Oxygen (%)			
NO_x, ppm, lb/h, kg/h lb/MBTU			
SO_2 (ppm, lb/h, kg/h lb/MBTU) kg/MJ			
Stack opacity (%)			
Ambient temp (°F) °C			
Barometric pressure (mmHg) mb			

preceding fields. The spiking on the opacity trace will be an overall indication of success. Next, double those values could be evaluated.

Plenty of time should be allowed between changes of rapper settings to allow the dust collection process to come to equilibrium. Several days is desirable because this can sometimes be a very slow process.

15C.6.6 Data collection

To assist the optimization efforts, system data should be collected at regular intervals. Initially, the data intervals are short (1/h typ.). This is due to the unknown response time and equilibrium times of the system. Once we have identified the reaction times, data acquisition can revert back to boiler house standards.

Table 15C.4 shows a typical data request. Boiler data and ESP data should be recorded on an hourly basis. FGC system data should be taken once per shift.

15C.7 Conclusions

In any ESP operation, dust collecting performance is often adversely affected by resistivity effects or by rapping and re-entrainment. Problems of this type reduce the performance level of the equipment compared with that which it is capable of attaining if the dust to be collected is treated to permit optimum operation of the collecting mechanisms. In many cases maximum economy may be attained by application of flue gas conditioning to an appropriately sized dust collector having fewer hoppers, less acreage, less power and much less maintenance than conventional units handicapped by being required to collect dusts which have not been properly prepared for efficient collection. Advantages obtainable by the use of flue gas conditioning will permit better emissions control with smaller, more reliable, less expensive ESP installations. There is a possibility, yet to be fully confirmed, that fabric filter performance may be similarly subject to improvement in certain cases.

References (15C)

1. White, H.I. (1963) *Industrial Electrostatic Precipitation.* Addison-Wesley Publishing Co., Reading, MA.
2. White, H.J. (1974) Resistivity problems in electrostatic precipitation. *J. APCA*, April, 313–38.
3. Oglesby, S. *et al.* (1970) *A Manual of Electrostatic Precipitation Technology*, Southern Research Institute, 25 August, SoRI Publishing, Birmingham, AL, USA.

4. Gooch, J.R. *et al.* (1973) A Mathematical Model of Characteristic Precipitation. Southern Research Institute Publication, Birmingham, AL, USA.
5. Bickelhaupt, R.E. (1979) A Technique for Predicting Fly Ash Resistivity. *EPA Report 600/7-79-204*, August.
6. Sproul, W.T. (1954) Laboratory performance of a special two-stage precipitator for collecting high-resistivity dust and fume. *Presented at American Chemical Society Meeting*, New York, NY, September.
7. Krigmont, H.V. and Coe, E.L. Jr. (1990) Experience in conditioning electrostatic precipitators. Presented at the *4th International Conference On Electrostatic Precipitation*, Beijing, 14–17 September, International Academic Publishers, Beijing, 1993.
8. Lee, G.K. and Whaley, H. (1983) Modification of combustion and fly ash characteristics by coal blending. *J. Inst. Energy*, December.
9. Frish, N.W. (1984) Analysis of air heater–fly ash–sulfur acid vapor interactions. In *Proceedings of the Fifth Symposium on the Transfer Utilization of Particulate Control Technology*, Kansas City, Missouri, August, EPRI, Palo Alto, CA, USA.
10. Coe, E.L., Jr and Krigmont, H.V. (1987) Prediction of SO_3 injection rates for fly ash conditioning systems. In *Proceedings of the Third International Conference on Electrostatic Precipitation*, Abano, University of Padova, Italy, 28 October.
11. Dismukes, E. (1975) Conditioning of fly ash with ammonia. *JAPCA*, **25(2)**, 152–6.
12. Dismukes, E.B. (1976) Fly ash conditioning. *Conference on Particulate Collection Problems in Converting to Low-Sulfur Coals*, Research Triangle Park, NC. 9–10 June, CPA.
13. Busby, H.G. and Darby, K. (1963) Efficiency of electrostatic precipitators as affected by the properties and combustion of coal. *J. Inst. Fuel*, **36**, 184–97.
14. Watson, K.S. and Blecher, K.J. (1963) Further investigation of electrostatic precipitators for large pulverized fuel-fired boilers. *Air Water Pollut. Int. J.*, **10**; 573–83.
15. Darby, K. and Heinrich D.O. (1966) Conditioning of boiler flue gases for improving efficiency of electrofilters. *Staub* **26** (11), 12–17, Nov.
16. Baxter, W.A. (1968) Recent electrostatic precipitators experience with ammonia conditioning of power boiler flue gases, *JAPCA*, **18** (12), 817–24.
17. Reese, J.T. and Greco, J. (1967) Experience with electrostatic fly-ash collecting equipment serving steam-electric generating plants. *ASME 67-WA/APC-3* (November).
18. Rendle, L.K. and Wilsdon, R.D. (1956) The prevention of acid condensation in oil fired boilers. *Combustion Mag.* July, 39–46.
19. Dahlin, R.S. *et al.* (1984) A field study of a combined NH_3/SO_3 conditioning system on a cold-side fly ash precipitator at a coal-fired power plant. Paper 84-96.3. In *Proceedings of the 76th Annual Meeting of the APCA*, June 1984, APCA, Pittsburgh, USA.
20. Krigmont, H.V. and Coe, E.L. Jr (1990) Dual flue gas conditioning, *The Eighth EPA/EPRI Symposium on Transfer and Utilization of Particulate Control Technology*, San Diego, California, 20–23 March, EPRI, Palo Alto, CA, USA.
21. Bickelhaupt, R.E. (1980) An interpretation of the deteriorative performance of hot side precipitators. *JAPCA*, **30**, 812.
22. Candeleria, R.B. and Johnson, J.S. (1983) Operating and maintenance of electrostatic precipitators at the Navajo generating station. *EPRI CS-2908 Electric Power Research Institute*, Palo Alto, CA, April, pp. 5–65.
23. Nelder, J.H. (1983) Experience with cold side precipitator on western coal at Pleasant Prarie power plant. *Proc. American Power Conference*, Chicago, IL, April.
24. Singer, J.G., Mellinger, G.A. and Cozza, A.J. (1979) Design for continuous ash removal: alternative concepts in ash handling. *Proc. American Power Conference*, Chicago, IL, April.
25. Crynack, R.R. (1985) Improving precipitator performance and saving energy with intermittent energisation—a case study. Paper 85-53.2, Presented at the *Annual APCA Meeting*, Detroit, MI, June, APCA, Pittsburgh, USA.
26. Box, G.E.R. *et al.* (1978) Statistics for Experiments. Chap. 15, pp. 510–39, John Wiley and Sons, New York, USA.

16 Possible future developments in the field of electrostatic precipitation

K.R. PARKER, C. RIEHLE and H. KRIGMONT

An attempt has been made in the previous chapters to review the major applications and practice of electrostatic precipitation engineering, such that it provides an update to the current state-of-the-art, which will provide the non-specialist and specialist alike with sufficient data and information to understand some of the finer points, of what in the past has been considered as the art, rather than the science, of precipitation.

The information should enable the non-specialist, who has electrostatic precipitators in his process clean-up train, to understand, not only the basics but, more important, some of the operational factors which will impact on the performance of the precipitator, which could result in emissions falling outside those demanded by regulation or legislation.

The specialist will hopefully find the contents useful in terms of a starting point for further investigations and measures to improve the performance, reliability and availability of the precipitator itself.

The precipitation industry has responded to the challenge of environmental pressures not only in controlling emissions using conventional precipitator designs and equipment, but like all modern technologies, is carrying out further research and development activities to further improve the precipitator's capability, in terms of performance and reliability of the equipment without hopefully increasing capital and operating costs.

The electrostatic precipitator, although mechanically not significantly different in basic design to that originally employed, has closely tracked the latest developments in a number of fields. The latest material science has led to improved fabrication techniques and materials, such that the mechanical reliability of the precipitator is significantly increased. It is not uncommon to find plants installed 20 years ago, still meeting the original performance figures with only normal routine maintenance activities being necessary.

As indicated in chapter 2, the electrostatic precipitation industry has always closely followed advances in the electrical field. The present developments in fast switching silicon technology and microelectronics, has not only led to improved control of the electrics, but the latest microprocessor-based controllers have simplified performance optimisation and will eventually lead to complete 'hands off' operation of the precipitator.

The latest programming now includes a self-diagnostic 'menu'-type window, which should assist the operators in quickly identifying any difficulty or problem effecting precipitator performance, whether it be electrical or mechanical in origin (see chapter 11 – Commissioning). These programs will also have 'intelligent' self-seeking optimisation memory programs, as reviewed in chapter 8. The Electrical Operation of Precipitators. Many of these programs are now available and should simplify the optimisation and reliability of the unit.

While it is not possible, because of commercial sensitivity, to include too many details of ongoing investigations and research, the following outlines some of the activities and developments that are being undertaken and some research topics which need, or are, being investigated by the various workers in the field.

16A Electrical developments
  (1) High frequency power converters – switch mode power supplies.
  (2) Nanosecond pulsing systems.

16B Use of natural sulphur dioxide in the flue gas as a feed stock for flue gas conditioning.

16C The application of high temperature and high pressure precipitator technology to advanced power generation particulate clean-up systems.

16D Precipitator sizing techniques based on computer-derived data.

In addition to the above items, which will be covered later in some detail, the following items are some personal ideas which are suggested for further investigation.

(a) While much has been accomplished in both the mechanical and electrical aspects of improving the performance, availability and reliability of electrostatic precipitators of the single type, where both charging and collection is achieved by the same set of electrodes, some useful investigations have been carried out, both in the laboratory and full-scale, to develop a hybrid precipitator having both single and part two-stage precipitation within the same casing. These trials were successful in minimising the problems of reverse ionisation, but suffered from rapping re-entrainment from the precipitation part of the two-stage section. Further work will no doubt be carried out where the precipitator will be further subdivided to include several hybrid stages in series, such that any uncharged particles leaving the precipitation stage will be recharged by a downstream single-stage section and thus minimise re-entrainment losses.

(b) As indicated elsewhere, if particulate re-entrainment could be eliminated from dry precipitators, the economics would significantly change. Even with a well-designed and controlled precipitator, measurements have shown that at least 30% of the emission arises from re-entrainment losses. Several different systems have been adopted, from total flow isolation during rapping to the use of internal dampers or air curtains to blank-off only the ducts being rapped. These approaches work in reducing the re-entrainment losses, but whether the additional complications, operating difficulties and costs can really be justified over a somewhat larger conventional plant, with fully optimised controlled rapping, minimising re-entrainment, is debatable.

Most dry precipitator applications, even when the rapping is fully optimised, tend to exhibit a re-entrainment factor of some 30%. In the past when higher emissions were acceptable and precipitators had only two series fields the re-entrainment losses favoured precipitator designs having moving belt-type collectors with scrapper mechanisms for removing the dust, but the mechanical integrity and high maintenance of these prevented general usage. Although the wet film flow collector design of precipitators offers the advantage of virtually zero re-entrainment, possible corrosion and water treatment costs restricted their application to special processes as indicated in chapter 13.

The importance of minimising re-entrainment was accepted by the precipitation industry as a challenge and on modern dry precipitators the rapping frequency is microprocessor or PLC controlled, such that no consecutive areas are simultaneously rapped, thus limiting the impact of re-entrainment on visible emission. Other controllers monitor and use the change in electrical operating conditions to initiate rapping on plant which have fairly consistent inlet conditions. Investigations into weighing of the collector plates using load cells and only rapping when a certain thickness/weight has been achieved, has proved promising, but again the high costs of monitoring instrumentation has restricted general usage.

(c) The use of CFD and associated systems will be developed to become fully acceptable as a means of achieving a uniform gas distribution within the precipitation field. This should produce data which are more representative of real plant operating conditions, since it should be possible to include space charge, ionic wind and other effects, which are impossible to simulate using a modelling approach, which cannot readily be energised with dust passing through.

(d) With the larger and faster computers which are becoming available, the field and current distributions within a precipitator duct should be readily determined, so that it should be possible to identify the best electrode arrangement for a particular application. The computer approach should also enable performance and sizing evaluations to be made with respect to operation under the various energisation systems that are becoming available as proven operational units.

(e) With the increasing concern over Pm 10 and smaller particles, particularly heavy metals, and their cumulative effect on health, there is a need to explore the collection of these particles to a much greater extent. While very low emissions have been determined from precipitators whose design and sizing have been based on total mass collection, in future, it may be necessary to base the design specifically addressed to Pm 10 and smaller particles.

Obviously one approach is to cool the gases to ensure the volatile metal components are all condensed at the operating temperature; this usually implies wet-type precipitation (chapter 13). The addition of a wet precipitator would, however, significantly add to the cost of the overall plant, even if the water treatment side could be incorporated into a wet acid gas scrubbing system; even then, one would need to precipitate the heavy metals separately from the general sludge effluent for 'safe disposal'; otherwise there is a risk of recycling and loss of the materials one is attempting to remove.

Even with wet precipitators there could be gaseous heavy metal elements still escaping, so one can envisage some form of target adsorption system being installed upstream of a dry precipitator. In this scenario it would be possible to avoid reducing the gas below saturation temperature, which should simplify collection and disposal methods.

The use of a target adsorption injection, upstream of the precipitator, has been successfully used in the incineration industry for the removal of numerous condensible phase materials. One advantage of this approach is that the target should remove predominantly the finer particle sizes, which both theoretically and from practical determinations show a higher penetration window on the fractional efficiency curve in the 0.5 μm area (see chapters 3 and 9). The addition of a target material would mean that a higher overall precipitator efficiency would be required, but the size could be less than the size increase necessary to capture a higher proportion of the 0.5 μm size particles.

16A Electrical developments

16A.1 High frequency power conversion or switched mode power supplies

For optimum performance, i.e. maximum collection efficiency, it is important to electrically operate the precipitator as near to breakdown/flashover condition as possible. This condition provides the maximum voltage and hence field strength, so that performance, which is theoretically proportional to the field strength squared, is maximised. Unfortunately, even on well-regulated and operated processes, changes in the inlet conditions invariably lead to flashover and hence discharging of the capacitive component of the precipitator, such that, during this time, the performance is compromised.

With conventional DC energisation, while one raises the operating voltage to the point of breakdown, as a resistance/capacitance load there is a significant ripple voltage which reduces the maximum field strength. Under flashover conditions, the minimum time frame to switch off the thyristor is half a cycle or 10 ms on 50 Hz supply (8.6 ms on 60 Hz), and at least a further half cycle to commence recharging the precipitator. On plants operating with multiple flashovers, the overall time lost can be very significant. This was recognised at an early date and attempts using three-phase energisation led to unresolvable problems in arc quenching within the field and three-phase investigations were discontinued.

Over the past two decades, the general availability of very fast switching devices in the form of high frequency power conversion devices widely used in PCs, welding machines and electroplating, etc., together with improved ferrite materials having low losses, has led to the development of a new approach to precipitation electrical energisation, by using high frequency power conversion (HFPC) or switch mode power supplies (SMPS).

The block diagram of such a system is shown in Figure 16A.1; in this the incoming three-phase supply feeds a power converter, an insulated gate bipolar transistor package (IGBT) operating at a frequency of some 10 kHz. This provides an initial DC link, which can be modulated, or 'chopped', by semi-conductor power switches, to any desired frequency, before feeding the ferrite cored HT transformer, whose output after rectification supplies the precipitator. The output from the HT transformer is controlled by varying the on/off times of the power switches using a feedback system which continuously monitors the output voltage and current levels to achieve optimum conditions on the precipitator.

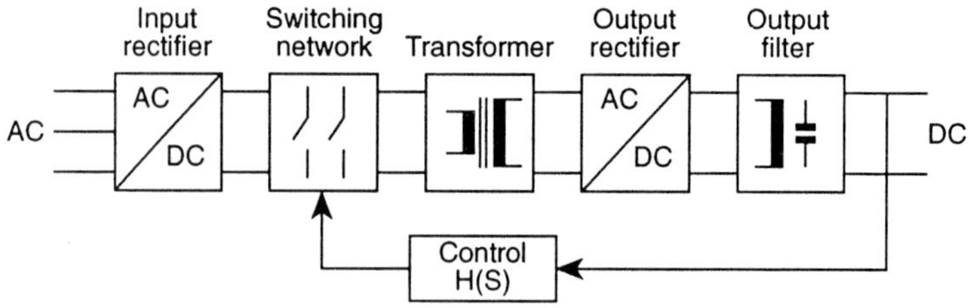

Figure 16A.1 General converter AC to DC (mains in, DC out).

The operational problems of interrupting the power supply, to quench an arc/flashover and re-establishing current to the precipitator, are considerably reduced by operation at high frequency, since at 10 kHz, the minimum switching time is 0.05 ms, compared with 10 ms at 50 Hz, and the recovery time, to recharge the precipitator, can be more controlled with the smaller time step increments, such as to rapidly re-establish precipitator performance.

Initial trials using SMPS units on a precipitator handling particulates from a wood chip installation as a load have been carried out and the results

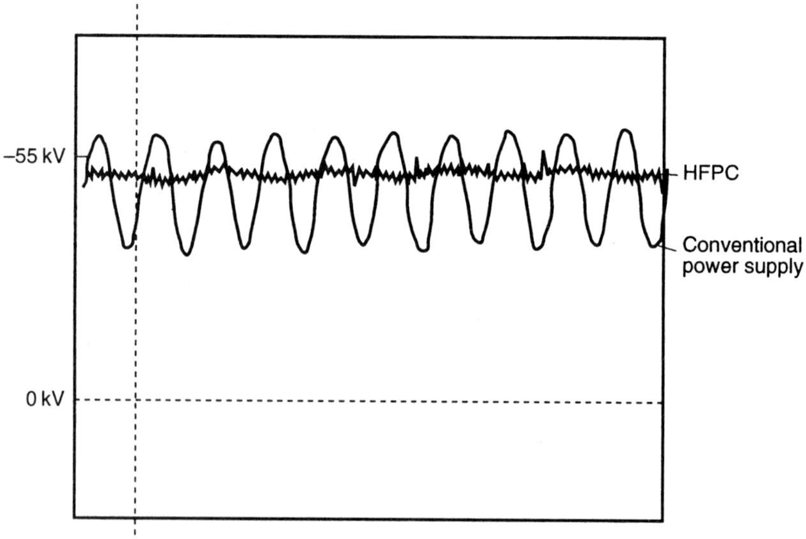

Figure 16A.2 ESP voltage, conventional and HFPC power supplies, same current.

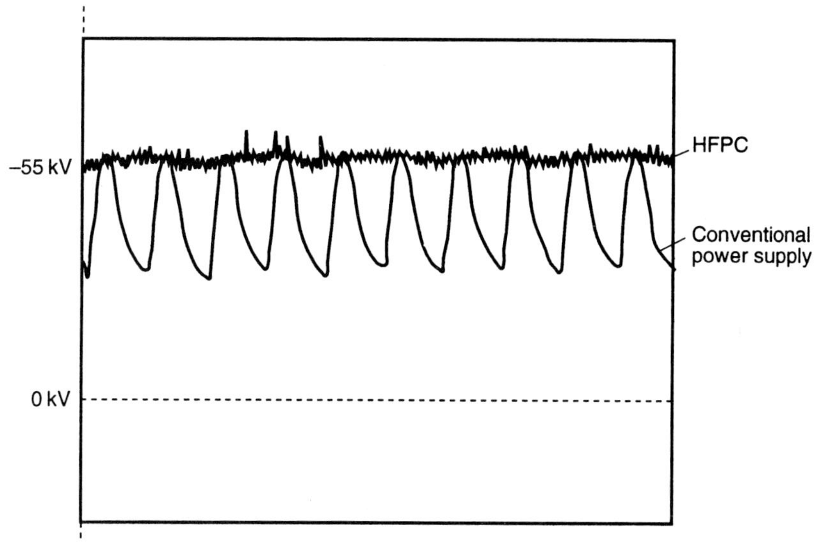

Figure 16A.3 ESP voltage, conventional and HFPC power supplies, same peak voltage.

compared with the same unit under conventional energisation. The results of these trials [1] are illustrated in the form of voltage oscillograms; Figure 16A.2 was obtained with both power supplies adjusted, without flashover arising, to give the same secondary currents of 200 mA. Under this condition, the conventional supply produced a peak voltage some 10 kV higher than the SMPS system, but at a lower mean voltage level. Figure 16A.3 was produced with the peak voltage on both adjusted to 55 kV; in this case, the currents for the conventional and SMPS systems were 110 and 230 mA, respectively. In Figure 16A.4, the power supplies were free running, with the current limited only by flashover/sparkover conditions; here the corresponding currents were 120 and 230 mA, respectively.

The above results indicate that SMPS energisation is capable of operating at a higher mean voltage and secondary current level, thereby providing significantly increased power inputs of up to three times that of a conventional DC system and consequently a much higher performance for a given size of precipitator.

This SMPS or HFPC development will probably have a greater impact on future precipitator technology than the pulse charger or flue gas conditioning systems, which, although ideal for improving precipitator performance for 'difficult dust' situations, does little for ordinary dusts typically met. The SMPS approach, on the other hand, appears to improve

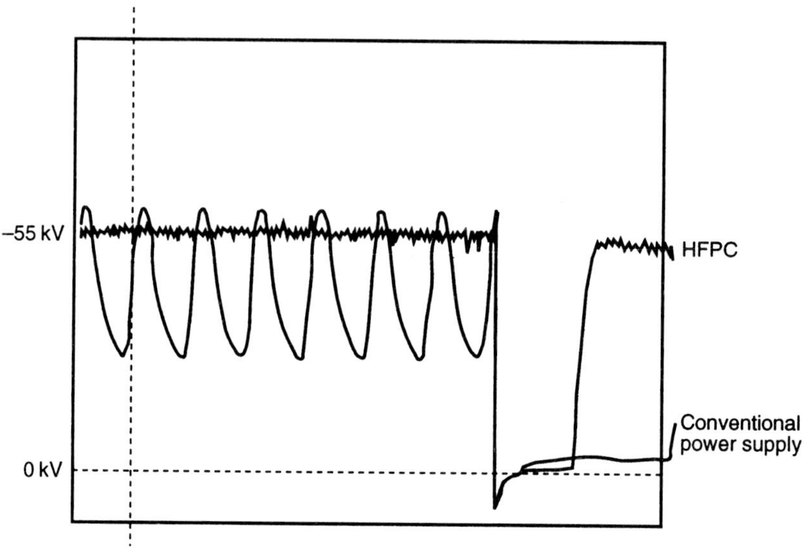

Figure 16A.4 ESP voltage, conventional and HFPC power supplies during normal operation, sparking.

all particulate collection efficiencies, as a result of the much higher power input and, for a given duty, should lead to a smaller size and hence lower cost precipitator.

In addition to the performance improvement, for the same power input, the primary current from each phase of the supply is lower than from a conventional single phase system. This has the advantage that all primary components, such as switch gear, control unit and all incoming cabling, are much smaller and hence there is a saving both in cost and in floor space. In some designs, the SMPS unit is a complete free-standing module, requiring only an incoming three-phase supply connection. With the much smaller and lighter power supply units, it is possible to consider dividing the precipitator into smaller bus sections, each with its own power supply, which would further improve the overall particulate collection efficiency, or lead to a reduction in size for a new unit.

16A.2 Nanosecond pulse operation and acid gas control

The initial object of operating a precipitator with nanosecond, rather than millisecond, pulse widths was to improve the performance of precipitators when handling 'difficult' fly ash.

Although such a system had merit in improving the collection efficiency of precipitators, there has been a move toward using this approach, under controlled precipitator operating conditions, to promote the formation of a plasma rather than a corona, such that the plasma gives rise to highly reactive gaseous radicles, for example O^+, which are then available for attachment to SO_2 molecules, to form initially SO_3 and then H_2SO_4, which can then be removed by the precipitator [2,3]. A similar reaction would occur with NOx molecules, so that the system is capable of controlling acid gas emissions in addition to fly ash from power stations.

Recently the system has been extended outside the power station field into waste incineration, where the plasma is being applied to break down VOCs and undesirable substances, such as dioxins and furans, and also to react with elemental mercury to simplify its collection as a condensed, rather than a gaseous, material.

This work is still in its experimental stage, but if successful, as indicated from laboratory and pilot field work, could change the present back end clean-up systems on a number of industrial process plant.

References (16A)

1. Ranstead, P. and Porle, K. High frequency power conversion: a new technique for ESP energisation, EPRI/DOE International Conference on Managing Hazardous and Particulate Air Pollutants, Aug. 15–17th, Toronto, Canada, in print.
2. Dinelli, G. (1987) Electrostatic precipitator performance by means of narrow pulse power energisation. 3rd International Conference on Electrostatic Precipitation. Abano, Italy, University of Padova, October.
3. Mazuda S. *et al.* (1993) High voltage nanosecond pulse discharge system for removing SO_2 and NO_x in flue gas. 10th EPRI Particulate Control Symposium and the 5th International Conference on Electrostatic Precipitation, Washington, USA, October, Paper 32.1–14, EPRI TR 103048 **2**, Palo Alto, CA, USA.

16B Use of natural sulphur dioxide as a feed stock for flue gas conditioning systems: flue gas conditioning today and tomorrow

The present status of flue gas conditioning systems is to use either elemental sulphur or liquid sulphur dioxide as a feed stock, which is converted to sulphur trioxide as the conditioning agent. This approach, while being completely satisfactory, has a cost element existing over the life of the plant. In many applications, particularly those firing carbonaceous fuels, there is usually sufficient sulphur dioxide present in the flue gas which, if converted, can provide the sulphur trioxide necessary for flue gas conditioning duties.

The following section, by Dr Henry Krigmont, explores and develops the various approaches that can be used to convert the naturally occurring sulphur dioxide as the feed stock for conditioning.

16B.1 Background

As discussed, typically, SO_3 is formed by a catalytic conversion of gaseous sulfur dioxide (SO_2) [1]. The SO_2 usually comes from an 'external' source (feed stock) by either evaporating liquid sulfur dioxide or by burning molten or solid sulfur. Conventional flue gas conditioning systems have two major drawbacks: they require an external continuous supply of feed stock and in the process of operation these systems slightly increase SO_2 emissions.

Another approach to create SO_3 is to utilize the 'native' SO_2 formed during combustion of sulfur contained in fossil fuels for a subsequent conversion to SO_3. The rate of such a process depends on temperature, concentrations of SO_2, O_2 and water vapor as well as the catalyst's properties. Various systems implementing this technique were proposed, but none were free of major deficiencies.

Figure 16B.1 depicts a summary of the SO_3 FGC technology [2]. A variety of designs could be offered to satisfy the preference and applications specifics of any customer. The WAHLCO designs substantially improve *in-situ* flue gas conditioning technology by providing simple yet reliable systems to furnish the required amount of sulfur trioxide for efficient electrostatic precipitator operation, without adding more sulfur oxides to the system and in a fully controlled manner. On the contrary, the amount of sulfur dioxide in the system is reduced.

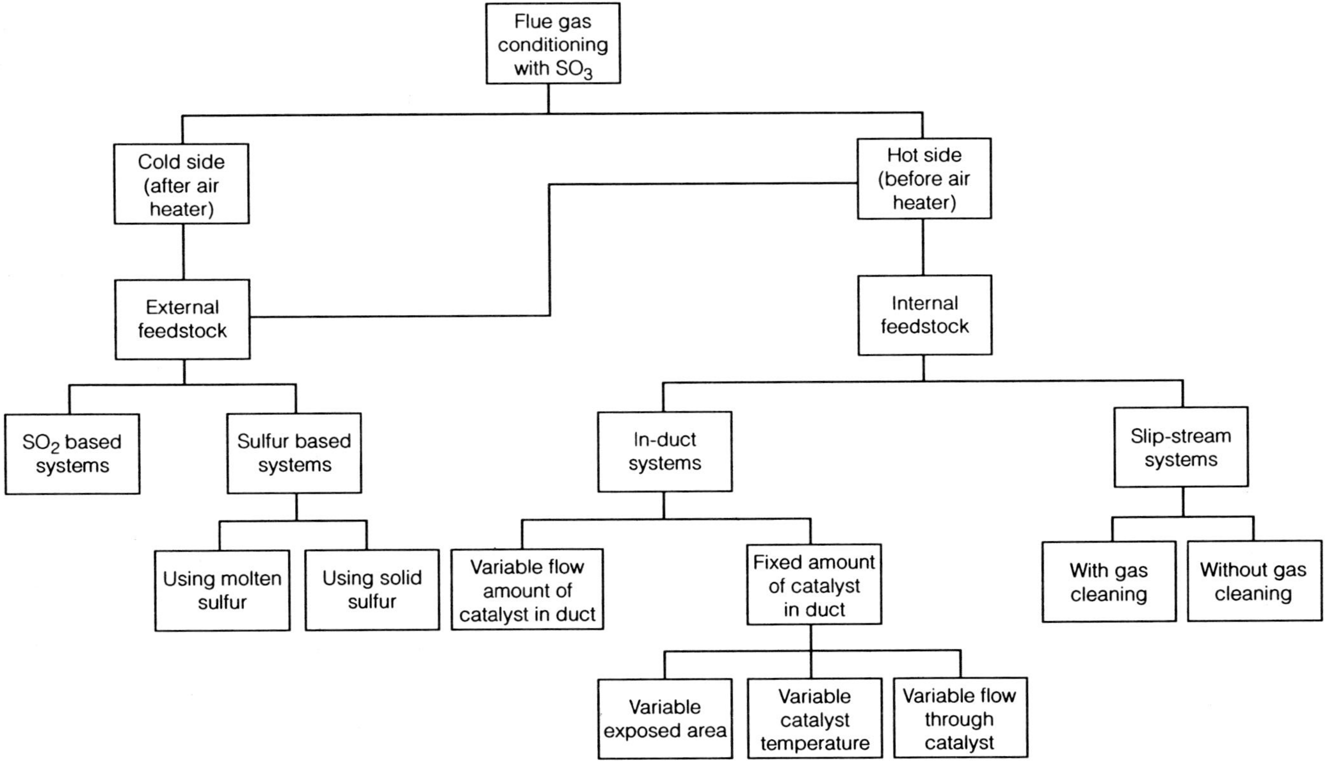

Figure 16B.1 FGC systems – classification.

16B.2 'Native' or 'internal' feed stock FGC technologies

Conventional FGC systems work well, and are widely used, but in some instances there are drawbacks. The catalytic conversion of sulfur dioxide to sulfur trioxide is not completely efficient, and additional sulfur dioxide is added to the flue gas flow. A constant supply of sulfur feed stock is required, and this feed stock must be safely handled. The components for burning, catalyzing, and injecting systems must be kept in good working order, and there is a power consumption associated with the process.

Another approach to create SO_3 is to utilize the 'native' SO_2 formed during combustion of sulfur contained in fossil fuels as a feed stock for a subsequent conversion to SO_3. Extensive experimentation at the Lehigh University [3] was conducted to investigate a new approach to fly ash conditioning based on conversion to SO_3 of a portion of the SO_2 normally present in flue gas even when low sulfur coals were burned.

The chemical equilibrium conditions predict that when the gas contains about 5% O_2, about 99% of the SO_2 can be oxidized to SO_3 at about 399 °C (750 °F) and about 90% at 510 °C (950 °F), with respectively more SO_3 being generated at lower temperatures and less at higher temperatures. Typically, however, not more than about 1–3% of the 'natural' SO_2 is being oxidized in SO_3, the rate of such reaction usually depending on the gas temperatures, the concentration of SO_2, O_2 and water vapors, as well as the possible catalytic properties of the boiler convection surfaces.

Kanowski and Coughlin [3] proved that it is possible to oxidize significant quantities of SO_2 present in flue gas at concentrations typical for commercial coal-fired boilers burning low sulfur coal, by inserting commercial catalysts in the SO_2-containing gas. Several systems implementing this technique were subsequently proposed, but none were free of major deficiencies.

16B.2.1 'Slip-stream' FGC systems

One earlier published approach suggested using a fan to draw a portion of the clean gas (taken from downstream of the precipitator) into a slip stream which is subsequently passed through a chamber with the catalyst, but gives no further details as to the amount of the flue gases, temperature and reheat requirements and how the amount of sulfur trioxide can be controlled. One can imagine that valving could be added to the slip stream to control its total flow, but such valves are complex, expensive, and difficult to fabricate and maintain.

Recently, a different approach was proposed and is being investigated as a possible way to avoid the purchase of sulfur or liquid SO_2 feed stock. A so-called 'EPRICON-Agentless Flue Gas Conditioning System' offers an approach in which a slip stream (about 3–5%) of the entire flue gas flow is drawn from the main flow ahead of the economizer and passed through the catalyst bed [4]. A portion of the sulfur dioxide in the slip stream is oxidized to sulfur trioxide, and the slip stream is merged back into the main flue gas flow downstream of the air preheater. Due to the relatively small portion of the flue gases used, very high efficiency catalytic conversion is required, which means strict control of the gas temperature and other operating parameters. While of interest, this approach has major drawbacks when implemented, such as reduced boiler thermal efficiency, because less heat is recovered, and there is typically insufficient mixing of the slip stream and the main flow at the point where they rejoin, due to a small pressure differential.

Furthermore, questions yet to be thoroughly investigated include control of SO_3 quantities to proper levels when a variety of different coals are burned producing ashes with differing conditioning requirements. Also, variation of flue gas temperatures as boiler loads change may lead to insufficient production of SO_3 at low loads, causing fouling of the ESP with high resistivity ash, which may interfere with emission compliance upon return to full load operation.

For economy of installation and simplicity of operation it is preferred that booster fans are not used, which means that the system should rely only on the relatively small pressure drop in the main system to provide the driving force for the side loop in which the catalytic conversion of SO_2 to SO_3 occurs. This may lead to big ducts, large volumes of catalysts, big injection probe sizes and difficulty in obtaining uniform distribution and mixing of SO_3 in the downstream gas. The expense of custom ductwork design and installation for such a system is not negligible. It is expected that a considerable amount of testing and demonstration work will be required to accurately assess the suitability of this system for commercial operation.

16B.2.2 In-duct FGC systems

In another design the catalytic converter is located in the boiler economizer outlet duct prior to the air heater [5]. By exposing most of the native sulfur dioxide in the gas stream to the catalyst, very low conversion efficiencies for the converter may suffice to produce SO_3 in the required quantities. The rate of conversion is proposed to be controlled by inserting or retracting the catalyst bed into the ductwork.

This approach also has several obvious deficiencies. Since the gas flow follows the path of least resistance, it could be difficult to force the gas into the catalyst bed presenting substantial resistance to the moving gases. This approach offers very little means for strict control of the SO_3 generation, for the catalyst bed will have to be in a continuous state of motion to provide the required SO_3 amounts in response to changes of boiler load, excess air (O_2), various boiler operating characteristics, etc. Therefore, this design offers only 'ON/OFF' type operation. Furthermore, this design will introduce yet another source of the possible air inleakage through the ductwork openings and create significant flow maldistribution.

16B.3 *In-situ* gas conditioning (IGC) approach

Recently, new solutions to these problems have been developed. These designs allow selective control of *in-situ* catalytic SO_3 generation. A sulfur trioxide conditioning system includes a catalytic converter that converts a portion of 'native' or naturally formed sulfur dioxide in a flow of flue gas to sulfur trioxide. In one design, the amount of the catalyzed surface exposed to the flow of flue gas is selectively varied to control the conversion of SO_2 to SO_3. Another approach provides the means to vary the catalyst bed temperature to selectively enhance or suppress the sulfur dioxide oxidation. In another design, variable amounts of flue gas are drawn through a fixed catalyst bed.

16B.3.1 Variable exposed area IGC system

Figure 16B.2 presents an artist's view of the Wahlco variable exposed area IGC system. This IGC system includes a catalytic converter consisting of a

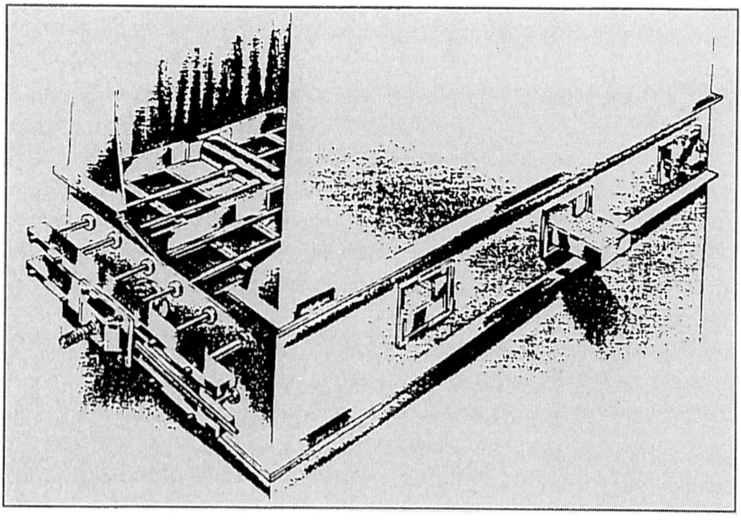

Figure 16B.2 The Wahlco variable exposed area IGC system.

bed of catalyst substrates distributed across the main duct, with the catalyst, for the *in-situ* oxidation of sulfur dioxide to sulfur trioxide. The bed is positioned in the flue gas stream at a location having the correct temperature to accomplish the catalyzed oxidation. The amount of sulfur trioxide produced is controlled by selectively adjusting the amount of catalyst-covered surface that is exposed to the flue gas stream.

To control the amount of sulfur trioxide generated, a shield is placed over a portion of the catalyst-covered surface. The shield prevents active contact between the flue gas flow and the covered portion of the catalyzed surface, so limiting the oxidation of sulfur dioxide. The greater the exposed surface area of catalyst, the greater the conversion of sulfur dioxide to sulfur trioxide. The shield also serves the important function of reducing erosion damage to the catalyst, and of permitting compensation for the effects of decreasing efficiency through ageing of the catalyst. This design can compensate for this change in the properties of the catalyst simply by exposing the required additional amount of catalyzed surface.

This design is easily automated. The operating parameters of the system are measured, such as, for example, boiler load, flue gas composition, electrostatic precipitator power consumption, fly ash properties, fly ash resistivity, and/or stack gas opacity. These parameters are provided to a computer, which determines whether any change in the sulfur trioxide level is required, and, if so, the nature of the change required. The shield is moved, if necessary, to change the amount of exposed catalyst surface area responsive to these measured operating parameters.

The gas flow distribution and sulfur trioxide distribution in the flue gas duct can even be improved by selection of the nature and distribution of the catalyst supports. For example, if there is an uneven temperature naturally in the duct, then the distribution of catalyst, or its exposed area, can be selected so that the production of sulfur trioxide matches the spatially varying requirements of the flue gas.

16B.3.2 Variable catalyst temperature IGC system

This sulfur trioxide flue gas conditioning system also includes a catalytic converter for converting a portion of the sulfur dioxide in the flue gas to sulfur trioxide. The catalytic converter consists of a bed with catalyst support distributed across the boiler back pass. In this design, the catalyst bed is furnished with a temperature controller to allow for the catalyst support heating or cooling to vary the conversion efficiency of the catalyst. The catalyst support temperature is maintained independently of the temperature of the flue gases, e.g. it could be electrically heated if the catalyst requires heating, or a cooling system, should the catalyst need to be cooled.

The main benefit of this design is that the catalyst bed could be located independent of the flue gas temperature, with the temperature controller providing optimum conditions for the IGC system operation. As an option, the catalyst bed could be positioned at the location where the gas temperature is always higher than that required; then the temperature controller may consist of only a cooler. Conversely, if the catalyst bed is positioned at the location where the gas temperature is always lower than the optimum the temperature controller may consist of only a heater.

The heating and/or cooling of the catalyst provides for a highly effective way to precisely control the extent of the SO_2 to SO_3 conversion.

16B.4 Variable flow IGC system

In this design the catalyst is enclosed in modules that are installed in a grid-like fashion in the flue duct (Figure 16B.3). Each module has an individual controllable aspirating device. Within the aspirator compressed air or stream is employed to induce the flue gas flow through the catalyst. The modules are positioned in the gas stream at a location having the proper temperature range to accomplish efficient catalytic oxidation of SO_2 to SO_3.

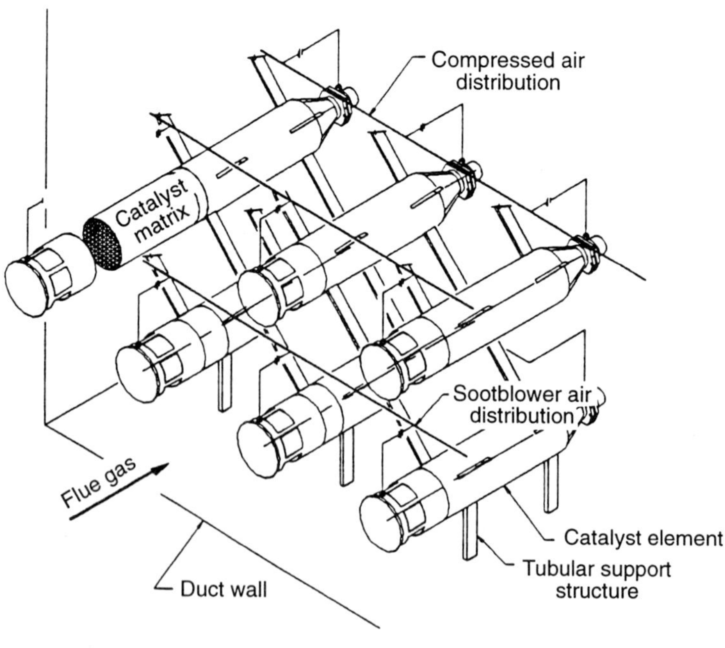

Figure 16B.3 Modular design for variable flow IGC.

The amount of SO_3 generated is controlled by varying the volume of flue gas that passes through the catalyst. By controlling the compressed air pressure or steam flow and, therefore, the velocity of the gas flow through the module, the design allows for precise control of the SO_3 generated and also limits catalyst wear. Only the minimal flow required for SO_3 generation is introduced into the catalyst module. The catalyst is shielded from direct erosion by the fly ash by a deflector located at the entrance to the module.

This design provides a method for a controllable amount of sulfur trioxide to be created and then uniformly distributed in the flue gas stream. The system is simple and durable in operation. Control of the process could be accomplished similar to the one described earlier using a logic proven with traditional external SO_3 generation systems.

The major benefits of this modular design are: installation is simple and easy and does not require any special tools and instrumentation; it provides for standardization and 'off-the-shelf' design; it allows for variable SO_3 distribution to match existing gas flow non-uniformity; only relatively minor duct modifications would be required; virtually no moving parts; maintenance is very easy, and periodic catalyst cleaning as well as change-out is simple to accomplish.

16B.5 Catalyst selection

The catalyst is the heart of any IGC system. The proper catalyst will improve system efficiency, longevity and economics. The major requirements for SO_2 to SO_3 oxidation catalysts are:

- High conversion efficiency
- Low 'light-off' temperature
- Activity in specified range of space velocity (SV)
- High stability and low sensitivity to poisoning
- Resistance to erosion/abrasion
- Cleanability

It is interesting that from a position of catalyst chemistry, the same types of catalyst developed for selective catalytic reduction (SCR) of NO_x are suitable for SO_2 oxidation. The only problem is that the catalyst manufacturers spent years sorcering with catalyst formulations to reduce SO_2 to SO_3 conversion down to only a few percentage points for a 'hot' catalyst and to a fraction of that for a 'normal' catalyst. For IGC applications, the goal is quite the opposite – to maximize SO_2 oxidation while still maintaining the mechanical integrity and chemical stability of the catalyst.

For the typical temperature range of 288–399 °C (550–750 °F), usually 316–321 °C (600–700 °F), two types of catalyst are applicable: based on base metal oxides and on precious metals. The first type usually has a maximum efficiency closer to the higher part of this temperature range. The noble metal catalyst is more active at lower temperatures.

The expected conversion efficiency is different for different types of IGC system:

- for variable exposed area IGC system 3–8%, with SV of 1×10^5–3×10^5 h^{-1}
- for variable catalyst temperature IGC system 1–3%, with SV of 3×10^5–6×10^5 h^{-1} and
- for variable flow IGC design 40–60%, with SV of 1×10^4–3×10^4 h^{-1}

Preliminary tests conducted in the simulated environment have shown that these target values are conservative, at least for a fresh catalyst.

Different types of substrates have been considered for this application: extruded monolith, metal foil, and plate type. The desired range of cell densities is 10–50 cpsi, with variable flow IGC at the higher side of this range. The catalysts, based on these substrates have specific features that may suit different IGC systems.

One of the important issues is the capability to withstand fly ash erosion. This may become critical for variable exposed area IGC systems, where the catalyst sees full particulate loading and the flue gas velocities are even higher than in non-obstructed ducts. The rate of erosion is usually proportional to the velocity to the power of 4 to 6 and at some point (approximately 10–15 m/s (35–50 ft/s)) it starts to increase dramatically. From this point of view the variable flow systems (like modular IGC) have definite advantages. Deflecting coarse particles from direct contact with the catalyst significantly reduces the danger of erosion. The smallest particles usually result in catalyst masking that may be effectively addressed by sootblowing. Some formulations, based on noble metals allow even washing the catalyst with chemical solutions and significantly extend its effective life by a properly planned maintenance program.

References (16B)

1. Krigmont, H.V. and Coe, E.L. (1992) Flue gas conditioning: key advances in recent years. *Power*, August, 30–6.
2. Ferrigen, J.J. *et al.* (1993) A naval approach to flue gas conditioning. EPRI/DOE International Conference on Managing Hazardous and Particulate Air Pollutants. Poster Session. August 16–18th, Toronto, Canada, in print.
3. Kanowski, S. and Coughlin, R.W. (1977) Catalytic conditioning of fly ash without addition of SO_3 from external sources. *Environ. Sci. Technol.*, **11(1)**, 67–70.
4. Bibbo, P.P. (1994) EPRICON: agentless flue gas conditioning for electrostatic precipitators. *Proc. POWERGEN America*, Orlando, Renwell. Inst. Houston, USA.
5. Woracek, D. *et al.* (1993) Catalytic conversion of native sulphur trioxide in flue gas for resistivity conditioning., 10th EPRI Particulate Control Symposium and 5th International Conference on Electrostatic Precipitators, October, Washington, pp. 13.1–16, EPRI TR-103048 **2**, Palo Alto, CA, USA.

16C High temperature/high pressure precipitators for advanced power generation systems

Worldwide environmental concern over 'green house' gases, particularly those arising from the power generation industry, has led to new designs for advanced power generation having far greater efficiencies than those obtainable from the conventional steam cycle of around 38%. Efficiencies of some 42–44% can be realised with pf units utilising supercritical steam conditions, mid 40s%, using pressurised fluid bed combustors' combined cycles and much higher using integrated gasification and combined cycle and hybrid approaches.

A schematic arrangement of an Air Blown Gasificaton Cycle is illustrated in Figure 16C.1 [1]. This employs an air blown gasifier stage, the gases from which drive a turbine, while the resultant char passes to a circulating fluid bed combustor in the steam cycle stage. Alternate approaches using oxygen, rather than air, for the gasifier result in improved efficiency, dispensing with the 'char' stage to produce a simpler steam-raising block.

In addition to the normal clean-up of the gases finally exiting the equipment, the gasifier gases need to be supercleaned to avoid erosion and high temperature corrosion difficulties arising with the turbine blades. Following the gasifier cyclone, indicated in Figure 16C.1, the emission is high in condensed alkali material, which needs to be removed to minimise adhesion and other problems. Cleaning as indicated is achieved by a ceramic filter operating at a temperature of some 400 °C and a pressure of 20 bar. An alternate approach would be to use high efficiency cyclones operating at a higher temperature, plus a 'ruggedised' turbine to accept a higher particulate loading.

In many high temperature but atmospheric pressure type applications, the increase in kinetic energy of the gas molecules leads to very conductive gases, this with the difficulties of suitable insulator materials, which have low resistance surface properties at high temperature, as indicated in Figure 4.9, has meant that electrostatic precipitation is not feasible for these processes. There are exceptions and precipitators have been successfully used at temperatures above 550 °C, in gases containing high levels of SO_2 and also at 800 °C, where the gases contain high levels of HF plus water vapour but basically no air.

With the new advanced power generation systems operating at high pressure and high temperature, the problems of the kinetic energy of the molecules disappear, and the difficulties of providing suitable insulator materials have been overcome by moving the suspension insulators outside the hot gas stream, so that they can operate, without breakdown, in a lower

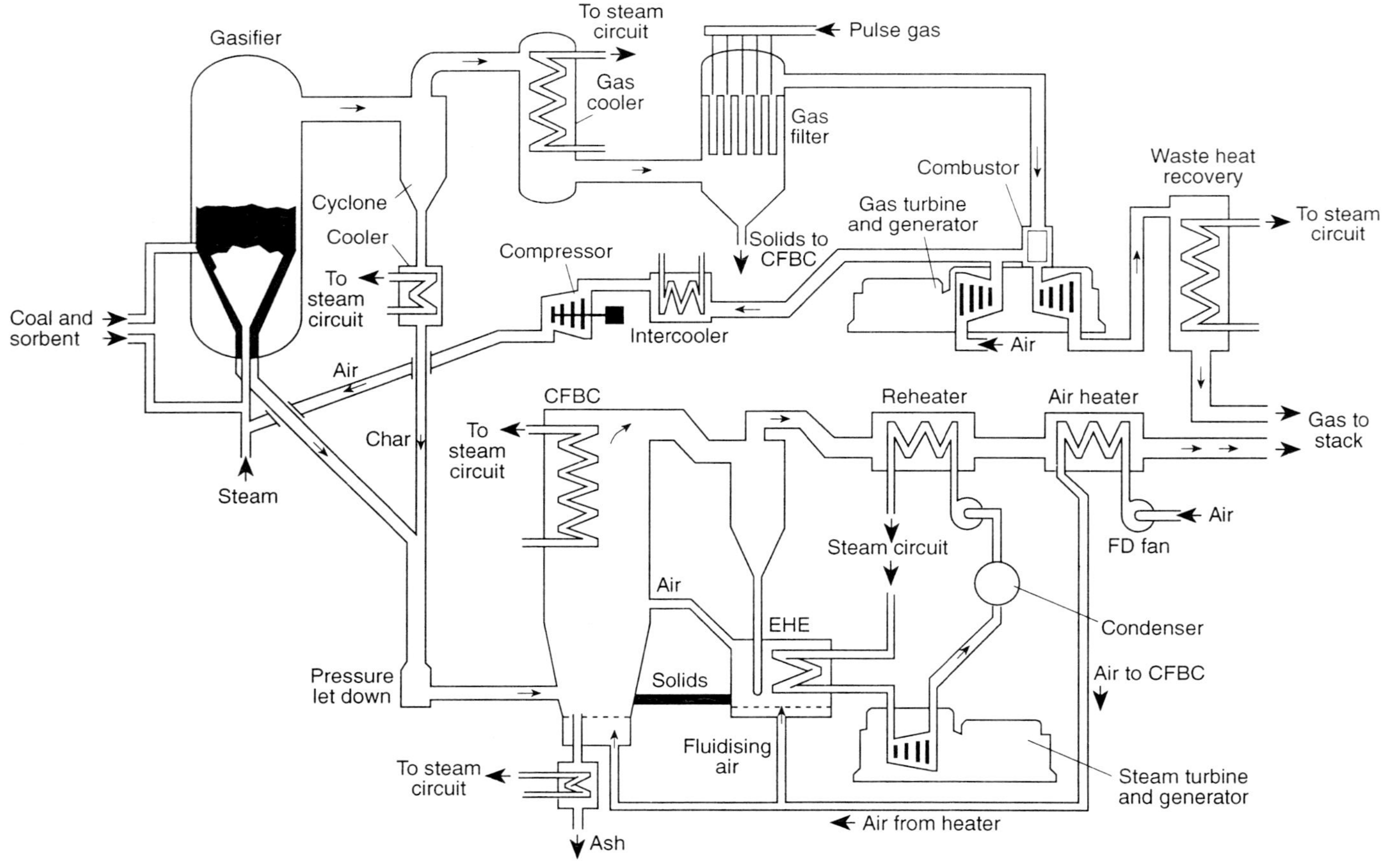

Figure 16C.1 Air blown gasification cycle [1].

temperature environment. While there will be engineering difficulties associated with the mechanical design of large precipitator units because of differential temperatures, these are probably no more onerous than those experienced by the ceramic filter and its cleaning system.

Over the next decade it is anticipated that there will be a significant increase in the demand for power worldwide, particularly with China, India and Russia becoming fully industrialised; the need for high efficiency power generation becomes of paramount importance to limit the discharge of carbon dioxide gases. To meet these opportunities a great deal of preemptive work has been carried out by the precipitator companies to clean up the high temperature/high pressure gases associated with these advanced power generation systems.

The theory and practical aspects of high temperature/high pressure precipitation are presented in the following resumé prepared by Dr Claus Riehle. This gives the case for the electrostatic precipitator together with some of the engineering difficulties which have to be resolved for its successful application in the advanced generation systems now being considered for implementation, both now and in the next century.

16C.1 Fundamentals

Changing gas temperature and/or changing gas pressure usually leads to a change in gas density. This can be easily seen when considering the equation of state for ideal gases (equation 16C.1).

$$\frac{p}{\rho} = R \cdot T \tag{16C.1}$$

For handling different states, it is convenient to introduce the so-called relative gas density δ, defined by equation (16C.2). This relates the gas density of a state 2 to the gas density of a known state 1. Obviously from Figure 16C.2 the relative gas density is a function of the pressure ratio and the reciprocal temperature ratio.

$$\delta = \frac{\rho_2}{\rho_1} = \frac{p_2}{p_1} \cdot \frac{T_1}{T_2} \tag{16C.2}$$

Usually temperature and pressure are related to normal conditions, i.e. $T_1 = T_0 = 273\text{K}$ and $p_1 = p_0 = 1\,\text{bar}$. Figure 16C.2 illustrates the value range of the relative gas density for temperatures from 300K up to 1500K and pressures from 1 bar up to 30 bar, represented by corresponding isolines.

A decreasing relative gas density corresponds to an increasing average distance between the gas molecules (which can be deduced from kinetic gas theory) or, in other words, comparing two different gas states, the mean free

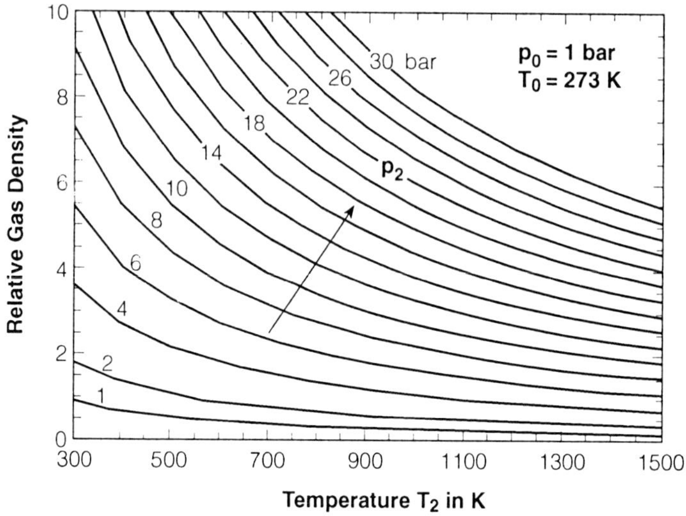

Figure 16C.2 Relative gas density as a function of temperature; isolines represent constant pressures.

path of the gas molecules behaves like the reciprocal of the relative gas density (equation (16C.3)). At normal conditions the mean free path $\lambda(p_0, T_0)$ is about 0.065 μm. Therefore, under normal conditions for particles smaller than 1 μm the mean free path of the gas molecules is of a similar order and has to be taken into account.

$$\lambda(p, T) = \frac{\lambda(p_0, T_0)}{\delta} \tag{16C.3}$$

Closely connected with the mean free path of the gas molecules is the mobility of the gas ions. The mobility of the charge carriers, b, is defined as the ratio of the mean drift velocity v of the charge carriers to the electrical field strength E generating the drift and a typical value for gas ions at normal conditions is given in equation (16C.4). The mean drift velocity of the charge carriers is determined by the collision frequency with other (neutral) gas molecules; obviously the collision frequency decreases with increasing intermolecular distances; thus the mobility increases in the same way as the mean free path of the gas molecules (equation (16C.5)). The mobility of gas ions influences the current–voltage relationship as will be discussed in the following sections.

$$b(T_0, p_0) = 2 \cdot 10^{-4} \frac{m^2}{Vs} \tag{16C.4}$$

$$b(p, T) = \frac{b(p_0, T_0)}{\delta} \tag{16C.5}$$

Finally we have to consider how the viscosity of the gas flow is affected by high temperatures and/or high pressures. It can be easily demonstrated that viscosity is not influenced by pressure but only by temperature, since the p, T-dependence of gas density and mean free path compensate each other (equation 16C.6). The well known $\sqrt{kT}$-dependence of the mean thermal velocity of a gas molecule with mass m remains valid (equation 16C.7). Therefore, viscosity is not modified in terms of relative gas density, but in terms of $\sqrt{(T_2/T_1)}$ (equation 16C.8).

$$\eta \sim \rho \cdot \lambda \cdot \langle v \rangle \tag{16C.6}$$

$$\langle v \rangle \sim \sqrt{\frac{kT}{m}} \tag{16C.7}$$

$$\eta(T) = \sqrt{\frac{T}{T_0}} \cdot \eta(T_0) \tag{16C.8}$$

Gas viscosity is decisive in the migration velocity of the particles which results as a balance between electrical and drag forces. Therefore, viscosity determines particle migration while relative gas density and mobility mainly govern the operating conditions. Finally, some particle properties important for the ESP process are also a function of temperature, e.g. electrical resistivity, particle–particle and particle–collector adhesive forces, as discussed elsewhere.

16C.2 Voltage and current

The corona initiation field strength E_0 was introduced in chapter 3. According to equation (3.1), Figure 16C.3 shows E_0 as function of wire radius and isolines for different relative gas densities. In general, an increase of wire radius leads to a decrease in electrical field strength; this trend softens for radii >1 mm. Relative gas densities >1 need higher electrical field strength values to initiate corona in the same geometry. This seems reasonable because in a denser gas the mean free path of molecules reduces and with the time available for acceleration; this can be compensated by higher electrical field values. Obviously, with relative gas densities <1 the opposite behaviour occurs.

The corona onset voltage for a tube-type precipitator can be estimated according to equation (3.3) and its dependence on relative gas density is plotted in Figure 16C.4. Generally for fixed geometries, an increasing relative gas density needs higher voltage levels to initiate corona. The thick line in Figure 16C.4 illustrates the onset condition for a wire radius of 1.0 mm and a tube radius of 150 mm. This figure also shows that a variation of tube radius has only small effects on voltage levels compared with

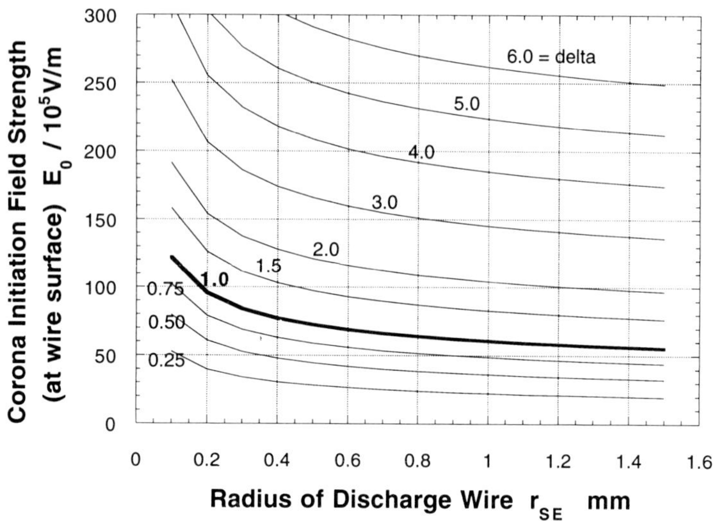

Figure 16C.3 Corona initiation field strength at wire surface as a function of wire radius; isolines represent constant relative gas densities.

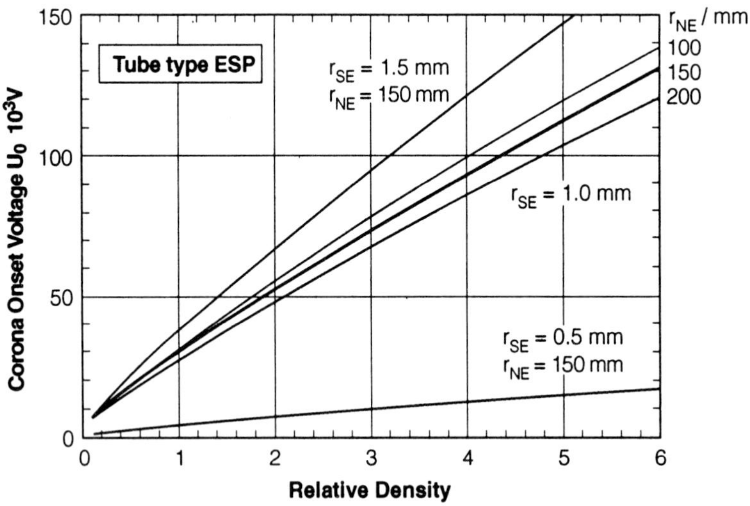

Figure 16C.4 Corona onset voltage as a function of relative gas density. For different tube and wire radii.

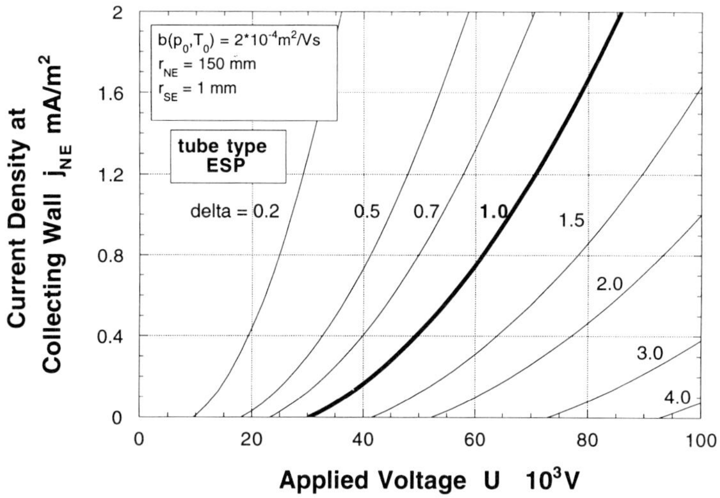

Figure 16C.5 Electrical current density at collecting wall as a function of applied voltage for different relative gas densities.

changes of discharge wire radius. It is interesting to note that the smaller the discharge wire, the weaker the dependence of onset voltage on gas density becomes.

In order to discuss the dependence of the current–voltage relationship on the relative gas density δ, one has to keep in mind that corona onset voltage U_0, and the mobility of gas ions b, are functions of δ. Figure 16C.5 shows calculations according to equation (3.13) for different relative gas densities for a tube type design of $r_{SE} = 1\,\text{mm}$ and $r_{NE} = 150\,\text{mm}$. The higher the relative gas density the flatter the curves become, or in other words, at higher relative gas densities, a constant applied voltage can force less gas ions through the gas phase.

The flow of charge carriers is limited by the electrical breakdown condition. This sparkover voltage plays an important role in the operation of electrical precipitators, because stable operation is only possible if an essential difference between corona onset and sparkover ($U_{crit} - U_0$) exists. As the experimental results in Figure 16C.6 show [2], the region of stable operation $U_{crit} - U_0$ in a tube-type ESP continuously decreases for increasing temperatures when the pressure is kept constant at 1 bar. However, when the pressure is also increased, then the region $U_{crit} - U_0$ widens and extends to higher temperatures. For example, at room temperature and 1 bar, $U_{crit} - U_0$ is about 25 kV (Figure 16C.6), whilst approximately the same voltage difference is observed at about 800 °C but at a pressure level

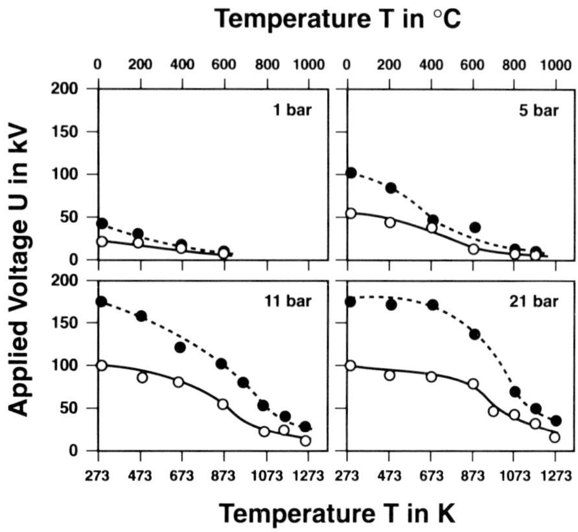

Figure 16C.6 Corona onset voltage (○) together with sparkover voltage (•) as a function of temperature for four different pressures. The measurements were done by Weber [2] in a tube-type ESP with $R_{NE}/R_{SE} = 62.5$.

of 21 bar. A stable operation of ESPs at high temperatures must, therefore, be accompanied by an appropriate high pressure level, i.e. high relative gas densities.

Furthermore, the high gas densities enable higher voltages to be applied, i.e. higher electrical fields can be sustained between the electrodes (Figure 16C.6), producing corresponding improvements in the particle collection, as will be shown. Since no reasonable forecast of the sparkover voltage can be made, no reliable estimates for the maximum electrical field strengths are possible and experiments are necessary. For example, Figure 16C.7 shows j_{SE}/U-curves experimentally determined by Bush *et al.* [3] in a laboratory-scale tube-type ESP covering temperatures in the range 533–1366K and pressures 3.4–35.5 bar ($R_{NE} = 3.63$ cm and $R_{SE} = 0.8$–1.6 mm). As other authors, they did not find a general limitation for corona discharge at high pressures and high temperatures in the range investigated. In most cases negative corona was found to be superior to positive corona. In Figure 16C.7 relative gas densities of about $\delta \geqslant 6$ enable voltage levels of about 100 kV; for a tube radius of about 3 cm, this corresponds to electric field strengths in the order of 30 kV/cm! For normal operating conditions the field strength is about 3 kV/cm, so this increase represents, obviously, an outstanding potential for successful particle collection at high temperatures and high pressures.

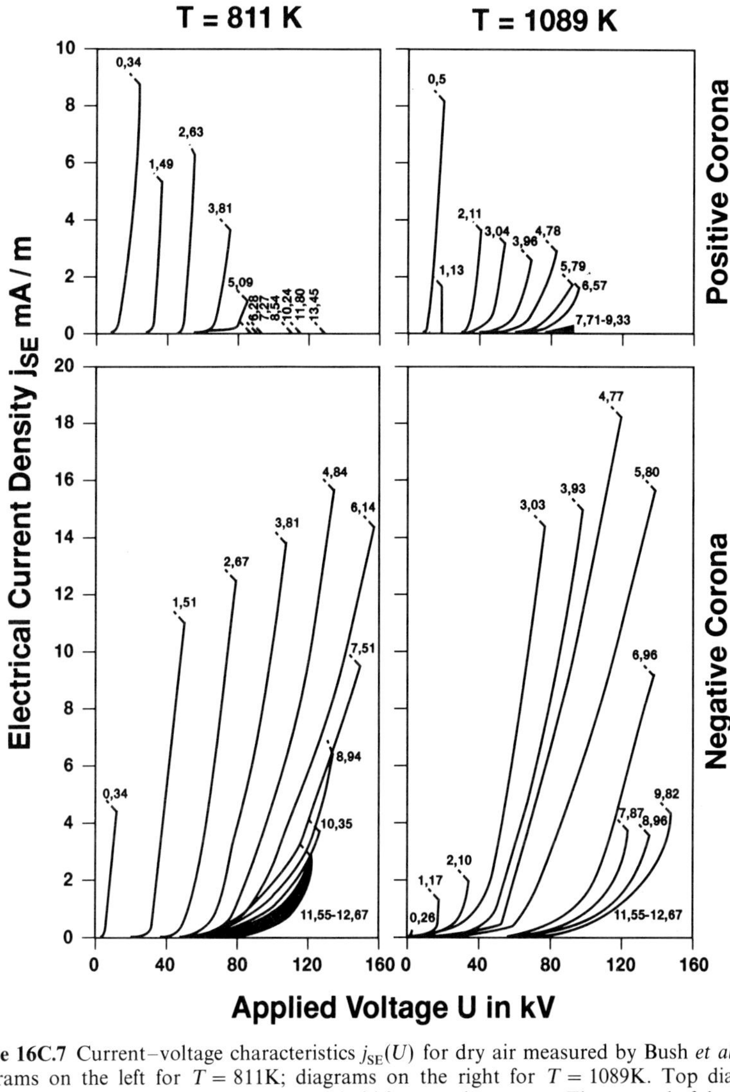

Figure 16C.7 Current–voltage characteristics $j_{SE}(U)$ for dry air measured by Bush *et al.* [13]. Diagrams on the left for $T = 811$K; diagrams on the right for $T = 1089$K. Top diagrams represent positive corona, bottom diagrams hold for negative corona. The top end of the curves corresponds to electrical breakdown. The value of the relative gas density is written along each curve. The tube radius was $R_{NE} = 3.63$ cm and that of the discharge wire $R_{SE} = 1.17$ mm.

16C.3 Particle charging

The consequences of temperature and pressure on the particle saturation charge can be calculated according to Cochet's equation given in section 3.3 (equation (3.56)). Figure 16C.8 shows the calculated influence of the relative

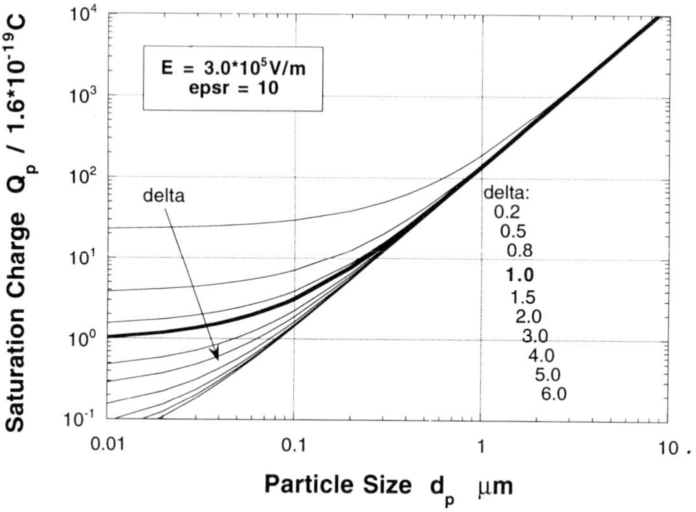

Figure 16C.8 Particle saturation charge according to Cochet [4] plotted against particle diameter. An electrical field strength of 3 kV/cm and a relative electrical permittivity of 10 are assumed for the calculations. Different lines correspond to different relative gas densities.

gas density on particle charge for an electrical field strength of 3 kV/cm and an electrical permittivity of the particle of $\varepsilon_r = 10$. With decreasing particle size, the achievable particle charge is generally reduced because of decreasing surface. This is why a straight line emerges for large particles on the log–log scale. At low relative gas densities, small sized particles can carry much more charge than at atmospheric conditions.

According to Figure 16C.2, gas densities <1 correspond to operation at atmospheric pressure and high temperature. To generate the charge carriers, however, additionally *high* pressure conditions are necessary, as previously discussed. Both factors do not correlate; thus, one could suppose at this stage, that ESPs will work less effectively in the fine particle region at high temperature and high pressure; on the other hand, the efficiency might be compensated by potentially higher electrical field strengths.

16C.4 Particle migration

Figure 16C.9 illustrates the theoretical migration velocity as a function of particle size under atmospheric pressure and raised temperatures, which corresponds to relative gas densities <1. The absolute value of the characteristic minimum in the migration velocity function is weakly lifted and its location is shifted with decreasing relative gas density to larger particle sizes.

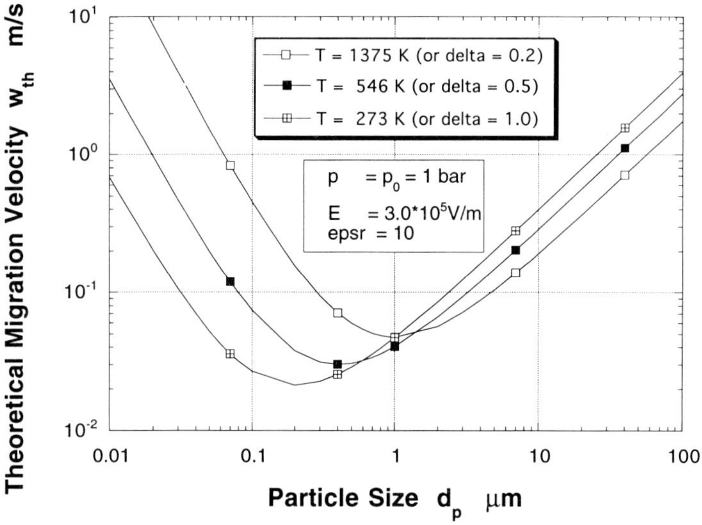

Figure 16C.9 Theoretical migration velocity calculated by equation (3.70) as a function of particle size for atmospheric pressure and three temperatures (or three relative gas densities, respectively). An electrical field strength of 3.0 kV/cm and a relative electrical permittivity of 10 are assumed for the calculations.

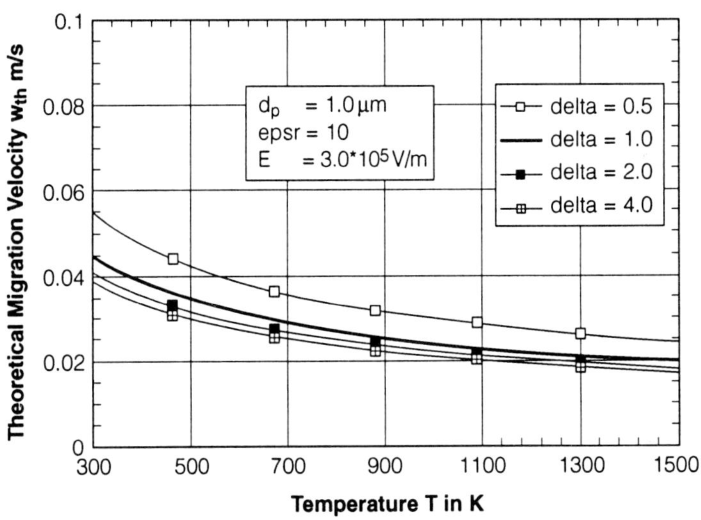

Figure 16C.10 Theoretical migration velocity of a 1.0 μm particle ($E = 3.0$ kV/cm; $\varepsilon_r = 10$) as a function of temperature; isolines correspond to constant relative gas densities.

For particles $>1\,\mu$m the influence of the dynamic viscosity η dominates the δ-dependence, i.e. at higher temperatures or relative gas densities <1, respectively, the gas phase becomes more viscous and, hence, the particles migrate more slowly. For particles $<1\,\mu$m this behaviour reverses: here the δ-dependence of particle charge and Cunningham correction dominate over the viscosity, i.e. at higher temperatures or relative gas densities <1, respectively, particles can penetrate the gas phase easier resulting in higher migration velocities.

The situation is more complex in an ESP, because for a corona discharge to develop the pressure level has to be raised (section 4.7.2). Therefore, it is necessary to look at migration velocities of particles with a fixed size as a function of temperature, keeping the relative gas density constant. This is illustrated in Figure 16C.10 for 1.0 μm particles. Here it can be seen that, for constant relative gas density, an increase in temperature will weakly decrease the migration velocity. For constant temperatures, doubling the relative gas density shows less effect on the theoretical migration velocity than reducing it by half.

The necessity of applying much higher electrical field strengths at high relative gas densities to increase the corona currents still has to be considered. An increase in the electrical field E results in an enormous increase of the migration velocity, because w_{th} *quadratically* depends on E (equation 3.70). This is illustrated in Figure 16C.11 for a 1.0 μm particle at a relative gas density of $\delta = 4.0$ for electrical field values of $E = 3/6/9/12\,\text{kV/cm}$.

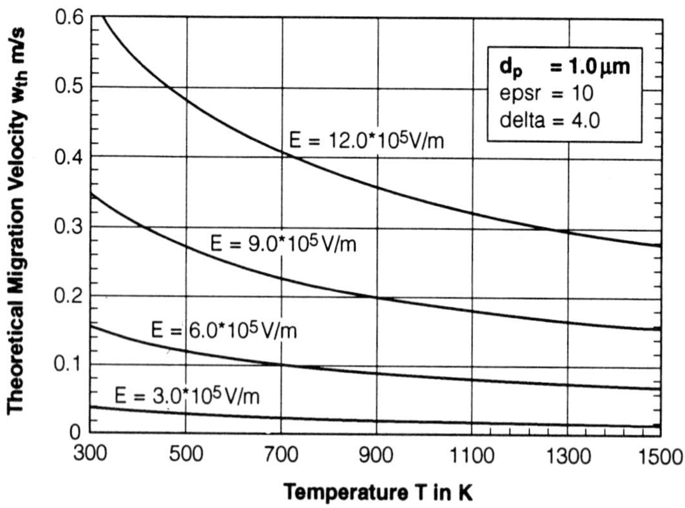

Figure 16C.11 Theoretical migration velocity of a 1.0 μm particle ($\varepsilon_r = 10$) for a relative gas density of 4.0 as a function of temperature; different lines correspond to different electrical field strengths.

16C.5 Grade efficiency

When looking for the T/p-dependence of the grade efficiency $T(d_p)$, the ESP design must first be specified. Here a tube-type precipitator was assumed with $r_{NE} = 150$ mm and $L_{NE} = 5.0$ m and operation conditions of $E = 3.0$ kV/cm and $v_0 = 1.0$ m/s. Grade efficiency was calculated according to the Deutsch equation (equation 3.87) for raised temperatures and atmospheric pressure which corresponds to relative gas densities < 1.

The results are illustrated in Figure 16C.12. The efficiency curves show, of course, the same tendencies as those of the migration velocities in Figure 16C.9; the absolute value of the typical minimum value is significantly raised and its location is shifted as before to larger particle sizes for decreasing relative gas densities. It can be seen that reducing the relative gas densities leads to a dramatic improvement in efficiency; however, this neglects the high pressure levels needed for corona generation.

Figure 16C.13 shows efficiency as a function of temperature with the relative gas density as parameter for 1.0 μm particles. Obviously, particles will be less efficiently collected at high temperatures when the relative gas density is kept constant and assuming a constant electric field strength.

However, when the potential higher electric field strengths at raised relative gas densities are considered, things look different. Figure 16C.14 shows the efficiency values of a 1.0 μm particle, calculated and plotted as a function of temperature as before, but for a relative gas density of $\delta = 4.0$

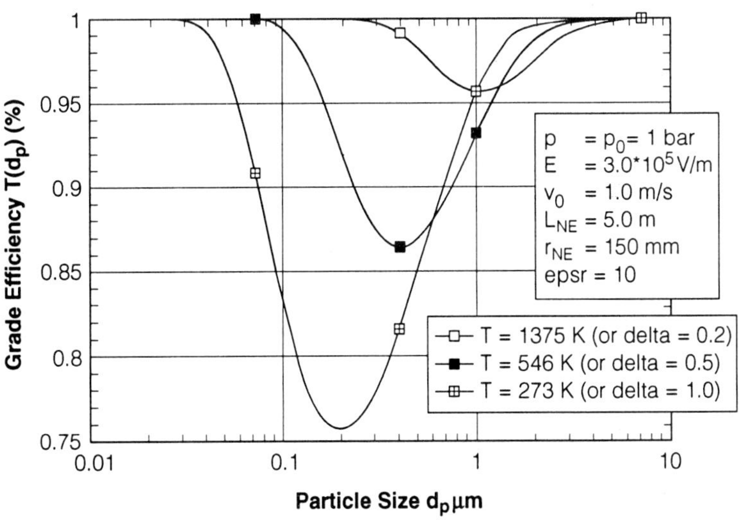

Figure 16C.12 Grade efficiencies at atmospheric pressure calculated according to Deutsch (equation 3.87) for three different temperatures; the ESP design was assumed to be $L_{NE} = 5.0$ m, $r_{NE} = 150$ mm and its operation $E = 3.0$ kV/cm and $v_0 = 1.0$ m/s.

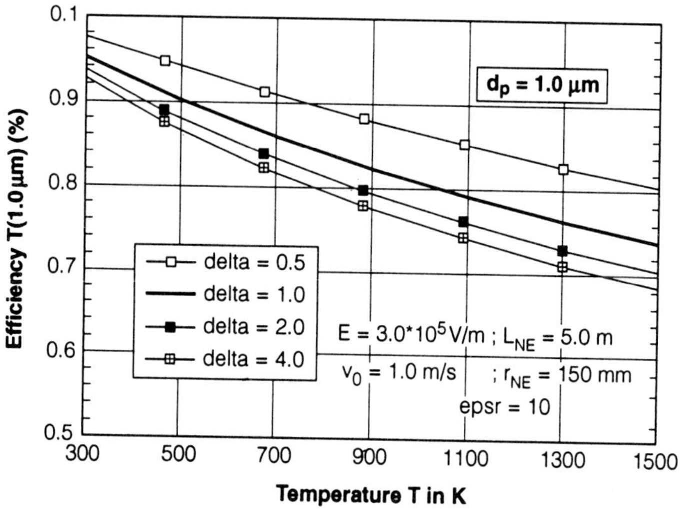

Figure 16C.13 Efficiencies according to Deutsch (equation 3.87) for a 1.0 μm particle as a function of temperature; different lines correspond to different relative gas densities. ESP design and operation conditions as before.

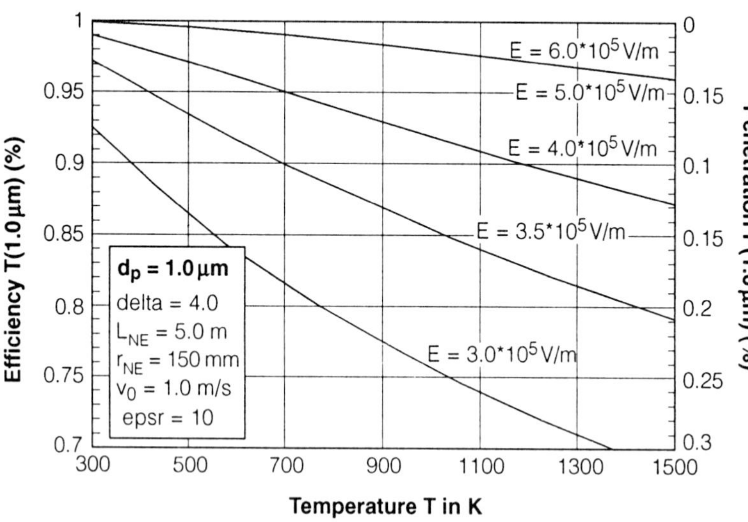

Figure 16C.14 Efficiency of a 1.0 μm particle ($\varepsilon_r = 10$) for a relative gas density of 4.0 as a function of temperature; different lines correspond to different electrical field strengths. ESP design and operation conditions as before.

with the applied electric field as parameter. Obviously, particle collection is extremely sensitive to electrical field strength; e.g. at a temperature level of 1300K, doubling the electrical field strength from 3 to 6 kV/cm increases the migration velocity by a factor 4 (compare Figure 16C.11), which reduces penetration (1 – efficiency) by more than a factor of 25!

Experimental results for total mass efficiencies (integrating over all particle sizes) have been published, e.g. by Feldman and Bush [5] and Rinard *et al.* [6]. As expected, they found a strong influence of the electrical field strength on particle collection. Feldman and Bush refer to a wire-pipe ESP of Union Carbide Olefins Co. operating at temperatures of 870–1000K, at pressures of 3–8 bar and gas velocities of 0.2–1.2 m/s. The ESP consisted of 19 pipes of 15.2 cm in diameter, 1.8 m in length and discharge wires of 2.1/3.4 mm in diameter. Rinard *et al.* refer to a test facility located at the Denver Research Institute with one tube, 30.5 cm in diameter and 2.1 m in length, operating at temperatures up to 1200K, at pressures up to 10 bar and a flow rate of 0.078 m^3/s at these conditions.

16C.6 Open questions

For a successful high temperature, high pressure ESP to be designed and developed the following factors need careful consideration.

16C.6.1 Electrical resistivity

The combination of low ash resistivity at high operating temperatures, together with the correspondingly high current densities, will not necessarily lead to back-corona since the high gas densities should suppress it. On the other hand electrical re-entrainment of particles might result from the low resistivity values.

16C.6.2 Mechanical stability of material

The suitability and integrity of the materials used at temperatures $\geqslant 1000\,^{\circ}C$ has to be considered carefully. For long-term applications at elevated temperatures, the creep behaviour of the material has to be taken into account. Furthermore, the flue gas and particularly the ash particles at high temperature are much more aggressive; therefore, corrosion could be a severe problem.

16C.6.3 Rapping

For some materials, at temperatures above 700 °C the region of forgeability starts. This can become a problem if rapping is done by conventional hammer systems. Cleaning by pulse jets, analogous to those used in bag houses, or by ultrasonic horns are under discussion.

16C.6.4 Electrical insulation

Insulator arrangements commonly used on the roofs of ESP housings operate at temperatures about 50 °C less than the gas and are stressed thermally, mechanically and electrically. Unfortunately the electrical resistivity of most insulator materials drastically decreases at temperatures above about 300 °C. An application of the so-called advanced ceramics for insulating material for the extreme requirements of both temperature and pressure, might be promising and should be investigated.

16C.6.5 Emptying of hoppers

To bring the collected dust out of the hoppers, it is necessary to overcome a pressure barrier of some 10 bar. In order to guarantee secure dust handling a carefully designed pressure sluice is absolutely essential.

16C.6.6 Electrical power consumption

The electric power consumption, as a fraction of the total electric output from a PCFB electric generating plant, has been observed to be between 1.5 and 2% [7]. If the ESP efficiency has to be increased, the collecting area has to be enlarged, which could lead to probably unacceptable high power consumption. A solution might be to use power-conserving means on the transformer rectifier sets, such as intermittent electrical energisation.

For successful ESP operation at high temperatures, correspondingly high pressure levels are necessary. Thus, the gas can withstand the electrical breakdown much better than at normal pressure conditions, leading to extremely high electrical field strengths. For the same reason back-corona is not expected to cause severe problems. The high electrical field strengths therefore suggest rather smaller specific collecting areas for efficiencies $\geqslant$99% than for 'normal' operation.

Tassicker [7] concludes that 'the data available would be sufficient for the commercial-scale of an ESP for conditions of 5–15 bar and 400–700 °C. The available data are less definite for a firm design at 850–900 °C. More pilot-plant work is needed before a commercial-scale plant in this range could be confidently sized'.

16C.7 Symbols

b mobility of gas ions

E	electric field strength
E_0	corona initiation field strength
j_{NE}	electrical current density at collecting wall
L_{NE}	length of collecting tube
Q_p^∞	particle saturation charge
p	pressure
r_{NE}	radius of collecting tube
r_{SE}	radius of discharge wire
T	temperature
$T(d_p)$	grade efficiency
U	applied voltage
U_0	corona onset voltage
U_{crit}	sparkover voltage
$\langle v \rangle$	mean thermal velocity of gas molecules
v	fluid velocity
w_{th}	theoretical migration velocity of an individual particle
δ	relative gas density
ε_r	(epsr) electrical permittivity of particle material
η	viscosity of fluid
λ	mean free path of gas molecules
ρ	density of fluid

References (16C)

1. Perrin, A.J. (1994) Clean Coal Technology — An Industry Perspective. Presented at Prospects for Clean Coal—A Contractors' Meeting, 1–2nd November, Nottingham, U.K. Sponsored by the DTI/DOE. RTSU Harwell, U.K.
2. Weber, E. (1984) Electrostatic precipitation under extreme conditions of temperature and pressure. *Proc. 2nd Int. Conf. Electrostatic Precipitation*, Kyoto, Japan, pp. 85–95, EPA, Pittsburgh, USA.
3. Bush, R.J., Feldman, P.L. and Robinson, M. (1979) High pressure, high temperature electrostatic precipitation. *J. Air Pollut. Control Assoc.*, **29**, 365–71.
4. Cachet, R. (1961) Lois charge des fines particules (submicroniques) etudes théoriques–controles récents spectre de particules. Coll. Int. la physique des forces electrostatiques et leurs application. Centre National de la Research Scientifique, **102**, 231–8, Paris, France.
5. Feldman, P.L. and Bush, J.R. (1980) Performance of electrostatic precipitators at high temperatures and high pressures. *VDI-Berichte*, **363**, 87–92.
6. Rinard, G., Rugg, D.E. and Yamamoto, T. (1987) High-temperature high-pressure electrostatic precipitator electrical characterization and collection efficiency. *IEEE Trans. Ind. Appl.*, **IA-23**, 114–19.
7. Tassicker, O.J. (1986) High temperature–pressure electrostatic precipitator for electric power generation technologies: an overview of the status. *IChemE Symp. Ser.*, **99**, 331–9.

16D Computer sizing of precipitators

Although at present, precipitators are normally sized by the supplier, using his databank knowledge and experience from similar processes, the advent of very high speed computers for solving Poisson and Laplace equations may eventually mean that sizing will be computer generated from first principles. A number of companies and research investigators are already working on this approach, as indicated in chapter 9, but for computer based sizing to be generally accepted by industry, a great deal of research and confirmatory investigations will have to be completed in the following areas to determine how they impact on performance from a theoretical standpoint for inclusion in the programming.

Some of these have been identified in previous chapters from various operational standpoints, but there is a need for them to be incorporated in the overall sizing programme. Some in fact will have their own subroutine to the major format on sizing if the approach is to prove acceptable.

1. Rapping and re-entrainment.
2. Effect of ionic wind on particle transport phenomena and on gas distribution under operational conditions.
3. Initiation conditions leading to reverse ionisation/back-ionisation from basic input information.
4. Influence of small variations in carrier gas analysis and temperature.
5. Improved usage of electrical power.
6. Effect of corona and collector plate current distribution.

Some of the above items must also be considered in their own right, as indicated in the introduction, since they can have considerable impact on precipitator performance and hence size and cost. In summary, although the electrostatic precipitator has been part of our industrial heritage for almost a century, there is a growing need for higher efficiency, improved reliability and availability of equipment, to ensure that future generations enjoy a clean environment, particularly free from man-made pollution in the form of particulate matter.

Index

Page numbers appearing in **bold** refer to figures and page numbers appearing in *italic* refer to tables.

Aerodynamic factors affecting performance 113–50
Aerosols 154
Agglomeration 157, 159, 162
Air inleakage hoppers 109
Alkali chlorides 177, 360, 363, 368
Applications of ESPs, *see* Contents v, ix, x, xi
Aspect ratio 188, *189*, 355
Automatic voltage control 22, 217–29
  analogue type 23
  microprocessor 23, 91, 217–29
  supervisory type 241–6

Back corona 169, 195, 226–8, 241, 265, 364
  onset of back ionisation 22
  severe back ionisation 216
Bag filters, *see* Fabric filters
Brownian motion 3, 154, 162, 184
Bus section/field 25

Carbon absorption 7, 301
Carbon particles 160, 353, 371, 379
Casings 97–100, 306–11, 324, 325, 387–90
  pressure testing of casings 318, 324–5
  purging of casings 340–1, 395, 415
Cenospheres 357
Cochet charging model 52–4, 84, 114, 509–10
Cohesion/cohesivity 163–6, 178, 262, 357, 452
Collector elements
  catch space **39**, 95, **313**
  concentric ring **405**
  FRP 97, 391, 393, 403, 408
  rolled channel 94
  tubular 15, 97, 403
Commissioning of precipitators 305–48
  electrical 326–339
  mechanical 306–326
  process 339–348
Composition of dust, effect of 17–28, **183**, 272–4
Computational fluid dynamics 74, 80, 142–8
Conditioning of gases and dusts 429–81
  ammonia 357, 445–52
  dual $SO_3 + NH_3$ 352, 452–4
  injection rate optimisation 471–80
  moisture 16, **169**, 184, 362, 365, 372, 425
  other reagents **426**
  sulphur trioxide 352, 429, 436–8
  SO_3 injection rate equation 440–5
  SO_3 injection rate prediction 438–40
Conditioning towers 362, 363, 369, 373
Corona discharge
  breakdown voltage 29, **30**, 31, 508
  corona power 193–5, 228–9
  corona suppression 91, 195, 358, 368, 376
  coaxial system 38–44
  current density 38, **40**, 42, **45**, 241
  initiation/onset voltage 33–6, **34**, 262, 505, **508**
  parallel plate 34–44, 246
  space charge 134–6, 195, 451
  streamers **30**, 170
  threshold voltage 31–44, 196–8
  see also Particle charging mechanisms
Cunningham correction factor 56–9
Cyclones/inertial separators 2, **3**, 4, 9

Design considerations 180–90
Detarring 17, 342, **406**, 411–15
Deutsch 20, 62–5, 84, 113–17, 182, 185, 255, 281–5, 514
Deutsch number 64, 116, **117**, 122
Diffusers 129, 317
Diffusion 114–17, 122
Diffusivity model 82–5
Discharge electrode spacing 34–6
Discharge electrode support 92–4, **187**
Dry precipitators 349–81
  applications 349–81
Dry scrubbing 17, 369, 379
Duct spacing 34–8, 40–4, 48–50, 185, 424

Earthing and grounding 335
Effective migration velocity 65–70, 182–6, 194, 434
Electron attachment 30
Electrical clearances
  external 110–11
  internal 110, 200, 309, 320

Electrical energization 192–46
  full wave 192, 201–10
  high frequency power conversion 192, 487–90
  intermittent energization 192, 210–17, 423
  pulse energization 121, 192, 230–41, 423
  transformer power ratings 206–8
Electric fields **49**, **50**
  distribution 44–52, 237, **238**
  plate precipitators 48, 49, **50**
  tube precipitators 46–8, **47**
Electrodes
  controlled emission type 90, 122
  discharge elements **38**, 89–92, **90**, 187, 315, 386
  high emission types 90
  mounting of electrodes 92–4, 187
  pubescent type **13**
Electropositive/negative gases 30
Extended Deutsch equation 256–8

Fabric filters **3**, 5, *9*, 35, 421
Flares (gas distribution) 317, **318**
Flue gas conditioning systems 429–81, 492–500
Fly ash composition, importance of 172–8, 183, 272–3
Fume 13, 370

Gas distribution 129–34, 138, 317, 355
CFD modelling 80, **81**, 142–8, 485
  field testing 148–9, 323
  flow bypass 109, 137
  gas distribution methods 129–31
  scale modelling 139–42
  visualization 122, 148
  see also Aerodynamic factors affecting performance

Heavy metals 298–303, 358, 369
High resistivity particles, *see* Particle resistivity
High temperature/high pressure precipitation 501–16
High voltage supplies, *see* Electrical energization
History of precipitation 11–23
  early designs and applications 11–20
  energization systems 20–3
Hoppers 108–110, 311
  air inleakage 109, 325, 377
  heating requirements 334, 351
  overfilling effects of 109
  pyramid type 316
  trough type 389
Hydrofiners 16

Inertial sampling 296–7
Insulators
  H.T. lead through 100–2, 322, 387, **409**, 516
  heating 334, 351, 387
  purging 387, 388
Ion mobility *37*, 504
Ionic space charge 46, 50, **51**
Isokinetic sampling 293–4
  see also Testing of precipitators

Lambert's law 161
Linear inductors 208–10

Mass flux 70–6
Materials of construction
  high temperature applications 515
  wet precipitators 391, *392*, 393
Maxwell equation 126, 246
Mechanical collectors 2, 4, 9
Mist precipitators
  applications 407–16
  design considerations 402–7
Modelling of precipitators 250–91
  Cañadas 286–7
  computer 274, **285**
  CSIRO 272
  full scale 275, **277**
  Padova University 288, **289**
  pilot scale 276
  Southern Research Institute 285
Modified Deutsch Formulae 20, 68, 250, 255, 256, 284, 420
  CSIRO 250, 256–74
  Matts Ohnfeldt 68, 250, 255, 284
  Petersen(FLS) 250, 255

Numerical flow model 126

Parallel plate nomenclatures 25, 34, 283
Particle charging mechanisms
  Cochet's model 52–4, 84, 114, 509
  collision charging 52
  ion diffusion 52
  saturation charge **53**, 54, 55, 510
Particle migration 54–85, 510–13
  exponential law 20
  practical consideration 59, 69
  theoretical 57, **81**
Particle re-entrainment 69, 128, 136–7, 356, 485
Particle resistivity 166–72, 263, **264**, 426, 430–6
  critical value 176
  effect on corona 170, 262

effect on performance 263–4, 362, 433, **434**, **435**
effect of temperature 167, 168, 263, 271, 427
measurement of 167
prediction of resistivity 432–6
Particle size
distributions 59, **60**, 153–4, 258, 259, 296
effect on performance **58**, **65**, 76–85, 184–5, **511**, **513**
grade efficiency relations 61, 63, 64, **66**
optical properties 161
shape 153, 160
Particle tracking model 76–82
Particle transport 28, 76–82, 114, 254, 510
see also Effective migration velocity
Péclet number 116, 121–6
Performance line 267–74
Performance testing
impactors inertial sampling 296
isokinetic sampling 293–4
methods and standards 292, 300
Pilot precipitators 81, 117, 263, 266, 276, 364, 508
Poisson's equation 45, **46**

Rapping 102–8, 346
collectors 103, 104, 314, **316**, 321
discharge elements 103, 104, 315, 321
Rapping optimization 107, 241, 245, 246, 346, 485
Re-entrainment effects 69, 103, 136, 425
Rectifier control methods
transductor/magnetic saturable reactors 22
thyristors 22, 201–46
see also Automatic voltage control
Residence time 131–4
Reynolds number 56, 126

Safety interlock systems 331
Sampling of gases 292–303
standards 292, **295**, 300
Scouring of dust from internals 137, 189
Scrubbers 2, **3**, 4, *9*, 396

Secondary flow in precips 113, 117, 122–6
Sectionalization 189, 199–201
Selective dust separation 15, 19, 382
Sliding bearings 97, **98**, **310**
Smog 1
Sneakage and sweepage 137
Sodium depletion 176, 464
Space charge effects 91, 134–6, 188, 386
Spray irrigation 322
Stokes–Cunningham correction 56, 162
Stokes' diameter 3, 159
Stokes' law 56, 258
Sulphate reducing bacteria 413
Supervisory AVCs 241–6

Temperature effect on performance 167, 183, **184**, **261–6**, 271–2, **353**, 427, **514**
Testing of precipitators 292–5
Theory of precipitation 25–87
Thermal expansion 344
Thermal insulation 317
Top housings 101, 308
Treatment time **410**, *420*, **424**
Trichel pulses 30
Tube type precipitators 15, 26, 33, 38, 46, 403, 406, 413
Turbulence 27–8, 76–85, 113–27
Two stage precipitation 25, **27**

Upgrading of precipitator performance 418–424

Velocity of gas 127–9, 138, 189, 355
Viscosity 28, 56, 113, 261, 505, 512
Volatiles 298, **299**, 367, 370, 501
Voltage/current relationships **29**, 42, **91**, 197, 337, **338**, 507, **509**

Water treatment 390, **391**
Wet precipitators 7, *9*, 8, 17, 25, 382–400
applications 394–401
collector film flow 19, 383, **384**
design considerations 383–94
spray irrigation **384**, 386
washdown spray system 322, 337, 385